Fatigue and Fracture of Adhesively-bonded Composite Joints

Related titles

Residual stresses in composite materials
(ISBN 978-0-85709-270-0)

Composite joints and connections
(ISBN 978-1-84569-990-1)

Failure mechanisms in polymer matrix composites
(ISBN 978-1-84569-750-1)

Fatigue life prediction of composites and composite structures
(ISBN 978-1-84569-525-5)

Woodhead Publishing Series in Composites
Science and Engineering: Number 52

Fatigue and Fracture of Adhesively-bonded Composite Joints

Behaviour, Simulation and Modelling

Edited by

A. P. Vassilopoulos

AMSTERDAM • BOSTON • CAMBRIDGE • HEIDELBERG
LONDON • NEW YORK • OXFORD • PARIS • SAN DIEGO
SAN FRANCISCO • SINGAPORE • SYDNEY • TOKYO
ELSEVIER   Woodhead Publishing is an imprint of Elsevier

Woodhead Publishing is an imprint of Elsevier
80 High Street, Sawston, Cambridge, CB22 3HJ, UK
225 Wyman Street, Waltham, MA 02451, USA
Langford Lane, Kidlington, OX5 1GB, UK

Copyright © 2015 Elsevier Ltd. All rights reserved.

No part of this publication may be reproduced, stored in a retrieval system or transmitted in any form or by any means electronic, mechanical, photocopying, recording or otherwise without the prior written permission of the publisher.

Permissions may be sought directly from Elsevier's Science & Technology Rights Department in Oxford, UK: phone (+44) (0) 1865 843830; fax (+44) (0) 1865 853333; email: permissions@elsevier.com. Alternatively, you can submit your request online by visiting the Elsevier website at http://elsevier.com/locate/permissions, and selecting Obtaining permission to use Elsevier material.

Notice
No responsibility is assumed by the publisher for any injury and/or damage to persons or property as a matter of products liability, negligence or otherwise, or from any use or operation of any methods, products, instructions or ideas contained in the material herein. Because of rapid advances in the medical sciences, in particular, independent verification of diagnoses and drug dosages should be made.

British Library Cataloguing-in-Publication Data
A catalogue record for this book is available from the British Library.

Library of Congress Control Number: 2014944400

ISBN 978-0-85709-806-1 (print)
ISBN 978-0-85709-812-2 (online)

For information on all Woodhead Publishing publications
visit our website at http://store.elsevier.com

Typeset by TNQ Books and Journals
www.tnq.co.in

Printed and bound in the United Kingdom

Contents

List of contributors	xi
Woodhead Publishing Series in Composites Science and Engineering	xiii

Part One Introduction to fatigue and fracture of adhesively-bonded composite joints   1

1 Investigating the performance of adhesively-bonded composite joints: standards, test protocols, and experimental design   3
A.J. Brunner
1.1   Introduction   3
1.2   Standards and test protocols for experimental fatigue and fracture testing of adhesively-bonded composite joints   7
1.3   Standards and test protocols for fatigue and fracture testing of pultruded glass-fiber reinforced polymer-matrix (GFRP) profiles   14
1.4   Standards and test protocols for determining environmental effects in fatigue, fracture, and durability testing   22
1.5   Standards and test protocols for modeling and simulation of fracture and fatigue behavior   24
1.6   Summary and future trends   26
1.7   Sources of further information and advice   28
     Acknowledgments   30
     References   30
     Appendix: list of abbreviations   42

2 Design of adhesively-bonded composite joints   43
L.F.M. da Silva, R.D.S.G. Campilho
2.1   Introduction   43
2.2   Factors affecting joint strength   48
2.3   Methods to increase joint strength   59
2.4   Hybrid joints   63
2.5   Repair techniques   66
2.6   Conclusions   68
     References   68

3	Understanding fatigue loading conditions in adhesively-bonded composite joints	73
	R. Sarfaraz	
	3.1 Introduction	73
	3.2 Fatigue data	75
	3.3 Tensile versus compressive fatigue	75
	3.4 Effects of fatigue loading parameters	76
	3.5 Future trends	86
	3.6 Sources of further information and advice	86
	References	86

Part Two Fatigue and fracture behaviour of adhesively-bonded composite joints 91

4	Mode I fatigue and fracture behaviour of adhesively-bonded carbon fibre-reinforced polymer (CFRP) composite joints	93
	R.D.S.G. Campilho, L.F.M. da Silva	
	4.1 Introduction	93
	4.2 Carbon fibre-reinforced polymer (CFRP) composite joints	96
	4.3 Preparation and testing of CFRP joints in mode I	98
	4.4 Fatigue characterization by the S–N approach	102
	4.5 Fatigue characterization by the fatigue crack growth (FCG) approach	104
	4.6 Fracture modes of CFRP joints in mode I	113
	4.7 Conclusions	116
	References	117

5	Mode I fatigue behaviour and fracture of adhesively-bonded fibre-reinforced polymer (FRP) composite joints for structural repairs	121
	J. Renart, J. Costa, C. Sarrado, S. Budhe, A. Turon, A. Rodríguez-Bellido	
	5.1 Introduction	121
	5.2 Configuration of the bonded joint	122
	5.3 Test generalities	124
	5.4 Fatigue testing	131
	5.5 Effect of waviness in crack growth rate curves	139
	5.6 Design and simulation approaches	141
	5.7 Conclusions	143
	Acknowledgements	144
	References	144

6	Mode I fatigue and fracture behavior of adhesively-bonded pultruded glass fiber-reinforced polymer (GFRP) composite joints	149

A.P. Vassilopoulos, M. Shahverdi, T. Keller

6.1	Introduction	149
6.2	Experimental investigation of adhesively-bonded pultruded glass fiber-reinforced polymer (GFRP) joints	155
6.3	Interpretation of the fatigue/fracture experimental results and discussion	158
6.4	Fracture mechanics data analysis	168
6.5	Fracture mechanics modeling	173
6.6	Conclusions	181
	References	182

7	Mixed-mode fatigue and fracture behavior of adhesively-bonded composite joints	187

M. Shahverdi, A.P. Vassilopoulos

7.1	Introduction	187
7.2	Mixed-mode fatigue and fracture experimental investigation	191
7.3	Fatigue and fracture data analysis	201
7.4	Results and discussion	209
7.5	Conclusions	215
	References	220

8	Fatigue and fracture behavior of adhesively-bonded composite structural joints	225

A.P. Vassilopoulos, T. Keller

8.1	Introduction	225
8.2	Experimental investigation of adhesively-bonded structural joints – experimental program description	227
8.3	Interpretation of quasi-static and fatigue/fracture experimental data	230
8.4	Analysis of the fracture mechanics measurements	243
8.5	Conclusions	252
	References	253

9	Block and variable amplitude fatigue and fracture behavior of adhesively-bonded composite structural joints	257

A.P. Vassilopoulos

9.1	Introduction	257
9.2	Experimental investigation of the block and variable amplitude fatigue behavior of adhesively-bonded joints	260
9.3	Experimental results and discussion of the effect of loading	269
9.4	Conclusions	283
	References	285

10	**Durability and residual strength of adhesively-bonded composite joints: the case of F/A-18 A–D wing root stepped-lap joint**	**289**
	W. Seneviratne, J. Tomblin, M. Kittur	
	10.1 Introduction	289
	10.2 Bonded joint applications in F/A-18	290
	10.3 Stress analysis of stepped-lap joints	291
	10.4 End-of-life residual strength evaluation of wing root stepped-lap joint	293
	10.5 Remaining life after fleet service	302
	10.6 Inner-wing full-scale fatigue test	315
	10.7 Conclusions	318
	Acknowledgments	319
	References	320

Part Three Modelling fatigue and fracture behaviour 321

11	**Simulating mode I fatigue crack propagation in adhesively-bonded composite joints**	**323**
	M.M. Abdel Wahab	
	11.1 Introduction	323
	11.2 Finite element (FE) modelling	324
	11.3 Fracture mechanics (FM) approach	329
	11.4 Cohesive zone model (CZM) approach	335
	11.5 Mixed CZM and FM approach	340
	11.6 Conclusions	342
	References	343
12	**Simulating the effect of fiber bridging and asymmetry on the fracture behavior of adhesively-bonded composite joints**	**345**
	M. Shahverdi, A.P. Vassilopoulos, T. Keller	
	12.1 Introduction	345
	12.2 Experimental investigation of asymmetry and fiber-bridging effects	348
	12.3 Finite element modeling	354
	12.4 Results and discussion of asymmetry and fiber-bridging effects	360
	12.5 Conclusions	364
	References	365
13	**Simulating the mixed-mode fatigue delamination/debonding in adhesively-bonded composite joints**	**369**
	A. Pirondi, G. Giuliese, F. Moroni, A. Bernasconi, A. Jamil	
	13.1 Introduction to the simulation of fatigue delamination/debonding	369

	13.2	Cohesive zone and virtual crack closure technique (VCCT) model formulation	376
	13.3	Comparison of cohesive zone and VCCT on fatigue delamination/debonding	391
	13.4	Conclusions	397
		References	397

14 Predicting the fatigue life of adhesively-bonded composite joints under mode I fracture conditions — 401
T.A. Hafiz, M.M. Abdel Wahab

	14.1	Introduction	401
	14.2	Characterization of fatigue in bonded joints	402
	14.3	Analytical approach to fatigue life prediction of adhesively-bonded joints	404
	14.4	Finite element analysis approach to fatigue life prediction of adhesively-bonded joints	412
	14.5	Validation of the finite element approach	415
	14.6	Conclusions	415
		References	416

15 Predicting the fatigue life of adhesively-bonded composite joints under mixed-mode fracture conditions — 419
P. Naghipour

	15.1	Introduction	419
	15.2	Diverse approaches to modeling fatigue life of composite materials	420
	15.3	Various cohesive zone models for cyclic delamination	421
	15.4	Cohesive zone model for cyclic delamination incorporating the Paris fatigue law	426
	15.5	Cohesive zone model for cyclic delamination incorporating the Paris fatigue law and a mixed-mode cohesive area	430
	15.6	Modeling cyclic mixed-mode delamination using the developed cohesive zone technique	431
	15.7	Conclusions and future trends	439
		References	439

16 Predicting the fatigue life of adhesively-bonded structural composite joints — 443
A.P. Vassilopoulos

	16.1	Introduction	443
	16.2	S–N formulations for composites and adhesively-bonded composite joints	448
	16.3	Comparison of existing fatigue models	456
	16.4	Discussion on the S–N formulations	465

	16.5	Constant life diagram (CLD) formulations for composites and adhesively-bonded composite joints	467
	16.6	Comparison of existing constant life diagram (CLD) formulations	478
	16.7	Conclusions	486
		Acknowledgments	487
		References	487
17		**Developing an integrated structural health monitoring and damage prognosis (SHM-DP) framework for predicting the fatigue life of adhesively-bonded composite joints**	**493**
		M. Gobbato, J.B. Kosmatka, J.P. Conte	
	17.1	Introduction	493
	17.2	Proposed reliability-based structural health monitoring and damage prognosis (SHM-DP) framework for fatigue damage prognosis	495
	17.3	Recursive Bayesian characterization of the current state of damage	500
	17.4	Probabilistic load hazard analysis	505
	17.5	Probabilistic mechanics-based debonding evolution analysis	507
	17.6	Probabilistic characterization of global system performance	511
	17.7	Damage prognosis analysis	513
	17.8	Effectiveness of proposed methodology in predicting the remaining time to failure	516
	17.9	Future trends	519
	17.10	Conclusions, recommendations, and additional sources of information	520
		References	521
Index			**527**

List of contributors

M.M. Abdel Wahab Ghent University, Zwijnaarde, Belgium

A. Bernasconi Politecnico di Milano, Milan, Italy

A.J. Brunner Swiss Federal Laboratories for Materials Science and Technology, Dübendorf, Switzerland

S. Budhe University of Girona, Girona, Spain

R.D.S.G. Campilho Instituto Politécnico do Porto, Porto, Portugal; Universidade Lusófona do Porto, Porto, Portugal

J.P. Conte University of California, La Jolla, CA, USA

J. Costa University of Girona, Girona, Spain

L.F.M. da Silva Universidade do Porto, Porto, Portugal

G. Giuliese Università di Parma, Parma, Italy

M. Gobbato Risk Management Solutions Inc., Newark, CA, USA

T.A. Hafiz University of Bristol, Bristol, UK

A. Jamil Politecnico di Milano, Milan, Italy

T. Keller École Polytechnique Fédérale de Lausanne (EPFL), Lausanne, Switzerland

M. Kittur Naval Air System Command, Patuxent River, MD, USA

J.B. Kosmatka University of California, La Jolla, CA, USA

F. Moroni Università di Parma, Parma, Italy

P. Naghipour NASA Glenn Research Center, Cleveland, OH, USA

A. Pirondi Università di Parma, Parma, Italy

J. Renart University of Girona, Girona, Spain

A. Rodríguez-Bellido AIRBUS Operations, Madrid, Spain

R. Sarfaraz École Polytechnique Fédérale de Lausanne (EPFL), Lausanne, Switzerland

C. Sarrado University of Girona, Girona, Spain

W. Seneviratne Wichita State University, Wichita, KS, USA

M. Shahverdi École Polytechnique Fédérale de Lausanne (EPFL), Lausanne, Switzerland

J. Tomblin Wichita State University, Wichita, KS, USA

A. Turon University of Girona, Girona, Spain

A.P. Vassilopoulos École Polytechnique Fédérale de Lausanne (EPFL), Lausanne, Switzerland

Woodhead Publishing Series in Composites Science and Engineering

1 Thermoplastic aromatic polymer composites
  F. N. Cogswell
2 Design and manufacture of composite structures
  G. C. Eckold
3 Handbook of polymer composites for engineers
  Edited by L. C. Hollaway
4 Optimisation of composite structures design
  A. Miravete
5 Short-fibre polymer composites
  Edited by S. K. De and J. R. White
6 Flow-induced alignment in composite materials
  Edited by T. D. Papthanasiou and D. C. Guell
7 Thermoset resins for composites
  Compiled by Technolex
8 Microstructural characterisation of fibre-reinforced composites
  Edited by J. Summerscales
9 Composite materials
  F. L. Matthews and R. D. Rawlings
10 3-D textile reinforcements in composite materials
   Edited by A. Miravete
11 Pultrusion for engineers
   Edited by T. Starr
12 Impact behaviour of fibre-reinforced composite materials and structures
   Edited by S. R. Reid and G. Zhou
13 Finite element modelling of composite materials and structures
   F. L. Matthews, G. A. O. Davies, D. Hitchings and C. Soutis
14 Mechanical testing of advanced fibre composites
   Edited by G. M. Hodgkinson
15 Integrated design and manufacture using fibre-reinforced polymeric composites
   Edited by M. J. Owen and I. A. Jones
16 Fatigue in composites
   Edited by B. Harris
17 Green composites
   Edited by C. Baillie
18 Multi-scale modelling of composite material systems
   Edited by C. Soutis and P. W. R. Beaumont

19 **Lightweight ballistic composites**
   Edited by A. Bhatnagar
20 **Polymer nanocomposites**
   Y-W. Mai and Z-Z. Yu
21 **Properties and performance of natural-fibre composite**
   Edited by K. Pickering
22 **Ageing of composites**
   Edited by R. Martin
23 **Tribology of natural fiber polymer composites**
   N. Chand and M. Fahim
24 **Wood-polymer composites**
   Edited by K. O. Niska and M. Sain
25 **Delamination behaviour of composites**
   Edited by S. Sridharan
26 **Science and engineering of short fibre reinforced polymer composites**
   S-Y. Fu, B. Lauke and Y-M. Mai
27 **Failure analysis and fractography of polymer composites**
   E. S. Greenhalgh
28 **Management, recycling and reuse of waste composites**
   Edited by V. Goodship
29 **Materials, design and manufacturing for lightweight vehicles**
   Edited by P. K. Mallick
30 **Fatigue life prediction of composites and composite structures**
   Edited by A. P. Vassilopoulos
31 **Physical properties and applications of polymer nanocomposites**
   Edited by S. C. Tjong and Y-W. Mai
32 **Creep and fatigue in polymer matrix composites**
   Edited by R. M. Guedes
33 **Interface engineering of natural fibre composites for maximum performance**
   Edited by N. E. Zafeiropoulos
34 **Polymer-carbon nanotube composites**
   Edited by T. McNally and P. Pötschke
35 **Non-crimp fabric composites: Manufacturing, properties and applications**
   Edited by S. V. Lomov
36 **Composite reinforcements for optimum performance**
   Edited by P. Boisse
37 **Polymer matrix composites and technology**
   R. Wang, S. Zeng and Y. Zeng
38 **Composite joints and connections**
   Edited by P. Camanho and L. Tong
39 **Machining technology for composite materials**
   Edited by H. Hocheng
40 **Failure mechanisms in polymer matrix composites**
   Edited by P. Robinson, E. S. Greenhalgh and S. Pinho
41 **Advances in polymer nanocomposites: Types and applications**
   Edited by F. Gao
42 **Manufacturing techniques for polymer matrix composites (PMCs)**
   Edited by S. Advani and K-T. Hsiao

43 Non-destructive evaluation (NDE) of polymer matrix composites: Techniques and applications
   Edited by V. M. Karbhari
44 Environmentally friendly polymer nanocomposites: Types, processing and properties
   S. S. Ray
45 Advances in ceramic matrix composites
   Edited by I. M. Low
46 Ceramic nanocomposites
   Edited by R. Banerjee and I. Manna
47 Natural fibre composites: Materials, processes and properties
   Edited by A. Hodzic and R. Shanks
48 Residual stresses in composite materials
   Edited by M. Shokrieh
49 Health and environmental safety of nanomaterials: Polymer nanocomposites and other materials containing nanoparticles
   Edited by J. Njuguna, K. Pielichowski and H. Zhu
50 Polymer composites in the aerospace industry
   Edited by P. E. Irving and C. Soutis
51 Biofiber reinforcement in composite materials
   Edited by O. Faruk and M. Sain
52 Fatigue and fracture of adhesively-bonded composite joints: Behaviour, simulation and modelling
   Edited by A. P. Vassilopoulos

Part One

Introduction to fatigue and fracture of adhesively-bonded composite joints

Investigating the performance of adhesively-bonded composite joints: standards, test protocols, and experimental design

A.J. Brunner
Swiss Federal Laboratories for Materials Science and Technology, Dübendorf, Switzerland

1.1 Introduction

1.1.1 General remarks

Adhesive bonding with polymeric or polymer-based adhesives (see, e.g., Awaja, Gilbert, Kelly, Fox, & Pigram, 2009 for a recent review of adhesion of polymers) has become an important joining technique in several areas of engineering (see, e.g., Baldan, 2004; Baldan, 2012; Banea & da Silva, 2009; Cognard, 2006). Notably, adhesive bonding is increasingly used for manufacturing parts and elements from fiber-reinforced polymer (FRP) composites for load-bearing structures (Kinloch, 1997). Important applications of adhesive bonding include the aerospace sector (e.g., Markatos et al., 2013; Park, Choi, Choi, Kwon, & Kim, 2010), automotive and general transportation industry (e.g., Feraboli and Masini, 2004; Tang, 2010) as well as ship building (e.g., Di Bella, Galtieri, Pollicino, & Borsellino, 2013; Diez de Ulzurrun, López, Herreros, & Suárez, 2007), but also civil and mechanical engineering (e.g., Brunner & Terrasi, 2008; de Castro & Keller, 2008a, 2008b; Keller, 2001; Vallée, Correia, & Keller, 2006; Vassilopoulos, Sarfaraz, Manshadi, & Keller, 2010). An important civil engineering application for adhesively-bonded joints from FRP composites is FRP bridges, discussed, e.g., by Cheng and Karbhari (2006) and Hizam, Manalo, and Karunasena (2013) who provide recent reviews. Since adhesively-bonded FRP composite load-bearing structures and elements are subject to quasi-static and cyclic or stochastic fatigue loads during service, assessment of their fatigue and fracture behavior is important for design, dimensioning, manufacturing, and processing, as well as for the determination of their service lifetime and of respective inspection intervals, including optimization of maintenance.

1.1.2 Objectives and current state of standardization

The focus of the present contribution is on fatigue, quasi-static fracture, and fatigue fracture behavior of adhesively-bonded joints made from pultruded glass-fiber-reinforced polymer (GFRP) profiles for civil engineering applications. However, most of the test methodology also applies to other types of adhesive joints made from FRP composites, e.g., carbon fiber–reinforced polymer (CFRP) materials, or hybrid joints with adherends made from different materials (see, e.g., Baldan, 2004; del Real, Ballesteros, Chamochin, Abenojar, & Molisani, 2011) as well as joints used in other areas of structural engineering, beside civil infrastructure, e.g., aerospace, automotive, and ship building industries.

So far, few standard test procedures for either fatigue or fracture behavior of adhesively-bonded joints made from FRP composites have been established, and none for either fatigue or for quasi-static or fatigue fracture of such joints made from pultruded GFRP profiles. One example of the former is the ISO 25217 standard on the determination of adhesive-fracture energy under the so-called Mode I, i.e., tensile opening loading that includes, among others, adhesively-bonded FRP composite beams as adherends. Other examples are standard fatigue tests that have been specified for metal–metal adhesive joints (e.g., ASTM D3166; EN 15190), but by scope these are not specified to be applicable for GFRP or CFRP composite adherends. It can be noted that EN 15190 includes environmental "loads," i.e., exposure, beside mechanical fatigue loading. Fatigue of FRP composites under tension–tension loading is standardized in, e.g., ASTM D3479/D3479M, but not explicitly specified for adhesively-bonded joints. A tensile shear fatigue standard for structural adhesives (ISO 9664) notes that the results are dependent on specimen geometry and hence cannot be used for design purposes. There is a guideline on the use of results from single-lap shear tests of adhesives (ASTM D4896) that specifies the range of applicability. In general, the use and the applicability of the test results are important issues that deserve detailed consideration. The transferability of laboratory-scale tests to engineering structures, to their design, or to estimating their service life is another issue, but this has not received much attention yet.

1.1.3 Overview of experimental work

Experimental tests reported on adhesively-bonded joints made from pultruded GFRP profiles in the literature are mainly based on test procedures developed for fatigue or quasi-static or fatigue fracture test procedures of FRP composite laminates or adhesive joints. Several procedures have been standardized or are in the process of standardization. For recent reviews on inter-laminar fracture properties of FRP composites on the one hand, see, e.g., Tay (2003), Brunner, Blackman, and Davies (2008), and Brunner (2014), and for details on tests developed for fracture of adhesively-bonded joints with arbitrary adherends, on the other, see, e.g., Blackman, Kinloch, Paraschi, and Teo (2003), Kinloch and Taylor (2004), and Blackman, Johnsen, Kinloch, and Teo (2008). Standardization agencies, such as the International Organization for Standardization (ISO), the American Society for Testing and Materials (ASTM) International, or the Comité Européen de Normalisation (CEN) and their websites (www.iso.ch, www.astm.org, www.cen.eu) are the main sources for standard tests that can potentially be adapted to the specific test requirements of adhesively-bonded joints made from pultruded GFRP profiles. These aspects will be discussed in more detail in the next section.

Experimental fatigue testing in general is based on applying a cyclic or more generally a dynamic (time-varying) load spectrum to the specimens. That can consist of constant or variable peak load or constant or variable peak displacement. The cases of constant peak load and peak displacement yield load and displacement controlled fatigue tests, respectively. The change of amplitude or displacement can be piece-wise linear in time (e.g., triangular), sinusoidal, intermittent, or almost instantaneous (e.g., square or rectangular load profiles as a function of time) and of slow, intermediate, or high rate or combine these in an emulation of real or in simulated service load spectra. Loads can be tensile, compressive, shear, or combinations of these and be applied unidirectionally, biaxially, or multi-directionally. The effects of these loads are then evaluated continuously or intermittently by, e.g., monitoring mechanical properties such as E-modulus, compliance, or strength of the specimens as a function of time or number of fatigue cycles. Experimental inter-laminar fracture testing of FRP composites is performed in one of the so-called basic Modes, i.e., tensile opening (called Mode I), in-plane shear (called Mode II), out-of-plane shear or twisting (called Mode III) or in combinations of these, i.e., various mixed modes. Loads are applied quasi-statically, at intermediate or high rates, or dynamically for fatigue fracture, see, again, e.g., Tay (2003), Brunner, Blackman, et al. (2008) or Brunner (2014) for reviews and more details on these test procedures and test-set-ups. These tests can be performed in a standard laboratory environment, e.g., conditioned to $+23$ °C and 50% relative humidity with defined variation limits according to ISO 291. Test standards usually prescribe (material-dependent) specimen conditioning and test climate to obtain comparable data because of the pronounced temperature and humidity dependence of polymer properties. Testing specimens at elevated temperature and humidity, or after and under exposure to fluid or gaseous media or to ultraviolet (simulating sunlight exposure) or higher energy electromagnetic radiation has been explored in research, but not standardized yet. Testing at (standard or non-standard) environmental conditions can also be performed after storage under constant or varying, combined environmental exposure for different durations. The environment can be chosen in an attempt at simulating specific service conditions or to accelerate aging (e.g., by storage at elevated temperature beyond the expected service temperature) for assessing service lifetime and durability of the joints. Environmental conditions can be combined with mechanical loads, but this may require elaborate test set-ups with monitoring and control of environment and of changes in the performance of the joints.

1.1.4 Overview of manufacturing and processing

Pultruded GFRP profiles can be manufactured in an essentially continuous process from various types of glass-fiber products, e.g., fiber strands, woven rovings, chopped strand mats, or non-woven veils that are embedded in a polymer matrix. In decreasing order of importance, polyester, vinyl ester, and epoxy are the main matrix materials for pultruded GFRP profiles (Mackerle, 2004). Bakis et al. (2002) provide a general and concise review of pultrusion manufacturing and the use of FRP composite materials in civil engineering that covers historical developments, state of the art (around the year 2000), and future challenges. Mackerle (2004) compiled a bibliography between 1985 and 2003 on finite element analysis and simulation of manufacturing and the mechanical properties of composites in general that includes a section on pultrusion. The challenges of the pultrusion process are discussed by, e.g., Miller, Dodds, Hale, and Gibson (1998) and in several of the references compiled by Mackerle (2004).

Pultrusion manufacturing can be set up to yield different sizes and shapes of the cross-sections of the resulting GFRP pultruded product, e.g., circular (rod), rectangular (beam or plate), I- or T-profile, or tube or box (circular and square or rectangular, hollow profiles, respectively). Selected examples of GFRP pultruded profiles in adhesively-bonded joints and their mechanical behavior, including strength, damage, and failure, are discussed, e.g., by Davalos, Salim, Qiao, Lopez-Anido, and Barbero (1996), Vallée, Keller, Fourestey, Fournier, and Correia (2009), Regel, van Hattum, and Dias (2013), and Feo and Mancusi (2013). In civil engineering, in particular, plate- or beam-type and I-shaped profiles have been used for test development and structural design with adhesively-bonded joints, respectively. The variety of pultruded products, in principle, yields a large range of different types of adhesively-bonded, structural joints. This variety is one factor that makes development and standardization of fatigue, quasi-static fracture, and fatigue fracture tests for such joints difficult.

Depending on the pultrusion process and its control, there can be significant variation in mechanical properties along the length of the profile or in the normal and transverse directions. The joints are then made from the profiles using suitable surface preparation techniques and polymeric or polymer-based adhesives. Polyurethane (PUR) or epoxy (EP)—based adhesives are the main choices for the adhesively-bonded joints. There is extensive literature dealing with performance and characterization of PUR adhesives used for manufacture of wooden joints, see, e.g., Clauß et al. (2011) or Sterley, Serrano, and Enquist (2013), which provides useful guidelines for GFRP adhesively-bonded joints. A general specification for epoxy adhesives is compiled in ASTM D6412/D6412M. Polymer-based adhesives, as polymers in general, may show rate-dependent properties, see, e.g., Yu, Crocombe, and Richardson (2001) discussing the case of epoxies and suitable modeling of properties. Improvement of epoxy adhesives for better adhesion performance has been reviewed by Ratna (2003). There are several guidelines on the preparation of adhesives joints, see, e.g., ASTM D2093, ASTM D6105, ASTM E1307, ISO 17212, and EN 13887. ISO 21368 provides a general guideline on adhesive joints in structures, which includes information on risk evaluation and respective reporting that may, in part, also be applicable to the specific joints discussed here.

1.1.5 Joint types and designs

In the literature, there is a variety of adhesively-bonded joint type designs, e.g., the simple adhesive bonding of beams or plates, single-lap joints (widely used in adhesives testing) or double-lap joints, scarf, bevel, and butt joints (including butt strap and double butt strap) and peel joints; these are discussed, e.g., by He (2011). Additional types, such as joggle lap joints and the L-section joints, are discussed by Taib, Boukhili, Achiou, Gordon, and Boukehili (2006) and stepped-lap joints by Zhang, Vassilopoulos, and Keller (2010a). Of course, for structural design, these basic joint types have to be adapted to the structure and its required performance. Test development for determining fatigue, fracture, or fatigue fracture behavior, on the other hand, is probably best based on simply-bonded, single-lap, stepped-lap, and double-lap joints, since these specimens conform to or at least provide similar aspect ratios as the standard specimens developed for other purposes, e.g., adhesives characterization or composites fracture and fatigue testing. The approaches and related problems for designing structural elements or structures with adhesively-bonded joints using data obtained from standard or non-standard testing of laboratory-scale or -size joint specimens are beyond the scope of this chapter It can, however, be noted that ASTM D5573 provides a base for classifying failure modes in FRP composite joints that may be useful in the interpretation of structural tests.

An important aspect of adhesively-bonded joints is the detailed geometric design of the joints, specifically of the adherends. Zhao, Adams, and da Silva (2011a, 2011b) investigated the effects of rounded corners of the adherends on the joint strength and determined respective stresses and strains, as well as a model for strength prediction of such joints. While the effects on the fatigue, on the fracture, and on the fatigue fracture behavior of the different joint designs have not been elaborated in detail, it is clear from the analysis that the design details will play an important role and hence have to be considered. It would be useful to extend the investigation into different adhesive types and to look into industrially reasonable manufacturing tolerances for the geometry of the adherend(s), of the adhesive (e.g., uniformity of thickness) as well as of the (relative) alignment and the respective effects on their fatigue and fracture behavior.

While there are standards on design, preparation, and manufacturing of adhesively-bonded FRP composite joints, there are no standardized procedures on experimental fatigue, fracture, and fatigue fracture testing of such elements manufactured for structural applications yet. The following section hence deals with standard tests or test procedures that have been established for other purposes, but have either been shown to be applicable to adhesively-bonded joints made from pultruded GFRP profiles or have potential applications in this area.

1.2 Standards and test protocols for experimental fatigue and fracture testing of adhesively-bonded composite joints

1.2.1 Basic considerations

As noted in the introduction, there are no specific standard test methods for fatigue or fracture testing of adhesively-bonded joints made from pultruded GFRP profiles. The relatively large variety of basic joint designs for engineering applications and civil engineering structures, e.g., the eight types described by He (2011) and the different cross-sections of pultrusion products, clearly make standardization difficult. One approach that can lead to standardization is to look for existing test standards for other types of adhesively-bonded joints, e.g., those developed for characterization of fatigue, fracture, or fracture fatigue properties (e.g., ASTM D3166; ASTM D5868; ASTM D1002; ISO 9664; EN 15190; ISO 9664) or for durability testing of adhesives (e.g., ASTM D2919; ISO 10354; ISO 14615). Another, alternative approach can be based on respective tests developed for fatigue or for inter-laminar, quasi-static, or fatigue fracture properties of FRP composite laminates (ASTM D3479/D3479M; ASTM D5528; ASTM D6115; ASTM D6671; ISO 15024).

In the case of fracture testing, standard tests and test methods published in the literature for inter-laminar fracture of FRP composites can, in principle, be easily adapted for investigating adhesively-bonded joints from pultruded GFRP profiles, as long as the profiles are beam-like and symmetrical with respect to the adhesive layer. The basic idea then is to regard the adhesive layer as analogous to the inter-laminar polymer layer between the fiber plies in continuous fiber laminates in which the crack or delamination is propagated, as discussed by, e.g., Tay (2003), Brunner, Blackman, et al. (2008), and Brunner (2014) for specimens used in inter-laminar fracture and fatigue fracture of FRP composites. Asymmetric specimens for inter-laminar fracture, with respect to thickness of the half beams and to half beams from different composite or laminate types manufactured into joints, are discussed in the respective literature, but not within the scope of standard test procedures yet.

Standard fracture tests developed for characterization of fracture properties of adhesives with either metal or composite adherends (ISO 25217; ASTM D5041; ASTM D3433) as well as fracture tests under development (e.g., ISO DIS 15114; ASTM WK 22949) can, therefore, be adapted for adhesive joints made from GFRP pultruded profiles. This also holds for fatigue fracture, where standardized test procedures are still under development and several issues are still debated, see, e.g., Brunner, Terrasi, Vallée, and Tannert (2009), Stelzer, Brunner, Argüelles, Murphy, and Pinter (2012), and Stelzer, Jones, and Brunner (2013) for detailed discussion. The so-called "tapered double cantilever beam (TDCB)" specimen developed for Mode I (tensile opening load) fracture of adhesives in ISO 25217 would not be useful for joints made from beam-like adherends with rectangular cross-section as the GFRP profiles, but the other specimen types defined in this standard can easily be adapted.

1.2.2 Fatigue testing

Fatigue testing of composites and adhesives (excluding fatigue fracture where a delamination or crack is extended under the action of cyclic loads) can be performed under different types of mechanical loading, e.g., tensile—tensile, tensile—compressive, compressive—compressive, or under more complex biaxial or multi-axial loads. The effect of such loads is typically a degradation of mechanical properties, sometimes yielding combined thermomechanical degradation from hysteretic heating of the material due to the cyclic loads. Also, environmental exposure can produce cyclic fatigue effects, e.g., from daily or seasonal temperature variations or from hygrothermal variations that lead to cyclic or stochastic thermal or hygrothermal expansion and contraction. These effectively result in cyclic mechanical stresses and strains, which, in the case of structural applications, are superposed on the applied mechanical service loads. The complexity of these load cases is further increased by possible internal stresses from manufacturing and processing, and by differences in the hygrothermal properties and behavior of the joint components (e.g., adhesive and pultruded adherend).

A comprehensive study of fatigue testing for adhesively-bonded joints made from pultruded GFRP profiles has been initiated by Keller and Tirelli (2004), Zhang and Keller (2008), Zhang, Vassilopoulos, and Keller (2008), and Zhang et al. (2010a) and continued by Sarfaraz, Vassilopoulos, and Keller (2011), Sarfaraz, Vassilopoulos, and Keller (2012), and Sarfaraz, Vassilopoulos, and Keller (2013a). These will be discussed in the next section that deals specifically with such joints.

Nevertheless, there are investigations on other types of composite joints that may also be considered in the development of fatigue test procedures for adhesively-bonded joints made from pultruded GFRP profiles. Chuang and Tsai (2013) compare the fatigue behavior of adhesively-bonded double lap joints made from CFRP with that of stepwise double-patched joints (also from CFRP), which show improved performance with respect to strength as well as improved fatigue life. This was attributed to the reduction in shear and peel stresses in the patched joints.

Fatigue performance characterization requires corresponding quasi-static tests for interpretation as well as for basic material properties for modeling and simulation. These have been investigated by, e.g., de Castro and Keller (2008a, 2008b), and Lee, Pyo, and Kim (2009). The latter use adhesively-bonded double-strap and single-lap joints made from pultruded GFRP vinyl ester profiles and investigated the effects of adhesive type, thickness of adhesive layer, and overlap length. The adhesive type was shown to have little effect, but joint strength decreased with increasing thickness of the adhesive layer and increased with overlap length. With respect to strength, the double-strap joints yielded better performance than the single-lap joints. This agrees in part with another study by Vallée, Correia, and Keller (2010) on quasi-static tensile experiments and modeling pultruded of GFRP double-lap joints bonded with a PUR adhesive where increasing overlap length yielded higher quasi-static joint strength, but in a range of 0.3—10 mm, the thickness of the adhesive layer yielded the highest strength for a thickness around 1 mm. However, the data show significant scatter in the plot presenting strength versus adhesive thickness. Campilho, Banea, Neto, and daSilva (2013) investigated modeling of single lap adhesive joints with two different epoxy adhesives using different types of cohesive zone models (CZMs) and compare their finite element analysis (FEA) results with experiments. The particular choice of the cohesive zone model is important for ductile adhesives and of essentially negligible effect for brittle adhesives.

1.2.3 Fracture testing

A recent review (Brunner, 2014) of quasi-static, high-rate, and fatigue fracture testing of FRP composites lists standard tests, tests under development for standardization, as well as specimen designs and test procedures reported in the literature for various types of loading and fracture Modes. All the specimen designs noted in the review are essentially beam- or plate-like, and most of the tests have been developed for inter-laminar or intra-laminar fracture (with a few exceptions dealing with translaminar fracture, which is defined as "fiber-breaking," see, e.g., Laffan, Pinho, Robinson, & McMillan,

2012). Therefore, the specimen types and test designs can be adapted for beam- or plate-like adhesively-bonded joints. Of course, test set-ups will have to be up-scaled for accommodating the typical size of the joints, but this is essentially a straightforward exercise as long as the strength and stiffness of the set-up ensures the applicability of linear-elastic fracture mechanics, which is the basis for essentially all fracture and fatigue fracture tests developed for FRP composites to date. One example of an up-scaled set-up is shown in Brunner, Blackman, et al. (2008) in Figure 9 for the so-called mixed mode bending I/II (MMB) fracture test set-up for GFRP composite profiles with woven reinforcement, which, however, have not been manufactured by pultrusion in this case.

For Mode I fracture, i.e., the tensile opening Mode, the so-called DCB specimen is used for the standard test under quasi-static load (ASTM D5528; ISO 15024; and the Japan Industrial Standard JIS K 7086). The same specimen is being used for the tensile (Mode I) fracture fatigue and high-rate Mode I tests under development (see, e.g., Brunner, Blackman, et al., 2008). Mode I fracture of adhesively-bonded T-joints under three-point bending has been modeled by Fernlund, Chaaya, and Spelt (1995). For Mode II, i.e., in-plane shear tests, at least two specimen types, end-loaded split (ELS) and end-notched flexure (ENF), compete for standardization, which for quasi-static loads is almost complete (JIS K 7086; ASTM WK 22949; ISO DIS 15114). Again, the same specimens are likely to be considered for fatigue loading (Brunner, Stelzer, Pinter, & Terrasi, 2013). So far, no test development has been attempted toward a high-rate version of Mode II, and it does seem rather unlikely that such a test will be developed. Mode III loading, i.e., out-of-plane shear or twist, has not been standardized for any loading case yet, even though the so-called edge-crack torsion (ECT) test has been explored in early quasi-static load developments (see, e.g., Browning, Carlsson, & Ratcliffe, 2010; Brunner, Blackman, et al., 2008; de Morais, Pereira, & de Moura, 2011; de Morais, Pereira, de Moura, & Magalhães, 2009). Mode III typically involves a plate-like composite specimen with a starter crack, which typically yields an intrinsic Mode II component (see, e.g., de Morais & Pereira, 2008 for a detailed discussion). Whether such a specimen with suitable aspect ratio could be manufactured by adhesively-bonding pultruded GFRP profiles will have to be investigated. Yoshihara (2006) examines the use of the so-called four point bending end-notched flexure test (4-ENF) originally developed as an alternative to the three-point bending ENF-test for Mode II fracture testing of FRP composites (see, e.g., Davidson, Sun, & Vinciquerra, 2007 for details) for determination of the Mode III resistance curve in wood. Since wood is similar to FRP composites in fracture testing, this may provide an approach to Mode III testing of adhesively-bonded joints, avoiding the complexity of the ECT test set-up and its special specimen design (Pereira, de Morais, & de Moura, 2011).

For mixed mode loading, several Mode combinations are feasible. Quasi-static mixed Mode I/II fracture has been standardized in ASTM D6671 and a fixed-ratio mixed Mode I/II test based on the ELS set-up, but with the specimen inverted compared to the Mode II test, has been explored in preliminary tests (see Brunner, Blackman, et al., 2008; Brunner, 2014). Mixed Mode I/II fracture and the basic Modes I and II of CFRP composite joints with a film adhesive (type FM 300M from Cyctec) have been investigated numerically and experimentally in a well-documented report by Balzani et al. (2011). They found that the critical energy release rate is the major parameter characterizing the adhesive fracture. Attempts at FEA modeling proved difficult and required some empirical fitting of critical energy release rates for obtaining them as a function of Mode mixity ratio.

Composite bonded joints with a ductile epoxy adhesive and CFRP epoxy adherends have been investigated under quasi-static Mode I loading using the DCB specimen by de Moura, Campilho, and Gonçalves (2008) and under quasi-static Mode II using the ENF specimen by the same group (de Moura, Campilho, & Gonçalves, 2009). The experimental tests were complemented by numerical analyses. For the Mode I case, a compliance-based beam method for determining crack length was compared with corrected-beam theory and reasonable agreement was obtained. This will prove useful for future testing, if visual determination of crack length can be replaced by the so-called equivalent crack length approach. The same holds for Mode II. There is potential for using the equivalent crack length concept also for fracture fatigue.

Mixed Mode I/III and mixed Mode II/III fracture (Fernlund, Lanting, & Spelt, 1995; Kondo, Sato, Suemasu, & Aoki, 2011) and other combinations, e.g., I-II-III fracture using a shear-torsion-bending test proposed by Davidson and Sediles (2011), have been investigated in quasi-static test development. Shindo, Miura, Takeda, Saito, and Narita (2011) and Miura, Shindo, Takeda, and Narita (2014) have explored fatigue fracture of GFRP composites at low (cryogenic) temperatures under mixed Mode I/II and Mode I/III, and maybe mixed Mode II/III will follow. mixed Mode I/II fatigue of multi-directional laminates with three different Mode mixity ratios has been investigated by Peng, Xu, Zhang, and Zhao (2012). They found that a modified Paris law approach yielded better results than the conventional Paris law, due to significant fiber bridging and crack branching, which occurred in the multi-directional laminates. Fiber bridging is frequently observed in Mode I tests, but combined with crack branching, the effects become even more complex and cannot be approached analytically anymore. Again, this is important for adhesively-bonded joints from pultruded GFRP profiles, since fiber bridging and crack branching, the latter at least to some extent, have been observed in Mode I tests (see, e.g., Shahverdi, Vassilopoulos, & Keller, 2011).

Composite joints from CFRP epoxy adherends and an epoxy adhesive have also been characterized under quasi-static and fatigue mixed Mode I/II loading, both experimentally and with phenomenological modeling as well as with a detailed damage model, by Carraro, Meneghetti, Quaresimin, and Ricotta (2013a, 2013b). They also performed quasi-static Mode I and Mode II tests for comparison with the mixed Mode I/II results. They found a dependence of the fatigue crack path on the mode mixity, going from the adherend—adhesive interface into the adhesive with increasing amount of Mode II contribution.

Fracture of adhesives under quasi-static Mode I loading has been investigated for aluminum adherends and thin adhesive layers by, e.g., Steinbrecher, Buchman, Sidess, and Sherman (2006). Numerical and experimental work tried to take the elasticity of the adhesive into account for calculating the fracture energy and a method for an improved determination of that was suggested, based on the results from a range of adhesives. da Silva, Esteves, and Chaves (2011) used asymmetric tapered DCB (ATDCB), single leg bending (SLB) and asymmetric DCB (ADCB) specimens with steel adherends for determining the quasi-static fracture toughness of a ductile epoxy adhesive under mixed mode loading, but performed pure Mode I and Mode II tests for comparison as well. They concluded that a small Mode II shear loading contribution reduced the total G compared with the pure Mode I tensile opening load.

Mode I fatigue fracture of composite joints (run at 4 Hz and an R-ratio of 0.1, i.e., a ratio of minimum to maximum load and displacement, respectively, with 50% of the static fracture load as maximum load) was investigated by Fernández, deMoura, daSilva, and Marques (2011) using CFRP adherends and an adhesive already characterized by Campilho, de Moura, and Domingues (2007) and used in a damage model for repaired CFRP laminates. They compared a compliance-based beam method (essentially an equivalent crack length method, see, e.g., Brunner, Blackman, et al., 2008 for details) for the analysis of the DCB adhesive joints that did not require crack length monitoring while simultaneously taking the effects of the fracture process zone into account with visual observation of crack length via a traveling microscope. The paper explicitly notes the necessity of a finite element model (FEM) for deducing the compliance calibration curve.

The problem of visual crack length detection in adhesively-bonded joints during quasi-static Mode I tests on DCB-specimens has been investigated by Richter-Trummer et al. (2011) who proposed using digital image correlation (DIC) and automated image analysis. They compared four different algorithms for crack detection that all worked reasonably well. The DIC method can be applied to Mode I fatigue fracture and hence provides an alternative to visual detection with a traveling microscope and compliance-based "equivalent crack" length analysis. The DIC method could hence provide a validation for compliance-based crack length determination and in the future maybe yield insight into the details of the fracture process zone development. Alternative methods for determination of crack lengths (e.g., crack gauges, video extensometers) have also been investigated (see, e.g., Zhang and Keller, 2008; Zhang et al., 2010a; Zhang, Vassilopoulos, & Keller, 2010b).

Mode I fracture of adhesively-bonded joints (essentially reporting the results leading to the standard ISO 25217) has been characterized by Blackman et al. (2003). They found a dependence of the resulting G_{IC} values on the type of adherend that could be traced to differences in glass-transition temperature of the adhesive in its final cured state in the joint. A further study by Blackman et al. (2008) highlighted the effects of pre-bond moisture in the CFRP adherends on the measured fracture toughness of the adhesive. These studies essentially point out processing and manufacturing effects that might crucially affect the performance of the adhesively-bonded joints. Whether these effects will also be relevant for GFRP adherends, and if so to what degree, is an open question. The amount of moisture uptake in GFRP may exceed that in CFRP epoxy composites, and it can hence be speculated that avoiding moisture effects, i.e., drying the adherends before bonding, may be advisable. Of course, guidelines on preparation of adherend surfaces before bonding (e.g., ASTM D2093) should also be consulted.

The rate dependence of the fracture of adhesively-bonded joints under Mode I, Mode II, and mixed Mode I/II loading has been investigated by Blackman, Kinloch, Rodriguez-Sanchez, and Teo (2012) from quasi-static to high rates (up to 15 m/s, i.e., typical impact speeds). They found a dependence of the fracture path (in the adhesive layer, cohesive or inter-laminar in the adherend) on the type of substrate, the type of adhesive, and on the type of loading Mode. The inter-laminar fracture path did seem to occur more frequently for mixed Mode I/II than either of the pure Modes.

Specific load cases for adhesively-bonded joints with CFRP and GFRP adherends—not necessarily manufactured via pultrusion—can be found in the literature, e.g., out-of-plane shear in a publication by Sohier, Cognard, and Davies (2013). This study intended to provide a basis for numerical models and identified composite properties and the bonding process as essential factors.

Fracture of adhesively-bonded single-lap joints with aluminum adherends has been reviewed and investigated by Kafkalidis and Thouless (2002). The study specifically looked into different Mode mixes and the problems of simulation of the behavior of such joints using cohesive zone elements. They found that linear elastic fracture mechanics is applicable, unless plastic effects become important, e.g., for high toughness adhesives, or low thickness or low yield stress of the adherend.

The literature summarized indicates that the test development for basic loading Modes and selected Mode mix for either FRP composite laminates or adhesive joints is under way, at least for the quasi-static load case. Fatigue fracture and high-rate fracture of composites are currently being explored in Mode I tests and, partially in Mode II and mixed Mode I/II tests. Standardization of these test procedures may still take considerable time, and reproducibility and accuracy may still be crucial issues to be solved. Once these procedures are established for FRP composites, it is expected that extension to adhesively-bonded FRP composite joints should be straightforward. As noted, Mode III and mixed mode test procedures involving Mode III (e.g., mixed Modes I/III, II/III, or I/II/III) may pose problems with respect to specimen shape (aspect ratio) and the resulting, effective Mode mixity.

Overall, there is a basis for exploring the applicability of these fracture standards and test procedures to the specific adhesively-bonded joints made from pultruded GFRP profiles. Test standards and test procedures reported in the literature for adhesively-bonded joints and FRP composites do seem adaptable to the specific test set-ups and test specimens. The next section will provide an overview of such tests and of the results obtained from them.

A recent investigation by Azari, Ameli, Papini, and Spelt (2013) deals with the effect of the thickness of aluminum adherends on the Mode I fracture fatigue of adhesive joints and reports a reduced fatigue performance, i.e., reduced threshold value and increased crack growth rate. The joints, however, were made from aluminum adherends and an epoxy adhesive with a thickness of 380 µm. The paper discusses other published investigations on the effect of adherend thickness, but it is not clear whether and, if so, how changing the thickness of the GFRP profiles would affect the performance of the adhesively-bonded joints made from them.

1.3 Standards and test protocols for fatigue and fracture testing of pultruded glass-fiber reinforced polymer-matrix (GFRP) profiles

1.3.1 Fatigue testing of adhesively-bonded joints made from pultruded GFRP profiles

So far, there are no standards on fatigue testing of adhesively-bonded joints made from pultruded GFRP profiles, but recent research literature reports testing procedures that, in the future, may become the basis for standardization. As expected, single-lap, stepped-lap, and double-lap joints represent the bulk of the specimens in this case. Table 1.1 is a synopsis of the currently available literature on fatigue testing. Keller and Tirelli (2004) investigated the tensile fatigue behavior of adhesively-bonded double lap joints made from pultruded GFRP profiles with epoxy adhesive. Quasi-static tests were performed for comparison with the fatigue behavior of the joints, but also for characterization of the properties of the components. Fatigue tests were run at amplitude ratios between 0.10 and 0.18 at frequencies between 1 and 10 Hz and no significant temperature increase was noted. The analysis showed a fatigue limit (above 10 million cycles) around 25% of the static failure strength. Failure, however, was always brittle with no sign of damage prior to failure.

An investigation of quasi-static strength and damage behavior of adhesively-bonded joints made from pultruded GFRP profiles (Zhang & Keller, 2008) showed the complexity of the damage occurring inside the pultruded profile. Following up on that, Zhang et al. (2008, 2010a), and Sarfaraz et al. (2011, 2012, 2013a, 2013b) also investigated the fatigue performance of such adhesively-bonded joints, both experimentally and with simulations, in an attempt at deriving models describing the fatigue performance. Zhang et al. (2008) investigated the fatigue performance of single- and double-lap joints (double butt strap joints in the terminology of He, 2011) with epoxy adhesives under tensile fatigue loading. They found a critical value of elongation and stiffness for of single- and double-lap joints, respectively, at which failure occurred independently of load level. Double-lap joints yielded almost linear stiffness degradation independent of load level, while single-lap joints showed such a behavior only in a limited range (around 20–80%) of crack propagation life. This behavior could best be described by a linear and a non-linear model, respectively.

The next step was an investigation of the fatigue life under constant amplitude loading (Zhang et al., 2010a). The fatigue crack curves showed a similar slope for both types of joint (stepped lap and double lap), but a different constant for the power law. This was tentatively explained by different Mode mixities in the fatigue fracture and by possible different crack propagation paths with different layup. It was further noted that the fracture mechanics data compared well with fatigue data (stiffness degradation). The latter, however, seemed to provide more conservative data for high cycle fatigue.

Zhang, Vassilopoulos, and Keller (2009) also investigated the effects of selected environmental parameters (temperature and humidity) on the fatigue performance

Table 1.1 **Literature summary of fatigue tests on adhesively-bonded joints made from pultruded GFRP profiles**

Test type	References	Adhesive type	GFRP profile	Remarks
Tensile–tensile fatigue	Keller and Tirelli (2004)	Two-component epoxy	E-glass in isophthalic polyester	Double-lap joint, R-ratio $= 0.10$–0.18, fatigue limit at 25% of the static failure load at 10 million cycles
	Zhang et al. (2008)	Epoxy SikaDur 330	E-glass in isophthalic polyester	Double-lap and stepped-lap joints, R-ratio $= 0.10$, 10 Hz
	Zhang et al. (2010a)	Epoxy SikaDur 330	E-glass in isophthalic polyester	Double-lap and stepped-lap joints, load control, R-ratio $= 0.10$, 10 Hz
	Sarfaraz et al. (2011)	Epoxy SikaDur 330	E-glass in isophthalic polyester	Double-lap joint, R-ratio $= 0.10$, 10 Hz
	Sarfaraz et al. (2012)	Epoxy SikaDur 330	E-glass in isophthalic polyester	Double-lap joint, R-ratio $= 0.5$ and 0.9
	Sarfaraz et al. (2013a)	Epoxy SikaDur 330	E-glass in isophthalic polyester	Double-lap joint, R-ratio $= 0.1$, 10 Hz, low-high and high low stress level sequences
	Sarfaraz, Vassilopoulos, and Keller (2013b)	Epoxy SikaDur 330	E-glass in isophthalic polyester	Modeling, double-lap joint, $R = 0.1$, 10 Hz

Continued

Table 1.1 Continued

Test type	References	Adhesive type	GFRP profile	Remarks
Compressive–compressive fatigue	Sarfaraz et al. (2011)	Epoxy SikaDur 330	E-glass in isophthalic polyester	Double-lap joint, R-ratio = 10.0, 10 Hz
	Sarfaraz et al. (2013a)	Epoxy SikaDur 330	E-glass in isophthalic polyester	Double-lap joint, R-ratio = 10.0, 10 Hz, low-high and high-low stress level sequences
	Sarfaraz et al. (2013b)	Epoxy SikaDur 330	E-glass in isophthalic polyester	Modeling, double-lap joint, R-ratio = 10.0, 10 Hz
Tensile-compressive fatigue	Sarfaraz et al. (2011)	Epoxy SikaDur 330	E-glass in isophthalic polyester	Double-lap joint, R-ratio = −1.0, 10 Hz
	Sarfaraz et al. (2012)	Epoxy SikaDur 330	E-glass in isophthalic polyester	Double-lap joint, R-ratio = −0.50
	Sarfaraz et al. (2013b)	Epoxy, SikaDur 330	E-glass in isophthalic polyester	Modeling, double-lap joint, R = −1.0, 10 Hz

of adhesively-bonded joints made from pultruded GFRP profiles joints that are again described as "double-lap joints," but effectively correspond to the "double butt joint" type described by He (2011). The authors also briefly review the older literature on environmental testing of adhesively-bonded composite joints (e.g., Ashcroft, Hughes, Shaw, Wahab, & Crocombe, 2001; Ferreira, Reis, Costa, & Richardson, 2002; Gilmore & Shaw, 1974; Gregory & Spearing, 2005). Most of that focused on aerospace rather than civil engineering applications, but these test procedures can, nevertheless, be adapted to adhesively-bonded joints for civil engineering applications. Stiffness degradation in fatigue seemed to be virtually linear up to failure independent of load level. At temperatures between −35 °C and +40 °C, high levels of humidity seemed to shift the failure type from adhesive to interfacial. Higher temperature yielded longer crack lengths due to softening of the adhesive. It can also be noted that crack gauges were used to monitor delamination lengths in these tests and they performed well in the range of environmental conditions that was investigated.

The effect of different load ratios (0.1, as in the previous test, −1 and +10) on the fatigue behavior of adhesively-bonded joints made from GFRP pultruded profiles was investigated by Sarfaraz et al. (2011). The experiments focused on double-lap joints and included quasi-static tensile and compressive tests, both at low and high loading rates for comparison. In fatigue, the failure occurred in a different layer, depending on the load ratio. This complex behavior also affects the modeling of the fatigue life. It is noted that linear models are not sufficient to describe the fatigue behavior and more sophisticated models will have to be developed, e.g., using the approach described by Vassilopoulos Manshadi, and Keller (2010) for FRP composite materials. The experimental and modeling investigation was continued (Sarfaraz, Vassilopoulos, & Keller, 2012) by focusing on the mean load effect on the fatigue behavior. The range of R-ratios was extended with seven ratio values between pure tension and compression. Again, the failure mode was found to depend on the load ratio, i.e., going from tensile to compression. The next step in the fatigue characterization involved block loading (Sarfaraz, Vassilopoulos, & Keller, 2013a). A significant effect was found for different sequences of two block loading patterns, yielding more damage for frequent changes of load levels. This was tentatively attributed to differences in crack propagation. It was noted that currently available models are inadequate to describe the experimentally observed behavior and this is hence an opportunity for future research. There are attempts at formulating tools for life prediction of adhesively-bonded joints made from pultruded GFRP profiles under fatigue loads, see, e.g., Vassilopoulos Sarfaraz, Manshadi, and Keller (2010), but the models can still be further refined and improved. The complex layup of the profiles combined with the complex stress states under realistic loading is still a challenge for predicting the long-term behavior. Even models for strength predictions are still being discussed and gradually improved (see, e.g., Vallée et al., 2006, 2009). All the models, however, do not yet take aging effects due to environmental exposure into account. The complexity of these effects will briefly be discussed in another section.

1.3.2 Fracture testing of adhesively-bonded joints made from pultruded profiles

Again, as in the case of fatigue tests, there are no standards on fracture or fatigue fracture testing of adhesively-bonded joints made from pultruded GFRP profiles, but recent research literature reports testing procedures that, in the future, may become the basis for standardization. The specimens are typically simply-bonded GFRP profiles or plates. Table 1.2 provides a synopsis of the currently available literature. With respect to quasi-static fracture of adhesively-bonded joints made from pultruded GFRP profiles, the tensile opening (Mode I), the in-plane shear (Mode II), and a range of mixed Mode I/II cases have been investigated, both experimentally and numerically by Zhang et al. (2010a, 2010b), Shahverdi et al. (2011), Shahverdi, Vassilopoulos, and Keller (2012a, 2012b), and Shahverdi, Vassilopoulos, and Keller (2013a, 2013b).

Fatigue under constant load amplitude producing crack initiation and propagation, i.e., fracture in double-lap and stepped-lap joints was investigated by Zhang et al. (2010a). Crack gauges were used to monitor the joints for crack initiation and propagation. The data were converted into fatigue crack growth curves that showed a difference between double- and stepped-lap joints. The double-lap joints yielded higher crack propagation or growth rates for a given value of maximum applied strain energy release rate. The two joints types also differed with respect to the shape of the compliance versus crack length curves, which hence required different fitting laws.

Taib et al. (2006) investigate adhesive layer thickness, fillet radius, and adherend stiffness on the quasi-static shear fracture (Mode II) of four different types of adhesively-bonded joints with composite adherends by means of tension test (based on ASTM D3165 for adhesives testing). Further, they also looked into the effects of humidity. Different failure modes were observed, but fracture mechanics data (e.g., energy release rates) were not evaluated from these tests.

Mode I and Mode II quasi-static fracture was investigated experimentally and with FEM by Zhang, Vassilopoulos, and Keller (2010b). Specimens used were the "standard" inter-laminar fracture types (DCB and ELS), but with aspect ratios adapted for the GFRP profiles. Crack lengths were monitored visually and with a video extensometer. The comparative analysis, including three-dimensional FEM simulations using the virtual crack closure technique (VCCT) confirmed the applicability of the test and analysis methods originally developed for FRP composite laminates. Only simple-beam theory analysis showed some differences in underestimating the energy release rates G_{IC} and G_{IIC}, but this can be attributed to neglecting essential corrections in this type of analysis.

Quasi-static and fatigue fracture with Mode I, Mode II, and several ratios of mixed Mode I/II was investigated by Shahverdi et al. (2011, 2012a, 2012b, 2013a, 2013b) in a series of papers. Following Zhang et al. (2010b), all tests were based on set-up and specimen designs originally developed (and partly standardized) for FRP composite laminates. Shahverdi et al. (2011) first investigated quasi-static Mode I fracture of simple bonded joints (emulating the DCB specimen developed for FRP composite

Table 1.2 Literature summary of fracture tests on adhesively-bonded joints made from pultruded GFRP profiles

Test type	References	Adhesive type	GFRP profile	Remarks
Quasi-static Mode I	Zhang et al. (2010b)	Epoxy SikaDur 330	E-glass in isophthalic polyester	Tensile opening DCB specimen
	Shahverdi et al. (2011)	Epoxy SikaDur 330	E-glass in isophthalic polyester	Tensile opening, DCB specimen, fiber bridging observed
	Shahverdi et al. (2013b)	Epoxy SikaDur 330	E-glass in isophthalic polyester	Tensile opening, asymmetric DCB specimen, 1 mm/min
	Shahverdi et al. (2013a)	Epoxy SikaDur 330	E-glass in isophthalic polyester	Tensile opening, DCB specimen
Quasi-static Mode II	Taib et al. (2006)	Two-component Hysol EA 9359.3	E-glass in vinyl ester Derakane Momentum 411-350	Shear through tensile load (ASTM D3165)
	Zhang et al. (2010b)	Epoxy SikaDur 330	E-glass in isophthalic polyester	In-plane shear ELS-specimen
	Shahverdi et al. (2013a)	Epoxy SikaDur 330	E-glass in isophthalic polyester	In-plane shear, ELS-specimen
Quasi-static Mode III	Not available			Out-of plane shear or twist
Quasi-static mixed Mode I/II	Shahverdi et al. (2013a)	Epoxy SikaDur 330	E-glass in isophthalic polyester	MMB-specimen, comparison with cyclic fatigue
	Not available			

Continued

Table 1.2 Continued

Test type	References	Adhesive type	GFRP profile	Remarks
Quasi-static mixed Mode I/III, II/III				
Cyclic fatigue Mode I	Zhang et al. (2010a)	Epoxy SikaDur 330	E-glass in isophthalic polyester	Tensile–tensile, double- and stepped-lap joint, R-ratio $= 0.1$, 10 Hz
	Shahverdi et al. (2011)	Epoxy SikaDur 330	E-glass in isophthalic polyester	Tensile opening, DCB specimen, 1 mm/min
	Shahverdi et al. (2012a)	Epoxy SikaDur 330	E-glass in isophthalic polyester	Tensile opening, DCB specimen, R-ratio $= 0.1$, 0.5, 0.8, also 0.3 and 0.65, constant amplitude, 5 Hz
Cyclic fatigue Mode II	Shahverdi et al. (2013b)	Epoxy SikaDur 330	E-glass in isophthalic polyester	Tensile opening, DCB specimen, R-ratio $= 0.5$, 5 Hz
	Shahverdi et al. (2013b)	Epoxy SikaDur 330	E-glass in isophthalic polyester	In-plane shear, ELS-specimen, R-ratio $= 0.5$, 5 Hz
Cyclic fatigue mixed Mode I/II	Shahverdi et al. (2013b)	Epoxy SikaDur 330	E-glass in isophthalic polyester	MMB-specimen displacement control, R-ratio $= 0.5$, 5 Hz
Cyclic fatigue mixed Mode I/III, II/III	Not available			

laminates) and deduced an empirical model for crack propagation. Since crack propagation or growth takes place in the GFRP profile rather than in the adhesive (a behavior observed in many types of adhesive joints with FRP composite or wood adherends, see, e.g., Brunner, Terrasi, Vallée, and Keller (2008), Brunner et al. (2009), Brunner, Pinter, and Murphy (2009) or Brunner and Terrasi (2008)) starting the crack at different depths in the profile yielded distinctly different values of the energy release rate depending on the local fiber type and layup. Fiber bridging plays an important role in crack propagation inside the GFRP profiles. This was further investigated with an FEM again using VCCT in a follow-up publication (Shahverdi, Vassilopoulos, & Keller, 2013a). This confirmed that the Mode II component introduced by the asymmetry in fracture of the joint contributed a negligible amount (around or less than 1%) in two out of three propagation paths; only the third (between the second chopped glass strand mat and the adjacent roving layer) yielded a contribution around 10%.

Mode I fatigue loading has been experimentally investigated for constant amplitude loading by Shahverdi, Vassilopoulos, and Keller (2012a) and modeled for different R-ratios by the same group of authors (Shahverdi, Vassilopoulos, & Keller, 2012b). Mode I fatigue fracture loading was performed under displacement control at 5 Hz with three load ratios (0.1, 0.5, and 0.8). Crack or delamination lengths were measured with crack gauges and visually. The crack path affected the results and yielded different crack curves, dependent on between which layers in the GFRP profile the delamination propagated. Higher R-ratios yielded higher slopes in the delamination rate versus applied load (G_{Imax}) curves. The data were then used for developing a phenomenological fatigue crack growth model for estimating the total fatigue life (Shahverdi et al., 2012b).

The next step investigated mixed Mode I/II fatigue fracture in the joints with different ratios between Mode I and Mode II contribution (Shahverdi, Vassilopoulos, & Keller, 2013b). The joints were tested with the DCB, ELS, and MMB set-ups developed for FRP composite laminates (see, e.g., Brunner, Blackman, et al., 2008 for details). The data were used to establish fatigue failure criteria and a fatigue life model. The fatigue crack growth curves deduced from the data yielded different slopes depending on the fatigue fracture Mode. Mode II components in mixed Mode I/II were observed to yield lower slopes than the pure Modes (I or II). The fatigue threshold values were observed to be lower than the strain energy release rates in quasi-static tests and different for Mode I and Mode II, respectively. This was attributed to different amounts of fiber bridging for the different Modes and Mode mixities, respectively.

In mixed Mode I/II loading, the determination of the total energy release rate $G_{I/IIC}$ is straightforward, but the distribution of the Mode I and Mode II components from that is still being debated, even for fracture of FRP composites (see, e.g., Harvey & Wang, 2012a, 2012b; Kinloch, Wang, Williams, & Yayla, 1993; Williams, 1988). Shahverdi in his Ph.D. thesis (2013) used FEM simulation and modified the so-called "global" method developed by Williams (Kinloch et al., 1993) which he called "extended global model" to achieve mixed mode separation for the joints. This represents an interesting development from which fracture analysis of FRP composites could benefit as well.

So far, the experimental evidence shows that test concepts and specimen designs developed for FRP composite laminates can be transferred and applied to fracture and fatigue fracture testing of adhesively-bonded joints made from pultruded GFRP profiles. The joints for these tests, so far, have been made from simply-bonded GFRP profiles. Other types of joints, e.g., based on the single leg four point bend proposed by Tracy, Feraboli, and Kedward (2003) designed for mixed mode testing, have not been investigated for fracture and fatigue fracture of the GFRP joints yet. It will be worthwhile to closely follow the developments in fracture and fatigue fracture testing of FRP composites in the future and to investigate the applicability of new concepts for the civil engineering structural designs using adhesively-bonded joints made from pultruded GFRP profiles.

1.4 Standards and test protocols for determining environmental effects in fatigue, fracture, and durability testing

As noted in the introduction, adhesively-bonded structural elements or structures in civil engineering applications made from pultruded profiles are also subject to a complex spectrum of environmental exposures. These range from daily and seasonal temperature variations (i.e., thermal fatigue) to humidity (e.g., from dew, rainfall, snow) in general leading to hygrothermal fatigue, and, depending on the application, possibly exposure to sunlight and ultraviolet (UV) radiation or contact with other media such as, e.g., oils, hydraulic fluids, acids, and gaseous chemicals. These effects are typically combined with mechanical service loads and possibly with additional loads generated by the environment, e.g., wind exposure, snow load, and hail impact. Fire exposure, even though not expected to occur regularly, nevertheless has to be considered as well (see, e.g., Mouritz et al., 2009). The service conditions via the associated respective mechanisms lead, in general, to aging of the adhesively-bonded joints and to a deterioration of their properties and, hence, of their performance.

Accelerated aging of GFRP profiles made from polyester and vinyl ester, but, however, not manufactured into adhesively-bonded joints, has recently been discussed by Cabral Fonseca, Correia, Rodrigues, and Branco (2012). Hygrothermal aging was performed in demineralized water and in salt water at temperatures between $+20$ °C and $+60$ °C for up to 18 months, under condensation conditions at $+40$ °C for up to 9 months, and under UV irradiation up to 3000 h. The effects of the different exposures were evaluated via mass change, dynamic mechanical analysis (DMA), mechanical behavior under tension, flexure and shear and chemical changes via Fourier transform infrared (FTIR) spectroscopy. Demineralized water caused larger changes than salt water, and UV irradiation affected the visual surface appearance, but not the viscoelastic (DMA) and mechanical properties. GFRP profiles made from vinyl ester in general performed better than those made from polyester.

Temperature effects on the fracture of adhesives under quasi-static Mode I loading have been investigated with numerical simulations using cohesive zone elements and experiments by Banea, da Silva, and Campilho (2011). In a second paper, the combined rate dependence and temperature effects were summarized (Banea, de Sousa, da Silva, Campilho, & Bastos de Pereira, 2011). For the adhesive used in the tests, the ductility increased significantly with temperature (up to $+150\,°C$). The strain-rate effects were considered minor compared with the temperature effects, but the tensile failure strain changed with temperature.

The temperature-dependent fatigue testing of adhesively-bonded joints made from pultruded GFRP profiles by Zhang et al. (2009) has already been noted. Fatigue fracture of composite joints at elevated temperature has been explored by, e.g., Newaz, Lustiger, and Yung (1989), using CFRP thermoplastic composites. While cyclic loads in load-controlled tests yielded increasing crack growth rates at room temperature, the rates were decreasing at $+93\,°C$. This was attributed to plasticizing effects that yielded higher ductility of the thermoplastic matrix and stabilized crack growth.

While exposure to specific environmental factors such as temperature, moisture, and even fire and the resulting effect on the adhesively-bonded GFRP joints can be simulated in laboratory tests, describing and evaluating the full complexity of causes and effects in real application environments is currently beyond the scope of such tests. The complexity arises in part from the different components used in the adhesively-bonded joints made from glass fibers, matrix polymers, fillers and additives, and adhesives, possibly again modified with fillers and additives, and their different and time-dependent behavior with respect to environmental exposure. A simple example of this is the effect of pre-bond moisture on the fracture behavior of adhesively-bonded GFRP composite joints (Blackman et al., 2008). ISO 9142 at least provides some guidelines for the selection of laboratory aging conditions for testing adhesively-bonded joints. Accelerating aging tests to some extent (see, e.g., Stewart & Douglas, 2012 for the example of epoxy FRP composites) become a challenging task for adhesively-bonded joints. The complexity posed by the multi-component joint increases further, as soon as more than a single environmental exposure factor shall be considered. Developing models for characterizing the complex interaction between different types of exposure and estimating realistic service lifetimes under combined mechanical and environmental loads can be formulated as the ultimate goal, but will require development of a conceptual framework and possibly of a step-by-step approach.

Civil engineering structures made from any material or combinations of materials also have to fulfill requirements with respect to fire behavior and fire resistance (e.g., according to ISO/TR 15655). The effects of exposure to fire on fatigue and fracture properties of adhesively-bonded joints determined under "normal conditions" (e.g., room or service temperature) also deserve consideration. The basic temperature-dependent properties of pultruded profiles have been determined and discussed with a combination of thermal or thermomechanical materials analysis by Bai, Post, Lesko, and Keller (2008). This however, does not provide information about behavior under exposure to fire. Specific investigations on fire exposure of GFRP columns (but without adhesive joint) following the international standard ISO 834 were performed by Bai, Hugi, Ludwig, and Keller (2011) and Bai and Keller (2011). In principle, the fatigue or fracture performance of adhesively-bonded GFRP joints after defined exposure to fire could be determined applying the test procedures for behavior compiled in previous sections. Considering the complex design and the complex processes occurring in GFRP under fire exposure (see, e.g., Mouritz, 2002), establishing validated test procedures is a promising area of research.

Another area, still largely unexplored for composite adhesive joints made from pultruded GFRP profiles (except for composite adhesive joints, see Blackman et al., 2012), is high-rate loading and impact that for fracture and fatigue fracture of FRP composites have recently been reviewed by Brunner (2014). A special aspect of that is impact of foreign objects that may play an important role in structural elements and structures made with adhesively-bonded joints. Examples that may deserve consideration are crashing automobiles, trees falling during storms, or hail impact. Explosive blast effects on special three-dimensionally reinforced GFRP composite laminate specimens have been investigated by Mouritz (2001), but to the best knowledge of the author no experience on such blast exposure for civil infrastructure or structural elements with adhesively-bonded joints is available.

A special environment for adhesively-bonded joints is exposure to ionizing radiation, for which a standard practice has recently been developed (ASTM D1879). This, however, is not a typical environment for civil engineering structures with adhesively-bonded composite joints. The GFRP bridge in Pontresina (Engadin valley in the southeastern Swiss Alps, see Keller, Bai, & Vallée, 2007; Keller & Tirelli, 2004 for details) possibly represents a case with somewhat higher exposure to ionizing radiation, both because of its altitude (around 1800 m above sea level), which entails higher cosmic radiation and also increased UV exposure, and its location in a region with a relatively high level of radioactive background radiation. The effects of long-term, relatively low-level ionizing radiation (comprising particles and electromagnetic radiation), however, have not been investigated for pultruded GFRP profiles yet. For FRP composites, there are a few investigations, mainly of effects from radiation levels in space environment on inter-laminar fracture (Funk & Sykes, 1986; Takeda, Tohdoh, & Takahashi, 1995). The reported results are somewhat contradictory in that improvement (Funk & Sykes, 1986) and degradation (Takeda et al., 1995) in inter-laminar fracture properties have been observed.

1.5 Standards and test protocols for modeling and simulation of fracture and fatigue behavior

Simulation and modeling require basic material properties, and these have to be determined either for each of the constituents or for representative adhesive joint specimens that yield the properties of the overall composite element or structure. A recent review by He (2011) summarizes basic issues in modeling a wide range of different types of adhesive joint designs (single or double lap, butt and butt strap, bevel and peel are included in the discussion) for various load cases (for static, fatigue, and dynamic loading, see, e.g., ASTM D3807) and considering damage modeling, fracture, and environmental exposures as well. The complexity of FEA in these cases, which show non-linear behavior, is clearly pointed out, especially with respect to predicting failure in adhesive or at the adhesive—adherend interface. For dynamic loading, it is explicitly noted that FEA predictions have to be validated by experimental work. This will be a task for future research and investigation. Modeling fracture of adhesive joints is, e.g., discussed by Kafkalidis and Thouless (2002) using a CZM for single-lap shear joints and by Steinbrecher et al. (2006) taking the elasticity of the adhesive into account, as well as more recently by Campilho et al. (2013). The modeling by Campilho et al. (2013) compared different types of cohesive zone elements and concluded that trapezoidal elements performed better than triangular in this case. However, the difference between the CZM shapes was much less pronounced for brittle adhesives.

Finite element modeling of fracture properties of adhesively-bonded joints from pultruded GFRP profiles includes the investigations by Zhang et al. (2010b), by Shahverdi et al. (2011, 2013a) already discussed previously, and by Taib, Boukhili, Achiou, and Boukehili (2006) who investigate the specific case of single- and joggle-lap joints.

FEA also plays a role in investigations of the properties of the constituents of the adhesively-bonded joints. Yu et al. (2001), for example, discuss the application of material models implemented in FEA codes for understanding the rate dependence of adhesives. It is concluded that viscoplastic models seem to be better suited for describing this behavior than the models implemented in FEA codes at the time (2001). For modeling fracture of adhesives, cohesive zone models (already briefly discussed in an earlier section) seem to become more and more popular, mainly because of the successful application of such models in inter-laminar fracture of FRP composite laminates.

Taib et al. (2006) note good agreement of their FEA model results with their experiments (Taib et al., 2006) up to a load level of about 2000 N (tensile load producing shear in the joint). For the joggle-lap joint, deformation behavior observed experimentally was well reproduced by the model, which assumed linear elastic behavior of the adherends and non-linear behavior for the adhesive.

Extensive numerical modeling by Gonçalves Teixeira and da Silva (2011) investigated the effects of the surface topography of substrates in adhesively-bonded joints. The numerical simulations indicated that surface roughness with regular patterns may reduce stress concentrations and simultaneously increase the contact area. The application of this concept to adhesively-bonded joints with pultruded GFRP profiles will require further studies on surface modification and experimental verification, however.

Sarfaraz, Vassilopoulos, and Keller (2013b) discuss different hypotheses for modeling the constant-amplitude fatigue behavior of adhesively-bonded double-lap joints with pultruded adherends, based on the experimental data obtained in earlier work (Sarfaraz et al., 2011, 2012, 2013a). Differences among different models were noted for extrapolation to the high cycle fatigue regime, while most models described the behavior in the experimental range of cycles sufficiently well. Nevertheless, the assumptions of the models and the data used have to be critically examined.

Modeling, not necessarily using FEA, is also required for extrapolating the properties of the components as well as those of the adhesive joints, under mechanical fatigue loading (see, e.g., Sarfaraz et al., 2013b) and aging effects, as well as of combinations of mechanical and environmental aging. An important issue is acceleration of aging tests for prediction of the long-term behavior of the adhesively-bonded composite joints. A recent review of accelerated aging of FRP composites used in civil engineering by Stewart and Douglas (2012) summarizes the potential pitfalls and discusses the major limitations, at least for epoxy-based FRP composites. It is essential not to activate additional damage mechanisms by increasing the aging or testing temperature too close to the relevant glass-transition temperature. Any modeling or prediction of long-term behavior based on accelerated tests hence requires careful analysis of the aged materials and of the resulting damage.

1.6 Summary and future trends

This contribution attempted to present a brief summary of available experimental test methods and perspectives for future development of such methods for characterizing the fatigue, the quasi-static fracture, and the fatigue fracture behavior of adhesively-bonded joints for structural applications in civil engineering made from pultruded GFRP profiles. At the time of writing (summer 2013), there are virtually no standardized test methods for this specific type of adhesive joint available. Published experimental research typically bases test specimens and test set-up on developments from fatigue and fracture of regular FRP composites or analogous methods for adhesively-bonded joints with adherends made from a range of different materials, including various composites. Of course, the size of test specimens and hence test set-ups may have to be up-scaled from the respective test standards to accommodate specimens that fully represent the behavior of commercially manufactured, pultruded profiles and of the adhesive joints made from them. As indicated by the data compiled in Tables 1.1 and 1.2 only a limited range of material types (essentially E-glass isophthalic polyester and two component epoxy SikaDur 330) have been used in experimental investigations so far. Hence, it would be straightforward to extend future work to other types of polymers and adhesives.

Most of the published literature on fatigue and fracture of adhesively-bonded joints made from pultruded GFRP profiles deals with laboratory-scale test specimens (typically centimeter to meter size). As noted in the introduction, the transferability of the test results to engineering structures still poses challenges, first because of size (up-scaling by at least on order of magnitude), but also because of the complex effective load spectrum experienced by the elements and structures.

Even though the emphasis here was on experimental aspects, concurrent theoretical work as well as modeling and simulation also play an important role. Examples of this are FEM investigations of the performance of up-scaled test set-up designs or of Mode mixity in quasi-static and in the future possibly also in cyclic fatigue fracture. These typically require property data of several materials as input and, hence, experimental determination of a range of (mainly mechanical) properties of the constituents and, in part, of the full adhesive joints. Modeling the long-term behavior of the joints under complex and variable environmental exposure again require materials data such as, e.g., temperature- and humidity-dependent behavior. In the near and intermediate future, the question of design optimization, as, e.g., discussed for FRP composite structures in general by Awad, Aravinthan, Zhuge, and Gonzalez (2012), will be important for further developments in adhesively-bonded joints as well. An example of this is the determination of the optimum thickness of the adhesive (Vallée et al., 2009). Exploring this field and establishing approaches and methods is a truly interdisciplinary activity that provides challenging research opportunities.

One area where extensive research on essentially all aspects of adhesive bonding of polymeric composites has been and still is being performed is research on adhesives for dental restoration materials, see, e.g., De Munck et al. (2005) or Milia, Cumbo, Cardoso, and Gallina (2012) for recent reviews. Most of the polymer-matrix composites (PMC) used in this field are, however, not fiber-reinforced and specimens are typically much smaller scale (i.e., around one or at most a few centimeters) than the GFRP joints used in civil engineering. The aspects of environmental exposure and its effect on the adhesive bond, specifically its durability, investigated for dental PMC (see, e.g., De Munck et al. (2005) for details), may nevertheless be useful as a guideline for future work on durability of adhesively-bonded GFRP pultruded composite joints. Closer to adhesively-bonded pultruded GFRP profiles are, on the other hand, adhesively-bonded timber joints, which also find widespread application in civil engineering (e.g., buildings and bridges). Fatigue and fracture of adhesively-bonded wood joints have been standardized or are described in the form of guidelines in, e.g., ASTM D1101, ASTM D2559, ASTM D4502, and ASTM D5574. Also, the tests discussed by Brunner and Terrasi (2008) show strong similarities with those of joints made from pultruded composite profiles. However, wood and wood products allow for more sophisticated designs of structural joints than GFRP, e.g., scarf or finger joints described in ASTM D7469, but with further development in FRP materials, more complex types of adhesively-bonded composite joints may become technically (and possibly also economically) feasible.

Again, environmental exposure and effects resulting from that also play an important role in timber joints (see, e.g., Custódio, Broughton, & Cruz, 2009) and essential aspects of that quite likely hold for GFRP adhesive joints as well.

An issue that affects structural performance of adhesively-bonded joints, but is rarely discussed, is manufacturing quality. In manufacturing and preparation of adhesively-bonded GFRP joints for structural applications, quality control and resulting durability of these joints under hygrothermomechanical loads have long been recognized to pose problems (see, e.g., Kinloch, 1979). The question of design values for FRP composites for civil infrastructure has been discussed, e.g., by Atadero and Karbhari (2009). Non-destructive testing (NDT) has hence been investigated for providing solutions to quality control and prediction of service life of FRP composites and adhesively-bonded joints (see, e.g., Allin, Cawley, & Lowe, 2003; Guyott, Cawley, & Adams, 1986; Hung et al., 2007; Karhnak & Duke, 1994). However, this still poses problems (see, e.g., Bossi, Housen, & Shepherd, 2002) even today as noted by Markatos et al. (2013), and new approaches or technical NDT methods are still welcome. In applications, the NDT methods (e.g., acoustic emission or ultrasonic-guided waves as discussed by Allin et al. (2003) or Brunner and Terrasi (2008), the use of fiber optics (see, e.g., Sans, Stutz, Renart, Mayugo, & Botsis, 2012; Stutz, Cugnoni, & Botsis, 2011) or modal analysis combined with FEA (Russo, 2013) may be implemented for continuous or intermittent monitoring of the adhesive joints in structural elements and structures under service loads. This relates to the general topic of structural health monitoring (SHM). In civil engineering, however, the cost for implementing or integrating and operating SHM systems in adhesively-bonded FRP structures may be prohibitive. Investigations of the economic feasibility of integrated SHM in other areas of structural engineering, e.g., aircraft operation and maintenance that has been investigated by Boller, Kapoor, and Goh (2007), indicate potential mainly for optimization of maintenance. The case of adhesively-bonded FRP composite structures has, to the best knowledge of the author, not been investigated yet in that respect.

Damage and fracture of complex shaped FRP composite elements under quasi-static or fatigue loads is an active area of research, see, e.g., Cartié, Dell'Anno, Poulin, and Partridge (2006), Cognard, Davies, Sohier, and Créac'hcadec (2006), or Hélénon, Wisnom, Hallett, and Trask (2012). The approach typically combines experiments and simulations (frequently based on FEA). Analogous investigations could be performed for adhesively-bonded structural elements made from pultruded profiles. Quite likely, this would provide useful information on structural behavior and damage accumulation of specific elements or parts. Except for a possible, future guideline on the general approach on structural element testing, these procedures and simulations are difficult to standardize. However, (standardized) tests on fatigue, fracture, and fatigue fracture performance of laboratory-size specimens (yielding essentially data on materials and elements) in combination with specific structural tests on selected elements and (maybe extensive) modeling and simulation will pave the way for future developments for understanding structures made from adhesively-bonded pultruded GFRP profiles.

If specialists from the area of fatigue and fracture of FRP composites and/or timber as well as specialists on adhesives and adhesive bond manufacturing continue to collaborate on test and specimen design and development, it will be fairly straightforward to establish validated and possibly standardized test procedures for fatigue, fracture, and fatigue fracture behavior of adhesively-bonded joints from pultruded GFRP profiles. Combined with research on the transferability of the results to design and manufacture of civil engineering structures, e.g., via structural element testing and FEM/FEA simulation, this will provide the basis for reliable and durable structures.

1.7 Sources of further information and advice

ASTM D1002-10 Standard test method for apparent shear strength of single-lap-joint adhesively bonded metal specimens by tension loading (metal-to- metal)
ASTM D1101-97a (2013) Standard test methods for integrity of adhesive joints in structural laminated wood products for exterior use
ASTM D1879-06 Standard practice for exposure of adhesive specimens to ionizing radiation
ASTM D2093-03 (2011) Standard practice for preparation of surfaces of plastics prior to adhesive bonding
ASTM D2559-12a Standard specification for adhesives for bonded structural wood products for use under exterior exposure conditions
ASTM D2919-01 (2007) Standard test method for determining durability of adhesive joints stressed in shear by tension loading

ASTM D3165-07 Standard test method for strength properties of adhesives in shear by tension loading of single-lap-joint laminated assemblies
ASTM D3166-99 (2012) Standard test method for fatigue properties of adhesives in shear by tension loading (metal/metal)
ASTM D3433-99 (2012) Standard test method for fracture strength in cleavage of adhesives in bonded metal joints
ASTM D3479/D3479M-12 Standard test method for tension—tension fatigue of polymer matrix composite materials
ASTM D3807-98 (2012) Standard test method for strength properties of adhesives in cleavage peel by tension loading (engineering plastics-to-engineering plastics)
ASTM D4502-92 (2011) Standard test method for heat and moisture resistance of wood-adhesive joints
ASTM D4896-01 (2008) Standard guide for use of adhesive-bonded single lap-joint specimen test results
ASTM D5041-98 (2012) Standard test method for fracture strength in cleavage of adhesives in bonded joint
ASTM D5528-01 (2007)e3 Standard test method for Mode I interlaminar fracture toughness of unidirectional fiber-reinforced polymer matrix composites
ASTM D5573-99 (2012) Standard practice for classifying failure modes in fiber-reinforced-plastic (FRP) joints
ASTM D5574-94 (2012) Standard test methods for establishing allowable mechanical properties of wood-bonding adhesives for design of structural joints
ASTM D5868-01 (2008) Standard test method for lap shear adhesion for fiber reinforced plastic (FRP) bonding
ASTM D6105-04 (2012) Standard practice for application of electrical discharge surface treatment (activation) of plastics for adhesive bonding
ASTM D6115-97 (2011) Standard test method for Mode I fatigue delamination growth onset of unidirectional fiber-reinforced polymer matrix composites
ASTM D6412/D6412M-99 (2012) Standard specification for epoxy (flexible) adhesive for bonding metallic and non-metallic materials
ASTM D6671/D6671M-06 Standard test method for mixed Mode I-Mode II interlaminar fracture toughness of unidirectional fiber reinforced polymer matrix composites
ASTM D7469-12 Standard test methods for end-joints in structural wood products
ASTM E1307-10 Standard practice for surface preparation and structural adhesive bonding of precured, non-metallic composite facings to structural core for flat shelter panels
ASTM WK22949 New test method for determination of the Mode II interlaminar fracture toughness of unidirectional fiber-reinforced polymer matrix composites using the end-notched flexure (ENF) test
EN 13887 Adhesives—test methods for fatigue properties of structural adhesives in tensile shear
EN 15190 Structural adhesives—test methods for assessing long-term durability of bonded metallic structures
ISO 291:2008 Plastics—standard atmospheres for conditioning and testing
ISO 834-12:2012 Fire resistance tests—elements of building construction—part 12: Specific requirements for separating elements evaluated on less than full scale furnaces
ISO 9142:2003 Adhesives—guide to the selection of standard laboratory aging conditions for testing bonded joints
ISO 9664:1993 Adhesives—test methods for fatigue properties of structural adhesives in tensile shear

ISO 10354:1992 Adhesives—characterization of durability of structural-adhesive-bonded assemblies—wedge rupture test
ISO 14615:1997 Adhesives—durability of structural adhesive joints—exposure to humidity and temperature under load
ISO 15024:2001 Fibre-reinforced plastic composites—determination of Mode I inter-laminar fracture toughness, G_{IC}, for unidirectionally reinforced materials
ISO 15114 Fibre-reinforced plastic composites—determination of the Mode II fracture resistance for unidirectionally reinforced materials using the calibrated end-loaded split (C-ELS) test and an effective crack length approach
ISO/TR 15655:2003 Fire resistance—tests for thermo-physical and mechanical properties of structural materials at elevated temperatures for fire engineering design
ISO 17212:2012 Structural adhesives—guidelines for the surface preparation of metals and plastics prior to adhesive bonding
ISO 21368:2005 Adhesives—guidelines for the fabrication of adhesively bonded structures and reporting procedures suitable for the risk evaluation of such structures
ISO 25217:2009 Adhesives—determination of the Mode I adhesive fracture energy of structural adhesive joints using double cantilever beam and tapered double cantilever beam specimens

Acknowledgments

Discussions of fracture and fatigue behavior, and of experimental and theoretical work, with members of Technical Committee 4 of the European Structural Integrity Society (ESIS), notably Prof. A.J. Kinloch and Dr B.R.K. Blackman, with Ph.D. students and researchers at the Composite Construction Laboratory (CCLab) of the Ecole Polytechnique Fédérale de Lausanne (EPFL), notably with Dr Ye Zhang and Dr Moslem Shahverdi, but also with Dr Aixi Zhou (now at the University of North Carolina Charlotte, USA), Dr Till Vallée (now at the Fraunhofer Institute for Manufacturing Technology and Advanced Materials IFAM, Germany) and Dr Anastasios P. Vassilopoulos, as well as discussions and experimental investigations of timber joints with Dr Thomas Tannert at The University of Applied Sciences Biel (now at the University of British Columbia, Canada), are gratefully acknowledged.

References

Allin, J. M., Cawley, P., & Lowe, M. J. S. (2003). Adhesive disbond detection of automotive components using first mode ultrasonic resonance. *NDT&E International, 36*(7), 503−514. http://dx.doi.org/10.1016/S0963-8695(03)00045-8.
Ashcroft, I. A., Hughes, D. J., Shaw, S. J., Wahab, M. A., & Crocombe, A. (2001). Effect of temperature on the quasi-static strength and fatigue resistance of bonded composite double lap joints. *Journal of Adhesion, 75*(1), 61−88. http://dx.doi.org/10.1080/00218460108029594.
ASTM D1002-10 Standard test method for apparent shear strength of single-lap-joint adhesively bonded metal specimens by tension loading (metal-to-metal).
ASTM D1101-97a (2013) Standard test methods for integrity of adhesive joints in structural laminated wood products for exterior use.
ASTM D1879-06 Standard practice for exposure of adhesive specimens to ionizing radiation.

ASTM D2093-03 (2011) Standard practice for preparation of surfaces of plastics prior to adhesive bonding.
ASTM D2559-12a Standard specification for adhesives for bonded structural wood products for use under exterior exposure conditions.
ASTM D2919-01 (2007) Standard test method for determining durability of adhesive joints stressed in shear by tension loading.
ASTM D3165-07 Standard test method for strength properties of adhesives in shear by tension loading of single-lap-joint laminated assemblies.
ASTM D3166-99(2012) Standard test method for fatigue properties of adhesives in shear by tension loading (metal/metal).
ASTM D3433-99 (2012) Standard test method for fracture strength in cleavage of adhesives in bonded metal joints.
ASTM D3479/D3479M-12 Standard test method for tension-tension fatigue of polymer matrix composite materials.
ASTM D3807-98 (2012) Standard test method for strength properties of adhesives in cleavage peel by tension loading (engineering plastics-to-engineering plastics).
ASTM D4502-92 (2011) Standard test method for heat and moisture resistance of wood-adhesive joints.
ASTM D4896-01 (2008) Standard guide for use of adhesive-bonded single lap-joint specimen test results.
ASTM D5041-98(2012) Standard test method for fracture strength in cleavage of adhesives in bonded joint.
ASTM D5528-01(2007)e3 Standard test method for mode I interlaminar fracture toughness of unidirectional fiber-reinforced polymer matrix composites.
ASTM D5573-99(2012) Standard practice for classifying failure modes in fiber-reinforced-plastic (FRP) joints.
ASTM D5574-94(2012) Standard test methods for establishing allowable mechanical properties of wood-bonding adhesives for design of structural joints.
ASTM D5868-01(2008) Standard test method for lap shear adhesion for fiber reinforced plastic (FRP) bonding.
ASTM D6105-04(2012) Standard practice for application of electrical discharge surface treatment (activation) of plastics for adhesive bonding.
ASTM D6115-97(2011) Standard test method for mode I fatigue delamination growth onset of unidirectional fiber-reinforced polymer matrix composites.
ASTM D6412/D6412M-99(2012) Standard specification for epoxy (flexible) adhesive for bonding metallic and nonmetallic materials.
ASTM D6671/D6671M-06 standard test method for mixed mode I-mode II interlaminar fracture toughness of unidirectional fiber reinforced polymer matrix composites.
ASTM D7469-12 Standard test methods for end-joints in structural wood products.
ASTM E1307-10 Standard practice for surface preparation and structural adhesive bonding of precured, nonmetallic composite facings to structural core for flat shelter panels.
ASTM WK22949 New test method for determination of the mode II interlaminar fracture toughness of unidirectional fiber-reinforced polymer matrix composites using the end-notched flexure (ENF) test.
Atadero, R., & Karbhari, V. M. (2009). Sources of uncertainty and design values for field-manufactured FRP. *Composite Structures*, *89*(1), 83–93. http://dx.doi.org/10.1016/j.compstruct.2008.07.001.
Awad, Z. K., Aravinthan, T., Zhuge, Y., & Gonzalez, F. (2012). A review of optimization techniques used in the design of fibre composite structures for civil engineering applications. *Materials and Design*, *33*, 534–544. http://dx.doi.org/10.1016/j.matdes.2011.04.061.

Awaja, F., Gilbert, M., Kelly, G., Fox, B., & Pigram, P. J. (2009). Adhesion of polymers. *Progress in Polymer Science*, *34*(9), 948−968. http://dx.doi.org/10.1016/j.progpolymsci. 2009.04.007.
Azari, S., Ameli, A., Papini, M., & Spelt, J. K. (2013). Adherend thickness influence on fatigue behavior and fatigue failure prediction of adhesively bonded joints. *Composites: Part A*, *48*, 181−191. http://dx.doi.org/10.1016/j.compositesa.2013.01.020.
Bai, Y., Hugi, E., Ludwig, C., & Keller, T. (2011). Fire performance of water-cooled GFRP columns. I: fire endurance investigation. *Journal of Composites for Construction*, *15*(3), 404−412. http://dx.doi.org/10.1061/(ASCE)CC.1943-5614.0000160.
Bai, Y., & Keller, T. (2011). Fire performance of water-cooled GFRP columns. II: postfire investigation. *Journal of Composites for Construction*, *15*(3), 413−421. http://dx.doi.org/ 10.1061/(ASCE)CC.1943-5614.0000191.
Bai, Y., Post, N. L., Lesko, J. J., & Keller, T. (2008). Experimental investigations on temperature-dependent thermo-physical and mechanical properties of pultruded GFRP composites. *Thermochimica Acta*, *469*(1−2), 28−35. http://dx.doi.org/10.1016/j.tca.2008. 01.002.
Bakis, C. E., Bank, L. C., Brown, V. L., Cosenza, E., Davalos, J. F., Lesko, J. J., et al. (2002). Fiber-reinforced polymer composites for construction—state-of-the-art review. *Journal of Composites for Construction*, *6*(2), 73−87. http://dx.doi.org/10.1061/(ASCE)1090- 0268(2002)6:2(73).
Baldan, A. (2004). Review adhesively-bonded joints and repairs in metallic alloys, polymers and composite materials: adhesives, adhesion theories and surface pretreatment. *Journal of Materials Science*, *39*(1), 1−49. http://dx.doi.org/10.1023/B: JMSC.0000007726. 58758.e4.
Baldan, A. (2012). Adhesion phenomena in bonded joints. *International Journal of Adhesion & Adhesives*, *38*, 95−116. http://dx.doi.org/10.1016/j.ijadhadh.2012.04.007.
Balzani, C., Wagner, W., Wilckens, D., Degenhardt, R., Busing, S., & Reimerdes, H.-G. (2011). *Adhesive joints in composite laminates − A combined numerical/experimental estimate of critical energy release rates*. Germany: Mitteilung 2, Karlsruher Institut für Technologie Institut für Baustatik.
Banea, M. D., & da Silva, L. F. M. (2009). Adhesively bonded joints in composite materials: an overview, proceedings of the institution of mechanical engineers, part L. *Journal of Materials Design and Applications*, *223*(1), 1−18s. http://dx.doi.org/10.1243/ 14644207JMDA219.
Banea, M. D., da Silva, L. F. M., & Campilho, R. D. S. G. (2011). Mode I fracture toughness of adhesively bonded joints as a function of temperature: experimental and numerical study. *International Journal of Adhesion & Adhesives*, *31*(5), 273−279. http://dx.doi.org/ 10.1016/j.ijadhadh.2010.09.005.
Banea, M. D., de Sousa, F. S. M., da Silva, L. F. M., Campilho, R. D. S. G., & Bastos de Pereira, A. M. (2011). Effects of temperature and loading rate on the mechanical properties of a high temperature epoxy adhesive. *Journal of Adhesion Science and Technology*, *25*(18), 2461−2474. http://dx.doi.org/10.1163/016942411X580144.
Blackman, B. R. K., Johnsen, B. B., Kinloch, A. J., & Teo, W. S. (2008). The effects of pre-bond moisture on the fracture behaviour of adhesively-bonded composite joints. *The Journal of Adhesion*, *84*(3), 256−276. http://dx.doi.org/10.1080/ 00218460801954391.
Blackman, B. R. K., Kinloch, A. J., Paraschi, M., & Teo, W. S. (2003). Measuring the mode I adhesive fracture energy, G_{IC}, of structural adhesive joints: the results of an international round-robin. *International Journal of Adhesion & Adhesives*, *23*(4), 293−305. http:// dx.doi.org/10.1016/S0143-7496(03)00047-2.

Blackman, B. R. K., Kinloch, A. J., Rodriguez-Sanchez, F. S., & Teo, W. S. (2012). The fracture behaviour of adhesively-bonded composite joints: effects of rate of test and mode of loading. *International Journal of Solids and Structures, 49*(13), 1434−1452. http://dx.doi.org/10.1016/j.ijsolstr.2012.02.022.

Boller, C., Kapoor, H., & Goh, W. T. (2007). *Structural health monitoring potential determination based on maintenance process analysis*. In L. Porter Davis, B. K. Henderson, & M. Brett McMickell (Eds.), *Proceedings of SPIE, industrial and commercial applications of smart structures technologies* (Vol. 6527). http://dx.doi.org/10.1117/12.720736, 65270C-1-65270C-10.

Bossi, R., Housen, K., & Shepherd, W. (2002). Using shock loads to measure bonded joint strength. *Materials Evaluation, 60*(11), 1333−1338.

Browning, G., Carlsson, L. A., & Ratcliffe, J. G. (2010). Redesign of the ECT test for mode III delamination testing. Part I: finite element analysis. *Journal of Composite Materials, 44*(15), 1867−1881. http://dx.doi.org/10.1177/0021998309356606.

Brunner, A. J. (2014). Fracture mechanics of polymer composites in aerospace, Chapter 10. In P. E. Irving & C. Soutis (Eds.), *Polymer composites in the aerospace industry*. Woodhead Publishing.

Brunner, A. J., Blackman, B. R. K., & Davies, P. (2008). A status report on delamination resistance testing of polymer−matrix composites. *Engineering Fracture Mechanics, 75*(9), 2779−2794. http://dx.doi.org/10.1016/j.engfracmech.2007.03.012.

Brunner, A. J., Pinter, G., & Murphy, N. (2009). Development of a standardized procedure for the characterization of interlaminar crack growth in advanced composites under fatigue mode I loading conditions. *Engineering Fracture Mechanics, 76*(18), 2678−2689. http://dx.doi.org/10.1016/j.engfracmech.2009.07.014.

Brunner, A. J., Stelzer, S., Pinter, G., & Terrasi, G. P. (2013). Mode II fatigue delamination resistance of advanced fiber-reinforced polymer−matrix laminates: towards the development of a standardized test procedure. *International Journal of Fatigue, 50*(1), 57−62. http://dx.doi.org/10.1016/j.ijfatigue.2012.02.021.

Brunner, A. J., & Terrasi, G. P. (2008). Acousto-ultrasonic signal analysis for damage detection in GFRP adhesive joints. *Journal of Acoustic Emission, 26*, 152−159.

Brunner, A. J., Terrasi, G. P., Vallée, T., & Keller, T. (2008). Acoustic emission analysis and acousto-ultrasonics for damage detection in GFRP adhesive joints. In K. Ono (Ed.), *Proceedings 28th European conference on acoustic emission, European working group on acoustic emission* (pp. 100−105). Kraków, Poland: EWGAE.

Brunner, A. J., Terrasi, G. P., Vallée, T., & Tannert, T. (2009). Monitoring and evaluation of damage behaviour of adhesively bonded joints under tensile loads. In *Proceedings 22nd international symposium Swiss bonding* (pp. 267−278). Switzerland: Swibotech GmbH Buelach.

Cabral Fonseca, S., Correia, J. R., Rodrigues, M. P., & Branco, F. A. (2012). Artificial accelerated ageing of GFRP pultruded profiles made of polyester and vinylester resins: characterisation of physical−chemical and mechanical damage. *Strain, 48*(2), 162−173. http://dx.doi.org/10.1111/j.1475-1305.2011.00810.

Campilho, R. D. S. G., Banea, M. D., Neto, J. A. B. P., & daSilva, L. F. M. (2013). Modelling adhesive joints with cohesive zone models: effect of the cohesive law shape of the adhesive layer. *International Journal of Adhesion & Adhesives, 44*, 48−56. http://dx.doi.org/10.1016/j.ijadhadh.2013.02.006.

Campilho, R., de Moura, M. F. S. F., & Domingues, J. J. (2007). Stress and failure analyses of scarf repaired CFRP laminates using a cohesive damage model. *Journal*

of Adhesion Science and Technology, *21*(9), 855−870. http://dx.doi.org/10.1163/156856107781061477.
Carraro, P. A., Meneghetti, G., Quaresimin, M., & Ricotta, M. (2013a). Crack propagation analysis in composite bonded joints under mixed-mode (I + II) static and fatigue loading: a damage-based model. *Journal of Adhesion Science and Technology*, *27*(13), 1393−1406. http://dx.doi.org/10.1080/01694243.2012.735901.
Carraro, P. A., Meneghetti, G., Quaresimin, M., & Ricotta, M. (2013b). Crack propagation analysis in composite bonded joints under mixed-mode (I + II) static and fatigue loading: experimental investigation and phenomenological modelling. *Journal of Adhesion Science and Technology*, *27*(11), 1179−1196. http://dx.doi.org/10.1080/01694243.2012.735902.
Cartié, D. D. R., Dell'Anno, G., Poulin, E., & Partridge, I. K. (2006). 3D reinforcement of stiffener-to-skin T-joints by Z-pinning and tufting. *Engineering Fracture Mechanics*, *73*(16), 2532−2540. http://dx.doi.org/10.1016/j.engfracmech.2006.06.012.
de Castro, J., & Keller, T. (2008a). Ductile double-lap joints from brittle GFRP laminates and ductile adhesives, part I: experimental investigation. *Composites: Part B*, *39*(2), 271−281. http://dx.doi.org/10.1016/j.compositesb.2007.02.015.
de Castro, J., & Keller, T. (2008b). Ductile double-lap joints from brittle GFRP laminates and ductile adhesives, part II: numerical investigation and joint strength prediction. *Composites: Part B*, *39*(2), 282−291. http://dx.doi.org/10.1016/j.compositesb.2007.02.016.
Cheng, L., & Karbhari, V. M. (2006). New bridge systems using FRP composites and concrete: a state-of-the-art review. *Progress in Structural Engineering and Materials*, *8*(4), 143−154. http://dx.doi.org/10.1002/pse.221.
Chuang, W. Y., & Tsai, J. L. (2013). Investigating the performances of step wise patched double lap joint. *International Journal of Adhesion & Adhesives*, *42*, 44−50. http://dx.doi.org/10.1016/j.ijadhadh.2013.01.005.
Clauβ, S., Dijkstra, D. J., Gabriel, J., Kläusler, O., Matner, M., Meckel, W., et al. (2011). Influence of the chemical structure of PUR prepolymers on thermal stability. *International Journal of Adhesion & Adhesives*, *31*(6), 513−523. http://dx.doi.org/10.1016/j.ijadhadh.2011.05.005.
Cognard, J. (2006). Some recent progress in adhesion technology and science. *Comptes Rendues Chimie*, *9*(1), 13−24. http://dx.doi.org/10.1016/j.crci.2004.11.016.
Cognard, J. Y., Davies, P., Sohier, L., & Créac'hcadec, R. (2006). A study of the non-linear behaviour of adhesively-bonded composite assemblies. *Composite Structures*, *76*(1), 34−46. http://dx.doi.org/10.1016/j.compstruct.2006.06.006.
Custódio, J., Broughton, J., & Cruz, H. (2009). A review of factors influencing the durability of structural bonded timber joints. *International Journal of Adhesion & Adhesives*, *29*(2), 173−185. http://dx.doi.org/10.1016/j.ijadhadh.2008.03.002.
Davalos, J. F., Salim, H. A., Qiao, P., Lopez-Anido, R., & Barbero, E. J. (1996). Analysis and design of pultruded FRP shapes under bending. *Composites: Part B*, *27*(3−4), 295−305. http://dx.doi.org/10.1016/1359-8368(95)00015-1.
Davidson, B. D., & Sediles, F. O. (2011). Mixed-mode I−II−III delamination toughness determination via a shear−torsion-bending test. *Composites: Part A*, *42*(6), 589−603. http://dx.doi.org/10.1016/j.compositesa.2011.01.018.
Davidson, B. D., Sun, X. K., & Vinciquerra, A. J. (2007). Influences of friction, geometric nonlinearities, and fixture compliance on experimentally observed toughnesses from three and four-point bend end-notched flexure tests. *Journal of Composite Materials*, *41*(10), 1177−1196. http://dx.doi.org/10.1177/0021998306067304.

De Munck, J., Van Landuyt, K., Peumans, M., Poitevin, A., Lambrechts, P., Braem, M., et al. (2005). A critical review of the durability of adhesion to tooth tissue: methods and results. *Journal of Dental Research, 84*(2), 118−132.

Di Bella, G., Galtieri, G., Pollicino, E., & Borsellino, C. (2013). Mechanical characterization of adhesive joints with dissimilar substrates for marine applications. *International Journal of Adhesion & Adhesives, 41*, 33−40. http://dx.doi.org/10.1016/j.ijadhadh.2012.10.005.

Diez de Ulzurrun, I., López, F., Herreros, M. A., & Suárez, J. C. (2007). Tests of deck-to-hull adhesive joints in GFRP boats. *Engineering Failure Analysis, 14*(2), 310−320. http://dx.doi.org/10.1016/j.engfailanal.2006.02.012.

EN 13887 Adhesives − test methods for fatigue properties of structural adhesives in tensile shear.

EN 15190 Structural adhesives − test methods for assessing long-term durability of bonded metallic structures.

Feo, L., & Mancusi, G. (2013). The influence of the shear deformations on the local stress state of pultruded composite profiles. *Mechanics Research Communications, 47*(1), 44−49. http://dx.doi.org/10.1016/j.mechrescom.2012.11.004.

Feraboli, P., & Masini, A. (2004). Development of carbon/epoxy structural components for a high performance vehicle. *Composites: Part B, 35*(4), 323−330. http://dx.doi.org/10.1016/j.compositesb.2003.11.010.

Fernández, M. V., deMoura, M. F. S. F., daSilva, L. F. M., & Marques, A. T. (2011). Composite bonded joints under mode I fatigue loading. *International Journal of Adhesion & Adhesives, 31*(5), 280−285. http://dx.doi.org/10.1016/j.ijadhadh.2010.10.003.

Fernlund, G., Chaaya, R., & Spelt, J. K. (1995). Mode I fracture load predictions of adhesive T-joints. *Journal of Adhesion, 50*(2−3), 181−190. http://dx.doi.org/10.1080/00218469508014365.

Fernlund, G., Lanting, H., & Spelt, J. K. (1995). Mixed mode II-mode III fracture of adhesive joints. *Journal of Composites Technology & Research, 17*(4), 317−330.

Ferreira, J. A. M., Reis, P. N., Costa, J. D. M., & Richardson, M. O. W. (2002). Fatigue behaviour of composite adhesive lap joints. *Composites Science and Technology, 62*(10−11), 1373−1379.

Funk, J. G., & Sykes, G. F. (1986). The effects of radiation on the interlaminar fracture toughness of a graphite/epoxy composite. *Journal of Composites Technology & Research, 8*, 92−97. http://dx.doi.org/10.1520/CTR10328J.

Gilmore, R. B., & Shaw, J. (1974). The effect of temperature and humidity on the fatigue behavior of composite bonded joints. Composite bonding. *ASTM STP, 1227*, 82−95.

Gonçalves Teixeira, F., & da Silva, L. F. M. (2011). Study of influence of substrate topography on stress distribution in structural adhesive joints. *The Journal of Adhesion, 87*(7−8), 671−687. http://dx.doi.org/10.1080/00218464.2011.596113.

Gregory, J. R., & Spearing, S. M. (2005). Constituent and composite quasi-static and fatigue fracture experiments. *Composites: Part A, 36*(5), 665−674. http://dx.doi.org/10.1016/j.compositesa.2004.07.007.

Guyott, C. C. H., Cawley, P., & Adams, R. D. (1986). The non-destructive testing of adhesively bonded structure: a review. *The Journal of Adhesion, 20*(2), 129−159. http://dx.doi.org/10.1080/00218468608074943.

Harvey, C. M., & Wang, S. (2012b). Experimental assessment of mixed-mode partition theories. *Composite Structures, 94*(6), 2057−2067. http://dx.doi.org/10.1016/j.compstruct.2012.02.007.

Harvey, C. M., & Wang, S. (2012a). Mixed-mode partition theories for one-dimensional delamination in laminated composite beams. *Engineering Fracture Mechanics, 96*, 737−759. http://dx.doi.org/10.1016/j.engfracmech.2012.10.001.

He, X. C. (2011). A review of finite element analysis of adhesively bonded joints. *International Journal of Adhesion & Adhesives, 31*(4), 248−264. http://dx.doi.org/10.1016/j.ijadhadh.2011.01.006.

Hélénon, F., Wisnom, M. R., Hallett, S. R., & Trask, R. S. (2012). Numerical investigation into failure of laminated composite T-piece specimens under tensile loading. *Composites: Part A, 43*(7), 1017−1027. http://dx.doi.org/10.1016/j.compositesa.2012.02.010.

Hizam, R. M., Manalo, A. C., & Karunasena, W. (2013). A review of FRP composite truss systems and its connection. In *22nd Australasian conference on the mechanics of structures and materials (ACMSM22): Materials to structures: Advancement through innovation, 11−14 December 2012*, Sydney, Australia. http://eprints.usq.edu.au/id/eprint/22535.

Hung, M. Y. Y., Chen, Y. S., Ng, S. P., Shepard, S. M., Hou, Y. L., & Lhota, J. R. (2007). Review and comparison of shearography and pulsed thermography for adhesive bond evaluation. *Optical Engineering, 46*(5). http://dx.doi.org/10.1117/1.2741277. Article Number 051007.

ISO 291:2008 Plastics—standard atmospheres for conditioning and testing.

ISO 834-12:2012 Fire resistance tests—elements of building construction—part 12: specific requirements for separating elements evaluated on less than full scale furnaces.

ISO 9142:2003 Adhesives—guide to the selection of standard laboratory ageing conditions for testing bonded joints.

ISO 9664:1993 Adhesives—test methods for fatigue properties of structural adhesives in tensile shear.

ISO 10354:1992 Adhesives—characterization of durability of structural-adhesive-bonded assemblies—wedge rupture test.

ISO 14615:1997 Adhesives—durability of structural adhesive joints—exposure to humidity and temperature under load.

ISO 15024:2001 Fibre-reinforced plastic composites—determination of mode I inter-laminar fracture toughness, G_{IC}, for unidirectionally reinforced materials.

ISO/DIS 15114 Fibre-reinforced plastic composites—determination of the mode II fracture resistance for unidirectionally reinforced materials using the calibrated end-loaded split (C-ELS) test and an effective crack length approach.

ISO/TR 15655:2003 Fire resistance—tests for thermo-physical and mechanical properties of structural materials at elevated temperatures for fire engineering design.

ISO 17212:2012 Structural adhesives—guidelines for the surface preparation of metals and plastics prior to adhesive bonding.

ISO 21368:2005 Adhesives—guidelines for the fabrication of adhesively bonded structures and reporting procedures suitable for the risk evaluation of such structures.

ISO 25217:2009 Adhesives—determination of the mode I adhesive fracture energy of structural adhesive joints using double cantilever beam and tapered double cantilever beam specimens.

Kafkalidis, M. S., & Thouless, M. D. (2002). The effects of geometry and material properties on the fracture of single lap-shear joints. *International Journal of Solids and Structures, 39*(17), 4367−4383. http://dx.doi.org/10.1016/S0020-7683(02)00344-X.

Karhnak, S. J., & Duke, J. C., Jr. (1994). Predicting performance of adhesively bonded joints based on acousto-ultrasonic evaluation. In D. J. Damico, T. L. Wilkinson, Jr., & S. L. F. Niks (Eds.), *Composites bonding, ASTM STP 1227* (pp. 60−67). Philadelphia: American Society for Testing and Materials.

Keller, T. (2001). Recent all-composite and hybrid fiber reinforced polymer bridges and buildings. *Progress in Structural Engineering and Materials, 3*(2), 132−140. http://dx.doi.org/10.1002/pse.66.

Keller, T., Bai, Y., & Vallée, T. (2007). Long-term performance of the Pontresina GFRP pedestrian bridge. In *Proceedings 3rd international conference on durability and field applications of FRP composites for construction (CDCC 2007)*, Quebec, Canada.
Keller, T., & Tirelli, T. (2004). Fatigue behavior of adhesively connected pultruded GFRP profiles. *Composite Structures, 65*(1), 55−64. http://dx.doi.org/10.1016/j.compstruct.2003. 10.008.
Kinloch, A. J. (1979). Interfacial fracture mechanical aspects of adhesive bonded joints − a review. *Journal of Adhesion, 10*(3), 193−219. http://dx.doi.org/10.1080/002184679 08544625.
Kinloch, A. J. (1997). Adhesives in engineering, proceedings of the institution of mechanical engineers, part G. *Journal of Aerospace Engineering, 211*(5), 307−335. http://dx.doi.org/ 10.1243/0954410971532703.
Kondo, A., Sato, Y., Suemasu, H., & Aoki, Y. (2011). Fracture resistance of carbon/epoxy composite laminates under mixed-mode II and III failure and its dependence on fracture morphology. *Advanced Composite Materials, 20*(5), 405−418. http://dx.doi.org/10.1163/ 092430411X568197.
Kinloch, A. J., & Taylor, A. C. (2004). *The use of fracture mechanics techniques to predict the service life of adhesive joints*. In D. R. Moore (Ed.), *The application of fracture mechanics to polymers, adhesives and composites* (Vol. 33) (pp. 187−192). Oxford: Elsevier, ESIS Publication.
Kinloch, A. J., Wang, Y., Williams, J. G., & Yayla, P. (1993). The mixed-mode delamination of fibre composite materials,. *Composites Science and Technology, 47*(3), 225−237. http:// dx.doi.org/10.1016/0266-3538(93)90031-B.
Laffan, M. J., Pinho, S. T., Robinson, P., & McMillan, A. J. (2012). Translaminar fracture toughness testing of composites: a review,. *Polymer Testing, 31*(3), 481−489. http:// dx.doi.org/10.1016/j.polymertesting.2012.01.002.
Lee, H. K., Pyo, S. H., & Kim, B. R. (2009). On joint strengths, peel stresses and failure modes in adhesively bonded double-strap and supported single-lap GFRP joints. *Composite Structures, 87*(1), 44−54. http://dx.doi.org/10.1016/j.compstruct.2007.12.005.
Mackerle, J. (2004). Finite element analyses and simulations of manufacturing processes of composites and their mechanical properties: a bibliography (1985−2003). *Computational Materials Science, 31*(3−4), 187−219. http://dx.doi.org/10.1016/j.commatsci. 2004.03.001.
Markatos, D. N., Tserpes, K. I., Rau, E., Markus, S., Ehrhart, B., & Pantelakis, Sp. (2013). The effects of manufacturing-induced and in-service related bonding quality reduction on the mode-I fracture toughness of composite bonded joints for aeronautical use. *Composites: Part B, 45*(1), 556−564. http://dx.doi.org/10.1016/j.compositesb.2012.05.052.
Milia, E., Cumbo, E., Cardoso, R. J. A., & Gallina, G. (2012). Current dental adhesives systems. A narrative review. *Current Pharmaceutical Design, 18*(34), 5542−5552.
Miller, A. H., Dodds, N., Hale, J. M., & Gibson, A. G. (1998). High speed pultrusion of thermoplastic matrix composites. *Composites: Part A, 29*(7), 773−782. http://dx.doi.org/ 10.1016/S1359-835X(98)00006-2.
Miura, M., Shindo, Y., Takeda, T., & Narita, F. (2014). Mixed-mode I/III fatigue delamination growth in woven glass/epoxy composite laminates at cryogenic temperatures. *Journal of Composite Materials, 48*(10), 1251−1259. http://dx.doi.org/ 10.1177/0021998313484951.
de Morais, A. B., & Pereira, A. B. (2008). Mixed mode II + III interlaminar fracture of carbon/ epoxy laminates. *Composites Science and Technology, 68*(9), 2022−2027. http:// dx.doi.org/10.1016/j.compscitech.2008.02.023.

de Morais, A. B., Pereira, A. B., & de Moura, M. F. S. F. (2011). Mode III interlaminar fracture of carbon/epoxy laminates using the six-point edge crack torsion (6ECT). *Composites: Part A, 42*(11), 1793−1799. http://dx.doi.org/10.1016/j.compositesa.2011.08.002.
de Morais, A. B., Pereira, A. B., de Moura, M. F. S. F., & Magalhães, A. G. (2009). Mode III interlaminar fracture of carbon/epoxy laminates using the edge crack torsion (ECT) test. *Composites Science and Technology, 69*(5), 670−676. http://dx.doi.org/10.1016/j.compscitech.2008.12.019.
de Moura, M. F. S. F., Campilho, R. D. S. G., & Gonçalves, J. P. M. (2008). Crack equivalent concept applied to the fracture characterization of bonded joints under pure mode I loading. *Composites Science and Technology, 68*(10−11), 2224−2230. http://dx.doi.org/10.1016/j.compscitech.2008.04.003.
de Moura, M. F. S. F., Campilho, R. D. S. G., & Gonçalves, J. P. M. (2009). Pure mode II fracture characterization of composite bonded joints. *International Journal of Solids and Structures, 46*(6), 1589−1595. http://dx.doi.org/10.1016/j.ijsolstr.2008.12.001.
Mouritz, A. P. (2001). Ballistic impact and explosive blast resistance of stitched composites. *Composites: Part B, 32*(5), 431−439.
Mouritz, A. P. (2002). Post-fire flexural properties of fibre-reinforced polyester, epoxy and phenolic composites. *Journal of Materials Science, 37*(7), 1377−1386. http://dx.doi.org/10.1023/A:1014520628915.
Mouritz, A. P., Feih, S., Kandare, E., Mathys, Z., Gibson, A. G., Des Jardin, P. E., et al. (2009). Review of fire structural modelling of polymer composites. *Composites: Part A, 40*(12), 1800−1814. http://dx.doi.org/10.1016/j.compositesa.2009.09.001.
Newaz, G. M., Lustiger, A., & Yung, J. Y. (1989). Delamination growth under cyclic loading at elevated temperature in thermoplastic composites. In G. M. Newaz (Ed.), *Advances in thermoplastic matrix composite materials, ASTM STP 1044* (pp. 264−278,). Philadelphia: American Society for Testing and Materials.
Park, S. Y., Choi, W. J., Choi, H. S., Kwon, H., & Kim, S. H. (2010). Recent trends in surface treatment technologies for airframe adhesive bonding processing: a review. *The Journal of Adhesion, 86*(2), 192−221. http://dx.doi.org/10.1080/00218460903418345.
Peng, L., Xu, J. F., Zhang, J. Y., & Zhao, L. B. (2012). Mixed mode delamination growth of multidirectional composite laminates under fatigue loading. *Engineering Fracture Mechanics, 96*, 676−686. http://dx.doi.org/10.1016/j.engfracmech.2012.09.033.
Pereira, A. B., de Morais, A. B., & de Moura, M. F. S. F. (2011). Design and analysis of a new six-point edge crack torsion (6ECT) specimen for mode III interlaminar fracture characterization. *Composites: Part A, 42*(2), 131−139. http://dx.doi.org/10.1016/j.compositesa.2010.10.013.
Ratna, D. (2003). Modification of epoxy resins for improvement of adhesion: a critical review. *Journal of Adhesion Science and Technology, 17*(12), 1655−1668. http://dx.doi.org/10.1163/156856103322396721.
del Real, J. C., Ballesteros, Y., Chamochin, R., Abenojar, J., & Molisani, L. (2011). Influence of surface preparation on the fracture behavior of acrylic adhesive/CFRP composite joints. *The Journal of Adhesion, 87*(4), 366−381. http://dx.doi.org/10.1080/00218464.2011.562114.
Regel, F., van Hattum, F. W. J., & Dias, G. R. (2013). A numerical and experimental study of the material properties determining the crushing behaviour of pultruded GFRP profiles under lateral compression. *Journal of Composite Materials, 47*(14), 1749−1764. http://dx.doi.org/10.1177/0021998312451297.
Richter-Trummer, V., Marques, E. A., Chaves, F. J. P., Tavares, J. M. R. S., da Silva, L. F. M., & de Castro, P. M. S. T. (2011). Analysis of crack growth behavior in a double cantilever

beam adhesive fracture test by different digital image processing techniques. *Materialwissenschaft und Werkstofftechnik*, *42*(5), 452−459. http://dx.doi.org/10.1002/mawe.201100807.

Russo, S. (2013). Damage assessment of GFRP pultruded structural elements. *Composite Structures*, *96*, 661−669. http://dx.doi.org/10.1016/j.compstruct.2012.09.014.

Sans, D., Stutz, S., Renart, J., Mayugo, J. A., & Botsis, J. (2012). Crack tip identification with long FBG sensors in mixed-mode delamination. *Composite Structures*, *94*(9), 2879−2887. http://dx.doi.org/10.1016/j.compstruct.2012.03.032.

Sarfaraz, R., Vassilopoulos, A. P., & Keller, T. (2011). Experimental investigation of the fatigue behavior of adhesively-bonded pultruded GFRP joints under different load ratios. *International Journal of Fatigue*, *33*(11), 1451−1460. http://dx.doi.org/10.1016/j.ijfatigue.2011.05.012.

Sarfaraz, R., Vassilopoulos, A. P., & Keller, T. (2012). Experimental investigation and modeling of mean load effect on fatigue behavior of adhesively-bonded pultruded GFRP joints. *International Journal of Fatigue*, *44*, 245−252. http://dx.doi.org/10.1016/j.ijfatigue.2012.04.021.

Sarfaraz, R., Vassilopoulos, A. P., & Keller, T. (2013a). Block loading fatigue of adhesively bonded pultruded GFRP joints. *International Journal of Fatigue*, *49*(1), 40−49. http://dx.doi.org/10.1016/j.ijfatigue.2012.12.006.

Sarfaraz, R., Vassilopoulos, A. P., & Keller, T. (2013b). Modeling the constant amplitude fatigue behavior of adhesively bonded pultruded GFRP joints. *Journal of Adhesion Science and Technology*, *27*(8), 855−878. http://dx.doi.org/10.1080/01694243.2012.727158.

Shahverdi, M. (2013). *Mixed-mode static and fatigue failure criteria for adhesively-bonded FRP joints* (Ph.D. thesis) No. 5728. Switzerland: Ecole Polytechnique Fédérale de Lausanne (EPFL).

Shahverdi, M., Vassilopoulos, A. P., & Keller, T. (2011). A phenomenological analysis of mode I fracture of adhesively-bonded pultruded GFRP joints. *Engineering Fracture Mechanics*, *78*(10), 2161−2173. http://dx.doi.org/10.1016/j.engfracmech.2011.04.007.

Shahverdi, M., Vassilopoulos, A. P., & Keller, T. (2012a). Experimental investigation of R-ratio effects on fatigue crack growth of adhesively-bonded pultruded GFRP DCB joints under CA loading. *Composites: Part A*, *43*(10), 1689−1697. http://dx.doi.org/10.1016/j.compositesa.2011.10.018.

Shahverdi, M., Vassilopoulos, A. P., & Keller, T. (2012b). A total fatigue life model for the prediction of the R-ratio effects on fatigue crack growth of adhesively-bonded pultruded GFRP DCB joints. *Composites: Part A*, *43*(10), 1783−1790. http://dx.doi.org/10.1016/j.compositesa.2012.05.004.

Shahverdi, M., Vassilopoulos, A. P., & Keller, T. (2013a). Modeling effects of asymmetry and fiber bridging on mode I fracture behavior of bonded pultruded composite joints. *Engineering Fracture Mechanics*, *99*, 335−348. http://dx.doi.org/10.1016/j.engfracmech.2013.02.001.

Shahverdi, M., Vassilopoulos, A. P., & Keller, T. (2013b). Mixed-mode fatigue failure criteria for adhesively-bonded pultruded GFRP joints. *Composites: Part A*, *54*(1), 46−55. http://dx.doi.org/10.1016/j.compositesa.2013.06.017.

Shindo, Y., Miura, M., Takeda, T., Saito, N., & Narita, F. (2011). Cryogenic delamination growth in woven glass/epoxy composite laminates under mixed-mode I/II fatigue loading. *Composites Science and Technology*, *71*(5), 647−652. http://dx.doi.org/10.1016/j.compscitech.2011.01.006.

da Silva, L. F. M., Esteves, V. H. C., & Chaves, F. J. P. (2011). Fracture toughness of a structural adhesive under mixed mode loadings. *Materialwissenschaft und Werkstofftechnik, 42*(5), 460−470.
Sohier, L., Cognard, J. Y., & Davies, P. (2013). Analysis of the mechanical behaviour of adhesively bonded assemblies of composites under tensile-shear out-of-plane loads. *Composites: Part A, 53*(1), 65−74. http://dx.doi.org/10.1016/j.compositesa.2013.05.008.
Steinbrecher, G., Buchman, A., Sidess, A., & Sherman, D. (2006). Characterization of the mode I fracture energy of adhesive joints. *International Journal of Adhesion & Adhesives, 26*(8), 644−650. http://dx.doi.org/10.1016/j.ijadhadh.2005.09.006.
Stelzer, S., Brunner, A. J., Argüelles, A., Murphy, N., & Pinter, G. (2012). Mode I delamination fatigue crack growth in unidirectional fiber reinforced composites: development of a standardized test procedure. *Composites Science and Technology, 72*(10), 1102−1107. http://dx.doi.org/10.1016/j.compscitech.2011.11.033.
Stelzer, S., Jones, R., & Brunner, A. J. (2013). Interlaminar fatigue crack growth in carbon fiber reinforced composites. In *Proceedings 19th international conference on composites, ICCM-19* (pp. 1689−1697,).
Sterley, M., Serrano, E., & Enquist, B. (2013). Fracture characterisation of green-glued polyurethane adhesive bonds in mode I. *Materials and Structures, 46*(3), 421−434. http://dx.doi.org/10.1617/s11527-012-9911-5.
Stewart, A., & Douglas, E. P. (2012). Accelerated testing of epoxy-FRP composites for civil infrastructure applications: property changes and mechanisms of degradation. *Polymer Reviews, 52*(2), 115−141. http://dx.doi.org/10.1080/15583724.2012.668152.
Stutz, S., Cugnoni, J., & Botsis, J. (2011). Studies of mode I delamination in monotonic and fatigue loading using FBG wavelength multiplexing and numerical analysis. *Composites Science and Technology, 71*(4), 443−449. http://dx.doi.org/10.1016/j.compscitech.2010.12.016.
Taib, A. A., Boukhili, R., Achiou, S., & Boukehili, H. (2006). Bonded joints with composite adherends. Part II. Finite element analysis of joggle lap joints. *International Journal of Adhesion & Adhesives, 26*(4), 237−248. http://dx.doi.org/10.1016/j.ijadhadh.2005.03.014.
Taib, A. A., Boukhili, R., Achiou, S., Gordon, S., & Boukehili, H. (2006). Bonded joints with composite adherends. Part I. Effect of specimen configuration, adhesive thickness, spew fillet and adherend stiffness on fracture. *International Journal of Adhesion & Adhesives, 26*(4), 226−236, 0.1016/j.ijadhadh.2005.03.015.
Takeda, N., Tohdoh, M., & Takahashi, K. (1995). Interlaminar fracture toughness degradation of radiation-damaged GFRP and CFRP composites. *Advanced Composite Materials, 4*(4), 343−354. http://dx.doi.org/10.1163/156855195X00195.
Tang, H. (2010). Latest advances in joining technologies for automotive body manufacturing. *International Journal of Vehicle Design, 54*(1), 1−25. http://dx.doi.org/10.1504/IJVD.2010.034867.
Tay, T. E. (2003). Characterization and analysis of delamination fracture in composites − a review of developments from 1990 to 2001. *Applied Mechanics Review, 56*(1), 1−23. http://dx.doi.org/10.1115/1.1504848.
Tracy, G. D., Feraboli, P., & Kedward, K. T. (2003). A new mixed mode test for carbon/epoxy composite systems. *Composites: Part A, 34*(11), 1125−1131. http://dx.doi.org/10.1016/S1359-835X(03)00205-7.
Vallée, T., Correia, J. R., & Keller, T. (2006). Probabilistic strength prediction for double lap joints composed of pultruded GFRP profiles − part II: strength prediction. *Composites Science and Technology, 66*(13), 1915−1930. http://dx.doi.org/10.1016/j.compscitech.2006.04.001.

Vallée, T., Correia, J. R., & Keller, T. (2010). Optimum thickness of joints made of GFPR pultruded adherends and polyurethane adhesive. *Composite Structures, 92*(9), 2102−2108. http://dx.doi.org/10.1016/j.compstruct.2009.09.056.

Vallée, T., Keller, T., Fourestey, G., Fournier, B., & Correia, J. R. (2009). Adhesively bonded joints composed of pultruded adherends: considerations at the upper tail of the material strength statistical distribution. *Probabilistic Engineering Mechanics, 24*(3), 358−366. http://dx.doi.org/10.1016/j.probengmech.2008.10.001.

Vassilopoulos, A. P., Manshadi, B. D., & Keller, T. (2010). Influence of the constant life diagram formulation on the fatigue life prediction of composite materials. *International Journal of Fatigue, 32*(4), 659−669. http://dx.doi.org/10.1016/j.ijfatigue.2009.09.008.

Vassilopoulos, A. P., Sarfaraz, R., Manshadi, B. D., & Keller, T. (2010). A computational tool for the life prediction of GFRP laminates under irregular complex stress states: influence of the fatigue failure criterion. *Computational Materials Science, 49*(3), 483−491. http://dx.doi.org/10.1016/j.commatsci.2010.05.039.

Williams, J. G. (1988). On the calculation of energy release rates for cracked laminates. *International Journal of Fracture, 36*(2), 101−119. http://dx.doi.org/10.1007/BF00017790.

Yoshihara, H. (2006). Examination of the 4-ENF test for measuring the mode III R-curve of wood. *Engineering Fracture Mechanics, 73*(1), 42−63. http://dx.doi.org/10.1016/j.engfracmech.2005.06.008.

Yu, X. X., Crocombe, A. D., & Richardson, G. (2001). Material modelling for rate-dependent adhesives. *International Journal of Adhesion & Adhesives, 21*(3), 197−210. http://dx.doi.org/10.1016/S0143-7496(00)00051-8.

Zhang, Y., & Keller, T. (2008). Progressive failure process of adhesively bonded joints composed of pultruded GFRP. *Composites Science and Technology, 68*(2), 461−470. http://dx.doi.org/10.1016/j.compscitech.2007.06.011.

Zhang, Y., Vassilopoulos, A. P., & Keller, T. (2008). Stiffness degradation and fatigue life prediction of adhesively-bonded joints for fiber-reinforced polymer composites. *International Journal of Fatigue, 30*(10−11), 1813−1820. http://dx.doi.org/10.1016/j.ijfatigue.2008.02.007.

Zhang, Y., Vassilopoulos, A. P., & Keller, T. (2009). Environmental effects on fatigue behavior of adhesively-bonded pultruded structural joints. *Composites Science and Technology, 69*(7−8), 1022−1028. http://dx.doi.org/10.1016/j.compscitech.2009.01.024.

Zhang, Y., Vassilopoulos, A. P., & Keller, T. (2010a). Fracture of adhesively-bonded pultruded GFRP joints under constant amplitude fatigue loading. *International Journal of Fatigue, 32*(7), 979−987. http://dx.doi.org/10.1016/j.ijfatigue.2009.11.004.

Zhang, Y., Vassilopoulos, A. P., & Keller, T. (2010b). Mode I and II fracture behavior of adhesively-bonded pultruded composite joints. *Engineering Fracture Mechanics, 77*(1), 128−143. http://dx.doi.org/10.1016/j.engfracmech.2009.09.015.

Zhao, X., Adams, R. D., & da Silva, L. F. M. (2011a). Single lap joints with rounded adherend corners: stress and strain analysis. *Journal of Adhesion Science and Technology, 25*(8), 819−836. http://dx.doi.org/10.1163/016942410X520871.

Zhao, X., Adams, R. D., & da Silva, L. F. M. (2011b). Single lap joints with rounded adherend corners: experimental results and strength prediction. *Journal of Adhesion Science and Technology, 25*(8), 837−856. http://dx.doi.org/10.1163/016942410X520880.

Appendix: list of abbreviations

ADCB	Asymmetric double cantilever beam (specimen)
ASTM	American Society for Testing and Materials
ATCB	Asymmetric tapered double cantilever beam (specimen)
CEN	Comité Européen de Normalisation
CFRP	Carbon fiber–reinforced polymer (material)
CZM	Cohesive zone model
DCB	Double cantilever beam (specimen)
DIC	Digital image correlation
DMA	Dynamic mechanical analysis
ECT	Edge-crack torsion (specimen or test)
ELS	End-loaded split (specimen)
EN	European Norm (standard)
ENF	End-notched flexure (specimen)
EP	Epoxy (polymer)
FEA	Finite element analysis
FEM	Finite element model (or) finite element modeling
FRP	Fiber-reinforced polymer (material)
FTIR	Fourier transform infrared (spectrometry)
GFRP	Glass fiber–reinforced polymer (material)
Hz	Hertz (unit of frequency)
ISO	International Organization for Standardization
JIS	Japan Industrial Standard
MMB	Mixed mode bending (usually combining Mode I and Mode II)
NDT	Non-destructive testing
PUR	Polyurethane
R-ratio	Ratio between minimum and maximum load and displacement, respectively (in cyclic fatigue tests)
SHM	Structural health monitoring
SLB	Single leg bending (specimen or test)
TDCB	Tapered double cantilever beam (specimen)
UV	Ultraviolet (radiation spectrum)

Design of adhesively-bonded composite joints

L.F.M. da Silva[1], R.D.S.G. Campilho[2,3]
[1] Universidade do Porto, Porto, Portugal; [2] Universidade Lusófona do Porto, Porto, Portugal; [3] Instituto Politécnico do Porto, Porto, Portugal

2.1 Introduction

Adhesively-bonded joints provide several benefits, such as more uniform stress distribution than conventional techniques such as fastening or riveting, high fatigue resistance and the possibility of joining different materials (da Silva, Öchsner, & Adams, 2011). The use of composite materials in industry is growing. In this chapter, the term 'composite material' is limited to classical fibre-reinforced plastics. Composite substrates in the form of fibre-reinforced plastics are not isotropic and several tests are necessary to determine all the mechanical properties of the material. One common form of manufacturing fibre-reinforced plastics is to use prepregs. A prepreg consists of a combination of a matrix (or resin) and fibre reinforcement. It is ready to use in the component manufacturing process. It is available in unidirectional (UD) form (one direction of reinforcement) or in fabric form (several directions of reinforcement). Technological improvements in composite materials have been accompanied by an improvement in structural adhesives. As a result, the use of bonded joints has begun to enhance or replace the use of traditional mechanical fasteners in composite and metallic structures. For example, the new airplane BOEING 787 Dreamliner has 50% composite materials in its structure. This higher use of composites combined with other technological improvements makes this airplane have fuel consumption 20% lower than other airplanes of a similar size (http://www.boeing.com/commercial/787family/background.html). The behaviour of composites is highly anisotropic in respect of both stiffness and strength. In the fibre direction, unidirectional composites can be very strong and stiff, whereas the transverse and shear properties are much lower.

Knowledge of the state of stresses inside the adhesive layer of an adhesively-bonded joint is essential for joint strength prediction and joint design. There are two methods for the stress analysis of lap joints, namely analytical and numerical methods. Analytical methods using closed-form algebra employ classical linear theories in which some simplifications are used. The finite element (FE) method, on the other hand, is a well-established numerical technique that can handle complex structures and non-linear material properties where classical methods generally fail to work. Cohesive zone models (CZMs) are increasingly being used in FE (da Silva & Campilho, 2012). This approach enables the complete response of structures up to the final point of failure to be modelled in a single analysis without the need for additional post-processing of FE analysis results. This is an emerging field, and the techniques for modelling damage can be divided into either local or continuum approaches. In the continuum approach, the damage is modelled over a finite region. In the local approach, the damage is confined to zero volume lines and surfaces in two and three dimensions, respectively, and is often referred to as the cohesive zone approach. In most of the CZMs, the traction−separation relations for the interfaces are such that with increasing interfacial separation, the traction across the interface reaches a maximum (crack initiation), then decreases (softening), and finally crack propagates, permitting a total debond of the interfaces. The whole failure response and crack propagation can thus be simulated. A CZM models the fracture process extending the concept of continuum mechanics by including a zone of discontinuity modelled by cohesive zones, thus using both strength and fracture mechanics concepts. CZMs may be used for adhesive debonding or for composite delaminations.

Although the closed-form solutions have their limitations, they are easy to use, especially for parametric studies. The FE method needs large computer power, especially if high accuracy is required, and experienced personnel. Consequently, the former is widely used for joint design and the latter for research or for complex geometries. There are many analytical models in the literature for obtaining stress and strain distributions. Many closed-form models are based on modified shear-lag equations, as proposed originally by Volkersen (1938). Reviews of these closed-form theories and their assumptions can be found in da Silva, das Neves, Adams, and Spelt (2009), da Silva, das Neves, Adams, Wang, and Spelt (2009) and Tong and Luo (2011). As the analytical equations become more complex (including factors such as stress variation through the adhesive thickness, plasticity, thermal effects, composite materials, etc.), there is a greater requirement to use computing power to solve for the stresses. Hart-Smith (1973) had a great influence on the methods used for stress analysis of adhesive joints. Versions of this method have been prepared as Fortran programmes and have been used extensively in the aerospace industry. Other analyses have been implemented in spreadsheets or as a programme for personal computers (PCs). The software packages assist in the design of efficient joints. A brief overview of commercial PC-based analysis/design software packages is given in Table 2.1. The main features of each software package are identified. As shown in Table 2.1, the existing software packages are very specific and most of them only cover one or two joint geometries. Moreover, only a limited number of models are suitable for composite materials. For most of the software, the choosing process is dependent on the previous experience of the designer, apart from those of da Silva, Lima, and Teixeira (2009) and Dragoni, Goglio, and Kleiner (2010).

The design philosophy associated with adhesive joints is radically different from that of other traditional methods of joining such as bolts or rivets. Therefore, it is not advised to take a joint initially designed for another type of joining method and modify it for adhesive bonding. The first point in design is to choose a suitable adhesive, which will depend on the type of loading, the adherends to bond and the environment (temperature and humidity). Adhesives, which are polymeric materials, are not as strong as the adherends they are joining, such as metals or composites. However, when they are loaded over a large area such as in a single-lap joint, they can provide a high

Table 2.1 Software packages available on the market

Name	Supplier	Application	Features
Bolt	G.S. Springer Stanford University, USA	Design of pin-loaded holes in composites	• Prediction of failure strength and failure mode • Three types of bolted joints: joints with a single hole, joints with two identical holes in a row • Joints with two identical holes in tandem • Applicable to uniform tensile loads and symmetric laminates
BISEPSLOCO	AEA Technology, UK	Closed form computer code for predicting stresses and strains in adhesively-bonded single-lap joints	• Tensile/shear/bending moment loading • Adhesive peel and shear stress predictions • Allowance for plasticity in adhesive layer • Thermal stress analysis
BISEPSTUG	AEA Technology, UK	Closed form computer code for predicting stresses and strains in adhesively-bonded coaxial joints	• Stepped and profiled joints • Orthotropic adherends • Torsional and axial loading • Allowance for plasticity in adhesive layer • Thermal stress analysis
CoDA	National Physical Laboratory, UK	Preliminary design of composite beams and panels, and bolted joints	• Synthesis of composite material properties (lamina and laminates for a range of fibre formats) • Parametric analyses • Panel and beam design • Bonded and bolted double shear joints • Bearing, shear-out, pin shear and by-pass tensile failure prediction
DLR	DLR-Mitteilung, Germany	Preliminary design of composite joints	• Adhesively-bonded and bolted joints • Linear-elastic and linear-elastic/plastic behaviour

Continued

Table 2.1 **Continued**

Name	Supplier	Application	Features
			• Tension and shear loading
• Symmetric and asymmetric lap joints
• Bearing, shear-out, pin shear and by-pass tensile failure prediction (washers and bolt tightening)
Feloco	AEA Technology, UK	Finite element module computer code for predicting stresses and strains in adhesively-bonded lap shear joints	• Stepped and profiled joints
• Tensile/shear/bending moment/pressure loading
• Linear and non-linear analysis
• Peel, shear and longitudinal stress predictions in adhesive layer and adherends
• Thermal stress analysis for adherend and adhesive
PAL	Permabond, UK	'Expert' system for adhesive selection	• Joined systems include:
• Lap and butt joints, Sandwich structures, Bushes/gears/bearings/shafts/pipes/threaded fittings
• Elastic analysis
• Creep/fatigue effects on joint stiffness (graphical)
RETCALC	Loctite, UK	Interactive Windows-based general-purpose software	• Joint strength
• Correction factors (temperature and fatigue)
ESDU	Engineering Science Data Unit, UK	Software for use in structural design	• ESDU 78042 shear stresses in the adhesives in bonded joints. Single-step double-lap joints loaded in tension
• ESDU 79016 Inelastic shear stresses and strains in the adhesives bonding lap joints loaded in tension or shear (computer program)
• ESDU 80011 elastic stresses in the adhesive in single step double-lap bonded joints

Table 2.1 **Continued**

Name	Supplier	Application	Features
			• ESDU 80039 elastic adhesive stresses in multi-step lap joints loaded in tension (computer program) • ESDU 81022 Guide to the use of Data Items in the design of bonded joints
Joint designer (da Silva, Lima, & Teixeira, 2009)	Faculty of Engineering, University of Porto, Portugal	Closed-form computer code for predicting stresses and strength in adhesively-bonded joints	• Accessible to non-experts • Lap joints • Adhesive peel and shear stress predictions • Joint strength prediction • Allowance for plasticity in adhesive layer and adherends • Orthotropic adherends • Thermal stress analysis
JointCalc (Dragoni et al., 2010)	Henkel AG, Germany	Closed-form computer code for predicting stresses and strength in adhesively-bonded joints	• Accessible to non-experts • Single- and double-lap joints, single- and double-strap joints, peel joints and cylindrical joints • Adhesive peel and shear stress predictions • Joint strength prediction

Source: Adapted from National Physical Laboratory (2007).

strength, sufficient to deform plastically the metal in some cases or to break the composite. That is why when designing an adhesive joint, the load must be spread over a large area and not concentrated in one point. Peel loads are the greatest enemy of the designer of bonded joints (Adams, Comyn, & Wake, 1997), especially for composite materials. The adhesive should be loaded in shear whenever possible and the peel and cleavage stresses should be avoided. However, there might be limitations in terms of manufacturing process, cost, consequences of failure, and desired final appearance that may complicate the designing process.

The strength of an adhesive joint in the absence of environmental factors is determined by the mechanical properties of the adhesive and adherends, the joint geometry and the residual internal stresses. In effect, localized stresses are not always apparent and may occur as a result of differential thermal expansion of the adhesive and adherends. Another cause is shrinkage of adhesive during cure. These factors are all discussed and simple design guidelines are given. The joint strength can be improved by a number of techniques such as fillets and adherend shaping. All these aspects are discussed in detail. Hybrid joints are increasingly being used for increasing adhesive joints. Adhesives can be used in conjunction with a rivet or a bolt. The advantages of such solutions are explained in this chapter. Repair techniques are also treated because joints may be damaged in some way and it is important to discuss methods to guarantee an efficient repair design.

2.2 Factors affecting joint strength

The major factors that affect the joint strength of lap joints are the material properties (adherends and adhesive) and the geometry (adherend and adhesive thickness, and overlap). Residual internal stresses due to thermal effects should also be taken into account. The stress distribution in adhesive joints is not uniform, and therefore the average shear stress (i.e. load divided by the bonded area) can be much lower than the local maximum stress. Failure always occurs at the stress concentrations and it is fundamental to decrease these stress peaks if a joint strength improvement is required. There are general guidelines whose objective is to increase the joint strength by minimising the stress concentrations:

- Use an adhesive with a low modulus and ductile behaviour.
- Use similar adherends, or if not possible, balance the stiffness.
- Use a thin adhesive layer.
- Use a large bonded area.

Each of these factors is discussed next and detailed design guidelines are given. The residual stresses caused by thermal effects are also discussed. Bonded joints experience peel loading, so the composite may fail in transverse tension before the adhesive fails. The composite adherend splits apart locally due to these peel stresses, thereby destroying the shear transfer capacity between the two adherends. In that case, the joint strength can be further improved by using adhesive fillets, adherend tapers, adhesive bands along the overlap or hybrid joints. These solutions are described in Section 2.3.

2.2.1 Adhesive properties

It is very important to distinguish between adhesive strength and joint strength. The joint strength may not increase if a stronger adhesive is used. The joint strength depends not only on the adhesive strength but also on its ductility and stiffness. Adhesives with high ductility and flexibility have generally a low strength. However, when used in a joint, their ability to distribute the stress uniformly along the overlap (low stiffness) and to deform plastically can give a joint strength much higher than with apparently strong adhesives that are less ductile (da Silva, Rodrigues, Figueiredo, de Moura, & Chousal, 2006). A low modulus adhesive gives a more uniform stress distribution in comparison to a stiff adhesive where there is a high stress concentration at the ends of the overlap (see Figure 2.1). A ductile adhesive is able to redistribute the

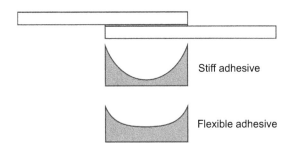

Figure 2.1 Effect of the adhesive modulus on the adhesive stress distribution along the overlap.

load and make use of the less stressed parts of the overlap, whereas a brittle adhesive concentrates the load at the ends of the overlap giving a low average shear stress (see Figure 2.2). Adhesives are either strong, brittle and stiff, or weak, ductile and flexible. The ideal would be to have a strong, ductile and flexible adhesive but this is very difficult to achieve, although the properties are independent. It is recommended to use ductile adhesives, but this also depends on the overlap, as will be seen in Section 2.2.4.

Ductile adhesives are also more resistant to crack propagation than brittle adhesives, giving a much higher toughness. The fatigue strength is generally lower for brittle adhesives. If the fatigue limit is measured in terms of a percentage of the static maximum joint strength, then the fatigue life of joints with ductile adhesives is considerably higher than that of joints with brittle adhesives. This is due to a more uniform stress distribution and also because of the higher damping energy associated with ductile adhesives. In the case of non-uniform loading such as peel, cleavage or thermal internal stresses, a joint with a ductile adhesive will also give a better response.

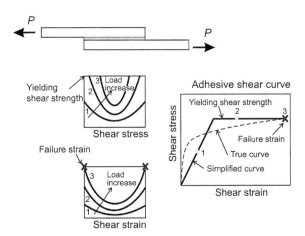

Figure 2.2 Effect of the adhesive ductility on the adhesive stress distribution along the overlap. Adapted from Hart-Smith (1973).

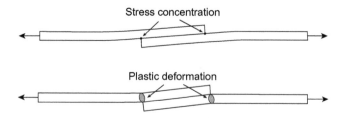

Figure 2.3 Adherend yielding in a single-lap joint.

2.2.2 Adherend properties

Adherend properties also have a huge impact on the joint strength. The most important are the adherend modulus and its strength. The higher the adherend modulus, the lower will be its deformation at the ends of the overlap, where the load transfer takes place, and the lower will be the effect of the differential straining in the adhesive (Volkersen, 1938). The adherend strength is also fundamental and can explain many joint failures. In the case of metallic adherends, the adherend yielding can cause a premature failure of the joint. As the load imposed on the joint increases, the stress at the edge of the overlap increases. When the stress reaches the yield point of the steel, large plastic strains appear, creating a plastic hinge, as shown in Figure 2.3. Although the stresses are limited, the strains associated with the stress in the plastic range are very large. As the maximum adhesive strain is limited, it therefore fails when the maximum adhesive strain is exceeded. In other words, it is the adherend yielding that controls failure.

In the case of composite laminate adherends, it is recommended to have the outer layers with a direction parallel to the loading direction to avoid intralaminar failure of these layers. In any case, the major problem is the low transverse strength (through the thickness) of composites that is of the same order or lower than the adhesive tensile strength. This is a major problem of adhesive joints with composites that tend to fail in an interlaminar manner due to the high peel stresses at the ends of the overlap, as shown in Figure 2.4. As an example, consider the case of a bismaleimide reinforced with a carbon fibre fabric (da Silva & Adams, 2008). The strength in the fibre direction was measured with a four-point flexure test as a function of temperature (Figure 2.5). The failure occurred between the inner rollers. Tests were performed at -55, 22, 100 and 200 °C. It is interesting to note that the strength increases with temperature. This is because the resin becomes tougher and more ductile as the temperature approaches T_g and is less sensitive to defects. It is widely acknowledged that the through-thickness strength of composites is a difficult property to measure. In this study (da Silva & Adams, 2008), the composite plate was cut and bonded to two steel blocks waisted to 12×15 mm². The steel and composite surfaces to bond were grit blasted. However, to avoid damaging the fibres, the composite was grit blasted with the gun of the shot blaster well away from the surface so as to decrease the blast pressure, and for a short period of time (approximately 5 s). Several designs were tested to make sure that failure occurs in the composite (design 3 in Figure 2.6). The adhesive used to bond the

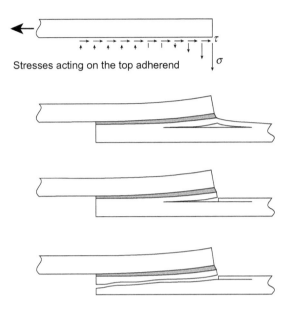

Figure 2.4 Interlaminar failure of the composite in adhesive joints. Adapted from Hart-Smith (1973).

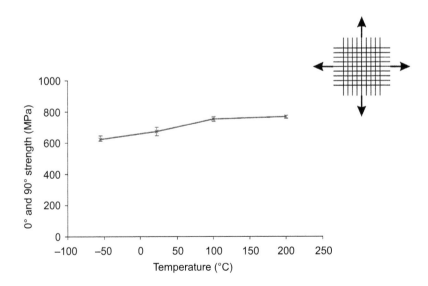

Figure 2.5 0 °C and 90 °C strength in flexure of (four-point flexure test) a bismaleimide composite.

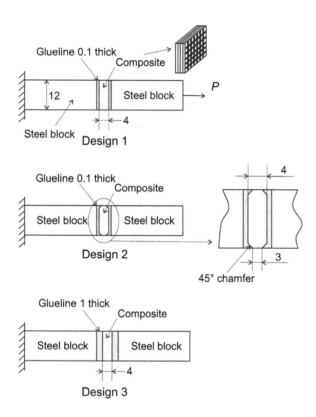

Figure 2.6 Designs studied for the determination of the transverse bismaleimide composite properties (dimensions in mm, not to scale).

composite to the steel blocks was an epoxy AV119 from Huntsman. Its T_g is 120 °C so that the composite was not tested at 200 °C. A jig was used to guarantee alignment during manufacture. The specimen was loaded normal to the plane of the fibres. It is necessary to ensure that the load is perfectly aligned and that there is no bending effect that would lead to a premature failure. This was achieved by loading the steel blocks through precisely aligned pins. Figure 2.7 shows that the strength obtained in the thickness direction is one order of magnitude lower than the longitudinal strength, comparable to the strength of an adhesive.

If failure occurs in the composite, a failure criterion in the adhesive (maximum stress and maximum strain) overestimates the joint strength (Figure 2.8). Predictions based on the composite transverse stress at the interface compare very well with the experiments (da Silva, das Neves, Adams, Wang, et al., 2009). One solution to decrease the peel stresses is by tapering the adherends, as will be discussed in Section 2.3.2.

The use of dissimilar adherends decreases the joint strength due to a non-uniform stress distribution (see Figure 2.9). To reduce this problem, the joint should be designed so that the longitudinal stiffness of the adherends to be bonded are equal,

Design of adhesively-bonded composite joints 53

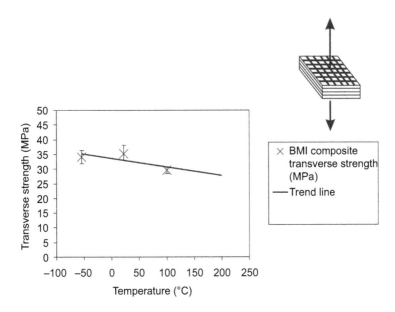

Figure 2.7 Transverse (through the thickness) strength of the bismaleimide composite.

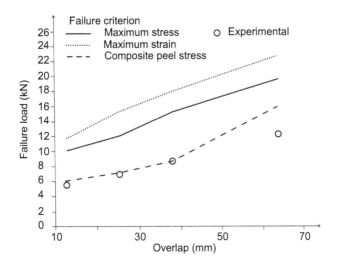

Figure 2.8 Aluminium-composite joints where failure initiates in the composite by peel. Adapted from Adams and Mallick (1993).

i.e. $E_1 t_1 = E_2 t_2$, where E is the longitudinal modulus, t the thickness and the subscripts (1, 2) refer to adherend 1 and adherend 2.

The composite material itself can be modified in order to increase its through-thickness strength. Many novel techniques have been developed. The most common

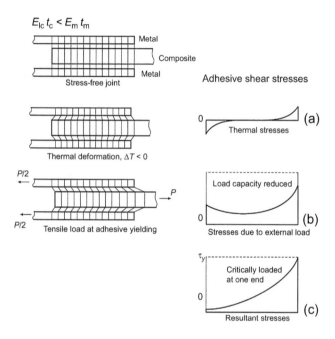

Figure 2.9 Adhesive shear stresses in a metal/composite double-lap joint for the case where the composite has a lower longitudinal stiffness than the metal ($E_{lc}\, t_c < E_m\, t_m$) under a tensile load and a thermal load, at adhesive yielding (τ_y).
Adapted from Hart-Smith (1973).

through-thickness reinforcement techniques are 3D weaving (Mouritz, 2007), stitching (Dransfield, Baillie, & Mai, 1994) and braiding (Tong, Mouritz, & Bannister, 2002, p. 1). However, the only technique capable of reinforcing prepreg laminates in the through-thickness direction in large commercial quantities is z-pinning. z-pins reinforced composites are composed of short small diameter rods inserted in the through-thickness direction (called the 'z-axis' in analytical models and hence the name 'z-pins') of the composite laminates. However, there are several concerns about the manufacture of the z-pinning process, particularly accurate insertion of the z-pins in the orthogonal direction, swelling of the laminate that reduces the fibre volume fraction, and fibre damage (Chang, Mouritz, & Cox, 2006).

Hybrid materials are rather new types of materials that have been developed in the last 20–30 years (Vlot & Gunnink, 2001). One type, the fibre metal laminates, has been developed at the Technical University of Delft, in cooperation with a number of partners. These material systems are created by bonding composite laminate plies to metal plies. It was found that the fatigue crack growth rates in adhesive bonded sheet materials can be reduced if they are built up by laminating and adhesively bonding thin sheets of the material, instead of using one thick monolithic sheet. The concept is usually applied to aluminium with aramid (ARALL) and glass fibres

(GLARE), but can also be applied to other constituents (Alderliesten, 2009; Vermeeren, 2003). In particular the GLARE material has been investigated intensively and has become one of the new materials used on the large Airbus A380 aircraft: two large sections of the fuselage and the leading edges of the horizontal tail planes are made of GLARE. Since metals and fibre-reinforced polymers have characteristic properties and features with respect to manufacturing, the manufacturing of hybrid materials has properties and features related to both material groups. This multilayer composition of the hybrid laminates also offer the opportunity to mix and combine constituent materials with the aim to optimize the component or substructure and have therefore a tailor-made material.

2.2.3 Adhesive thickness

The effect of the bondline thickness on single-lap joints is well documented in the literature. Most of the results are for typical structural adhesives and show that the lap joint strength decreases as the bondline increases (Adams & Peppiatt, 1974; da Silva et al., 2006). Experimental results show that for structural adhesives, the optimum joint strength is obtained with thin bondlines, in the range of 0.1–0.2 mm. However, the classical analytical models such those of Volkersen (1938) or Goland and Reissner (1944) predict the opposite. There are many theories that attempt to explain this fact and this subject is still controversial. Adams and Peppiatt (1974) explained that an increase in the bondline thickness increases the probability of having internal imperfection in the joint (voids and microcracks), which will lead to premature failure of the joints. Crocombe (1989) shows that thicker single-lap joints have a lower strength by considering the plasticity of the adhesive. An elastic analysis shows that the stress distribution of a thin bondline is more concentrated at the ends of the overlap than a thicker bondline that has a more uniform stress distribution. A thin bondline will therefore reach the yielding stress at a lower load than a thick bondline. However, when yielding does occur in a thicker joint, there is a less 'elastic reserve' to sustain further loading and thus yielding spreads more quickly (see Figure 2.10). Gleich, van Tooren, and Beukers (2001) showed with a FE analysis on single-lap joints that increases in the interface stresses (peel and shear) as the bondline gets thicker cause the failure load of a bonded joint to decrease with increasing bondline thickness. They found that for the low bondline thickness range, an optimum distribution of stresses along the joint interface exists for maximum joint strength. Grant, Adams, and da Silva (2009) found a reduction in joint strength with increasing the bondline thickness when testing single-lap joints for the automotive industry with an epoxy adhesive. The strength reduction was attributed to the higher bending moments for the lap joints with thick bondlines due to the increase in the loading offset.

The earlier analyses are in general for metal adherends. When composites are used, a decrease in the adhesive stiffness increases the peel stress at the ends of the overlap and might trigger composite delamination. Therefore, the benefits of using a thin bondline might be reduced.

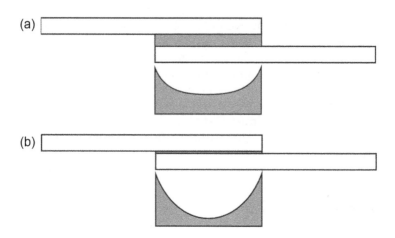

Figure 2.10 Stress distribution in a thick (a) and a thin (b) adhesive layer.

2.2.4 Overlap

Increasing the joint width increases the strength proportionally. However, the effect of the overlap length depends on the type of adhesive (i.e. ductile or brittle) and on the type of adherend. For metal adherends, three cases should be considered: (1) elastic adherends (e.g. high strength steel) and ductile adhesive, (2) elastic adherend and brittle adhesive and (3) adherends that yield. For elastic adherends and ductile adhesives (more than 20% shear strain to failure), the joint strength is approximately proportional to the overlap. This is because ductile adhesives can deform plastically, redistribute the stress as the load increases and make use of the whole overlap. In this case, the failure criterion is the global yielding of the adhesive. For adhesives with intermediate ductility, the adhesive fails because the strain in the adhesive at the ends of the overlap reaches the adhesive shear strain to failure (see Figure 2.11). For elastic adherends and brittle adhesives, the joint strength is not proportional to the overlap and a plateau is reached. This is because the stress is concentrated at the ends of the overlap and a longer overlap does not alter the stress distribution along the overlap. For adherends that yield, the failure is dictated by the adherend yielding, and again a plateau is reached corresponding to the adherend yielding.

The effect of the overlap for composite adherends is mainly dictated by the composite transverse strength. Recently, Neto, Campilho, and da Silva (2012) studied this matter with carbon fibre-reinforced epoxy composites. The main objectives of this work were the characterization of the failure process and strength of adhesive joints with composites, bonded with different adhesives and from short to long overlaps, and the validation of different predicting methods. This work allowed to conclude that for single-lap joints with composites bonded with a ductile adhesive (SikaForce® 7888) the failure load increases as the overlap increase (Figure 2.12) and the failure was cohesive in the adhesive for all overlaps (Figure 2.13(a)). In the case of the brittle

Design of adhesively-bonded composite joints 57

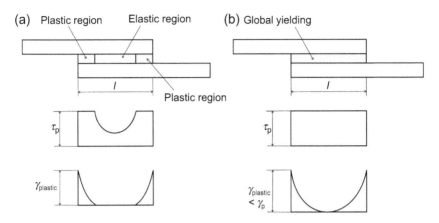

Figure 2.11 Failure due to adhesive shear strain (a) and due to global yielding (b). τ_p is the yielding shear stress and γ_p is the plastic shear strain at failure.

adhesive (AV138), as the overlap increased a plateau was reached (Figure 2.12), since for the overlap of 30 mm, the experimental failure observed was interlaminar and therefore the failure was dictated by the composite (Figure 2.13(b)). Analytical and numerical methods were used to predict the strength of the joints. For the joints with the brittle adhesive, the analytical model of Hart-Smith can predict the failure load in the composite using the peel stress. For the ductile adhesive, the best prediction was obtained with the global yielding criterion (Figure 2.11):

$$P_{GY} = \tau_y \cdot b \cdot l \tag{2.1}$$

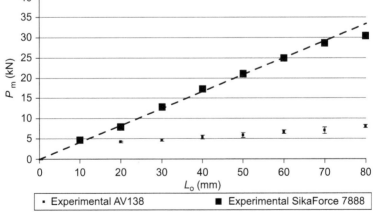

Figure 2.12 Experimental failure loads for composite single-lap joints with a brittle adhesive AV138 and a ductile adhesive SikaForce® 7888.

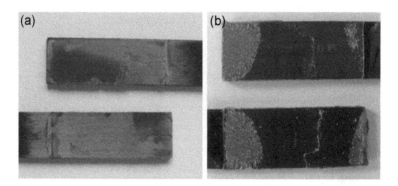

Figure 2.13 (a) Cohesive failure of joints with ductile adhesive SikaForce® 7888 for overlap 50 mm. (b) Composite failure for joints with brittle adhesive AV138 for an overlap of 40 mm.

where P_{GY} is the failure load of the adhesive due to global yielding, τ_y is the yield strength of the adhesive, b is the joint width and l is the overlap length. The numerical methods used (CZM) returned satisfactory values with the brittle adhesive, but with the ductile adhesive these models did not work with the same precision. This probably happened because a triangular law was used and the behaviour of the ductile adhesive is closer to a trapezium shape. Also the determination of the cohesive law properties can be improved with an inverse method or a direct determination method. Nevertheless, the numerical models were capable to simulate the failure initiation and propagation observed in the tests.

2.2.5 Residual stresses

One of the main advantages of using adhesive bonding is the possibility to bond dissimilar materials, such as carbon fibre-reinforced plastics (CFRP) to aluminium in many aeronautical applications. However, dissimilar adherends may have very different coefficients of thermal expansion (CTE). Thus, temperature changes may introduce thermal stresses in addition to the externally applied loads. The adhesive curing and the resulting thermal shrinkage may also introduce internal stresses. Deformations or even cracks can appear. It is important to consider thermal effects because these generally lead to a joint strength reduction, even though in some cases the opposite happens (da Silva, Adams, & Gibbs, 2004). Several authors have found that the stresses caused by adhesive shrinkage have much less effect on the lap joint strength than those generated by the adherend thermal mismatch. Thermal loads are especially important when bonding adherends with different CTEs (Hart-Smith, 1973). For metal/composite joints for example, the metal tends to shrink as the temperature is decreased from the cure value (generally a high temperature) and this is partially resisted by the composite (lower CTE), thereby inducing residual bond stresses especially at the ends of the joint. One end has positive residual shear stresses and the other end has negative residual shear stresses (see Figure 2.9(a)). The thermal stresses are beneficial at one end of the joint but have the reverse effect on the other side of the joint. The thermal load ΔT is given by Eqn (2.2):

$$\Delta T = T_O - T_{SF} \tag{2.2}$$

where T_O is the operating temperature and T_{SF} is the stress-free temperature. It is reasonable to consider the stress-free temperature as the normal cure temperature of the adhesive for most cases.

2.3 Methods to increase joint strength

2.3.1 Fillets

Various authors have shown that the inclusion of a spew fillet at the ends of the overlap reduces the stress concentrations in the adhesive and the substrate (Adams, Atkins, Harris, & Kinloch, 1986; Adams & Harris, 1987; Adams & Peppiatt, 1974; Crocombe & Adams, 1981; Dorn & Liu, 1993; Lang & Mallick, 1998; da Silva & Adams, 2007a, 2007b; Tsai & Morton, 1995). The load transfer and shear stress distribution of a single-lap joint with and without fillet are schematically represented in Figure 2.14. It can be seen that there is a stress concentration at the ends of the overlap for the single-lap joint with a square end. Modification of the joint end geometry with a spew fillet spreads the load transfer over a larger area and gives a more uniform shear

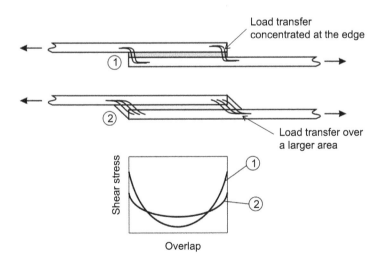

Figure 2.14 Load transfer and shear stress distribution in single-lap joints with and without fillet.

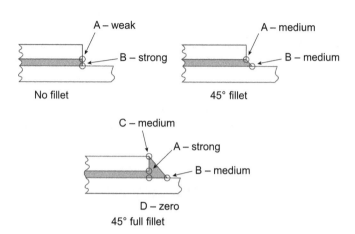

Figure 2.15 Stress intensity factors in adhesive lap joints with different spew fillet geometry.

stress distribution. The fillet not only gives a smoother load transfer but also alters the stress intensity factors, as shown in Figure 2.15.

Adams and Peppiatt (1974) found that the inclusion of a 45° triangular spew fillet decreases the magnitude of the maximum principal stress by 40% when compared to a square-end adhesive fillet. Adams and Harris (1987) tested aluminium/epoxy single-lap joints with and without fillet and found an increase of 54% in joint strength for the filleted joint. Adams et al. (1986) tested aluminium/CFRP single-lap joints and found that the joint with a fillet is nearly two times stronger than the joint without a fillet. Crocombe and Adams (1981) did similar work but included geometric (overlap length, adhesive thickness and adherend thickness) and material (modulus ratio) parameters. The reduction in peel and shear stresses was greatest for a low modulus ratio (low adhesive modulus), a high adhesive thickness and a low adherend thickness.

Dorn and Liu (1993) investigated the influence of the spew fillet in plastic/metal joints. The study included an FE analysis and experimental tests, and they concluded that the spew fillet reduces the peak shear and peel adhesive stresses and decreases stress and strain concentrations in the adherends in the most critical regions. They also studied the influence of different adhesive and different metal adherends. A ductile adhesive and a more balanced joint (aluminium/plastic instead of steel/plastic) give a better stress distribution.

Tsai and Morton (1995) studied the influence of a triangular spew fillet in laminated composite single-lap joints. The FE analysis and the experimental tests (Moire interferometry) proved that the fillet helps to carry part of the load, thus reducing the shear and peel strains.

The above analyses are limited to triangular geometry. Lang and Mallick (1998) investigated eight different geometries: full and half triangular, full and half rounded, full rounded with fillet, oval and arc. They showed that shaping the spew to provide a smoother transition in joint geometry significantly reduces the stress concentrations. Full rounded with fillet and arc spew fillets give the highest percent reduction in

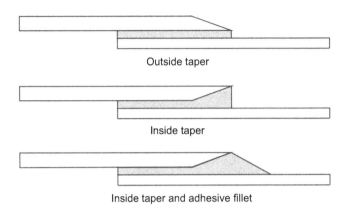

Figure 2.16 Adherend shaping.

maximum stresses, whereas half rounded fillets give the least. Furthermore, increasing the size of the spew also reduces the peak stress concentrations.

2.3.2 Adherend shaping

Adherend shaping is a powerful way to decrease the stress concentration at the ends of the overlap. Figure 2.16 presents typical geometries used for that purpose. Some analytical models were proposed to have a more uniform stress distribution along the overlap (Cherry & Harrison, 1970). However, the FE method is more appropriate for the study of adherend shaping. The concentrated load transfer at the ends of the overlap can be more uniformly distributed if the local stiffness of the joint is reduced. This is particularly relevant for adhesive joints with composites due to the low transverse strength of composites (see Section 2.2.2). Adams et al. (1986) addressed this problem. They studied various configurations of double-lap joints where the central adherend is CFRP and the outer adherends are made of steel. They found with FE and experiments that the inclusion of an internal taper and an external fillet can triple the joint strength. Later da Silva and Adams (2007b) tested joints with an internal taper and an adhesive fillet (see the geometry in Figure 2.16), which were manufactured and tested with the epoxy adhesive Supreme 10HT (Master Bond) at 22 °C. The failure load is higher than for the joint without a taper and an adhesive fillet (basic design) but the increase is very small (see Figure 2.17). The increase in strength obtained by Adams et al. (1986) was three times. They also showed that, with an adhesive fillet and an internal taper, the loading is predominantly tensile in the adhesive, and that the locus of failure is at the outer surface of the adhesive fillet close to the outer adherend corner. The Supreme 10HT joints also failed at the outer adherend corner, as shown in Figure 2.18. The tensile strength of Supreme 10HT at room temperature is 46 MPa (da Silva & Adams, 2005), whereas that of the adhesive used by Adams et al. (1986) was 82 MPa. Therefore, another adhesive with a higher tensile strength than Supreme 10HT was used to check if a joint with an internal taper and an adhesive fillet can give a higher strength increase. The epoxy

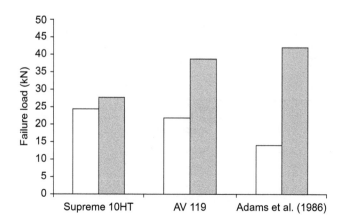

Figure 2.17 Basic design (open bars) versus design with taper (grey bars) and adhesive fillet failure loads for various adhesives at 22 °C (titanium/composite double-lap joints with the composite as the inner adherend).

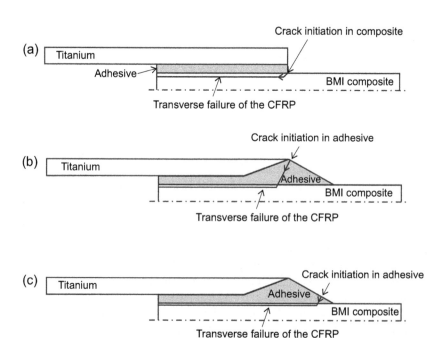

Figure 2.18 Failure mode of titanium/composite double-lap joints: (a) in basic design; (b) with an internal taper and an adhesive fillet at 22 °C; (c) with an internal taper and an adhesive fillet at −55 °C (schematic representation).

AV119 (Huntsman) was selected, as its tensile strength at room temperature is 67 MPa. The strength increase compared to the basic design is now 1.8 times (see Figure 2.17). The locus of failure was also at the outer adherend corner (see Figure 2.18(b)). These results show that the use of an internal taper and adhesive fillet is not necessarily beneficial and will depend on the adhesive properties.

Hildebrand (1994) did similar work on SLJs between fibre-reinforced plastic and metal adherends. The optimisation of the SLJs was done by modifying the geometry of the joint ends. Different shapes of adhesive fillet, reverse tapering of the adherend, rounding edges and denting were applied in order to increase the joint strength. The results of the numerical predictions suggest that, with a careful joint-end design, the strength of the joints can be increased by 90−150%.

The use of internal tapers in adherends in order to minimize the maximum transverse stresses at the ends of bonded joints has also been studied by Rispler, Tong, Steven, and Wisnom (2000). An evolutionary structural optimisation method (EVOLVE) was used to optimize the shape of adhesive fillets. EVOLVE consists of an iterative FE analysis and a progressive removal of elements in the adhesive that are low stressed.

Other examples of joint-end modifications for joint transverse stress reduction but using external tapers are those of Sancaktar and Nirantar (2003) and Kaye and Heller (2005). Kaye and Heller (2005) used numerical optimization techniques in order to optimize the shape of the adherends. This is especially relevant in the context of repairs using composite patches bonded to aluminium structures (see Section 2.5) due to the highly stressed edges.

Tapers (internal or external) or more complex adherend shaping are excellent methods to reduce the peel stresses at the ends of the overlap, and therefore to increase the joint strength. Internal tapers with a fillet seem to be the more efficient way to have a joint increase, especially with brittle adhesives and when composites are used. The FE method is a convenient technique for the determination of the optimum adherend geometry; however, the complexity of the geometry achieved is not always possible to realize in practice.

2.4 Hybrid joints

Joints with different methods of joining are increasingly being used. The idea is to gather the advantages of the different techniques, leaving out their problems. Another possibility is to use more than one adhesive along the overlap or varying the adhesive and/or adherend properties. All these cases have been grouped here under a section called 'hybrid joints' (see Figure 2.19). Such joints are particularly difficult to simulate using analytical models for obvious reasons. The FE method is the preferred tool to investigate the application of such techniques and find design guidelines.

2.4.1 Mixed adhesive joints

Mixed modulus joints have been proposed in the past to improve the stress distribution and increase the joint strength of high-modulus adhesives. The stiff, brittle adhesive

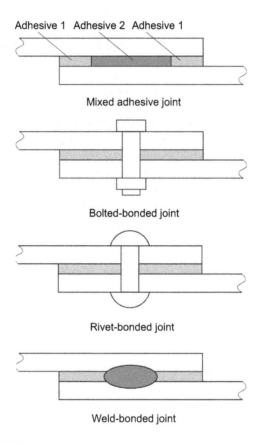

Figure 2.19 Hybrid joints.

should be in the middle of the overlap, while a low-modulus adhesive is applied at the edges prone to stress concentrations. Sancaktar and Kumar (2000) used rubber particles to toughen the part of the adhesive located at the ends of the overlap and increase the joint strength. The concept was studied with the FE method and proved experimentally. Pires, Quintino, Durodola, and Beevers (2003) also proved with an FE analysis and experimentally with two different adhesives that the mixed adhesive method gives an improvement in joint performance. Temiz (2006) used an FE analysis to study the influence of two adhesives in double-lap joints under bending and found that the technique greatly decreased the stresses at the ends of the overlap. Bouiadjra, Fekirini, Belhouari, Boutabout, and Serier (2007) used the mixed modulus technique for the repair of an aluminium structure with a composite patch. The use of a more flexible adhesive at the edge of the patch increases the strength performance of the repair. das Neves, da Silva, and Adams (2009a, 2009b) have developed an analytical model that takes into account two adhesives along the overlap and permits determination of the best combination of adhesives and the optimum geometric factors (e.g. overlap) to have the maximum joint strength. The authors showed that the technique is more efficient for single-lap joints than for double-lap joints. da Silva and Lopes (2009) have studied single-lap adhesive joints maintaining the same brittle adhesive in the middle of the overlap and using three different ductile adhesives of increasing ductility at the ends of the overlap. A simple joint strength prediction is proposed for mixed adhesive joints. The mixed adhesive technique gives joint strength improvements in relation to a brittle adhesive alone in all cases. For a mixed adhesive joint to be stronger than the brittle adhesive and the ductile adhesive used individually, the load carried by the brittle adhesive must be higher than that carried by the ductile adhesive. Marques and da Silva (2008) studied mixed adhesive joints with an internal taper and a fillet. They show that the use of a taper and a fillet have little effect on the strength of mixed adhesive joints. The ductile adhesive at the ends of the overlap is sufficient to have an improved joint strength in relation to a brittle adhesive alone. The taper and the fillet are only useful when the brittle adhesive is used alone. One of the problems associated with the mixed adhesive technique is the adhesive proper separation. The best way to control the process is to use film adhesives. However, it is difficult to find compatible adhesives in the film form. This is a problem for the manufacturers to solve. Meanwhile, the adhesive separation can be done by the use of silicone strips even though a small portion of the load-bearing area is reduced.

The technique of using multi-modulus adhesives has been extended to solve the problem of adhesive joints that need to withstand low and high temperatures by da Silva and Adams (2007b, 2007c). At high temperatures, a high-temperature adhesive in the middle of the joint retains the strength and transfers the entire load, while a low-temperature adhesive is the load-bearing component at low temperatures, making the high-temperature adhesive relatively lightly stressed. The authors studied various configurations with the FE method and proved experimentally that the concept works, especially with dissimilar adherends (titanium/CFRP).

2.4.2 Adhesive joints with functionally graded materials

The mixed adhesive joint technique can be considered to be a rough version of a functionally graded material. The ideal would be to have an adhesive functionally modified with properties that vary gradually along the overlap, allowing a true uniform stress distribution along the overlap. Gannesh and Choo (2002) and Apalak and Gunes (2007) have used functionally graded adherends instead of functionally graded adhesives. Gannesh and Choo (2002) used a braided preform with continuously varying braid angle and the variation of the braid angle measured to realistically evaluate the performance of adherend modulus grading in single-lap bonded joints. An increase of 20% joint strength was obtained due to a more uniform stress distribution. Apalak and Gunes (2007) studied the flexural behaviour of an adhesively-bonded single-lap joint with adherends composed of a functionally gradient layer between a pure ceramic (Al_2O_3) layer and a pure metal (Ni) layer. The study is not supported with experimental results, and the adhesive stress distribution was not hugely affected. To the author's knowledge, the work of Gannesh and Choo (2002) and Apalak and Gunes (2007) are the only studies that deal with the application of functionally graded materials to adhesive joints. There has never been an attempt to modify the adhesive, which might be easier and more logical than modifying the adherends. Only Sancaktar and Kumar (2000) used rubber particles to locally modify the adhesive at the ends of the overlap, but that is not a gradually modified adhesive. This is an area that is being intensively studied and where modelling at different scales is essential. More recently, Stapleton, Waas, and Arnold (2011) used the same idea by strategically placing glass beads within the adhesive layer at different densities along the joint with composite adherends. Kumar (2009) studied numerically the joint behaviour of a functionally graded joint with different degrees of grading in the adhesive stiffness along the overlap. However, several practical concerns impede the actual use of such adhesives. Carbas, da Silva, and Barbosa (2011) are developing an apparatus that allows varying gradually the adhesive properties (stiffness, strength and ductility) along the overlap by using a differential cure.

2.4.3 Rivet-bonded joints

Liu and Sawa (2001) investigated, using a three-dimensional FE analysis, rivet-bonded joints and found that for thin substrates, bonded, riveted joints, adhesive joints and rivet-bonded joints gave similar strengths whilst for thicker substrates the rivet-bonded joints were stronger. They proved this experimentally. Later, the same authors (Liu, Liu, & Sawa, 2004) proposed another technique similar to rivet-bonded joints: adhesive joints with adhesively-bonded columns. Strength improvements are also obtained in this case. The advantage of this technique of that the appearance is the that joint is maintained in relation to an adhesive joint. Grassi, Cox, and Zhang (2006) studied through-thickness pins for restricting debond failure in joints. The pins were simulated by tractions acting on the fracture surfaces of the debond crack. Pirondi and Moroni (2009) found that the adhesive layer strongly increases the maximum load and the initial stiffness in comparison to a joint with a rivet alone. When failure of the adhesive occurs, the joint behaves similar as with rivet alone.

2.4.4 Bolted-bonded joints

Chan and Vedhagiri (2001) studied the response of various configurations of single-lap joints, namely bonded, bolted and bolted-bonded joints by three-dimensional FE method and the results were validated experimentally. The authors found that for the bonded-bolted joints, the bolts help to reduce the stresses at the edge of the overlap, especially after the initiation of failure. The same type of study was carried out by Lin and Jen (1999).

2.5 Repair techniques

Adhesively-bonded repairs are generally associated with complex geometries and the FE method has been extensively used for the design and optimization of the repair, especially with composites. The literature review of Odi and Friend (2002) about

Design of adhesively-bonded composite joints 67

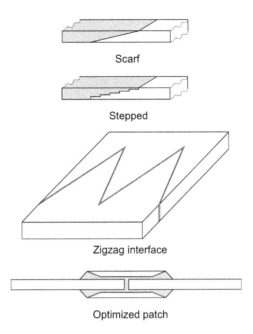

Figure 2.20 Repair techniques.

repair techniques illustrates clearly this point. Typical methods and geometries are presented in Figure 2.20. Among the various techniques available, bonded scarf or stepped repairs are particularly attractive, because a flush surface is maintained that permits a good aerodynamic behaviour. Gunnion and Herszberg (2006) studied scarf repairs and carried out an FE analysis to assess the effect of various parameters. They found that the adhesive stress is not much influenced by mismatched adherend lay-ups and that there is a huge reduction in peak stresses with the addition of an over-laminate. Campilho, de Moura, and Domingues (2007) studied scarf repairs of composites with a CZM and concluded that the strength of the repair increased exponentially with the scarf angle reduction.

Bahei-El-Din and Dvorak (2001) proposed new design concepts for the repair of thick composite laminates. The regular butt-joint with a patch on both sides was modified by the inclusion of pointed inserts or a "zigzag" interface in order to increase the area of contact and improve the joint strength.

Soutis and Hu (1997) studied numerically and experimentally-bonded composite patch repairs to repair cracked aircraft aluminium panels. The authors concluded that the bonded patch repair provides a considerable increase in the residual strength.

Tong and Sun (2003) developed a pseudo-3D element to perform a simplified analysis of bonded repairs to curved structures. The analysis is supported by a full 3D FE analysis. The authors found that external patches are preferred when the shell is under an internal pressure, while internal patches are preferred when under an external pressure.

2.6 Conclusions

Simple design rules for single-lap joints were proposed as a function of the main variables that influence joint strength: adhesive and adherend properties, adhesive thickness, overlap and residual stresses. The main rules are:

- Use an adhesive with a low modulus and high ductility.
- Use similar adherends whenever possible.
- Use a thin adhesive layer.
- Use a large bonded area.

Several methods to increase the joint strength were discussed for lap joints: fillets and adherend shaping. The designer should always bear in mind that when designing an adhesive joint, the load must be spread over a large area and not concentrated in one point. Peel loads are the greatest enemy of the designer of bonded joints, especially with composites due to the low transverse strength.

Adhesive in conjunction with other methods of joining (rivet, bolt) were explained. The idea is to get a synergetic effect and combine the advantages of two methods. Mixed adhesive joints and functionally graded joints are also very promising techniques.

Finally, various types of repairs are discussed in order to obtain the maximum strength recovery. The scarf joint is particularly efficient and aesthetically attractive.

References

Adams, R. D., Atkins, R. W., Harris, J. A., & Kinloch, A. J. (1986). Stress analysis and failure properties of carbon-fibre-reinforced-plastic/steel double-lap joints. *Journal of Adhesion*, 20, 29–53.

Adams, R. D., Comyn, J., & Wake, W. C. (1997). *Structural adhesive joints in engineering* (2nd ed.). London: Chapman & Hall.

Adams, R. D., & Harris, J. A. (1987). The influence of local geometry on the strength of adhesive joints. *International Journal of Adhesion and Adhesives*, 7, 69–80.

Adams, R. D., & Mallick, V. (1993). The effect of temperature on the strength of adhesively-bonded composite-aluminum joints. *Journal of Adhesion*, 43, 17–33.

Adams, R. D., & Peppiatt, N. A. (1974). Stress analysis of adhesive-bonded lap joints. *Journal of Strain Analysis*, 9, 185–196.

Alderliesten, R. (2009). On the development of hybrid material concepts for aircraft structures. *Recent Patents on Engineering*, 3, 25–38.

Apalak, M. K., & Gunes, R. (2007). Elastic flexural behaviour of an adhesively bonded single lap joint with functionally graded adherends. *Materials & Design*, 28, 1597–1617.

Bahei-El-Din, Y. A., & Dvorak, G. J. (2001). New designs of adhesive joints for thick composite laminates. *Composites Science and Technology*, 61, 19–40.

Bouiadjra, B. B., Fekirini, H., Belhouari, M., Boutabout, B., & Serier, B. (2007). Fracture energy for repaired cracks with bonded composite patch having two adhesive bands in aircraft structures. *Computational Materials Science*, 40, 20–26.

Campilho, R. D. S. G., de Moura, M. F. S. F., & Domingues, J. J. M. S. (2007). Stress and failure analyses of scarf repaired CFRP laminates using a cohesive damage model. *Journal of Adhesion Science and Technology*, 21, 855–870.

Carbas, R. J. C., da Silva, L. F. M., & Barbosa, A. Q. (2011). Influence of graded cure on the strength of adhesive joints. In *International conference on structural adhesive bonding 2011, Porto, Portugal, 7–8 July*.
Chan, W. S., & Vedhagiri, S. (2001). Analysis of composite bonded/bolted joints used in repairing. *Journal of Composite Materials, 35*, 1045–1061.
Chang, P., Mouritz, A. P., & Cox, B. N. (2006). Properties and failure mechanisms of z-pinned laminates in monotonic and cyclic tension. *Composites, Part A, 37*(10), 1501–1513.
Cherry, B. W., & Harrison, N. L. (1970). The optimum profile for a lap joint. *Journal of Adhesion, 2*, 125–128.
Crocombe, A. D. (1989). Global yielding as a failure criteria for bonded joints. *International Journal of Adhesion and Adhesives, 9*(3), 145–153.
Crocombe, A. D., & Adams, R. D. (1981). Influence of the spew fillet and other parameters on the stress distribution in the single lap joint. *Journal of Adhesion, 13*, 141–155.
Dorn, L., & Liu, W. (1993). The stress state and failure properties of adhesive-bonded plastic/metal joints. *International Journal of Adhesion and Adhesives, 13*, 21–31.
Dragoni, E., Goglio, L., & Kleiner, F. (2010). Designing bonded joints by means of the JointCalc software. *International Journal of Adhesion and Adhesives, 30*(5), 267–280.
Dransfield, K., Baillie, C., & Mai, Y.-W. (1994). Improving the delamination resistance of CFRP by stitching – a review. *Composites Science and Technology, 50*, 305–317.
Gannesh, V. K., & Choo, T. S. (2002). Modulus graded composite adherends for single-lap bonded joints. *Journal of Composite Materials, 36*, 1757–1767.
Gleich, D. M., van Tooren, M. J. L., & Beukers, A. (2001). Analysis and evaluation of bond line thickness effects on failure load in adhesively bonded structures. *Journal of Adhesion Science and Technology, 15*, 1091–1101.
Goland, M., & Reissner, E. (1944). The stresses in cemented joints. *Journal of Applied Mechanics, 66*, A17–A27.
Grant, L. D. R., Adams, R. D., & da Silva, L. F. M. (2009). Experimental and numerical analysis of single lap joints for the automotive industry. *International Journal of Adhesion and Adhesives, 29*, 405–413.
Grassi, M., Cox, B., & Zhang, X. (2006). Simulation of pin-reinforced single-lap composite joints. *Composites Science and Technology, 66*, 1623–1638.
Gunnion, A. J., & Herszberg, I. (2006). Parametric study of scarf joints in composite structures. *Composite Structures, 75*, 364–376.
Hart-Smith, L. J. (1973). *Adhesive-bonded double-lap joints*. NASA CR-112235.
Hildebrand, M. (1994). Non-linear analysis and optimization of adhesively bonded single lap joints between fibre-reinforced plastics and metals. *International Journal of Adhesion and Adhesives, 14*, 261–267.
Kaye, R., & Heller, M. (2005). Through-thickness shape optimisation of typical double lap-joints including effects of differential thermal contraction during curing. *International Journal of Adhesion and Adhesives, 25*, 227–238.
Kumar, S. (2009). Analysis of tubular adhesive joints with a functionally modulus graded bondline subjected to axial loads. *International Journal of Adhesion and Adhesives, 29*, 785–795.
Lang, T. P., & Mallick, P. K. (1998). Effect of spew geometry on stresses in single lap adhesive joints. *International Journal of Adhesion and Adhesives, 18*, 167–177.
Lin, W.-H., & Jen, M.-H. R. (1999). Strength of bolted and bonded single-lapped composite joints in tension. *Journal of Composite Materials, 33*(7), 640–666.
Liu, J., Liu, J., & Sawa, T. (2004). Strength and failure of bulky adhesive joints with adhesively-bonded columns. *Journal of Adhesion Science and Technology, 18*, 1613–1623.

Liu, J., & Sawa, T. (2001). Stress analysis and strength evaluation of single-lap adhesive joints combined with rivets under external bending moments. *Journal of Adhesion Science and Technology, 15*, 43−61.

Marques, E. A. S., & da Silva, L. F. M. (2008). Joint strength optimization of adhesively bonded patches. *Journal of Adhesion, 84*, 917−936.

Mouritz, A. P. (2007). Review of z-pinned composite laminates. *Composites, Part A, 38*, 2383−2397.

National Physical Laboratory (2007). http://www.npl.co.uk.

Neto, J. A. B. P., Campilho, R. D. S. G., & da Silva, L. F. M. (2012). Parametric study of adhesive joints with composites. *International Journal of Adhesion and Adhesives, 37*, 96−101.

das Neves, P. J. C., da Silva, L. F. M., & Adams, R. D. (2009a). Analysis of mixed adhesive bonded joints − Part I: theoretical formulation. *Journal of Adhesion Science and Technology, 23*, 1−34.

das Neves, P. J. C., da Silva, L. F. M., & Adams, R. D. (2009b). Analysis of mixed adhesive bonded joints − Part II: parametric study. *Journal of Adhesion Science and Technology, 23*, 35−61.

Odi, R. A., & Friend, C. M. (2002). A comparative study of finite element models for the bonded repair of composite structures. *Journal of Reinforced Plastics and Composites, 21*, 311−332.

Pires, I., Quintino, L., Durodola, J. F., & Beevers, A. (2003). Performance of bi-adhesive bonded aluminium lap joints. *International Journal of Adhesion and Adhesives, 23*, 215−223.

Pirondi, A., & Moroni, F. (2009). Clinch-bonded and rivet-bonded hybrid joints: application of damage models for simulation of forming and failure. *Journal of Adhesion Science and Technology, 23*, 1547−1574.

Rispler, A. R., Tong, L., Steven, G. P., & Wisnom, M. R. (2000). Shape optimisation of adhesive fillets. *International Journal of Adhesion and Adhesives, 20*, 221−231.

Sancaktar, E., & Kumar, S. (2000). Selective use of rubber toughening to optimize lap-joint strength. *Journal of Adhesion Science and Technology, 14*, 1265−1296.

Sancaktar, E., & Nirantar, P. (2003). Increasing strength of single lap joints of metal adherends by taper minimization. *Journal of Adhesion Science and Technology, 17*, 655−675.

da Silva, L. F. M., & Adams, R. D. (2005). Measurement of the mechanical properties of structural adhesives in tension and shear over a wide range of temperatures. *Journal of Adhesion Science and Technology, 19*(2), 109−142.

da Silva, L. F. M., & Adams, R. D. (2007a). Adhesive joints at high and low temperatures using similar and dissimilar adherends and dual adhesives. *International Journal of Adhesion and Adhesives, 27*, 216−226.

da Silva, L. F. M., & Adams, R. D. (2007b). Joint strength predictions for adhesive joints to be used over a wide temperature range. *International Journal of Adhesion and Adhesives, 27*, 362−379.

da Silva, L. F. M., & Adams, R. D. (2007c). Techniques to reduce the peel stresses in adhesive joints with composites. *International Journal of Adhesion and Adhesives, 27*, 227−235.

da Silva, L. F. M., & Adams, R. D. (2008). Effect of temperature on the mechanical and bonding properties of a carbon reinforced bismaleimide. *Proceedings of the Institution of Mechanical Engineers, Part L: Journal of Materials: Design and Applications, 222*, 45−52.

da Silva, L. F. M., Adams, R. D., & Gibbs, M. (2004). Manufacture of adhesive joints and bulk specimens with high temperature adhesives. *International Journal of Adhesion and Adhesives, 24*, 69−83.

da Silva, L. F. M., & Campilho, R. D. S. G. (2012). *Advances in numerical modelling of adhesive joints*. Heidelberg: Springer.

da Silva, L. F. M., Lima, R. F. T., & Teixeira, R. M. S. (2009). Development of a computer program for the design of adhesive joints. *Journal of Adhesion*, *85*, 889−918.
da Silva, L. F. M., & Lopes, M. J. C. Q. (2009). Joint strength optimization by the mixed adhesive technique. *International Journal of Adhesion and Adhesives*, *29*, 509−514.
da Silva, L. F. M., das Neves, P. J. C., Adams, R. D., & Spelt, J. K. (2009). Analytical models of adhesively bonded joints − Part I: literature survey. *International Journal of Adhesion and Adhesives*, *29*, 319−330.
da Silva, L. F. M., das Neves, P. J. C., Adams, R. D., Wang, A., & Spelt, J. K. (2009). Analytical models of adhesively bonded joints − Part II: comparative study. *International Journal of Adhesion and Adhesives*, *29*, 331−341.
da Silva, L. F. M., Öchsner, A., & Adams, R. D. (Eds.). (2011). *Handbook of adhesion technology*. Heidelberg: Springer.
da Silva, L. F. M., Rodrigues, T., Figueiredo, M. A. V., de Moura, M., & Chousal, J. A. G. (2006). Effect of adhesive type and thickness on the lap shear strength. *Journal of Adhesion*, *82*, 1091−1115.
Soutis, C., & Hu, F. Z. (1997). Design and performance of bonded patch repairs of composite structures. *Proceedings of the Institution of Mechanical Engineers, Part G*, *211*, 263−271.
Stapleton, S. E., Waas, A. M., & Arnold, S. M. (2011). *Functionally graded adhesives for composite joints*. NASA/TM-2011-217202.
Temiz, S. (2006). Application of bi-adhesive in double-strap joints subjected to bending moment. *Journal of Adhesion Science and Technology*, *20*, 1547−1560.
Tong, L., & Luo, Q. (2011). Analytical approach. In L. F. M. da Silva, A. Öchsner, & R. D. Adams (Eds.), *Handbook of adhesion technology*. Heidelberg: Springer.
Tong, L., Mouritz, A. P., & Bannister, M. K. (2002). *3D fibre reinforced polymer composites*. Boston: Elsevier. pp. 1−12.
Tong, L., & Sun, X. (2003). Nonlinear stress analysis for bonded patch to curved thin-walled structures. *International Journal of Adhesion and Adhesives*, *23*, 349−364.
Tsai, M. Y., & Morton, J. (1995). The effect of a spew fillet on adhesive stress distributions in laminated composite single-lap joints. *Composite Structures*, *32*, 123−131.
Vermeeren, C. A. J. R. (2003). An historic overview of the development of fibre metal laminates. *Applied Composite Materials*, *10*, 189−205.
Vlot, A., & Gunnink, J. W. (2001). *Fibre metal laminates − an introduction*. Dordrecht: Kluwer Academic Publisheas.
Volkersen, O. (1938). Die nietkraftoerteilung in zubeanspruchten nietverbindungen mit konstanten loschonquerschnitten. *Luftfahrtforschung*, *15*, 41−47.

Understanding fatigue loading conditions in adhesively-bonded composite joints

R. Sarfaraz
École Polytechnique Fédérale de Lausanne (EPFL), Lausanne, Switzerland

3.1 Introduction

New load-bearing structures made of fiber-reinforced polymer (FRP) composites comprise adhesively-bonded joints, which are components vulnerable to fatigue failure (Figure 3.1). These structural components are frequently subjected to complex cyclic loading histories during their service life. Therefore, identification of the loading parameters that influence the fatigue life and quantification of their effects for reliable prediction of fatigue life under realistic loading patterns is essential.

The source of fatigue loading can be mechanical or due to residual effects. The residual stresses are developed mainly due to the difference between the hygrothermal

Figure 3.1 Pontresina composite bridge, Switzerland (adhesively-bonded joints connect the glass fiber-reinforced polymer (GFRP) profiles).
Courtesy of Composite Construction Laboratory, EPFL, Switzerland.

behavior, thermal, and moisture expansion coefficients of adhesive and adherend. The fatigue mechanical loads are developed during the function of structures. Different types of loads occur on a structure in service that together comprise the fatigue loading spectrum. In a bridge, for example, the structure is loaded by each passage of a vehicle, and the loading spectrum depends on the frequency of their passage. In addition, the thermomechanical stresses due to the environmental conditions and the mechanical stresses by the wind forces coming from different directions and with varying intensity should be taken into account to have a realistic estimation of the loading spectrum (Figure 3.2). For a motorcar, the load spectrum depends mainly on the condition of the roads, loading, and the driver, and therefore it may vary for all cars of the same type. A relatively severe load spectrum, therefore, should be considered for fatigue analysis of a car to avoid fatigue failures. For an airplane, the load spectrum is fairly complex and can be divided to the loads produced during different stages of a flight, i.e., taxi, take-off, climb, cruise, landing, etc.

The scientific efforts to study the fatigue behavior of adhesively-bonded FRP joints are mainly focused on constant amplitude fatigue loading, and many loading parameters involved in the realistic loading spectrums have not usually been investigated. The aim of this chapter is to highlight their significant influence on the fatigue life.

Figure 3.2 Composite bridge deck, Bex, Switzerland (adhesive bonds used to assemble the composite deck).
Courtesy of Composite Construction Laboratory, EPFL, Switzerland.

3.2 Fatigue data

The fatigue data that are usually collected in the laboratory depend on the approach used for the fatigue analysis. Two main approaches, namely phenomenological and fracture mechanics approaches, are common in this domain. In the phenomenological approach, the loading condition and the corresponding fatigue life are used for the fatigue analysis. The fatigue data are usually reported as $S-N$ curves presenting the stress level (S) versus fatigue life (number of cycles, N) for a specific loading condition. The fatigue data are usually reported as $S-N$ curves presenting the stress level versus fatigue life for a specific loading conditions. The effects of loading parameters are reflected in the $S-N$ curve as, e.g., change of slope or shift of the curves.

In a fracture mechanics approach, acquiring complementary data regarding the crack propagation during fatigue life is necessary. Using these data, a relationship for the rate of fatigue crack propagation can be expressed in terms of fracture mechanics parameters such as strain energy release rate or stress intensity factor. The typical outputs of this approach are the fatigue crack growth (FCG) curves, which, similar to $S-N$ curves, are derived for specific loading conditions, and their changes represent alterations in loading conditions.

It is worthwhile to mention that in all fatigue analysis approaches, a thorough investigation of fracture surface is essential. The comparison of fracture surfaces associated with different loading conditions can provide invaluable information regarding the likely change of failure mechanisms and consequently change in fatigue life.

3.3 Tensile versus compressive fatigue

To date, experimental studies performed on bonded joints are usually based on tensile fatigue loads mainly because they have focused on joints with metal substrates (Ashcroft & Shaw, 2002), in which a cohesive or an interfacial failure is exhibited. However, the fatigue behavior of the joints is not necessarily the same under compressive loads, a loading pattern that is very common during the operation of a structure. This phenomenon is more pronounced for joints made of FRPs with anisotropic material properties. In composite joints, cracks in the adherend—which are generally multidirectional composite laminates—may lead the failure of fiber tear type. This type of failure has been observed by several researchers (Zhang, Vassilopoulos, & Keller, 2008; Zhang, Vassilopoulos, & Keller, 2009). Therefore, due to the variable architecture of the composite laminates as the adherents, the failure modes under tensile and compressive loading are not necessarily the same and depend on the composite lay-up. The difference between tensile and compressive failure modes of a typical adhesively-bonded glass fiber–reinforced polymer (GFRP) composite joint is shown in Figures 3.3 and 3.4 (Sarfaraz, Vassilopoulos, & Keller, 2011).

In this specific case, the composite laminates comprise two mat layers on each side and a roving layer in the middle. Each mat layer comprises a woven fabric stitched to a chopped strand mat (CSM). Under tensile loading, the crack initiated from the joint

Figure 3.3 Failure mode of adhesively-bonded GFRP joint under tensile loading.

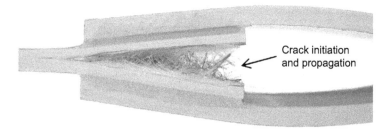

Figure 3.4 Failure mode of adhesively-bonded GFRP joint under compressive loading.

corner of one of the bond lines—the upper in this figure—between the adhesive and the inner laminate and then shifted deeper, between the first and the second mat layers of the inner laminate, and remains along this path up to failure. Under compression loading, as shown in Figure 3.4, the crack initiated and propagated from the right side of the inner laminate inside the roving layer.

This example clearly shows that both tensile and compressive components of the fatigue spectrum must be taken into account to avoid any unpredictable fatigue failure. Apparently when the joints are subjected to tensile-compressive loading cases, both failure modes may appear simultaneously (Sarfaraz et al., 2011).

3.4 Effects of fatigue loading parameters

As described previously, bonded joints are subjected to fairly complex loading spectra. There are several features in these spectra such as variation of mean loads and transitions of load levels. The effects of these features are discussed below by comparing the acquired fatigue data under different loading conditions. For example, changes in the slopes of $S-N$ curves are used to identify the effects of loading parameters.

3.4.1 Effect of mean load

The constant amplitude cycling loading is characterized by two parameters, i.e., the mean load (F_{mean}) and load amplitude (F_{amp}), minimum and maximum loads (F_{min} and F_{max}), or load range (ΔF) and load ratio ($R = F_{min}/F_{max}$), which are practically equivalent (Figure 3.5). In the case of realistic loading, the first set of parameters, mean and amplitude stress, may be preferred for the fatigue life investigation.

The load ratio is usually used to specify the loading type; $0 < R < 1$ expresses tension–tension (T–T) fatigue, $1 < R < +\infty$ represents compression–compression (C–C) fatigue, and $-\infty < R < 0$ denotes mixed tension–compression (T–C) fatigue loading that can be tension or compression dominated. It is well documented that for a given maximum stress in a tension–tension case, the fatigue life of the composite increases with increasing magnitude of R. In compression–compression loading, increasing the magnitude of R reduces the fatigue life of the examined composite (Abd Allah, Abdin, Selmy, & Khashaba, 1997; Ellyin & El-Kadi, 1994; Mallick & Zhou, 2004).

The influence of the mean load on the fatigue behavior of composite materials has been the subject of numerous investigations in the past. For laminates exhibiting significantly higher tensile strength than compressive strength, e.g., unidirectional carbon/epoxy laminates (Kawai & Koizumi, 2007), small components of compressive load drastically deteriorate the fatigue strength, whereas for materials exhibiting higher compressive than tensile fatigue properties, e.g., short-fiber composites (Silverio Freire, Dória Neto, De Aquino, 2009), the tensile components of the fatigue spectrum strongly influence the fatigue performance.

In spite of the considerable amount of information that exists regarding the mean load effect on the fatigue life of composite laminates, there is little literature regarding

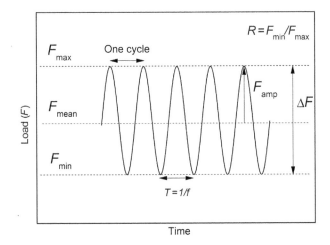

Figure 3.5 Schematic representation of constant amplitude loading spectrum.

similar investigations for adhesively-bonded structural joints. Experimental studies on joints are based on tensile fatigue loads because, as discussed in Section 3.3, they focus on joints exhibiting a cohesive or an interfacial failure. Nevertheless, as shown previously, different failure modes can be observed depending on the adherend materials and the joint configuration.

Significant mean load effects have been observed for pultruded GFRP joints in which cracks in the adherend lead to the fatigue failure. Moreover different failure modes are observed under tension and compression fatigue, with a transition of the failure mode from tensile to compressive as the mean load was decreased from zero to negative values (Sarfaraz et al., 2011).

The increase of the mean load under constant amplitude loading leads to an increase in the tensile and compressive fatigue life of the examined joints. The slopes of the $S-N$ curves were decreased by increasing the mean load level. Figure 3.6 shows variation of fatigue life as the R-ratio is changing, representing the transition from tensile fatigue to tensile-dominated fatigue loading. Figure 3.7 provides similar data for compressive and compression-dominated fatigue loading (Sarfaraz, Vassilopoulos, & Keller, 2012). The experimental results in this investigation show high dependency of the fatigue strength on the mean load, and therefore the necessity of appropriate understanding and modeling of this effect to avoid extensive experimental programs and to obtain a reliable estimation of fatigue life time.

The influence of mean load is usually assessed by using constant life diagrams (CLD). Constant life diagrams reflect the combined effect of mean stress and material anisotropy on fatigue life, and can be used for estimation of the fatigue life of the

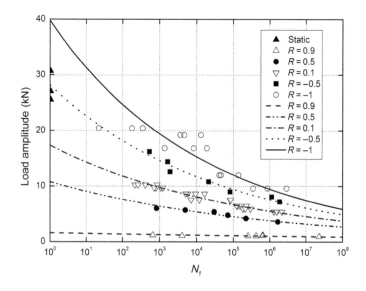

Figure 3.6 $S-N$ data (cyclic load versus number of cycle) for tension and tension-dominant fatigue loading.

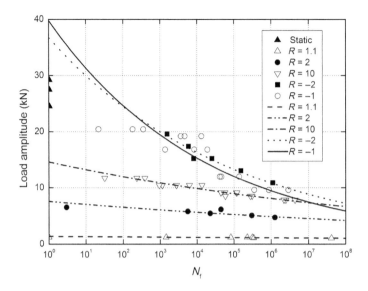

Figure 3.7 $S-N$ data (cyclic load versus number of cycle) for compression and compression-dominant fatigue loading.

material under loading patterns for which no experimental data exist. The main parameters that define a CLD are the cyclic mean stress, stress amplitude, and number of fatigue cycles. For more information on modeling of mean load effect, readers are referred to a comprehensive review recently performed by Vassilopoulos, Manshadi, and Keller (2010a).

3.4.2 Creep–fatigue interaction

The cyclic loading at R-ratios close to 1 cannot be considered as fatigue loading but, rather, as creep of the material (constant static load over a short or long period with a very small load amplitude).

The increase in mean strain of short E-glass–reinforced polyamide 6.6 under constant amplitude fatigue was attributed to the creep caused by the constant mean stress applied during cycling loading (Mallick & Zhou, 2004). Petermann and Schulte (2002) stated the damage evolution of $(\pm 45)_{2S}$ carbon/epoxy laminates at high stress ratio and maximum stresses below the endurance limit dominated by creep.

This type of loading is very common in service loading, as it frequently happens that the structure is loaded with a small load amplitude oscillating around a high mean load. This pattern can involve a major part of the loading spectrum. However, since performing this type of experiment is very time consuming, the behavior is approximated by interpolation of fatigue data at smaller R-ratios and the quasi-static strength of materials.

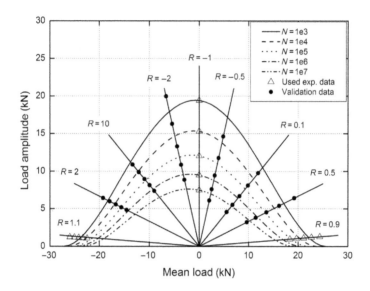

Figure 3.8 Typical constant life diagram for a composite bonded joint.

It has been shown that this simplification can significantly reduce the accuracy of fatigue life prediction. Few modifications to take the time-dependent material strength into account have been introduced (Mallick & Zhou, 2004), and yet none of these modifications can provide a general model to characterize the fatigue—creep interaction in composite materials.

Figure 3.8 shows several sets of constant amplitude fatigue data for a typical GFRP bonded join at different R-ratios. Apparently the creep—fatigue interaction is significant at high mean load levels ($R = 0.9$ and 1.1). The iso-life curves shown in Figure 3.8 represent a new model taking into account this effect in the form of a constant life diagram (Sarfaraz et al., 2012).

3.4.3 Effect of loading sequence on fatigue life

A simplified form of a realistic loading pattern can be assumed as a sequence of constant amplitude loading blocks. Therefore, investigation of the effect of different loading sequences on the fatigue performance provides very useful information to fill the gap between constant amplitude fatigue behavior and the fatigue response under realistic loading patterns. Experimental studies on the fatigue behavior of composite laminates show their sensitivity to the loading sequence. These experiments are usually composed of two blocks of constant amplitude loading changing from a low load level to a higher load level (L—H sequence, Figure 3.9(a)) or vice versa (H—L sequence, Figure 3.9(b)).

The results obtained from these experiments are not consistent, showing a greater damaging effect due to the L—H sequence (Broutman & Sahu, 1972; Hosoi, Kawada,

Understanding fatigue loading conditions in adhesively-bonded composite joints 81

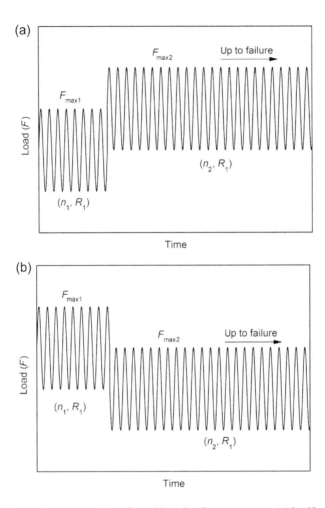

Figure 3.9 Schematic representation of two-block loading sequences: (a) L—H sequence and (b) H—L sequence.

& Yoshino, 2006; Yang & Jones, 1980) or the opposite behavior (Gamstedt & Sjögren, 2002; Hwang & Han, 1987; Lee & Jen, 2000), depending on the material and loading parameters.

Several different loading parameters such as the R-ratio and the cyclic load levels govern the sequence effects (Adam, Gathercole, Reiter, & Harris, 1994; Bonnee, 1996; Harris, Gathercole, Reiter, & Adam, 1997; Lee & Liu, 1994). Tension or compression loading blocks can cause different damage compared to mixed tension—compression blocks. The difference between the applied load levels in a two-block loading sequence is also an important parameter, as the sequence effect can be magnified when the difference between two load levels is increased (Lee & Liu, 1994).

Despite numerous publications dedicated to the study of the sequence effect, explicit explanations regarding the involved failure mechanisms are very limited (Found & Quaresimin, 2003; Gamstedt & Sjögren, 2002; Plumtree, Melo, & Dahl, 2010). The activation of competing failure mechanisms, like initiation mechanisms vs. progressive failure mechanisms or resin cracking vs. fiber breakage or delamination under different stress levels, was considered for the explanation of the sequence effects observed for different types of composite materials (Found & Quaresimin, 2003; Gamstedt & Sjögren, 2002). For instance, transverse cracking dominates the failure of cross-ply laminates under high stress levels, whereas delamination is activated under lower stress levels (Gamstedt & Sjögren, 2002). Therefore, the H—L sequence results in shorter fatigue lifetimes than the L—H sequence, since the transverse cracks, created under a high stress level, are potential places for the initiation of delamination. A reverse effect was observed after an experimental investigation of multidirectional carbon/epoxy laminates and is explained by the assumption that since most of the applied load is borne by the matrix under low stress levels and by the fibers under high stress levels, the damage involves mainly the growth of microcracks in the matrix throughout the specimen under lower stress levels, which can induce rapid failure in the following high stress level stage (Found & Quaresimin, 2003). The balance between the damage state and the stress levels and its effect on stress intensity was also proposed as a way of interpretation of sequence effects in angle ply laminates. The longer life of $[\pm 45]_{2s}$ carbon/epoxy laminates under both L—H and H—L sequences compared to the expected life, characterized by the Palmgren—Miner sum, was thus attributed to the decrease in local stress intensity due to a large number of well-distributed matrix cracks when the stress level is decreased (Plumtree et al., 2010).

In addition to the loading sequence effect, the significant influence of loading transition and its frequency of occurrence on the fatigue life has been discussed in several investigations (Filis, Farrow, & Bond, 2004; Schaff & Davidson, 1997; Van Paepegem & Degrieck, 2002). The effect of frequent transition of cyclic load level on fatigue life was found to be more significant than the loading sequence effect (Van Paepegem & Degrieck, 2002). The transition effect, as defined by the term "cycle mix," was introduced for modeling the damage accumulated under block and variable amplitude loading in Schaff & Davidson (1997) and Filis et al. (2004).

The load sequence also affects the fatigue behavior of (FRP) composite joints, although only a limited number of works exist concerning this phenomenon. Similar to composite laminates, several parameters were found to contribute to the loading sequence effects on the fatigue life (Ashcroft, 2004; Erpolat, Ashcroft, Crocombe, & Abdel-Wahab, 2004a, 2004b). A significant load interaction effect (overloads and loading sequence effect) was identified (Erpolat et al., 2004b) for adhesively-bonded double-lap joints composed of carbon/epoxy laminates and a single-part epoxy adhesive. The crack growth acceleration due to the load interaction was put forward as the main reason for the shorter fatigue life exhibited by the joints under investigation. The cycle mix effect and the variation in mean stress were also investigated, and it has been proved that they both affect crack growth acceleration, whereas overloads were shown to increase the likelihood of fatigue crack initiation.

The recent investigation on GFRP bonded joints (Sarfaraz et al., 2013a) shows block loading sequences does not change the failure mode. However a significant loading sequence effect is identified by applying two-block loading sequences. The results shows that the sequence effect is a function of type of loading and the applied load levels. The L—H sequences were found to be more damaging than the H—L sequences under tensile loading, whereas under compressive loading this trend is reversed.

The effect of loading sequence on the fatigue life of the examined joints was associated with the crack growth rate during the applied loading blocks. The H—L tensile loading blocks led to retardation of crack growth rate, whereas acceleration was observed under L—H sequences. In contrast, the crack growth rate under L—H compressive loading blocks did not increase significantly when the load level was increased, and led to longer fatigue life. However, under the H—L sequences, the first loading block did not affect the expected crack propagation rate under the second loading block. The difference in sequence effects under tension and compression was attributed to the difference in failure modes as explained in Section 3.3.

It has been also shown that the frequent change of load levels (Figure 3.10(a) and (b) presenting two spectra having the same number of cycles but different number of transitions) has a very strong damaging effect compared to the sequence effect and therefore is more critical than the sequence effect on the fatigue life (Sarfaraz et al., 2013a).

Review of the previous studies highlights the significant influence of load interaction on the fatigue behavior of materials and structures under realistic loading patterns. It also shows that such interaction strongly depends on the materials as well as the applied loading spectrum. A well-organized experimental program including different loading sequences and simultaneously monitoring the crack propagation in the joints can provide clear insight into the effect of load interactions on the fatigue life of bonded joints.

3.4.4 Variable amplitude loading

Variable amplitude loading is representative of the actual loading patterns that engineering structures experience during their service life. Despite the vast number of researches concerning the fatigue of FRP composite joints under constant amplitude loading, due to the complicated and time consuming testing process, only a small number of studies have been performed on the variable amplitude (VA) fatigue behavior of FRP bonded joints. These studies are mainly related to bonded joints exhibiting cohesive or adhesive failure and not fiber tear failure of composite adherends (Erpolat et al., 2004b; Jeans, Grimes, & Kan, 1983; Sarkani, Michaelov, Kihl, & Beach, 1999; Shenoy, Ashcroft, Critchlow, & Crocombe, 2010).

Regardless of the adherend material, the load interaction effects such as load transition, load sequence, and overload effects have been reported in different investigations. The overloads can accelerate the fatigue crack initiation (Erpolat et al., 2004b) or increase the damaging effect of the following cycles of lower amplitude, although their effect is reduced when the number of low amplitude cycles following the overloads is increased (Nolting, Underhill, & DuQuesnay, 2008). The change of mean load can also accelerate the crack growth in bonded composite joints

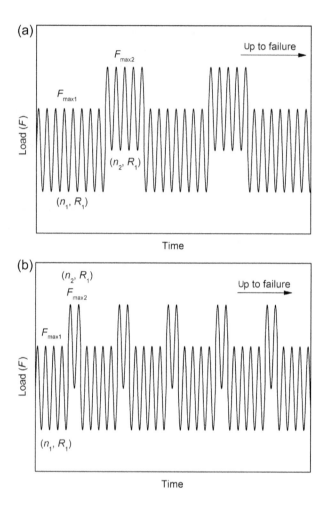

Figure 3.10 Schematic representation of multiblock loading sequences: (a) low transition and (b) high transition.

(Erpolat et al., 2004b). The significant damaging effect of introducing a small number of cycles at a higher mean load has also been addressed (Shenoy et al., 2010). The load interaction effect was also observed for pultruded GFRP joints in which cracks in the multidirectional laminate (adherend) lead the failure process. The investigation of two-block loading sequences under tension loading demonstrated a retardation effect under high−low and an acceleration effect under low−high loading sequences. However, the damaging effect of frequent load transitions in a spectrum dominated the load sequence effect (Sarfaraz et al., 2011). Several experimental investigations of the fatigue behavior of composite laminates show also their sensitivity to the loading sequence. The results obtained from these experiments show the sequence effects depend on the material and loading parameters, as discussed in Section 3.4.3.

The aforementioned retardation or acceleration of the fatigue crack growth rate due to load interactions is common for metals, where one dominant crack mainly governs the fracture behavior. In composite materials, which exhibit several contributing fatigue failure mechanisms, identification of a single dominant crack for this investigation is difficult. The situation is less complicated for adhesively-bonded lap joints under cyclic loading, however, since experimental observation in previous investigations (Sarfaraz et al., 2011; Zhang, Vassilopoulos, & Keller, 2010) showed that in several cases, only one dominant crack led to final failure even for joints composed of composite materials. Therefore, for adhesively-bonded joints, the load interaction effects can be correlated with the acceleration or retardation in the propagation rate of the dominant crack, consequently providing explanations regarding the fatigue behavior under VA loading.

Study of adhesively-bonded pultruded GFRP double-lap joints under variable amplitude loading conditions (i.e., WISPEREX standard spectrum; see Figure 3.11) and comparison with constant amplitude fatigue behavior show that variable nature of the loading pattern does not change the failure mode and the crack propagation trend. These results confirm the use of constant amplitude fatigue data for prediction of variable amplitude fatigue life. However it has been shown that the load interaction effects that are introduced in the variable amplitude loading spectrum, particularly the load transition effect, must be taken into account for an accurate prediction (Sarfaraz et al., 2013b).

To date, investigations on fatigue behavior of composite bonded joints show that several parameters significantly affect fatigue behavior under variable amplitude loading. A systematic investigation of the effect of these parameters is therefore necessary to establish reliable fatigue life prediction method.

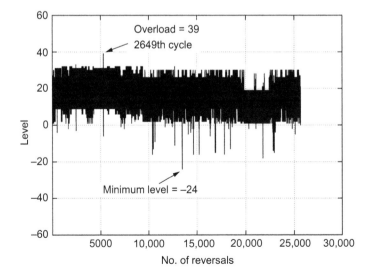

Figure 3.11 WISPERX spectrum.

3.5 Future trends

The new developments in strain measuring systems can provide more detailed information concerning the strain state in the load-bearing components such as bonded FRP joints. The fast advancement in optical sensors technology facilitates continuous multipoint strain measurement in composite joints (Canal et al., 2014). A more precise definition of the loading spectra that structures bear is accordingly obtained, and allows the researchers and engineers to take into account all load variations in the actual spectrum.

The increase of new powerful test frames capable of applying user-defined loading spectrum will also facilitate more investigations on influencing loading parameters on fatigue behavior and possibly exploring new loading parameters. The rapid progress in this domain from the 1970s, when the first variable amplitude experiments were performed, supports this trend.

Nevertheless, to avoid time-consuming and expensive experimental programs for different materials and structures, a systematic modeling of these parameters are essential. New models taking into account the effects of these parameters will significantly assist this progress.

3.6 Sources of further information and advice

For more details on the loading parameters influencing the fatigue response and also their modeling, readers are referred to the following articles. The modeling of mean load effect in bonded composite joints is discussed in Sarfaraz et al. (2012), and a comprehensive review of composite materials is presented in Vassilopoulos et al. (2010a). Also, further discussions on modeling for different composite material systems are provided (Andersons & Paramonov, 2011; Bond & Ansell, 1998; Fernando, Dickson, Adam, Reiter, & Harris, 1988; Towo & Ansell, 2008; Vassilopoulos, Manshadi, & Keller, 2010b). More information on fatigue—creep interaction is available in Guedes (2011). It is also helpful to see some of the articles on block and variable amplitude loading in composite materials (Bartley-Cho, Lim, Hahn, & Shyprykevich, 1998; Found & Kanyanga, 1996; Jen, Kau, & Wu, 1994; Otani & Song, 1997; Wahl, Mandell, & Samborsky, 2001; Yang & Jones, 1983).

References

Abd Allah, M. H., Abdin, E. M., Selmy, A. I., & Khashaba, U. A. (1997). Effect of mean stress on fatigue behavior of GFRP pultruded rod composites. *Composites, Part A, 28*(1), 87–91.

Adam, T., Gathercole, N., Reiter, H., & Harris, B. (1994). Life prediction for fatigue of T800/5425 carbon-fibre composites: II. Variable amplitude loading. *International Journal of Fatigue, 16*(8), 533–547.

Andersons, J., & Paramonov, Yu (2011). Applicability of empirical models for evaluation of stress ratio effect on the durability of fiber-reinforced creep rupture-susceptible composites. *Journal of Materials Science, 46*(6), 1705–1713.

Ashcroft, I. A. (2004). A simple model to predict crack growth in bonded joints and laminates under variable-amplitude fatigue. *Journal of Strain Analysis, 39*(6), 707−716.

Ashcroft, I. A., & Shaw, S. J. (2002). Mode I fracture of epoxy bonded composite joints: 2. Fatigue loading. *International Journal of Adhesion and Adhesives, 22*(2), 151−167.

Bartley-Cho, J., Lim, S. G., Hahn, H. T., & Shyprykevich, P. (1998). Damage accumulation in quasi isotropic graphite/epoxy laminates under constant-amplitude fatigue and block loading. *Composites Science and Technology, 58*(9), 1535−1547.

Bond, I. P., & Ansell, M. P. (1998). Fatigue properties of jointed wood composites: part II life prediction analysis for variable amplitude loading. *Journal of Materials Science, 33*(16), 4121−4129.

Bonnee, W. J. A. (1996). NLR investigation of polyester composite materials. In C. W. Kensche (Ed.), *Fatigue of materials and components for wind turbine rotor blades* (pp. 39−70). German Aerospace Establishment, European Commission-EUR 16684.

Broutman, L. J., & Sahu, S. (1972). A new theory to predict cumulative fatigue damage in fibre-glass reinforced plastics. In *Composite materials: Testing and design (2nd conference)* (pp. 170−188). Philadelphia: ASTM STP 497, American Society for Testing and Materials.

Canal, L. P., Sarfaraz, R., Violakis, G., Botsis, J., Michaud, V., & Limberger, H. G. (2014). Monitoring strain gradients in adhesive composite joints by embedded fiber Bragg grating sensors. *Composite Structures, 112*(1), 241−247.

Ellyin, F., & El-Kadi, H. (1994). Effect of stress ratio on the fatigue of unidirectional glass fibre/epoxy composite laminae. *Composites, 25*(10), 917−924.

Erpolat, S., Ashcroft, I. A., Crocombe, A. D., & Abdel-Wahab, M. M. (2004a). Fatigue crack growth acceleration due to intermittent overstressing in adhesively bonded CFRP joints. *Composites, Part A: Applied Science and Manufacturing, 35*(10), 1175−1183.

Erpolat, S., Ashcroft, I. A., Crocombe, A. D., & Abdel-Wahab, M. M. (2004b). A study of adhesively bonded joints subjected to constant and variable amplitude fatigue. *International Journal of Fatigue, 26*(11), 1189−1196.

Fernando, G., Dickson, R. F., Adam, T., Reiter, H., & Harris, B. (1988). Fatigue behaviour of hybrid composites: part I carbon/Kevlar hybrids. *Journal of Materials Science, 23*(10), 3732−3743.

Filis, P. A., Farrow, I. R., & Bond, I. P. (2004). Classical fatigue analysis and load cycle mix-event damage accumulation in fibre reinforced laminates. *International Journal of Fatigue, 26*(6), 565−573.

Found, M. S., & Kanyanga, S. B. (1996). The influence of two-stage loading on the longitudinal splitting of unidirectional carbon-epoxy laminates. *Fatigue & Fracture of Engineering Materials & Structures, 19*(1), 65−74.

Found, M. S., & Quaresimin, M. (2003). Two-stage fatigue loading of woven carbon fibre reinforced laminates. *Fatigue & Fracture of Engineering Materials & Structures, 26*(1), 17−26.

Gamstedt, E. K., & Sjögren, B. A. (2002). An experimental investigation of the sequence effect in block amplitude loading of cross-ply laminates. *International Journal of Fatigue, 24*(2−4), 437−446.

Guedes, R. M. (Ed.). (2011). *Creep and fatigue in polymer matrix composites.* Cambridge: Woodhead Publishing Limited.

Harris, B., Gathercole, N., Reiter, H., & Adam, T. (1997). Fatigue of carbon-fibre reinforced plastics under block-loading conditions. *Composites, Part A: Applied Science and Manufacturing, 28*(4), 327−337.

Hosoi, A., Kawada, H., & Yoshino, H. (2006). Fatigue characteristic of quasi-isotropic CFRP laminates subjected to variable amplitude cyclic loading of two-stage. *International Journal of Fatigue, 28*(10 Spec. Iss.), 1284−1289.

Hwang, W., & Han, K. S. (1987). Statistical study of strength and fatigue life of composite materials. *Composites, 18*(1), 47–53.
Jeans, L. L., Grimes, G. C., & Kan, H. P. (1983). Fatigue sensitivity of composite structure for fighter aircraft. *Journal of Aircraft, 20*(2), 102–110.
Jen, M. H. R., Kau, Y. S., & Wu, I. C. (1994). Fatigue damage in a centrally notched composite laminate due to two-step spectrum loading. *International Journal of Fatigue, 16*(3), 193–201.
Kawai, M., & Koizumi, M. (2007). Non-linear constant fatigue life diagrams for carbon/epoxy laminates at room temperature. *Composites, Part A, 38*(11), 2342–2353.
Lee, C. H., & Jen, M. H. R. (2000). Fatigue response and modelling of variable stress amplitude and frequency in AS-4/PEEK composite laminates, composite laminates, part 1: experiments. *Journal of Composite Materials, 34*(11), 906–929.
Lee, B. L., & Liu, D. S. (1994). Cumulative damage of fiber-reinforced elastomer composites under fatigue loading. *Journal of Composite Materials, 28*(13), 1261–1286.
Mallick, P. K., & Zhou, Y. (2004). Effect of mean stress on the stress-controlled fatigue of a short E-glass fiber reinforced polyamide–6.6. *International Journal of Fatigue, 26*(9), 941–946.
Nolting, A. E., Underhill, P. R., & DuQuesnay, D. L. (2008). Variable amplitude fatigue of bonded aluminum joints. *International Journal of Fatigue, 30*(1), 178–187.
Otani, N., & Song, D. Y. (1997). Fatigue life prediction of composite under two-step loading. *Journal of Materials Science, 32*(3), 755–760.
Petermann, J., & Schulte, K. (2002). The effects of creep and fatigue stress ratio on the long-term behavior of angle-ply CFRP. *Composite Structures, 57*(1–4), 205–210.
Plumtree, A., Melo, M., & Dahl, J. (2010). Damage evolution in a [± 45]$_{2S}$ CFRP laminate under block loading conditions. *International Journal of Fatigue, 32*(1), 139–145.
Sarfaraz, R., Vassilopoulos, A. P., & Keller, T. (2011). Experimental investigation of the fatigue behavior of adhesively-bonded pultruded GFRP joints under different load ratios. *International Journal of Fatigue, 33*(11), 1451–1460.
Sarfaraz, R., Vassilopoulos, A. P., & Keller, T. (2012). Experimental investigation and modeling of mean load effect on fatigue behavior of adhesively-bonded pultruded GFRP joints. *International Journal of Fatigue, 44*, 245–252.
Sarfaraz, R., Vassilopoulos, A. P., & Keller, T. (2013a). Block loading fatigue of adhesively-bonded pultruded GFRP joints. *International Journal of Fatigue, 49*, 40–49.
Sarfaraz, R., Vassilopoulos, A. P., & Keller, T. (2013b). Variable amplitude fatigue of adhesively-bonded pultruded GFRP joints. *International Journal of Fatigue, 55*, 22–32.
Sarkani, S., Michaelov, G., Kihl, D. P., & Beach, J. E. (1999). Stochastic fatigue damage accumulation of FRP laminates and joints. *Journal of Structural Engineering, 125*(12), 1423–1431.
Schaff, J. R., & Davidson, B. D. (1997). Life prediction methodology for composite structures. Part II – spectrum fatigue. *Journal of Composite Materials, 31*(2), 158–181.
Shenoy, V., Ashcroft, I. A., Critchlow, G. W., & Crocombe, A. D. (2010). Fracture mechanics and damage mechanics based fatigue lifetime prediction of adhesively bonded joints subjected to variable amplitude fatigue. *Engineering Fracture Mechanics, 77*(7), 1073–1090.
Silverio Freire, R. C., Dória Neto, A. D., & De Aquino, E. M. F. (2009). Comparative study between ANN models and conventional equation in the analysis of fatigue failure of GFRP. *International Journal of Fatigue, 31*(5), 831–839.
Towo, A. N., & Ansell, M. P. (2008). Fatigue of sisal fibre reinforced composites: constant-life diagrams and hysteresis loop capture. *Composites Science and Technology, 68*(3–4), 915–924.

Van Paepegem, W., & Degrieck, J. (2002). Effects of load sequence and block loading on the fatigue response of fibre-reinforced composites. *Mechanics of Advanced Materials and Structures, 9*(1), 19–35.

Vassilopoulos, A. P., Manshadi, B. D., & Keller, T. (2010a). Influence of the constant life diagram formulation on the fatigue life prediction of composite materials. *International Journal of Fatigue, 32*(4), 659–669.

Vassilopoulos, A. P., Manshadi, B. D., & Keller, T. (2010b). Piecewise non-linear constant life diagram formulation for FRP composite materials. *International Journal of Fatigue, 32*(10), 1731–1738.

Wahl, N. W., Mandell, J. F., & Samborsky, D. D. (2001). Spectrum fatigue lifetime and residual strength for fiberglass laminates in tension. In *ASME wind energy symposium, AIAA-2001-0025, ASME/AIAA*.

Yang, J. N., & Jones, D. L. (1980). Effect of load sequence on the statistical fatigue of composite. *AIAA Journal, 18*(12), 1525–1531.

Yang, J. N., & Jones, D. L. (1983). Load sequence effects on graphite/epoxy $[\pm 35]_{2s}$ laminates. In *Long-term behavior of composites* (pp. 246–262). Philadelphia: ASTM STP 813, American Society for Testing and Materials.

Zhang, Y., Vassilopoulos, A. P., & Keller, T. (2008). Stiffness degradation and fatigue life prediction of adhesively-bonded joints for fiber-reinforced polymer composites. *International Journal of Fatigue, 30*(10–11), 1813–1820.

Zhang, Y., Vassilopoulos, A. P., & Keller, T. (2009). Environmental effects on fatigue behavior of adhesively-bonded pultruded structure joints. *Composites Science and Technology, 69*(7–8), 1022–1028.

Zhang, Y., Vassilopoulos, A. P., & Keller, T. (2010). Fracture of adhesively-bonded pultruded GFRP joints under constant amplitude fatigue loading. *International Journal of Fatigue, 32*(7), 979–987.

Part Two

Fatigue and fracture behaviour of adhesively-bonded composite joints

Mode I fatigue and fracture behaviour of adhesively-bonded carbon fibre-reinforced polymer (CFRP) composite joints

R.D.S.G. Campilho[1,2], L.F.M. da Silva[3]
[1] Instituto Politécnico do Porto, Porto, Portugal; [2] Universidade Lusófona do Porto, Porto, Portugal; [3] Universidade do Porto, Porto, Portugal

4.1 Introduction

Fibre-reinforced composites are becoming increasingly popular in many sectors of industry as a result of some attractive characteristics over typical construction materials (Rudawska, 2010). Carbon fibre-reinforced polymers (CFRPs) in particular are widely used nowadays in high-end industries that require high specific strength and stiffness (i.e. weight normalized values), allowing reducing the weight of components, while keeping the necessary strength and stiffness to withstand the imposed loads. These specific properties are partly responsible for the increasing use of high-performance CFRPs in automotive, marine, military, aeronautical and aerospace applications. Among these applications, it was the aerospace and defence industries that initiated using CFRPs, because of the weight reductions that resulted on higher performance and payloads. Unidirectional CFRP composites are also ideal for reinforcement purposes because they do not significantly change the structures stiffness, and they are corrosion resistant. As an application example, fibre-reinforced composite materials make up 22% of the total weight of the Airbus A380, including the centre wing box, the tail cone, the pressure bulkheads, and the vertical and horizontal tails (Kolesnikov, Herbeck, & Fink, 2008), and this tendency is likely to increase in the near future (the expected use of composites in the Boeing 787 should be around 50%). Despite these advantages, the fracture behaviour and related mechanisms of these advanced materials, either under static of fatigue loads, are still being currently studied, because of the multitude of possible failure mechanisms as a result of their complexity. One of the major issues in composite structures is the phenomenon of delamination, due to the fact that composites are made of stacked layers. This is particularly true if the composites are made of layers with different fibre orientations. As for the loading conditions, the dynamic regime is particularly critical, due to associated testing difficulties (Coronado, Argüelles, Viña, Mollón, & Viña, 2012). Regarding fabrication of composite structures, even though current manufacturing methods offer the possibility to reduce structural coupling to a minimum, by means of integral design and special manufacturing techniques, post-fabrication interconnections are still required due to some issues such as the typical size of the components, and design, technological, cost, fabrication simplification and logistic limitations (Hou & Liu, 2003). Other problems related to repair, maintenance, inspection and handling requirements may also bring the need to connect different parts. A static analysis is the starting point of any design process, especially if the structures to analyse will mainly sustain static loads. Nonetheless, in many structural components, and more specifically in aerospace structures, time-dependent loads (of which fatigue is a particular case) can be found. Ashcroft, Hughes, Shaw, Abdel-Wahab, and Crocombe (2001) experimentally tested CFRP double-lap bonded joints under tensile fatigue loadings, considering multidirectional and unidirectional lay-ups. The maximum load that the specimens could endure 10^6 cycles was measured (considering a load ratio, R, of 0.1). The authors found out that this load was between 26% and 62% of the quasi-static strength. Thus, it is clear that cyclical (e.g. fatigue) loads must be taken into account during the design process when designing bonded CFRP joints. This makes it possible not only to predict fatigue life but also to establish suitable inspection intervals so that fatigue cracks can be found before they become critical or the component exceeds its residual strength. Moreover, this is especially important for structures designed with a damage-tolerant method (Pirondi & Moroni, 2010). The fatigue strength of structures is generally characterized by the number of cycles to failure, N_f, which can be expressed as the cyclic count for crack nucleation (first phase), stable crack growth (second phase) and catastrophic failure (third phase). A reliable approach for the fatigue design of joints is to keep service loads below the limit of crack nucleation, so that little or no crack propagation occurs. However, in some scenarios this design criterion may be considered too conservative. If this is the case, the applied loads may be based on a sufficiently small crack growth rate to ensure the joint will fulfil its service life (Azari, Papini, Schroeder, & Spelt, 2010). In general, the complete fatigue characterization covers both the threshold and higher crack growth rate regimes.

Despite the design and fabrication flexibility of CFRP, most structures must be assembled from individual parts. This requires a suitable method of joining that does not bring a significant weight penalty (Ashcroft & Shaw, 2002). Independently of the method, joining increases structural complexity and local stress concentrations, but can also reduce the lightweight advantage of composite design. As a result, advanced joining technologies must be developed to bring the corresponding weight penalties to a minimum. Adhesive bonding of CFRP structures is used in several fields of industry, equally to the use of more traditional joining methods. Adhesive bonding is a permanent joining process in which an adhesive bonds the components after solidification/curing. Co-curing CFRP components can also be viewed as a bonding method, by using the resin of the composite matrix as an adhesive that bonds parts together. Composite curing and joining are thus achieved simultaneously (Campilho et al., 2013). However, a significant increase of the bond strength, when compared with conventional bonded joints, cannot be expected. Notwithstanding the bonding method, the static and dynamic joint strength of joints in CFRP structures highly depends on design parameters such as the lay-up of the composite adherend, geometry and surface roughness (for adhesive bonding). Adhesive bonding is being extensively studied to join CFRP components. It enables the fabrication of complex shaped structures that could not be manufactured in a single piece, and it can be used for reinforcement/repairing purposes. Adhesive joints exceed welded or fastened/riveted joints in engineering applications because of large investigation and optimization efforts over the last decades that, together with the continuous developments in adhesives, have made this process highly efficient (Campilho, Banea, Pinto, da Silva, & de Jesus, 2011). Currently, they provide many advantages over mechanical methods, such as lower structural weight, lower fabrication cost, ease of fabrication, improved damage tolerance and design flexibility. Stresses are also more evenly distributed, which highly benefits the joint strength, particularly under fatigue loads. The structural integrity is preserved, which does not occur with bolted or riveted joints, because of cutting of fibres and hence the introduction of stress concentrations. Because of the aforementioned characteristics and advantages of adhesive bonding, it is thus easy to accept that the application of bonding in structures involving fibre-reinforced composites has significantly increased in recent years. Actually, this technology is presently chosen in various areas from high-technology industries such as aeronautics, aerospace, electronics and automotive to traditional industries such as construction, sports and packaging (da Silva et al., 2011).

Despite these facts, bonded joints cannot be disassembled without damage, they are very sensitive to environmental factors such as humidity and temperature, and they tend to collapse abruptly, rendering damage monitoring not feasible. Concurrently, it is known that the structural integrity of composite structures is usually determined by the strength and durability of their joints (Messler, 1993). However, the limited understanding of the behaviour of bonded CFRP structures over their life (Yuan & Xu, 2008) and the lack of well-established failure criteria (de Morais, Pereira, Teixeira, & Cavaleiro, 2007) prevent their prompt usage in industry applications. Actually, the lack of plasticity characteristic of composite materials does not allow the redistribution of stresses at the loci of stress concentrations, such as sharp edges or regions of fabrication-induced defects, requiring a deep knowledge for a safe design (Lee & Soutis, 2008). Because of these issues, a large effort has been made to measure the mechanical properties and to validate predictive methodologies for the mechanical behaviour of bonded structures, either subjected to static or fatigue loadings. In addition, depreciation of the mechanical properties of bonded structures by contact with adverse environments (e.g. temperature and moisture) also requires quantification for the advantages of bonded joints to be fully exploited. However, the most serious handicap of adhesive bonding is the designers' uncertainty regarding long-term structural integrity.

This work deals with mode I fatigue behaviour of bonded CFRP structures, considering both secondary bonding (i.e. with a structural adhesive) and co-cured parts. CFRP joints are initially discussed and weighed against more traditional bonding methods such as bolting or riveting, followed by a brief state-of-the-art overview of mode I, mode II and mixed-mode fatigue characterization, allowing categorizing the present work. Current preparation and testing techniques for mode I fatigue testing are then discussed. Fatigue characterization of CFRP composites in mode I will be further detailed with recent application examples, focusing on the two major categories: stress-fatigue life and fatigue crack growth (FCG) techniques. Finally, the fracture mechanisms under mode I fatigue are described, showing the multitude of failure scenarios in these joints.

4.2 Carbon fibre-reinforced polymer (CFRP) composite joints

The bonded joints of CFRP structures can be subjected to static or fatigue in-service loadings, and these loads can induce pure tensile (mode I; although this is not recommended, as adhesives should work under shear loads), pure shear (mode II) or mixed loads (mode I + mode II). Although this work only deals with mode I fatigue behaviour, the three loading scenarios are briefly described next to provide an overview of fatigue characterization. The discussions that follow are valid for both static and fatigue loadings. Actually, as will be detailed further in this work and depending on the method, the fracture behaviour under fatigue loads can rely on static characterization principles.

4.2.1 Mode I characterization

Mode I characterization of adhesive bonds is widespread for bonded joints and composites. The double-cantilever beam (DCB) is undeniably the most popular test geometry for fatigue characterization (Abdel-Wahab, Ashcroft, Crocombe, & Smith, 2004). The DCB specimens consist of two cantilever arms with uniform width (B) and thickness (h), bonded together by an adhesive layer or co-cured between adherends with a starter crack at one end (Figure 4.1). a_0 is the initial crack length, t_A the adhesive thickness and L the specimen length. The load (F) or displacement (δ) is applied to provide a purely mode I loading (opening mode). The DCB is a useful joint for studying crack propagation and to estimate the fatigue constants appearing in the Paris law, since the test coupons are simple to fabricate and the available analytical methods such as beam theories are well established (the most relevant ones are described later in Section 4.5.1). Despite these facts, the crack can grow unstably (da Silva & Campilho, 2011) and the specimen fabrication requires some attention to ensure the visibility of the crack during propagation. Otherwise,

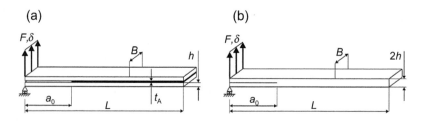

Figure 4.1 Geometry and characteristic dimensions of the bonded (a) and co-cured (b) DCB specimens.

measurement errors of the crack length (a) can reduce the measurement accuracy of the critical strain energy release rate or fracture toughness (generally G_c; G_{Ic} or G_{IIc} for mode I or II components, respectively). Apart from these issues, when testing adhesive bonds with ductile adhesives, a large amount of energy dissipates at the fracture zone, which implies that beam theory-based methods without any corrections will underestimate G_{Ic} (Campilho et al., 2012). The tapered double-cantilever beam (TDCB) is also used for metals (e.g. Hadavinia, Kinloch, Little, & Taylor, 2003) because measurement of a is not required by the standard fracture formulae, but its application in CFRP joints is not practical because of machining of the adherends. The peel test was used by Blackman, Hadavinia, Kinloch, and Williams (2003) with the same purpose.

4.2.2 Mode II characterization

Fracture characterization under mode II, either static or fatigue, is still not well addressed owing to some inherent features of the most popular tests: end-notched flexure (ENF), end-loaded split (ELS) and four-point end-notched flexure (4ENF). Between these, the ENF test is the most suited for mode II fracture characterization of adhesively-bonded structures and composites (Leffler, Alfredsson, & Stigh, 2007). The ENF specimen is similar to the DCB, although loading occurs by bending, induced by the loading cylinder at the specimen mid-length, while the edges are supported. The resulting load creates an almost pure shear stress state at the crack tip, provided that the specimen is designed so that the adherends deform elastically, which allows for shear characterization. However, unstable crack growth and crack monitoring during propagation are issues to be solved in this test method, although the former can be minimized by the proper choice of a_0 and by slightly loading the specimens in mode I, prior to mode II experiment to ensure a sharp pre-crack (Argüelles, Viña, Fernández-Canteli, Viña, & Bonhomme, 2011). In addition, the classical data reduction schemes, based on beam theory analysis and compliance calibration, require the monitoring of a during propagation. Several works are available detailing the fatigue characterization of CFRP composites in mode II in terms of initiation and propagation by acoustic emission damage monitoring (Roy & Elghorba, 1988) and delamination rate versus strain energy release rate or fracture energy (generally G; G_I or G_{II} for mode I or II components, respectively), presenting the corresponding Paris plots (Beghini, Bertini, & Forte, 2006). Although the ENF test is the most widespread, 4ENF, ELS, end-notched cantilever beam (ENCB) or central cut ply (CCP) tests were also successfully applied to fatigue characterization of CFRP composites in mode II (Brunner, Stelzer, Pinter, & Terrasi, 2013). Regarding bonded joints between CFRP adherends, available data are scarce (Fernández, de Moura, da Silva, & Marques, 2013).

4.2.3 Mixed-mode characterization

Practically all of the CFRP joints (either adhesively-bonded or co-cured) are loaded under mixed-mode conditions. Despite this fact, one of the stress components, through-thickness normal or shear, may be much larger than the other and used alone for strength prediction. However, in predicting fatigue damage in these joints, typically both modes I and II require consideration. To this end, the total strain energy release rate, G_T, can be adopted (Abdel-Wahab et al., 2004):

$$G_T = G_I + G_{II}. \tag{4.1}$$

Alternative approaches exist, as the one proposed by Quaresimin and Ricotta (2006b), based on the definition of an equivalent strain energy release rate, G_{eq}:

$$G_{eq} = G_I + \frac{G_{II}}{G_I + G_{II}} G_{II}. \tag{4.2}$$

These two options were inclusively compared in the work of Quaresimin and Ricotta (2006a), and for composite single-lap joints with thin adherends, the two methods gave identical predictions. To make this possible, the values of G_I and G_{II} in the joints were calculated numerically by the virtual crack closure technique (VCCT). Bernasconi, Jamil, Moroni, and Pirondi (2013) fatigue-tested tapered CFRP lap joints, with a pre-crack at each of the overlap ends, while G_I and G_{II} were taken from the VCCT technique applied through finite element (FE) simulations. The plot of G_{II}/G_I-a showed that crack growth initiated already in mixed mode, but quickly became mode II-dominant as the crack continued to grow. On account of this, both expressions (4.1) and (4.2) were applied, and the plots of a versus number of cycles (N) between these two methods were compared with the equivalent experimental curves. From the results, N_f was best approximated by the model of expression (4.1), whereas the model of G_{eq} was more suited for the simulation of the initial stages of crack growth.

4.3 Preparation and testing of CFRP joints in mode I

In the fatigue characterization of CFRP parts, particular attention must be paid to the fabrication method and testing procedure. The present section describes general principles for a correct bonding between CFRP composites, with emphasis on the adherends preparation and bonding procedure (either adhesive bonding or co-curing). Testing of the specimens in mode I is also discussed.

4.3.1 Specimen fabrication

The adherends fabrication method largely depends on the material. For CFRP composites, the structures/adherends are usually made of stacked plies according to a predefined lay-up and set to their final dimensions by machining. Grinding stones at a high rotational speed (>2000 rpm) and small linear feed (<50 mm/min) are suggested. Milling tools are not recommended based on their fast wear rate (carbide or carbide-coated mills) or high cost (diamond-coated mills). Particular attention must be given to the curing conditions of the composite resin, more specifically the temperature, curing time and required pressure for the bonded set. Adherend curing or CFRP joint co-curing can occur at room temperature or at high temperature and/or pressure, depending on the matrix characteristics. If applicable, a hot plates press (for flat components) (Owens & Sullivan, 2000) or an autoclave (da Silva & Adams, 2007) are the most straightforward approaches to induce the specified thermal/pressure cycle. For pressure application, vacuum-bagging is another alternative, although the final quality is poor compared to the use of an autoclave. Ultrasonic phased array C-scanning can confirm the bond quality before testing. Coronado et al. (2012) tested co-cured CFRP DCB specimens under static and fatigue conditions. The sequentially piled prepreg plies were cured in an autoclave, assuring a volume fraction of 63%. Ashcroft and Shaw (2002) performed fatigue tests on bonded CFRP specimens (DCB test geometry). The adherends were cut from thin plates composed of unidirectional lay-ups of 16 plies, cured in an autoclave at 182 °C during 2 h, with a pressure of 0.6 MPa. After curing, the plates were checked for defects and voids by ultrasonic scanning. Adhesive bonding the CFRP components to make up the structure is highly relevant. When considering paste adhesives, pouring of the adhesive in the bonding surfaces is carried out with a low-moisture atmosphere to avoid moisture absorption and preferably in both of the adherends, which is essential for fast curing adhesives to prevent weakening at the upper adhesive/component interface due to adhesive curing before assembly of the upper component. Preparation of the bonding surfaces should be performed by manual abrasion, followed by cleaning with acetone or an equivalent degreaser (Fernández, de Moura, da Silva, & Marques, 2011). An alignment and curing mould should be considered for a correct alignment, uniform application of pressure and achieving the design value of adhesive thickness. To this end, the placement of calibrated steel spacers at the outer periphery of the bonding length is recommended (Sugiman, Crocombe, & Katnam, 2011). Alternatively, calibrated glass beads can be mixed with the adhesive in amount up to 0.5% of the total weight (Owens & Sullivan, 2000). Bernasconi et al. (2013) fatigue-tested CFRP DCB joints and assured a uniform bond line thickness of 0.2 mm using calibrated copper wires inserted in the adhesive in the longitudinal direction. Rudawska (2010) used manual abrasion and a solvent cleaner by Loctite® (reference 7063) for cleaning and degreasing the composite (aramid/epoxy) parts before bonding with an epoxy adhesive, which proved to be quite effective in providing a strong bond. Film adhesives can also be used to bond the adherends. Ashcroft and Shaw (2002) used a modified epoxy, supplied as a 0.2 mm thick film, with a non-woven nylon carrier. When considering the co-curing technique, the specimens can be manufactured by piling prepreg plies. For the starter cracks, a thin Teflon® film (e.g. 20 μm) can be placed at the laminate mid-plane during fabrication (Stelzer, Brunner, Arguelles, Murphy, & Pinter, 2012). For mode I loading in particular, piano hinges or aluminium blocks can be bonded to the specimen edges of the DCB specimens. Although the placement of hinges does not affect the stiffness characteristics of the specimens (provided that they are bonded outside the active area of the beams), if using the blocks, a correction to the G_{Ic} formulae is recommended (ISO 15024 standard, 2001).

4.3.2 Testing of the specimens

Bonded specimens can be loaded under cyclic loads in servo hydraulic testing machines that apply the established fatigue cycle. Data to be recorded in these tests include a, F and δ as function of N. The G-parameter, used from this point, is the generic term to address G_{Ic} or any quantity from fracture mechanics to be measured for fatigue characterization. When speaking about constant-amplitude fatigue testing, it is important to realize that covering the desired range of the G-parameter during a test can be performed in two ways, depending on the control mode of the applied load (Figure 4.2): displacement or load control. On one hand, displacement control supposes the application of a displacement-based fatigue cycle. As the crack continuously grows, the measured G-parameter gradually diminishes because of the corresponding reduction of load. On the other hand, and opposing to this behaviour, by load control the G-parameter increases with the crack length, as the crack tip region is increasingly loaded. Because the G-parameter has a direct relation to the crack growth rate (Pirondi & Moroni, 2010), testing under displacement control conditions gives an initially fast crack propagation and decrease with the testing time. Thus, each test begins with the maximum desired crack growth rate and finishes when the crack reaches the threshold value of the G-parameter, G_{th}, which represents the lower limit of absence of fatigue damage. Standardized determination of this sole parameter can be found in ASTM D6115-97 (2011). Displacement controlled fatigue tests are usually faster and easier to control than load controlled tests. Oppositely, in load control the crack initially grows slowly, but it steadily increases up to complete failure of the specimen. As a result, crack growth is typically initiated in the threshold region and is

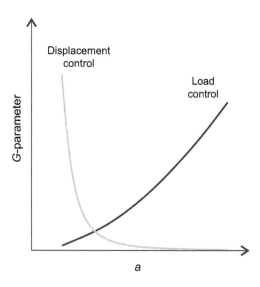

Figure 4.2 Schematic representation of G-parameter–a plots under displacement and load control conditions.

progressively accelerated to the maximum desired crack speed. Fernández et al. (2011) fatigue tested CFRP composite DCB joints in mode I, aiming to evaluate different data reduction schemes to obtain G_I. The tests were carried out in load control with $R = 0.1$, with the maximum load being defined as 50% of the static failure load of identical specimens. Figure 4.3 shows the maximum strain energy release rate (G_{max}) with the growth of a for three calculation methods, all of these showing the increase of G_{max} with a due to the increasing loading applied to the crack tip. The variation of the growth rate during the tests (for both control methods) has the advantage of allowing building the complete fatigue curve from one specimen, by covering the complete range of G-parameter. Previous investigations showed that results are identical whether load or displacement control is used (Mall, Ramamurthy, & Rezaizdeh, 1987). However, displacement control allows making more than one test per specimen (controlling the test parameters such that the crack stops in a position that another test can initiate in the same specimen). In the work of Ashcroft and Shaw (2002), the displacement control technique was chosen for fatigue testing in mode I of adhesively-bonded CFRP DCB specimens. A frequency of 5 Hz and a displacement ratio of 0.1 were used. Each test was run up to a growth rate of 0.02 mm/day (assumption of crack arrest). This limit value was then considered to define G_{th}. After each test, the maximum displacement was increased and another test initiated, until no usable length was left for another test to initiate (Figure 4.4 shows an example of G-parameter versus a by different calculation methods under displacement control). Depending on the method used for fatigue characterization, data of a versus N may be required for further processing and construction of the fatigue laws. For this to be accomplished, a few methods can be used. Stelzer et al. (2012) measured the delamination length in co-cured CFRP DCB specimens by an optical microscope. The measurement resolution was 50 μm, which corresponded to 500 fatigue cycles, and it was enough for a

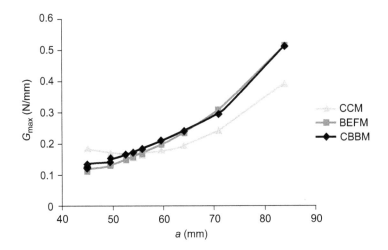

Figure 4.3 G_{max} versus a by different calculation methods under load control (Fernández et al., 2011).

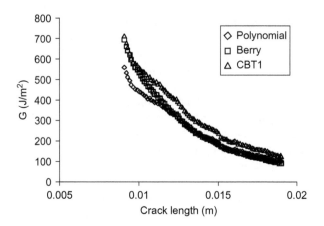

Figure 4.4 G-parameter versus a by different calculation methods under displacement control (Ashcroft & Shaw, 2002). Berry represents the Berry method for G calculation; CBT, corrected beam theory.

good quality in the resulting fatigue plots. In the work of Bernasconi et al. (2013), the value of a of adhesively-bonded CFRP DCB specimens was continuously measured during the tests by special clip gauges that measured the opening at the loading points, and correlation with F allowed estimation of a by analytical formulae developed by Krenk (1992). Ashcroft and Shaw (2002) used a Krak gauge/Fractomat system. Krak gauges were bonded to the specimen sides and a constant current was applied by the Fractomat. Crack growth was detected by the local tearing of the gauge and respective increase of electrical resistance of the gauge. This variation was measured and converted to a value of a. The Fractomat data were merged with the test machine data, allowing the building of the fatigue laws.

4.4 Fatigue characterization by the $S-N$ approach

Fatigue characterization of bonded structures (independently of the substrate materials) has been studied for a long time. At first, it is important to stress that G_c obtained from static testing is not the most representative parameter for fatigue analyses, because of the viscoelastic nature of adhesives or composite matrices. This causes heat dissipation and plastic deformation resulting in micro-cracking, cavitations and voiding during cyclic loading. Therefore, most of the fatigue life prediction techniques are based on empirical formulae derived from experimental data (Abdel-Wahab et al., 2004). Mainly two major categories can be defined for fatigue life prediction: stress-fatigue life ($S-N$) and FCG techniques. These two techniques are described independently, since they correspond to two different lines of analysis that can be followed, with varying degrees of complexity and procedure. Theoretical background of the two methods is also presented, together with relevant studies in the field of CFRP structures (either secondary bonding or co-curing).

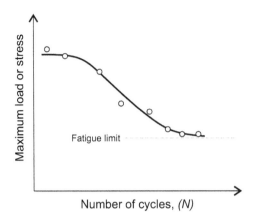

Figure 4.5 Schematic representation of an $S-N$ curve.

Characterization of the fatigue behaviour of structures by $S-N$ techniques is performed by testing, preferably under constant amplitude cycles, of several test coupons at different maximum fatigue loads (Bathias, 2006). The $S-N$ plot is built, for a specific set of test geometry-material, from S (equivalent stress, stress component or load) versus N. Figure 4.5 gives an example of a typical $S-N$ curve. The value of fatigue limit (or G_{th}) gives an indication of the S value below which the material has an infinite fatigue life. In adhesively-bonded CFRP joints, the existence of an endurance limit indicates that a threshold load or stress value exists below which cracks do not propagate in the adhesive layer, along either of the adhesive/adherend interfaces or as delaminations (Abdel-Wahab et al., 2004). In the application of these techniques to bonded joints, there is some uncertainty on how to define the value of S for a fatigue test. For mode I loading, peel stresses can be used. The problem of peel in lap joints is that it is not easily defined, and depending on the existence of stress singularities, stress calculations can be mesh-dependent in FE analyses. Thus, the most convenient way to present results by this approach is to use the applied load. However, by the non-normalization of the applied load over any parameter (e.g. bonded area) that somehow accounts for the geometry, the obtained law can only be applied to structures with an identical geometry. The $S-N$ approach is more viable for high-cycle fatigue (Ashcroft & Shaw, 2002). For low-cycle fatigue, the consideration of a strain-life plot is recommended since it allows accounting for the material plasticity (Bannantine, Comer, & Handrock, 1990). Compared to the FCG technique, this methodology suffers from not allowing the total fatigue life to be separated in fatigue initiation and propagation phases, since only the cyclic count to complete failure is available in the $S-N$ plot. This is an important handicap, as the proportion between the two phases is highly dependent on a number of factors: test geometry, materials, defects and magnitude of the applied load. The $S-N$ approach is an established fatigue prediction technique that is used in many fields of application. Nonetheless, it is applied to the actual structures whose behaviour needs to be determined, rather than using pure-mode tests for parameter identification and extrapolation to other geometries (as it is the case of FCG approaches).

4.4.1 Results of testing by the S–N approach

Since pure-mode loads are practically inexistent in real-life structures, no significant data is currently available regarding pure-mode I characterization by this method. Instead, many works deal with specific mixed-mode geometries, even though in some cases mode I data is used in some manner to aid with the procedure (Abdel-Wahab et al., 2004; Curley, Jethwa, Kinloch, & Taylor, 1998). In the work of Curley et al. (1998), FCG curves from bonded DCB joints were used to predict S–N curves of mixed-mode joints. Two assumptions were made for the prediction procedure to be possible: (1) the initiation phase is ignored in predicting the total fatigue life and (2) pure-mode data can predict mixed-mode failure. A similar approach was followed by Abdel-Wahab et al. (2004), considering adhesively-bonded single- and double-lap joints, whose S–N behaviour was predicted by DCB test FCG data. This was performed by numerical integration of the FCG law from an initial to a final crack size. In the S–N curves, load was used instead of stresses, as these are misleading on account of the non-uniform nature of stresses across the bonded regions. The value of G_{th} was defined as the load not inducing damage in 10^6 cycles. The estimated fatigue behaviour was compared to experiments, and a fair agreement was found. Despite this fact, the validity of this procedure is not unanimous (de Goeij, van Tooren, & Beukers, 1999), because of differences in geometry, boundary conditions and edge effects. Thus, if this extrapolation from mode I to mixed-mode data is rejected, the relevant S–N information must come from the actual structure under analysis, which requires using prototypes for obtaining the data. Fatigue design will thus occur at a late phase of the design, where changes to the geometry are expensive to carry out.

4.5 Fatigue characterization by the fatigue crack growth (FCG) approach

Techniques based on the FCG approach are more commonly used nowadays to characterize adhesive bonds in general, on account of a more faithful representation of the fatigue damage, which led to the implementation of numerical procedures for the prediction of the structures behaviour. In these methods, damage variables are established to depreciate the material properties, updated based on the cyclic count and FCG laws, allowing fatigue damage modelling (Turon, Costa, Camanho, & Dàvila, 2007). For the crack growth characterization, suitable test configurations with a pre-crack are selected (e.g. DCB), and only the crack growth phase is analysed. The tests allow quantifying the growth rate of fatigue cracks, defined by da/dN, as function of parameters from fracture mechanics related to the material (G or J-integral) or geometry (stress intensity factor, K). Metals are often characterized by this method, using log da/dN–log ΔK (Bannantine et al., 1990). For bonded joints and fibre-reinforced composites, a G-related quantity is considered the most adequate to characterize fatigue crack propagation (Dessureault & Spelt, 1997), resulting on a da/dN-G-parameter plot. Actually, it is complicated to apply K to bonded joints, because of the adhesive deformations being constrained by the adherends, which makes the definition of stresses around the crack tip difficult (Erpolat, Ashcroft, Crocombe, & Abdel-Wahab, 2004). Between the available works on this matter,

G_{max} or the strain energy release rate amplitude ($\Delta G = G_{max} - G_{min}$; G_{min} is the minimum strain energy release rate) are used. Nonetheless, G_{max} is more suited for adhesives/co-cured interfaces, since in bonded joints the vicinity of the adherend leads to the overestimation of G_{min}, which in turn wrongly reduces ΔG (Ashcroft & Shaw, 2002). Moreover, it was shown by Mall et al. (1987), when testing DCB specimens with varying values of R, that no distinction existed between these conditions when reducing the data to the $da/dN-\Delta G$ form. On the other hand, by considering $da/dN-G_{max}$, a clear increase of the fatigue performance was found by increasing R. Figure 4.6 shows a typical fatigue characterization curve, whose generic behaviour is representative of most materials. The fatigue behaviour can be divided into three well-distinguishable parts, related to different damage phases of materials. The first or crack nucleation phase is associated with load patterns (and corresponding values of G-parameter) for which crack growth is nil or negligible. The physical significance of G_{th} is also shown in Figure 4.6. This parameter largely depends on the loading configuration and environmental effects, and it is paramount when designing structures for avoidance of crack growth. It was shown that in bonded joints and co-cured CFRP (i.e. delaminations), the value of da/dN is sensitive to changes in the load, which gives large variations between tested specimens in the $da/dN-G$-parameter curves. As a result of this inconsistency, Ashcroft and Shaw (2002) recommend to base designs on the value of G_{th} rather than allowable crack growths. In the first phase, da/dN varies between zero (up to G_{th}) and the leftmost value of the second phase. Shivakumar, Chen, Abali, Le, and Davis (2006) proposed the following law for CFRP delaminations, where G_{max} is the driving fracture parameter (although extrapolation to bonded joints is perfectly possible):

$$\frac{da}{dN} = C \left(\frac{G_{max}}{G_c}\right)^m \cdot \left[1 - \left(\frac{G_{th}}{G_{max}}\right)^{D_1}\right], \quad (4.3)$$

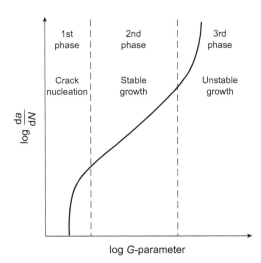

Figure 4.6 Hypothetical plot of log da/dN−log G-parameter.

where C, m and D_1 are the material constants determined by best fitting the data. Fatigue behaviour in the second phase (also called the stable growth or linear phase) is ruled by the G-parameter, and it reports a linear behaviour between this value and $\mathrm{d}a/\mathrm{d}N$ (in the double logarithmic plot). Under these conditions, a simple Paris-like law can fit the experimental data (Paris, Gomez, & Anderson, 1961):

$$\frac{\mathrm{d}a}{\mathrm{d}N} = C(G\text{-parameter})^m, \tag{4.4}$$

where C and m are the same material constants of expression (4.3). The G-parameter is typically $G_{\max}$, but other quantities such as ΔG, ΔK or ΔJ are eligible as well. The value of m in particular is the slope of the linear regime of the second phase, and it relates to the dependency of $\mathrm{d}a/\mathrm{d}N$ with the chosen G-parameter. Typical values of this parameter can be found in the works of Mangalgiri, Johnson, and Everett (1987) for bonded joints or Turon et al. (2007) for composites. In the work of Bernasconi et al. (2013), the value of ΔG was considered in expression (4.4) as the G-parameter to extract the fatigue law from bonded CFRP DCB specimens, and its application revealed accurate in the extrapolation for mixed-mode joints. Shivakumar et al. (2006) used a normalized G-parameter for fatigue growth in CFRP DCB specimens. When structures to be analysed have a pre-crack, the first phase is neglected, as beginning of the loading falls directly into the second phase (Fernández et al., 2011). For bonded joints and composite delaminations, the second phase is usually dominant over nucleation (first phase) and failure (third phase), as it was shown in the work of Bernasconi, Beretta, Moroni, and Pirondi (2010). In this case, the first and third phases can be neglected in the fatigue life prediction and expression (4.4) can be integrated, giving a simplification of N_f:

$$N_f = \int_{a_i}^{a_f} \frac{\mathrm{d}a}{C(G\text{-parameter})^m}, \tag{4.5}$$

where a_i and a_f are naturally the initial and final values of a. The third phase of the plot (unstable region) corresponds to fast crack growth, where the G-parameter enters the vicinity of the respective critical value (e.g. when $G_{\max}$ approaches the static G_c). The critical G-parameter/G_{th} ratio is addressed as the sensitivity to fatigue, with bigger values indicating that a large difference between static and fatigue strengths. In some situations, the third phase can be excluded from the fatigue life prediction because of its reduced effect on N_f. If this cannot be performed without compromising the accuracy, proposed laws are available, such as the following for CFRP delaminations, considering $G_{\max}$ as the G-parameter (Shivakumar et al., 2006):

$$\frac{\mathrm{d}a}{\mathrm{d}N} = C\left(\frac{G_{\max}}{G_c}\right)^m \cdot \frac{1}{1 - (G_{\max}/G_c)^{D_2}}, \tag{4.6}$$

where C, m and D_2 are the material constants, and C and m common to those of expressions (4.3) and (4.4). It should be mentioned that a set of parameters C, m, D_1 and

D_2 fully defines the da/dN–G-parameter law for a given material under the tested fatigue conditions by expressions (4.3), (4.4) and (4.6), or equivalent ones. The estimation process consists of fitting C and m to the second-phase curve (expression (4.4)), followed by insertion of these parameters in expression (4.3) for fitting of D_1, and finally using C and m in expression (4.6) to define D_2. These parameters are also specific for the mode-ratio in which they were determined (Turon et al., 2007). Laws can also be found in the literature that cover the entire spectrum of the G-parameter for general-purpose materials (Ewalds, 1984). In the definition of the da/dN law for co-cured CFRP components, it is also important to bear in mind that fibre-bridging events may take place, which increases the G-parameter as it tends to oppose crack opening. Non-consideration of this fact will invariably lead to under-predictions of the fatigue life. A more detailed description of these phenomena and predictive methods can be found in the work of Shivakumar et al. (2006). For bonded joints in particular, the crack nucleation or first phase, considering a specific loading applied to the structure, highly depends on the presence and shape of a spew fillet (Hadavinia et al., 2003), while the duration of the propagation phases (stable and catastrophic) depends on the length and shape of the joint (Pirondi & Moroni, 2010). In the work of Quaresimin and Ricotta (2006a), by comparing square-edge and filleted-bonded lap joints, a significant improvement of the nucleation phase was found by using the fillet (up to 75% of N_f). For theoretically defect-free joints, the first phase can be studied by the local stresses in the adhesive layer, notch-stress intensity factor (N-SIF) or stress singularity at the sharp corners of the overlap edges (Quaresimin & Ricotta, 2006a). If no fillet exists or in presence of defects due to lack of polymerization or adhesion, the nucleation phase should be negligible compared to propagation. On the other hand, simulation of general structures without pre-crack may not correlate well with data from cracked specimens such as the DCB, with under predictions of the damage accumulation rate. Nonetheless, the problem can be surpassed by considering the limiting scenario of pre-crack in the model, assuming some kind of fabrication defect. Predictions from these analyses are very conservative since they neglect the first phase, which can be a significant part of the total fatigue life (May & Hallett, 2010).

On account of the described method to characterize the fatigue behaviour, besides the values of a, F and δ obtained from testing, two addition quantities are required: (1) the value of G-parameter during the test and (2) the value of da/dN. For both of these, several theories can be applied, depending on a number of factors. The most relevant ones applied to bonded and co-cured CFRP joints are discussed.

4.5.1 Estimation of the G-parameter in mode I

In fatigue testing, the G-parameter is function of a, and it can be evaluated analytically for simple geometries like the DCB specimen, or in more complex cases by numerical methods such as FE (Pirondi & Moroni, 2009). The FE techniques to determine the G-parameter rely on the calculation of the J-integral, the virtual crack extension technique (VCET) or the VCCT for a given crack position. After it has been evaluated, the crack is propagated by a predefined amount by moving the mesh or by debonding nodes at the crack tip, giving another measurement. This process is then repeated until the area of interest is addressed. This section describes analytical techniques to obtain G_I and

J-integral, since these are the most common G-parameters applicable to bonded and co-cured specimens (Pirondi & Moroni, 2010). The proposed methods are also valid for the DCB specimen only. It should be noted that in the co-cured specimens, these parameters relate to the CFRP interlaminar fracture, while in the bonded specimens, fracture occurs along the adhesive bondline. In the first scenario, due to the typical brittleness, fracture occurs without significant plasticization and, as a result, it can be categorized under the scope of conventional linear elastic fracture mechanics (LEFM)-based methods. When considerable plasticization occurs in the adhesive layer, as it occurs in the second scenario with modern toughened adhesives, LEFM methods are rendered inaccurate (Fernández et al., 2011) and the well-known path-independent J-integral is a feasible option as G-parameter. The Compliance Calibration Method (CCM) is based on the Irwin–Kies equation (Kanninen & Popelar, 1985):

$$G_\text{I} = \frac{F^2}{2B}\frac{\text{d}C}{\text{d}a}, \tag{4.7}$$

where C is the compliance ($C = \delta/F$). Cubic polynomials ($C = C_3 a^3 + C_2 a^2 + C_1 a + C_0$) can be used to fit the $C = \text{f}(a)$ curves required for the definition of $\text{d}C/\text{d}a$ (Banea, da Silva, & Campilho, 2012). The equation can then be differentiated to obtain $\text{d}C/\text{d}a$ as function of a, which makes possible the calculation of G_I for all recorded values of a. The Berry method can also be applied to fit the $C = \text{f}(a)$ curves (Ashcroft & Shaw, 2002), by plotting in a log–log chart the C–a curve and making a straight line fit of the form:

$$C = Y a^n, \tag{4.8}$$

where Y is the axis intercept and n the slope. Beam theories are also widely used. The Direct Beam Theory (DBT), based on elementary beam theory, gives (de Moura, Campilho, & Gonçalves, 2008):

$$G_\text{I} = \frac{12 a^2 F^2}{B^2 h^3 E_\text{x}}, \tag{4.9}$$

with E_x representing the Young's modulus of the adherends in the longitudinal direction. The DBT is known to underestimate the measured value of G_I, since it is assumed that the specimen beams are built-in at the crack tip. The corrected beam theory (CBT) takes this feature into account, and G_I is obtained using (Robinson & Das, 2004):

$$G_\text{I} = \frac{3F\delta}{2B(a + |\Delta|)}, \tag{4.10}$$

where Δ is the crack length correction for crack tip rotation and deflection. In fact, in the beam theory it is assumed that each arm of the DCB specimen is a clamped beam with the length equal to the value of a, which does not reflect the real conditions of the DCB test. Δ corrects the value of a, being determined by a linear regression of $C^{1/3} = \text{f}(a)$. This can be performed by slightly loading the specimen with three different initial crack lengths to define the $C^{1/3} = \text{f}(a)$ linear regression. The value of

Δ is extracted from $C^{1/3} = 0$ (de Moura et al., 2008). A variation of this theory was used by Bernasconi et al. (2013), which included the effect of shear upon the transverse deformation. This was performed by considering a Timoshenko beam scheme for the adherends and a Timoshenko beam on elastic foundation model for the DCB specimen. Pirondi and Nicoletto (2004) proposed a model based on beam theory that includes the adhesive influence in the bonded structure, considering the joint modelled as a beam on an elastic foundation:

$$G_I = \frac{F^2 a^2}{2BE_x I} \cdot (1 + \lambda_\sigma)^2, \quad \text{with} \quad \lambda_\sigma^4 = \frac{6}{h^3 t_A} \cdot \frac{E_a'}{E_x} \quad \text{and} \quad E_a' = \frac{E_a}{(1-\nu^2)}. \tag{4.11}$$

where I is the second moment of area of the adherends, E_a the adhesive Young's modulus, E_a' the adhesive Young's modulus considering plane strain conditions and ν the Poisson's coefficient. The compliance-based beam method (CBBM) only depends on the specimen compliance during the test (de Moura et al., 2008). G_I can be obtained by the following expression:

$$G_I = \frac{6F^2}{B^2 h} \left(\frac{2 a_{eq}^2}{t_P^2 E_f} + \frac{1}{5 G_{xy}} \right), \tag{4.12}$$

where a_{eq} is an equivalent crack length obtained from the experimental compliance and accounting for the fracture process zone (FPZ) at the crack tip, E_f is a corrected flexural modulus to account for all phenomena affecting the $F-\delta$ curve, such as stress concentrations at the crack tip and stiffness variability between specimens, and G_{xy} is the shear modulus in the xy plane. An identical method that does not require measurement of a was developed by Biel and Stigh (2008).

The J-integral is valid for the non-linear elastic behaviour of materials, but it remains applicable in the presence of a plastic but monotonic applied loading to characterize the cohesive separation and plastic dissipation in the adhesive (Ji, Ouyang, Li, Ibekwe, & Pang, 2010). Based on the fundamental expression for J defined by Rice (1968), it is possible to derive an expression for the value of G_I applied to the DCB specimen from the concept of energetic force and also the beam theory for this particular geometry, as follows (the following formulae are developed assuming that the J-integral gives a measurement of G_I) (Banea, da Silva, & Campilho, 2010):

$$G_I = 12 \frac{(F_w a)^2}{E h^3} + F_w \theta_o \tag{4.13}$$

or

$$G_I = F_w \theta_p, \tag{4.14}$$

where F_w represents the value of F per unit width, θ_o the relative rotation of the adherends at the crack tip and θ_p the relative rotation of the adherends at the loading line (Figure 4.7).

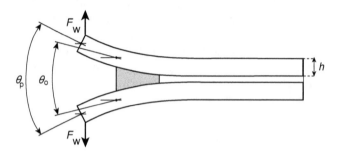

Figure 4.7 DCB specimen under loading, with description of the analysis parameters.

4.5.2 Estimation of da/dN

The final step towards characterizing the fatigue behaviour is the calculation of da/dN, based on the test data (value of a as function of N). Standardized procedures are available for this process (e.g. ASTM E647, 2011), and a few of these are described, based on their past application and suitability to adhesively-bonded and co-cured CFRP joints. A straightforward method to be applied is the secant method, recommended in the standard ASTM E647 (2011), which considers a discrete number of measurements during the fatigue test. The value of da/dN between two consecutive measurements (i and $i+1$) of the test is the slope of the $a = f(N)$ curve (Stelzer et al., 2012):

$$\left(\frac{da}{dN}\right)_{\bar{a}} = \frac{a_{i+1} - a_i}{N_{i+1} - N_i}, \quad \text{with} \quad \bar{a} = \frac{a_{i+1} + a_i}{2}. \tag{4.15}$$

The secant method is very simple to apply and accurately extracts da/dN from the data, although the estimation of da/dN is highly affected by fluctuations or small measurement errors in the test data. Another possibility, which minimizes the scatter in the calculation of da/dN, consists on fitting, for each value of N, a second-order polynomial function of a set of ($2n-1$) points (Ashcroft & Shaw, 2002). The most suited value of n will consider the sufficient but minimum number of points for an accurate representation. The polynomial function can be expressed as:

$$a = b_0 + b_1\left(\frac{N - C_1}{C_2}\right) + b_2\left(\frac{N - C_1}{C_2}\right)^2, \tag{4.16}$$

where the parameters b_i ($i = 0, 1, 2$) are estimated by regression techniques to the data set such as the least squares. C_1 and C_2 are the auxiliary parameters that normalize the test data, and aim to prevent numerical problems:

$$C_1 = \frac{1}{2(N_{i-n} + N_{i+n})}; \quad C_2 = \frac{1}{2}(N_{i+n} - N_{i-n}); \quad \text{with} \quad i = (n+1).$$

$$\tag{4.17}$$

Stelzer et al. (2012) used an $n = 7$ approximation for CFRP DCB specimens, which provided a good fit to the data. After having approximated $a = f(N)$ by the polynomial of expression (4.16), differentiation promptly gives da/dN as:

$$\frac{da}{dN} = \frac{b_1}{C_2} + 2b_2 \left(\frac{N - C_1}{C_2^2} \right). \quad (4.18)$$

4.5.3 Results of testing by the FCG approach

Many works are currently available on FCG methodologies to characterize the fatigue behaviour of CFRP joints. Beginning with co-cured joints, Turon et al. (2007) studied the delamination behaviour in mode I by fatigue DCB testing, using a Paris-like law and ΔG normalized to G_c as the G-parameter. An accurate fit was found between the test data and the obtained law. Stelzer et al. (2012) addressed the co-cured configuration on CFRP DCB specimens under fatigue loads. Load control was selected for the tests, performed with $R = 1$, until a crack growth rate of nearly 10^{-6} mm/cycle was reached. The value of G_{max} was chosen as the G-parameter for the Paris law characterizing the stable propagation phase, estimated by the modified calibration method (MCC), while the da/dN values were taken from point−wise (secant) approximations or seven point polynomial fit. Figure 4.8(a) represents the da/dN−G_{max} plot for three laminates (C1 and C2 are CFRP, while G1 is a $S2$ glass composite). Results showed an approximate linear trend (in the double logarithmic plot), although the experimental scatter made difficult to distinguish trends amongst the laminated systems. To avoid this complication, the same data were plotted (Figure 4.8(b)), but considering values of a estimated from the $C = \delta/F$ data, instead of the visual inspection. As a result, much of the scatter disappeared and clear trends could be found. Figure 4.9 reports the fitting laws for composite systems C1 (a) and C2 (b), also representing the non-linear (NL) and maximum or 5% (MAX/5%) initiation values from quasi-static testing

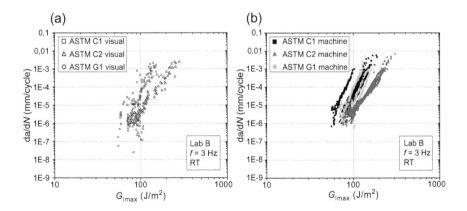

Figure 4.8 da/dN−G_{max} curves for DCB specimens of three laminated systems (a) and identical plot with improved method for the measurement of a (b) (Stelzer et al., 2012).

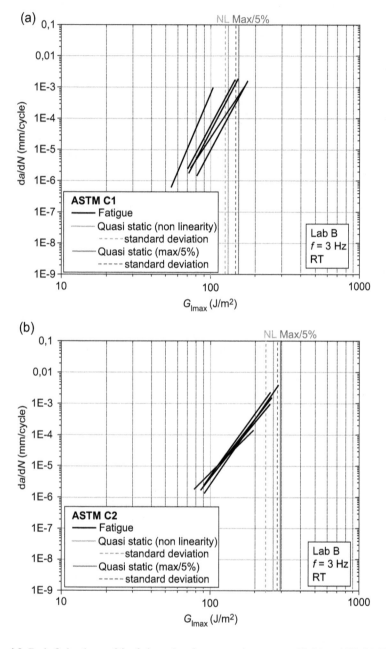

Figure 4.9 Paris fitting laws of the fatigue data for composite systems C1 (a) and C2 (b) (Stelzer et al., 2012).

of specimens of these laminates. The reported results showed that laminate C1 had a bigger scatter than C2 and laminate C2 is more resistant to crack growth than C1. For laminate C1, all fitting curves except one reached the average MAX/5% initiation value determined from quasi-static tests, whereas for C2 only one fit exceeds the average NL value from the same tests. Also, between the tested laminates, the highest delamination rates were found for laminate C2, on the order or approximately 5×10^{-3} mm/cycle. Fatigue characterization at different temperatures in adhesively-bonded CFRP DCB specimens was carried out by Ashcroft and Shaw (2002) in displacement control, considering both secant and polynomial methods to calculate da/dN. For the polynomial method, the $a = f(N)$ curves were fit making $n = 3$ (giving 7 points). Comparison between both methods (Figure 4.10(a)) revealed a fair agreement, although some scatter was found with the secant method. As a result, the authors chose to apply the polynomial method from this point on. Figure 4.10(b) shows FCG curves for specimens tested at room temperature at different maximum displacements, with evidence of good repeatability and no obvious effect of the maximum displacement on the behaviour. The curves exhibit a shape close to that of Figure 4.6, with a clear distinction between the three crack growth phases. The first phase emphasizes G_{th}, with a mean value of 120 J/m^2 (values from both axes are converted to logarithmic equivalents). The second phase shows a linear behaviour, corresponding to fast crack growth between 400 and 500 J/m^2, at comparable values to the static value of G_{Ic} for crack initiation defined in a previous work (Ashcroft, Hughes, & Shaw, 2001). Similar tests were carried out at -50 and 90 °C, with variations in behaviour that were properly discussed, showing in all cases the robustness of the chosen method to characterize the joints in mode I. Erpolat et al. (2004) carried out a similar study, but additionally considering variable amplitude (VA) fatigue loading. For constant amplitude fatigue, values of $R = 0.1$ and 0.5 were considered, and the 7-point polynomial fit allowed the estimation of da/dN. A beam on elastic foundation model was used to calculate G_I. The value of G_{th} was much lower for $R = 0.1$ than 0.5; however, when da/dN was plotted as function of ΔG, the two curves were practically coincident. This was justified by facial interference of the adhesives on the debonding surfaces that prevented cracks from closing, giving an artificially high value of G_{min} and thus affecting the results. For VA fatigue loading, a two-stage block loading was applied, with overloads every 20 cycles. During testing, crack growth occurred steadily, but after approximately 250 overloads sudden crack growth took place. From this point on, the steady growth was resumed up to the test ending. The da/dN behaviour for VA fatigue was predicted by integration of the constant amplitude crack growth data, which showed a slight under-prediction.

4.6 Fracture modes of CFRP joints in mode I

The fracture mechanisms under fatigue loadings acquire special relevancy in CFRP structures because of the complexity of the crack path, arising from comparable characteristics between the adhesive bond and the composite matrix. On account of this, several crack deviations can occur, together with delaminations between

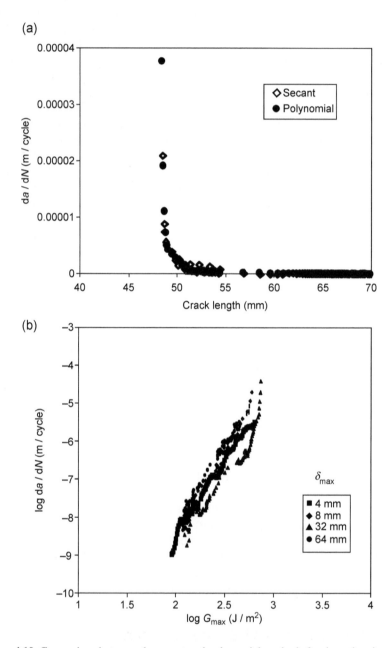

Figure 4.10 Comparison between the secant and polynomial methods for the estimation of da/dN (a) and FCG curves for different maximum displacements (b) (Ashcroft & Shaw, 2002).

composite plies. Damage uptake and crack path progression in the structures can be recorded during testing with high-speed cameras focused on a small region around the damage region. The correlation of the damage events with the N data is possible by the elapsed time from the beginning of the test and testing frequency. Analysis of the fractured surfaces can be accomplished by optical or scanning electron microscopy (SEM). In SEM, samples of reduced size need to be cut from the specimens, preferably with a diamond saw for a clean cut, mounted on the analysis stubs and gold coated to ensure the electrical conductivity of the sample.

The fracture surfaces of co-cured CDRP DCB specimens were characterized with SEM by Coronado et al. (2012), after fatigue testing at temperatures between -60 and 90 °C. In general, the surfaces showed 'river markings', indicating matrix plasticity (Figure 4.11). This behaviour was more evident at high temperatures, with clear matrix deformation and resin adhesion to the fibres, while at low temperatures failure was more brittle. Bigger magnifications emphasized the presence of voids near the fibre—matrix interfaces, which potentiated delaminations, and also broken fibres resulting from fibre-bridging events. Identical findings were reported by Argüelles et al. (2011) by fatigue-testing DCB specimens made of CFRP with matrices of different toughness. The work of Ashcroft and Shaw (2002) described the fatigue behaviour of DCB specimens with CFRP adherends (testing temperatures between -60 and 90 °C). At room temperature, crack growth began predominantly cohesively in the adhesive layer with longitudinal striations, indicative of stick-slip behaviour. After a length of crack growth, the failure path shifted to one of the adherends, but still with small regions of cohesive failure of the adhesive. At high temperatures, the dominant

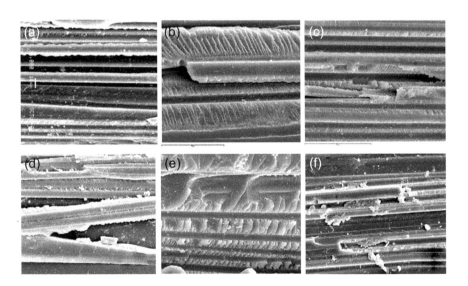

Figure 4.11 Fracture surfaces of the DCB specimens at 20 °C (a), 50 °C (b), 90 °C (c), 0 °C (d), -30 °C (e) and -60 °C (f), 1000× magnification (Coronado et al., 2012).

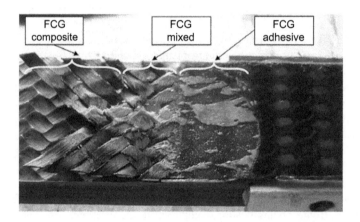

Figure 4.12 Fracture surface of a bonded CFRP DCB specimen after fatigue testing (Bernasconi et al., 2013).

failure was cohesive in the adhesive, although few longitudinal bands indicated composite failure. At low temperatures, fracture was mostly on the composite, although with minor cohesive spots. Fernández et al. (2011) fatigue tested CFRP DCB joints, bonded with the ductile epoxy Araldite® 2015 (Huntsman), for fatigue characterization of the adhesive in mode I. Although the bonding surfaces were carefully prepared, due to comparable peel characteristics between the adhesive and CFRP in the transverse direction, a few specimens failed by delamination between the two nearest plies to the adhesive. These tests were discarded for the analysis, and only the specimens that suffered cohesive failure were considered. The study of Bernasconi et al. (2013) consisted on the fatigue analysis of adhesively-bonded CFRP DCB and lap joints. The parameters of the Paris law were derived by fracture mechanics principles applied to the DCB specimens. Complex failure paths were observed on both test geometries. In the DCB specimens, this was clearly visible in the non-continuous shape of the $da/dN - \Delta G$ curve, as two distinct lines interpolating the experimental data were identified by a least squares analysis. Crack path initially developed in the adhesive bond and subsequently propagated to CFRP delaminations near the adhesive. Figure 4.12 reports to a microscope record of the fracture surface, showing initial propagation at the CFRP-adhesive interface with small regions of cohesive failure, followed by mixed crack growth and finally between the first and the second woven plies in one of the CFRP adherends.

4.7 Conclusions

This chapter dealt with mode I fatigue behaviour of bonded CFRP structures (considering secondary bonding and co-curing). The use of CFRPs and bonding to produce advanced structures was discussed regarding the comparative advantages over more conventional materials and joining methods. An overview of the possible loading scenarios (modes I, II and mixed-mode) showed that fatigue characterization can be performed under each one of these conditions. Mode I characterization is well established, with the DCB test as the most widespread. Mode II, on the other hand, is still not well addressed and, inclusively, mode II fatigue characterization for bonded joints is extremely scarce. Mixed-mode characterization can take advantage of specific criteria to merge modes I and II data to analyse mixed-mode failures. Alternatively, fatigue characterization can also be performed under mixed-mode conditions and then applied to the structures. Fabrication and fatigue testing were also discussed for mode I loadings, showing the importance of fabrication techniques, testing procedures and equipment, necessary for the fatigue tests to work properly. Special attention is required for the measurement of a during propagation, and a few techniques were described to this end. The two main categories for fatigue characterization were described: $S-N$ and FCG approaches. The $S-N$ based methods suffer from requiring testing at different values of S for building the curve, uncertainties on how to define S for a particular application and no allowance of the separation between fatigue initiation and propagation. Because of this, the FCG method is generally chosen for fatigue description of structures, relying on plotting da/dN as function of a suitable G-related parameter. Related literature seems to find unanimous that G_{max} is the most suited G-parameter for this purpose, although ΔG was considered by some authors. With this data, it is also possible to implement numerical procedures for fatigue strength prediction. Fracture modes in CFRP structures, either adhesively-bonded or co-cured, are highly complex on account of the different mechanisms possible to occur. In bonded CFRP joints, delaminations can occur because of comparable properties between the adhesive and composite matrix. Deviations from the original fracture path (e.g. to the adherends) prevent the measured laws from working for a specific mode of failure over the entire range of G-parameter values (i.e. from nucleation to catastrophic failure).

References

Abdel-Wahab, M. M., Ashcroft, I. A., Crocombe, A. D., & Smith, P. A. (2004). Finite element prediction of fatigue crack propagation lifetime in composite bonded joints. *Composites Part A, 35*, 213–222.
Argüelles, A., Viña, J., Fernández-Canteli, A., Viña, I., & Bonhomme, J. (2011). Influence of the matrix constituent on mode I and mode II delamination toughness in fiber-reinforced polymer composites under cyclic fatigue. *Mechanics of Materials, 43*, 62–67.
Ashcroft, I. A., Hughes, D. J., & Shaw, S. J. (2001). Mode I fracture of epoxy bonded composite joints: Part 1. Quasi-static loading. *International Journal of Adhesion and Adhesives, 21*, 87–99.
Ashcroft, I. A., Hughes, D. J., Shaw, S. J., Abdel-Wahab, M., & Crocombe, A. (2001). Effect of temperature in the quasi-static strength and fatigue resistance of bonded composite double lap joints. *Journal of Adhesion, 75*, 61–88.
Ashcroft, I. A., & Shaw, S. J. (2002). Mode I fracture of epoxy bonded composite joints 2. Fatigue loading. *International Journal of Adhesion and Adhesives, 22*, 151–167.
ASTM D6115-97. (2011). *Standard test method for mode I fatigue delamination growth onset of unidirectional fiber-reinforced polymer matrix composites.*

ASTM E647. (2011). *Standard test method for measurement of fatigue crack growth rates.*
Azari, S., Papini, M., Schroeder, J. A., & Spelt, J. K. (2010). Fatigue threshold behavior of adhesive joints. *International Journal of Adhesion and Adhesives, 30*, 145−159.
Banea, M. D., da Silva, L. F. M., & Campilho, R. D. S. G. (2010). Temperature dependence of the fracture toughness of adhesively bonded joints. *Journal of Adhesion Science and Technology, 24*, 2011−2026.
Banea, M. D., da Silva, L. F. M., & Campilho, R. D. S. G. (2012). Effect of temperature on tensile strength and mode I fracture toughness of a high temperature epoxy adhesive. *Journal of Adhesion Science and Technology, 26*, 939−953.
Bannantine, J. A., Comer, J. J., & Handrock, J. L. (1990). *Fundamentals of metal fatigue analysis.* Englewood Cliffs: Prentice Hall.
Bathias, C. (2006). An engineering point of view about fatigue of polymer matrix composite materials. *International Journal of Fatigue, 28*, 1094−1099.
Beghini, M., Bertini, L., & Forte, P. (2006). Experimental investigation on the influence of crack front to fiber orientation on fatigue delamination growth rate under mode II. *Composites Science and Technology, 66*, 240−247.
Bernasconi, A., Beretta, S., Moroni, F., & Pirondi, A. (2010). Local stress analysis of the fatigue behaviour of adhesively bonded thick composite laminates. *Journal of Adhesion, 86*, 480−500.
Bernasconi, A., Jamil, A., Moroni, F., & Pirondi, A. (2013). A study on fatigue crack propagation in thick composite adhesively bonded joints. *International Journal of Fatigue, 50*, 18−25.
Biel, A., & Stigh, U. (2008). Effects of constitutive parameters on the accuracy of measured fracture energy using the DCB-specimen. *Engineering Fracture Mechanics, 75*, 2968−2983.
Blackman, B. R. K., Hadavinia, H., Kinloch, A. J., & Williams, J. G. (2003). The use of a cohesive zone model to study the fracture of fibre composites and adhesively-bonded joints. *International Journal of Fracture, 119*, 25−46.
Brunner, A. J., Stelzer, S., Pinter, G., & Terrasi, G. P. (2013). Mode II fatigue delamination resistance of advanced fiber-reinforced polymer-matrix laminates: towards the development of a standardized test procedure. *International Journal of Fatigue, 50*, 57−62.
Campilho, R. D. S. G., Banea, M. D., Pinto, A. M. G., da Silva, L. F. M., & de Jesus, A. M. P. (2011). Strength prediction of single- and double-lap joints by standard and extended finite element modeling. *International Journal of Adhesion and Adhesives, 31*, 363−372.
Campilho, R. D. S. G., Moura, D. C., Gonçalves, D. J. S., da Silva, J. F. M. G., Banea, M. D., & da Silva, L. F. M. (2013). Fracture toughness determination of adhesive and co-cured joints in natural fibre composites. *Composites: Part B, 50*, 120−126.
Coronado, P., Argüelles, A., Viña, J., Mollón, V., & Viña, I. (2012). Influence of temperature on a carbon-fibre epoxy composite subjected to static and fatigue loading under mode-I delamination. *International Journal of Solids and Structures, 49*, 2934−2940.
Curley, A. J., Jethwa, J. K., Kinloch, A. J., & Taylor, A. C. (1998). The fatigue and durability behaviour of automotive adhesives. Part III: predicting the service life. *Journal of Adhesion, 66*, 39−59.
Dessureault, M., & Spelt, J. K. (1997). Observations of fatigue crack initiation and propagation in an epoxy adhesive. *International Journal of Adhesion and Adhesives, 17*, 183−195.
Erpolat, S., Ashcroft, I. A., Crocombe, A. D., & Abdel-Wahab, M. M. (2004). Fatigue crack growth acceleration due to intermittent overstressing in adhesively bonded CFRP joints. *Composites Part A, 35*, 1175−1183.
Ewalds, H. L. (1984). *Fracture mechanics.* London: Edward Arnold.

Fernández, M. V., de Moura, M. F. S. F., da Silva, L. F. M., & Marques, A. T. (2011). Composite bonded joints under mode I fatigue loading. *International Journal of Adhesion and Adhesives, 31*, 280−285.

Fernández, M. V., de Moura, M. F. S. F., da Silva, L. F. M., & Marques, A. T. (2013). Characterization of composite bonded joints under pure mode II fatigue loading. *Composite Structures, 95*, 222−226.

de Goeij, W. C., van Tooren, M. J. L., & Beukers, A. (1999). Composite adhesive joints under cyclic loading. *Materials and Design, 20*, 213−221.

Hadavinia, H., Kinloch, A. J., Little, M. S. G., Taylor, A. C. (2003). The prediction of crack growth in bonded joints under cyclic-fatigue loading I. Experimental Studies. *International Journal of Adhesion and Adhesives, 23*, 449−461.

Hou, L., & Liu, D. (2003). Size effects and thickness constraints in composite joints. *Journal of Composite Materials, 37*, 1921−1938.

ISO 15024. (2001). *Fibre-reinforced plastic composites − Determination of mode I interlaminar fracture toughness, G_{Ic}, for unidirectionally reinforced materials*.

Ji, G., Ouyang, Z., Li, G., Ibekwe, S., & Pang, S. S. (2010). Effects of adhesive thickness on global and local mode-I interfacial fracture of bonded joints. *International Journal of Solids and Structures, 47*, 2445−2458.

Kanninen, M. F., & Popelar, C. H. (1985). *Advanced fracture mechanics*. Oxford: Oxford University Press.

Kolesnikov, B., Herbeck, L., & Fink, A. (2008). CFRP/titanium hybrid material for improving composite bolted joints. *Composite Structures, 83*, 368−380.

Krenk, S. (1992). Energy release rate of symmetric adhesive joints. *Engineering Fracture Mechanics, 43*, 549−559.

Lee, J., & Soutis, C. (2008). Measuring the notched compressive strength of composite laminates: specimen size effects. *Composites Science and Technology, 68*, 2359−2366.

Leffler, K., Alfredsson, K. S., & Stigh, U. (2007). Shear behaviour of adhesive layers. *International Journal of Solids and Structures, 44*, 530−545.

Mall, S., Ramamurthy, G., & Rezaizdeh, M. A. (1987). Stress ratio effect on cyclic debonding in adhesively bonded composite joints. *Composite Structures, 8*, 31−45.

Mangalgiri, P. D., Johnson, W. S., & Everett, R. A. (1987). Effect of adherend thickness and mixed mode loading on debond growth in adhesively bonded composite joints. *Journal of Adhesion, 23*, 263−288.

May, M., & Hallett, S. R. (2010). A combined model for initiation and propagation of damage under fatigue loading for cohesive interface elements. *Composites Part A, 41*, 1787−1796.

Messler, R. W. (1993). *Joining of advanced materials*. Stoneham: Butterworths/Heinemann.

de Morais, A. B., Pereira, A. B., Teixeira, J. P., & Cavaleiro, N. C. (2007). Strength of epoxy adhesive-bonded stainless-steel joints. *International Journal of Adhesion and Adhesives, 27*, 679−686.

de Moura, M. F. S. F., Campilho, R. D. S. G., & Gonçalves, J. P. M. (2008). Crack equivalent concept applied to the fracture characterization of bonded joints under pure mode I loading. *Composites Science and Technology, 68*, 2224−2230.

Owens, J. F. P., & Sullivan, P. L. (2000). Stiffness behaviour due to fracture in adhesively bonded composite-to-aluminium joints II. Experimental. *The International Journal of Adhesion and Adhesives, 20*, 47−58.

Paris, P., Gomez, M., & Anderson, W. (1961). A rational analytical theory of fatigue. *Trend in Engineering, 13*, 9−14.

Pirondi, A., & Moroni, F. (2009). An investigation of fatigue failure prediction of adhesively bonded metal/metal joints. *International Journal of Adhesion and Adhesives, 29*, 796−805.

Pirondi, A., & Moroni, F. (2010). A progressive damage model for the prediction of the fatigue crack growth in bonded joints. *Journal of Adhesion, 86*, 501−521.

Pirondi, A., & Nicoletto, G. (2004). Fatigue crack growth in bonded DCB specimens. *Engineering Fracture Mechanics, 71*, 859−871.

Quaresimin, M., & Ricotta, M. (2006a). Life prediction of bonded joints in composite materials. *International Journal of Fatigue, 28*, 1166−1176.

Quaresimin, M., & Ricotta, M. (2006b). Stress intensity factors and strain energy release rate in single lap bonded joints in composite materials. *Composites Science and Technology, 66*, 647−656.

Rice, J. R. (1968). A path independent integral and the approximate analysis of strain concentration by notches and cracks. *Journal of Applied Mechanics, 35*, 379−386.

Robinson, P., & Das, S. (2004). Mode I DCB testing of composite laminates reinforced with z-direction pins: a simple model for the investigation of data reduction strategies. *Engineering Fracture Mechanics, 71*, 345−364.

Roy, C., & Elghorba, M. (1988). Monitoring progression of mode-II delamination during fatigue loading through acoustic-emission in laminated glass−fiber composite. *Polymer Composites, 9*, 345−351.

Rudawska, A. (2010). Adhesive joint strength of hybrid assemblies: titanium sheet-composites and aluminium sheet-composites − experimental and numerical verification. *International Journal of Adhesion and Adhesives, 30*, 574−582.

Shivakumar, K., Chen, H., Abali, F., Le, D., & Davis, C. (2006). A total fatigue life model for mode I delaminated composite laminates. *International Journal of Fatigue, 28*, 33−42.

da Silva, L. F. M., & Adams, R. D. (2007). Adhesive joints at high and low temperatures using similar and dissimilar adherends and dual adhesives. *International Journal of Adhesion and Adhesives, 27*, 216−226.

da Silva, L. F. M., & Campilho, R. D. S. G. (2011). *Advances in numerical modelling of adhesive joints.* Springer, Heidelberg.

da Silva, L. F. M., Öchsner, A., & Adams, R. D. (Eds.). (2011). *Handbook of adhesion technology.* Springer, Heidelberg.

Stelzer, S., Brunner, A. J., Arguelles, A., Murphy, N., & Pinter, G. (2012). Mode I delamination fatigue crack growth in unidirectional fiber reinforced composites: development of a standardized test procedure. *Composites Science and Technology, 72*, 1102−1107.

Sugiman, S., Crocombe, A. D., & Katnam, K. B. (2011). Investigating the static response of hybrid fibre-metal laminate doublers loaded in tension. *Composites Part A, 42*, 1867−1884.

Turon, A., Costa, J., Camanho, P. P., & Dàvila, C. G. (2007). Simulation of delamination in composites under high-cycle fatigue. *Composites Part A, 38*, 2270−2282.

Yuan, H., & Xu, Y. (2008). Computational fracture mechanics assessment of adhesive joints. *Computational Materials Science, 43*, 146−156.

Mode I fatigue behaviour and fracture of adhesively-bonded fibre-reinforced polymer (FRP) composite joints for structural repairs

5

J. Renart[1], J. Costa[1], C. Sarrado[1], S. Budhe[1], A. Turon[1], A. Rodríguez-Bellido[2]
[1] University of Girona, Girona, Spain; [2] AIRBUS Operations, Madrid, Spain

5.1 Introduction

In recent years, the use of composite components in structures has been continuously evolving. In particular, the use of composites in the current generation of aircraft has reached a volume of 50% or more (i.e. Airbus A350 or Boeing 787). Fibre-reinforced composite material is used in primary structural components such as the fuselage, wings, tail stabilizers and doors. Some of these uses have highlighted a need to develop new manufacturing approaches, which include bonded joints. In addition, components already in service need to be maintained and repaired. Repairs by means of patches rely on the bonded joints transmitting loads, and thus they are the crucial aspect in the quality of the repair.

The application of bonded repairs in composite components is well established and widespread in the aeronautical industry, especially for minor accidental damage during the operational life of the aircraft. Bonded joints are also used to rework small defects incurred during the manufacturing process. The frequency of minor accidental damage during the operational life of the aircraft is high and their repair has a significant impact on maintenance costs.

Severe damage, although less frequent, also has a significant influence on maintenance costs as they require costly and time-consuming repairs that might also entail stopping the aircraft's normal operation. In spite of their feasibility, the use of bonded repairs for severe damage is restricted by failsafe criteria limits (the repaired structure must sustain limit load in the event of a complete loss of the bonded repair). Overcoming this restriction is linked to the availability of reliable non-destructive inspection (NDI) techniques that are capable of demonstrating the quality of the repair (i.e. detecting contaminants in the adherent surface, detecting weak bonded joints, etc.).

Once the repair size is restrained to that capable of sustaining a limit load residual strength with failure or partial failure of the bond line, implementing bonded repairs for highly loaded primary carbon-fibre-reinforced polymer (CFRP) components also requires (via tests, or through analysis substantiated by tests) demonstrating that the full strength, stiffness, damage tolerance and fatigue durability in all operational conditions have been restored. This capability relies on the use of a proper repair design, repair materials and processes.

The use of original production materials is usually prevented for in-field repairs. Indeed, original materials require curing in an autoclave to ensure their best performance, and this is not a suitable process for repairing a wing, a fuselage, stabilizers or any other primary structures. Even for components that can be easily disassembled, like doors, access panels, fairings, etc., the use of original production materials is not a common practice either due to, once again, needing to use an autoclave. Curing in autoclaves requires the appropriate tooling to avoid component deformation and can harm the component as a consequence of heating it to a high temperature (the prospect of skins possibly blowing up from the honeycomb around the repair in sandwich panels is an example of a quite frequent and typical problem).

In view of this situation, materials specifically dedicated to repair purposes have therefore been developed. All of them share their suitability for achieving good properties (low porosity) when cured under vacuum pressure only. Due to the fact that these materials are different from those used and certified for production, a complete set of tests, covering all levels of the certification test pyramid, must be developed. These tests should include the patch itself, as well as the bonded joint, which is the most critical part of the repair. In order to reduce the effort at the higher levels of the test pyramid, it is essential to develop reliable procedures to assess the fatigue and damage tolerance at a coupon level. In the case of fatigue, determining crack growth behaviour at a coupon level (onset and crack growth rates) would permit an effective repair system screening and would help identify their performance when compared to the original unrepaired material.

This chapter gives an overview of the methodologies used to analyse fatigue-induced damage in bonded joints under cyclic loads. The work is focused on mode I peel tests at a coupon level. First, to provide a proper background, a description of the different bonded joint configurations is presented. Second, a revision of the static and fatigue tests is introduced. That section describes the static tests that have to be performed prior to fatigue testing, and then it concentrates on how to obtain the curves for fatigue onset and for fatigue propagation data (crack growth rate versus loading). Third, the fracture modes and their effect on the test results are described. Finally, the chapter includes a concise review of the current numerical techniques available to simulate the behaviour of the bonded joint, accentuating their main advantages and drawbacks.

5.2 Configuration of the bonded joint

There are different choices when manufacturing bonded joints between composite panels. Secondary bonding and co-bonding are the most common processes (see Figure 5.1). Co-curing or co-bonding are usually preferred over secondary bonding because the number of parts and/or curing cycles is reduced. However, for large and

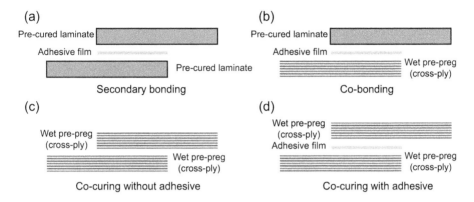

Figure 5.1 Drawing of the most common manufacturing processes used to produce bonded joints between composite adherents.

complex assemblies, secondary bonding can provide stronger joints. Failure processes, failure modes and joint strength are all influenced by the bonding method (Kim, Yoo, Yi, & Kim, 2006; Song et al., 2010). Therefore, in order to have the best performance of the bonded joint, it is important to choose the appropriate bonding method for each particular case.

Here, we omit manufacturing details such as the use of a 'peel ply' on the manufacturing of the adherent to protect their surfaces from environmental degradation, to generate a constant roughness pattern, and to prevent the contamination of the surface, thus leading to a proper strength of the bonded joint (Kanerva & Saarela, 2013).

The bond line thickness also plays an important role in the strength of the bonded joint (Mall & Ramamurthy, 1989; Marzi, Biel, & Stigh, 2011). Several works concluded that the bond line thickness should be within the range of 0.2–0.4 mm to attain a high joint strength. This is often achieved by embedding a textile membrane into the adhesive film, called carrier, or by using gages during the curing process. A bonded joint produced with an adhesive film with a carrier maintains a constant-bond line thickness as compared to that obtained with resin. Figure 5.2 shows the fractured surface of an adhesive joint where the footprint of the peel ply and that of the carrier can be observed. The carrier is used to control the bond line thickness and the adhesive bleeding during the curing phase. However, in co-bonding or co-curing processes involving woven fabrics, a constant adhesive thickness is difficult to achieve due to the geometry of the fabric itself, resulting in fairly significant differences in thickness.

The stacking sequences in the adherents also play a role in the strength of the joints in composite laminates (Matthews & Tester, 1985).

Poor surface preparation can cause a complete disbanding of the adherents during service at virtually zero stress (Hart-Smith, 1999). Finally, the in-service environmental conditions of the bonded joint (temperature, moisture, chemical agents) can influence strength and durability (Caminero et al., 2013; Whittingham, Baker, Harman, & Bitton, 2009).

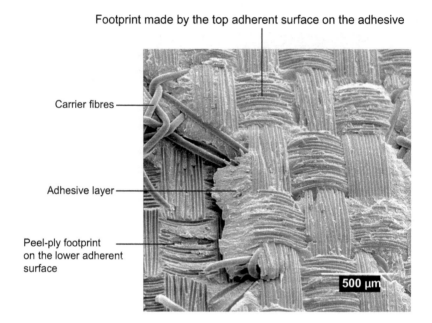

Figure 5.2 Fractured surface of a secondary bonded joint between two unidirectional (UD) pre-cured adherents and an adhesive film with carrier. The bonding surfaces of both adherents were prepared with a peel-ply layer that was removed just before the adhesive joint.

In order to gain a complete understanding of the behaviour of bonded joints, it is of the utmost importance to establish the relationship between the joint strength (often determined by means of the fracture energy, G_C) and the particular mechanism responsible for the failure. As Figure 5.1 shows, there are several interfaces where failure can potentially occur. What the fracture energy associated to each of them is and how it is influenced by fatigue loads or environmental parameters is the knowledge required to optimize the bonded joint performance and to predict its behaviour in service.

5.3 Test generalities

5.3.1 Static tests prior to fatigue testing

Peel tests have been proven to be one of the most reliable methods in determining the quality of an adhesive joint. They measure the adhesive joint strength under opening displacements, or mode I. The double cantilever beam (DCB) test is widely used in the aircraft industry as the peel test characterizing the quality of the adhesive joint between two composite adherents. The bond strength is quantified by means of a single parameter, the energy fracture toughness (G_{IC}). The principal advantage of this test over other alternative tests is that it allows bonded joints to be tested between two rigid adherents. It is routinely applied to any kind of bonded joint: secondary bonded, co-bonded or co-cured. In addition, the same test configuration is used to characterize bonded joints under cyclic loading. In this case, preliminary quasi-static DCB tests have to be conducted in order to define fatigue test parameters.

The DCB test was formerly conceived for delamination in composites (ASTM D5528, 2013; ISO 15024, 2001). While the scope of the American standard (ASTM D5528, 2013) includes the characterization of bonded joints, it does not introduce any refinement in the theory different to that of the delamination tests. In 2009, the International Standard Organization published a specific standard for bonded joints: ISO 25217 (2009). It was based on the British Standard BS 7991 (2001) and the testing protocol developed by Blackman and Kinloch in 2001. The data reduction methods used to obtain the energy fracture toughness (G_{IC}) are similar to those proposed in the delamination tests.

The DCB test consists of opening the adherents of the bonded joint to cause the propagation of an already existing crack in the mid-plane of the specimen (see Figure 5.3). The test is performed under controlled crosshead displacement. The displacement (δ), load to open the specimen arms (P) and the crack length (a) are recorded during the test.

Load and displacement are obtained directly from the test machine. The crack length should be obtained from visual observation at the specimen side. According to ISO 25217 (2009), the crack length is measured every 1 mm from 1 to 10 mm and from 60 to 65 mm, and then every 5 mm from 10 to 60 mm. An artificial crack is created at one of the ends of the specimen by introducing a Teflon insert in the adhesive mid-plane during the manufacturing phase. As a result of the insert, the crack tip is blunt and does not represent the real shape of possible defects, for example, created during an impact event. Therefore, a short propagation is performed in order to obtain a sharp tip (pre-crack). A propagation test is then conducted from that pre-crack.

Two values of the fracture energy G_{IC} are determined: an initiation and a propagation value. Furthermore, there are several methods for determining the initiation values: NL (point at which the load-displacement curve becomes nonlinear), VIS (point at which the crack propagation is visually observed) or 5%/MAX

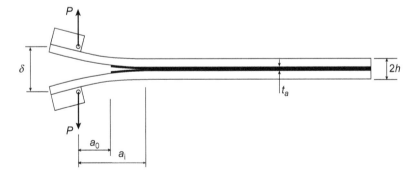

Figure 5.3 Schema of a DCB test of a bonded joint. P is the applied load, δ the deflection of the specimen arms, a the delamination length, a_0 the initial delamination length, t_a the adhesive thickness and $2h$ the total thickness of the specimen.

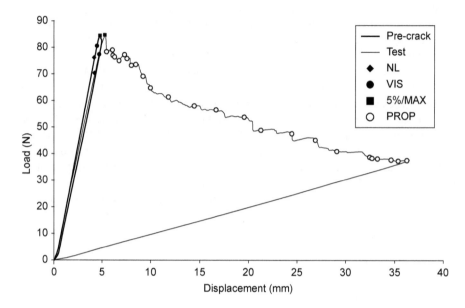

Figure 5.4 Load–displacement curves of a DCB test: pre-crack and propagation test curves including initiation (NL, VIS, 5%/MAX) and propagation (PROP) points.

(5% offset and maximum load) (ISO 25217, 2009). Therefore, from each test initiation, values of G_{IC} from the pre-crack, and initiation and propagation values from the propagation test are measured (see Figure 5.4).

Apart from this, there are several methods to calculate G_{IC}. All of them are based on the assumptions of linear elastic fracture mechanics (LEFM). In particular, the Irwin Kies approach relates the energy release rate (G_I) to the derivative of the compliance in function of the crack length (dC/da). The modified beam theory (MBT) is also of specific interest, as it too is used in fatigue tests. According to this method, the energy fracture toughness is determined by Eqn (5.1):

$$G_{IC} = \frac{3P\delta}{2B(a + \Delta)} \quad (5.1)$$

where δ is the displacement of the specimen arms, P the load, B the specimen width, a the crack length and Δ a crack length correction parameter that is determined from the linear regression of the cube root of the compliance ($C^{1/3}$) against the crack length (a) (see Figure 5.5).

5.3.2 Stick-slip effects on adhesive joints

In many cases, when testing bonded joints in mode I, the crack growth does not evolve continuously but rather proceeds as a succession of rapid growth and arrest phases

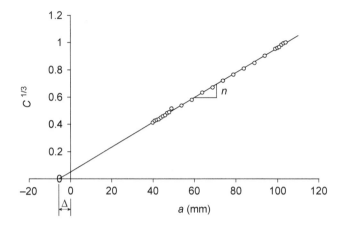

Figure 5.5 Linear regression curve to obtain the crack length correction factor $|\Delta|$.

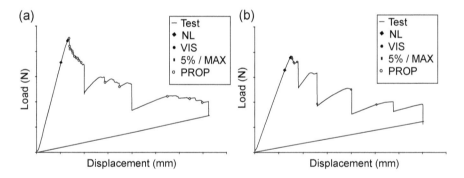

Figure 5.6 Load–displacement curves from DCB tests: (a) continuous propagation with some regions of stick-slip and (b) stick-slip during the entire test.

(Ashcroft, Hughes, & Shaw, 2001). This is commonly referred to as stick-slip growth. It can be easily identified by looking at the load–displacement curve, because when the rapid crack growth occurs there is a sudden drop in the force (see Figure 5.6). During the arrest phases, the load increases linearly according to the specimen stiffness. The stick-slip can occur sporadically in between stages of constant crack propagation (see Figure 5.6(a)) or during the entire propagation test (Figure 5.6(b)).

The stick-slip propagation can be observed on the specimen's fractured surfaces (see Figure 5.7). The darker bands correspond to the zones of fast crack growth, whereas the lighter zones correspond to constant propagation or arrest phases. If the stick-slip occurs during the entire propagation test, the amplitude of the arrest phases is small, only a few mm, compared to the size of the fast crack growth regions.

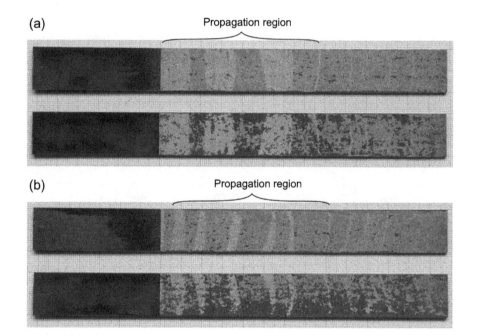

Figure 5.7 Fractured surface of (a) load—displacement curve of Figure 5.6(a), lighter bands correspond to zones of continuous crack growth; (b) load—displacement curve of Figure 5.6(b), lighter bands correspond to arrest phases.

When stick-slip occurs, two values of the fracture energy are obtained, one from the initiation of the fast crack growth (G_{ICi}) and the other from the arrest propagation points (G_{Ia}). Each of them corresponds to the maximum and minimum points of the load—displacement curve. The initiation values are directly related to the bonded joint fracture properties. In contrast, the arrest values include dynamic effects and should not be used to assess the quality of the adhesive joint.

Standard ISO 25217 (2009) suggests calculating both initiation (G_{ICi}) and arrest (G_{Ia}) values with the simple beam theory (SBT) method. The values of crack length needed to calculate G_{ICi} can be measured from the arrest bands of the fractured surface. However, in some cases they are difficult to observe (i.e. bonded joints with a resin layer or woven fabric adherents). If this is the case, it is preferable to use an optical device to video-record the crack front during the test and measure the value of the force and displacement at the initiation of fast crack propagation.

An alternative, and simpler, way to calculate G_{IC} is the area method. The fracture energy is obtained from the ratio between the area under the load—displacement curve and the fractured surface (specimen width multiplied by the crack extension). This method has the advantage of measuring the crack only at the beginning and at the end of the propagation. However, in situations of crack growth with stick-slip behaviour, a region of the curve is underestimated, as a consequence of the jumps in the load—displacement curve, and this results in a conservative value of G_{IC}.

5.3.3 Failure modes and fractography

Basically, there are three types of failure modes in bonded composite joints: adhesive failure at the interface between the adhesive and the adherent, cohesive failure inside the adhesive layer and substrate failure. In composite repairs, it is common to have different stacking sequences for each of the adherents, and that the repair patch consists of fabrics bonded to the structure by a wet lay-up operation. Therefore, the internal test procedure AITM1-0053, which was developed by Airbus, takes into account six failure modes:

- Substrate failure inside the pre-cured panel: delamination inside the pre-cured adherent.
- Inter adherent/peel ply resin failure: failure at the resin layer from the pre-cured peel ply.
- Adhesion failure: failure at the interface between the adhesive layer and the resin layer form the pre-cured peel ply.
- Cohesive failure: failure inside the adhesive.
- Adhesion failure in the wet–wet interphase: failure at the interface between the adhesive layer and the co-cured substrate in co-bonded joints.
- Substrate failure inside the co-cured panel: delamination inside the co-cured adherent.

Many research works have provided evidence of the influence of the failure mode on the strength of bonded joints. It has been found that different adhesives exhibit different failure modes despite having the same adherent (Parker, 1983), and that temperature and moisture strongly influence the locus of failure (Ashcroft et al., 2001). Therefore, bonded joint failure is still difficult to predict because, depending on the bonding method, adhesive used, temperature, moisture and other parameters, the failure modes are different.

With this in mind, fractography analysis can provide key information about the causes of failure in bonded joints, failure mode and the source of failure location. It is important that fracture surface analysis is conducted in the sequence outlined in Figure 5.8; otherwise vital information may be lost. Specifically, it is important to collect all the information from the surface before dissection.

Visual examination of the fractured surface provides information, without damaging the fracture surface, about the failure mode, crack growth direction, distribution of failure mode along the fracture surface and fracture pattern. The roughness of the fractured surfaces provides a qualitative indication of fracture toughness.

Microscopy is used to analyse fractured surfaces in detail. There are two main techniques: optical microscopy and scanning electron microscopy (SEM). Selection of fracture surface area in optical or SEM samples is very important, especially when a mixed type of failure occurs. Optical microscopy allows transverse and longitudinal sections of the specimen to be analysed, whereas SEM permits the observation of small portions of the fractured surfaces at high levels of magnification (see Figure 5.9). The sketch in Figure 5.10 shows the sample preparation process for optical microscopic and SEM observation of fractured specimens. The sample preparation stages for optical microscopic observation involve cutting, embedding and polishing, while a gold coating is applied to increase the imaging capability of samples for SEM observation.

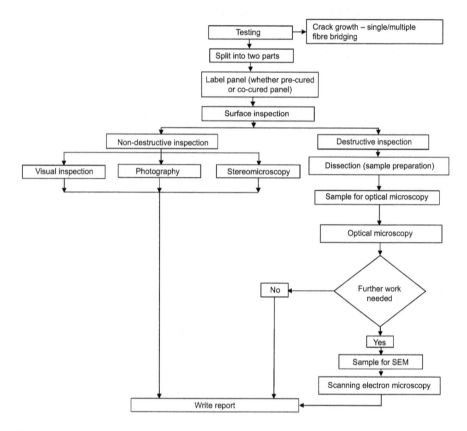

Figure 5.8 General procedure for fracture surface analysis.

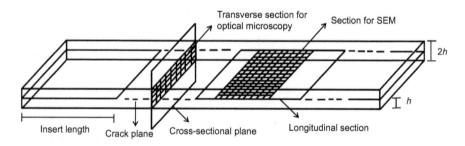

Figure 5.9 Configuration of DCB specimen and sample extraction for microscopic inspections.

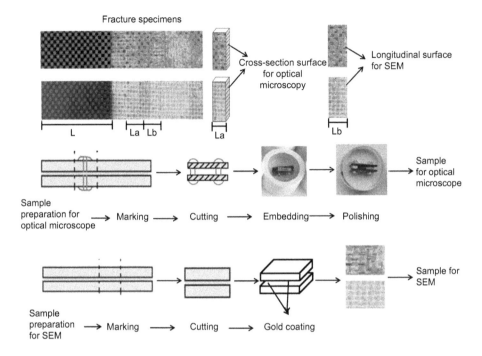

Figure 5.10 Sample preparation process for optical microscopic observation and SEM.

5.4 Fatigue testing

Most structural components of transport vehicles are subjected to cyclic loads during service. Fatigue tests aim to reproduce the behaviour of the bonded joint during service conditions. As in quasi-static tests, DCB specimens are used to analyse the quality of the bonded joint. The assessment of bonded joint fatigue behaviour consists of determining:

1. The number of cycles at which the crack starts to propagate (onset of crack propagation) for a given load level (usually expressed by means of the energy release rate).
2. In the event of crack propagation:
   a. The crack growth rate dependence on the energy release rate (crack growth rate curves).
   b. The value of the energy release rate at which the crack growth rate becomes null or practically immeasurable (threshold value of the energy release rate).

In spite of the practical importance of this topic in composite design, there is only one standard available, ASTM D6115-97 (2011), which focuses on determining onset curves for mode I fatigue loading. The standard was developed to analyse delamination in unidirectional composite specimens. Nevertheless, the methodology can be applied to bonded joints. Although currently there is no standard for fatigue delamination propagation in composites or bonded joints, an effort (round-robin test campaigns) to develop a standard test for delamination propagation is being made (Brunner, Murphy, & Pinter, 2009).

5.4.1 Test procedure

DCB fatigue tests are performed with the same specimens and load configurations as those in quasi-static tests. The test consists of opening the specimen arms cyclically, thus causing subcritical crack propagation in the adhesive. The required loads are lower than those applied in static tests. Loading can be introduced by controlling the load or the displacement. DCB fatigue tests are mainly displacement controlled because the propagation is stable in this case, and unstable under load control.

The loading cycle is sinusoidal in the tests, and it is defined by two parameters: the amplitude (R-ratio) and the level of the energy release rate (ERR) with respect to the critical fracture toughness (percentage or the energy release rate, %ERR). The R-ratio is defined by:

$$R = \frac{\sigma_{min}}{\sigma_{max}} \tag{5.2}$$

where σ_{max} and σ_{min} are the maximum and minimum values of the cyclic stress, respectively. For linear elasticity and small deflections ($\delta/a < 0.4$ being the crack length), the displacement ratio ($\delta_{min}/\delta_{max}$) is identical to the R-ratio (ASTM D6115-97, 2011):

$$R = \frac{\delta_{min}}{\delta_{max}} \tag{5.3}$$

The maximum displacement (δ_{max}) is related to %ERR. If the specimen geometry for the quasi-static DCB test is identical to those for the fatigue tests (same material and geometry: width, thickness and crack length), δ_{max} obeys:

$$\left(\frac{\delta_{max}}{\delta_{cr}}\right)^2 = \frac{G_{Imax}}{G_{IC}} = \%ERR \tag{5.4}$$

where δ_{cr} is the critical displacement for quasi-static delamination growth obtained from the quasi-static test (ASTM D6115-97, 2011).

If the value of ERR is equal to 1, it means that $G_{Imax} = G_{IC}$, and the crack would fracture statically during the first loading cycle. Therefore, in fatigue tests, %ERR < 1 (it generally varies between 0.1 and 0.9). With the R-ratio, it is a common practice to test at $R = 0.1$. This involves the cycle amplitude being almost equal to the maximum displacement (the worst load case in fatigue), and ensures that there is no contact between the specimen arms. Figure 5.11 shows the displacement cycles for three different test configurations.

In fatigue tests performed under displacement control, δ_{max} and δ_{min} are kept constant during the test. Depending on the load level, after a certain number of cycles the

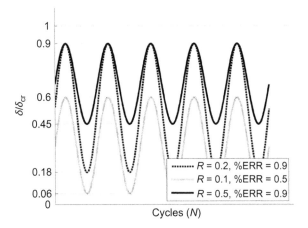

Figure 5.11 Displacement cycles of a DCB fatigue test.

crack eventually propagates at a crack growth rate, which decreases as the crack extends (the applied energy release rate decreases with the crack extension for a given displacement). Due to the stability of the test, the crack growth rate becomes immensurable after a certain crack extension (usually after a large number of cycles).

During the test, displacement, load and crack length are monitored. From these data a curve of the compliance against the number of cycles is obtained (see Figure 5.12). If the test is controlled by displacement the compliance increases rapidly during the first cycles and then tends to stabilize. The same occurs with the crack length. In the opposite case where the test is controlled by load, the compliance increases exponentially

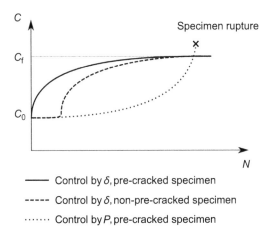

Figure 5.12 Evolution of the compliance against the number of cycles.

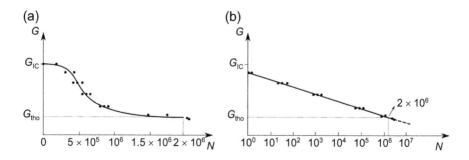

Figure 5.13 $G-N$ crack growth onset curve: (a) linear representation of N and (b) logarithmic representation of N.

with the number of cycles until the specimen breaks completely. Again, the same occurs with the crack length. Whereas, the evolution of the compliance (thus, crack length) is different if the specimen is pre-cracked than if the fatigue test is performed from the insert. If the crack propagation initiates at the insert, where there is a blunt crack tip, more cycles are needed to initiate propagation. The curve of the compliance has a flat region during the initial cycles.

The onset of crack propagation may be determined either by visual observation of crack growth or by an increase in specimen compliance (direct evidence of crack growth). The method based on monitoring compliance is preferred because of its objectivity. An initial value of the compliance is measured during the initial cycles (C_0). When compliance has increased by a certain percentage from the initial value, the test is stopped and the compliance and number of cycles recorded. ASTM D6115-97 (2011) suggests taking the points where compliance increases by a 1% or a 5% as onset values. Based on detailed visual observation at the crack tip, other studies consider that the onset of crack growth is when there is an increment of 2% in the compliance (Martin & Murri, 1990).

The onset curve is obtained by plotting the maximum energy release rate (G_{Imax}) against the number of cycles (see Figure 5.13). G_{Imax} is calculated from the maximum values of force and displacement (P_{max} and δ_{max}) from the first cycle ($N = 1$) with the following equation (ASTM D6115-97, 2011):

$$G_{Imax} = \frac{3P_{max}\delta_{max}}{2B(a_0 + |\Delta|_{av})} \tag{5.5}$$

where a_0 is the initial crack length (from the insert in ASTM D6115-97, 2011), B the specimen width, and $|\Delta|_{av}$ the average of Δ from the quasi-static tests (see Figure 5.5).

The test is repeated using different levels of the applied energy release rate. In most cases a level of energy release rate at which the crack does not propagate even for a large number of cycles is reached. This value is the onset threshold (G_{tho}). To obtain it, a maximum run-out of 2 million cycles is defined (ASTM D6671). If there is no crack propagation after this number of cycles, G_{Imax} is commonly considered to be under the threshold value. However, it has been observed that there is crack growth beyond the threshold defined according to the standard (Brunner et al., 2009). Therefore, threshold values in fatigue experiments cannot be identified accurately. A consensus in the number of cycles at which a fatigue test should be stopped is still to be achieved (Argüelles, Viña, Canteli, & Bonhomme, 2010; Argüelles, Viña, Canteli, Castrillo, & Bonhomme, 2008; Kenane, Azari, Benmedakhene, & Benzeggagh, 2011). What is commonly adopted, especially in propagation tests, is a crack growth rate limit. However there is a large disparity between the values proposed by the authors (Hojo, Tanaka, Gustafson, & Hayashi, 1987; Martin & Murri, 1990; Stelzer, Brunner, Argüelles, Murphy, & Pinter, 2012; Stelzer et al., 2014).

The analysis of the crack growth rate in fatigue propagation is done from the measurement of the crack length versus the number of cycles. The crack length is differentiated against the number of cycles required to obtain the crack growth rate ($dа/dN$). And the crack growth rate is represented against the energy release rate measured every certain number of cycles (see Figure 5.14).

The crack growth rate curve can be divided into three regions: the energy threshold where the crack growth becomes immeasurable (I); the region close to static failure (III); and in between, the linear propagation zone (II), well described by the Paris law (Paris & Erdogan, 1963; Paris, Gomez, & Anderson, 1961):

$$\frac{dC}{da} = K(G_{Imax})^m \qquad (5.6)$$

The key parameters obtained from the curve are the value of the energy release rate threshold (G_{th}) and the slope of the Paris law relationship (m). Design methodologies accounting for fatigue damage in bonded joints make use of one or both of these parameters.

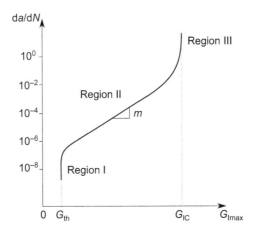

Figure 5.14 Typical crack growth rate curve.

5.4.2 Test set-up

Before conducting the fatigue test campaign, a static characterization must be performed. Two parameters are used from the quasi-static campaign to set up the fatigue test: G_{IC} to determine the percentage of energy release rate for the test (%ERR from Eqn (5.4)) and $|\Delta|_{av}$ (Figure 5.5) to calculate G_{Imax}. In the case that a specimen with the same configuration is used for both the static and fatigue test, the percentage of energy release rate is directly related to the maximum displacement; see Eqn (5.4). If the specimen configuration is different (i.e. the initial crack length is not the same), compliance has to be included when calculating maximum displacement (ASTM D6115-97, 2011).

Another test parameter to be taken into account is the test frequency. The duration of the test depends on this parameter. However, high frequencies can cause heating at the crack tip, in turn leading to unexpected results (Brunner et al., 2009). Test standards such as ASTM 6115 recommend testing at frequencies from 1 to 10 Hz. In general, mode I fatigue tests are performed at 5 Hz, this being a good compromise between the duration of the test and the capacities of the acquisition systems of current testing machines.

Finally, in fatigue tests involving propagation, it is important to predict the extension of crack propagation during the test. For example, in order to achieve the highest precision when measuring a crack length, the crack length should be as large as possible. Based on linear elastic fracture mechanics (LEFM), there is a proportional relationship between the crack growth extension ($a_f - a_0$) and the initial crack length (a_0):

$$a_f - a_0 = a_0 \left[\left(\frac{G_{Imax}^0}{G_{Ith}} \right)^{1/4} - 1 \right] \tag{5.7}$$

where G_{Imax}^0 is the value of the energy release rate at the first cycle and G_{Ith} the estimated threshold value at which the crack will practically arrest, a_f. According to Eqn (5.7), a large initial crack length, a_0, has a positive effect as it results in large crack propagation ($a_f - a_0$). However, the amplitude of the displacement ($d_{max} - d_{min}$) required to obtain a given energy release rate has to be increased too, which might not be attainable with the testing machine. Therefore, a compromise between crack length extension and displacement amplitude, which in most cases depends on the capacity of the testing machine, should be established.

5.4.3 Dynamic compliance

The standard definition of the compliance of any elastic structural component is the ratio between displacement and load, $C = \delta/P$ (see Figure 5.15). Assuming perfect linear behaviour of the material (curve 1), and that the load—displacement curve passes through the origin, the specimen compliance is the inverse of the slope of the load—displacement curve. This slope remains constant until any damage occurs on the material (i.e. crack growth), when the slope then changes. However, the geometry and manufacture of the fixture tools usually introduce non-linearities (see Figure 5.15; curve 2) or even initial offsets at small displacements (curve 3). The slope of the

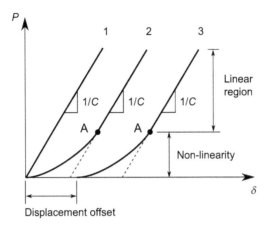

Figure 5.15 Effects of the initial non-linearity and displacements offsets on the load–displacement curve. Curve 1: a perfect linear behaviour of the material is assumed. Curve 2: the load–displacement curve has an initial non-linearity. Curve 3: there is an initial non-linearity and displacement offset. Point A determines the beginning of the linear region of the curve.

load–displacement curve then increases from the initial zone (related to the load introduction system rather than to the specimen) until it reaches a constant value corresponding to the stiffness/compliance of the specimen tested (beyond point A). Plays between pieces of the fixture, an inaccurate load introduction on the specimen or the material properties of the samples are the principal causes of these non-linearities (Blackman, Kinloch, Paraschi, & Teo, 2003).

Due to these non-linearities, the traditional definition of compliance during the fatigue test generates problems in the interpretation of compliance and its relation to the crack length (Renart, Vicens, Budhe, Costa, & Mayugo, submitted for publication). In a fatigue test controlled by displacement, the displacement is usually sinusoidal and the load response is also sinusoidal. In Figure 5.16, the displacements δ_A and δ_B correspond to points A and B respectively in the load–displacement curve. If there is a non-linearity at the beginning of the curve, the slope varies from point A to point B. Thus, compliance during a fatigue test does not remain constant and varies from a minimum value, C_A, to a maximum, C_B, being $C_A \neq C_B$.

To avoid the errors induced by the initial non-linearity, compliance must be determined at the linear region of the load–displacement curve. The calculated compliance (C^*) is the slope of the line that connects points A and B and corresponds to the ratio of displacement and load amplitude (δ_a and P_a).

Normalized procedures for obtaining fatigue onset curves require the test being stopped periodically in order to evaluate the specimen compliance using static tests on the lineal zone of the material (Argüelles et al., 2008; Argüelles et al., 2011; ASTM D6115, 2011). Therefore, accuracy in determining the onset point depends on the frequency of interruptions in the fatigue tests. Unavoidably, the results exhibit a large scatter.

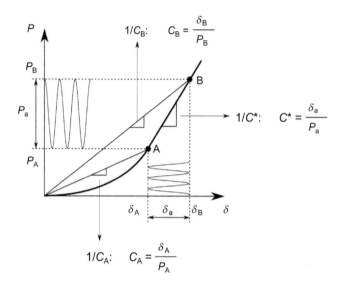

Figure 5.16 Dynamic compliance obtained from the linear region of the load–displacement curve.

A continuous representation of the compliance against the number of cycles is obtained if the data of the whole fatigue cycle (load and displacement) is recorded at a higher frequency than the test. At least between 15 and 20 points per cycle are required (i.e. if the fatigue test is performed at 5 Hz, the data acquisition system should be above 75 or 100 Hz). The large data files that are obtained (several Gbytes) have to be processed once the test is finished. Therefore, the method is not practical in determining the onset curves because the data cannot be processed in real time, and so it is not possible to know when compliance has increased by a certain percentage. As a consequence, the tests have to be performed until the run-out number of cycles is reached.

In order to reduce testing time, a new method of monitoring compliance in real time was developed (Renart et al., submitted for publication). This method is based on the internal calculations of the testing machine that provide the value of the dynamic compliance at every cycle. Therefore, a continuous curve of $C(N)$ is acquired and the onset of crack propagation can be precisely determined in one cycle. This method means being able to stop the test once onset criterion is satisfied, i.e. without the need to reach the maximum number of cycles.

5.4.4 Compliance calibration as an alternative to the visual inspection of the crack length

Crack growth rate curves are obtained by measuring crack length during the fatigue test. Crack tip positions are commonly determined by means of visual methods such as traveling cameras or microscopes (Argüelles et al., 2008; Hojo, Matsuda, Tanaka, Ochiai, & Murakami, 2006; Stelzer et al., 2012). Such methods introduce two main uncertainties. On the one hand, visually monitoring the crack length does not constitute a robust method, as it is considered 'operator dependent' (Brunner et al., 2009). On the other hand, the fact that the crack length is measured at the edge of the coupons adds imprecision to the measurement because the crack front is not straight (Sans, Stutz, Renart, Mayugo, & Botsis, 2012).

When using optical devices, the crack length is measured every certain number of cycles, which introduces scattering into the results. This scattering increases with the differentiation of the crack length against the number of cycles. In addition, as the crack growth rate decreases during the test, the crack front needs to be determined with even greater precision as the test evolves in order to produce reliable data near the threshold.

An alternative approach relies on the estimation of the crack length by means of compliance. The relationship between the compliance and the crack length should be calibrated before or after the test (Sans et al., 2013). This method is widely used in mode II 3-point end-notched flexure tests (3ENF) (Davies, 1992). The compliance calibration consists of opening the specimen arms at different crack lengths (without damaging the specimen) to measure the compliance. This results in a representation of compliance against the crack length. This representation is equivalent to the linear relationship between $C^{1/3}$ and the crack length that is used for the MBT data reduction method in quasi-static tests. In addition to the intersection of the curve with the abscissa (Δ), the slope of the curve (n) is also measured (see Figure 5.5). Using Eqn (5.8), a value of crack length (a_i) is obtained from each value of compliance (C_i). Therefore, a continuous curve of $a(N)$ can be represented if $C(N)$ is measured continuously:

$$a_i = \frac{C_i^{\frac{1}{3}} - \Delta}{n} \qquad (5.8)$$

5.5 Effect of waviness in crack growth rate curves

The crack growth of a bonded joint under cyclic loading is a subcritical growth, so it is mostly a smooth growth (stick-slip is not likely to occur). However, if the bond line thickness is not constant, the crack growth rate curve may exhibit undulations in the region of linear propagation (region II).

Figure 5.17 shows a da/dN curve of a co-bonded joint between a pre-cured adherent and a wet lay-up patch for repair purposes. The adherents were bonded with a structural adhesive film. Compliance was monitored in real time and the crack length was estimated from a compliance calibration conducted before the fatigue tests. Therefore, continuous curves for both compliance and crack length were obtained. In Figure 5.17, the linear propagation zone presents undulations.

To analyse the response, it is more informative to plot the crack growth rate against the crack length (Figure 5.18). This plot reveals the periodic character of the undulation, which appears every 2 mm. The fact that the undulation is periodic suggests that it is caused by the geometry of the adhesive joint.

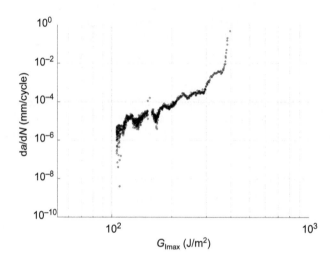

Figure 5.17 Crack growth rate curve of a bonded specimen with variable adhesive thickness.

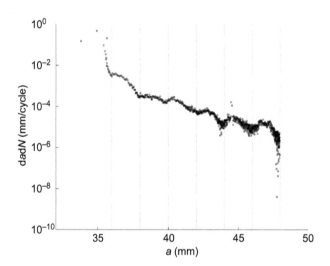

Figure 5.18 Crack growth rate against crack length.

Figure 5.19 shows the transverse section of the specimen as observed with the optical microscope. The image is taken at mid-plane of the bonded joint. The central line corresponds to the adhesive film; the pre-cured panel is placed on top of the adhesive film and the co-cured adherent underneath it. Both pre-cured and co-cured

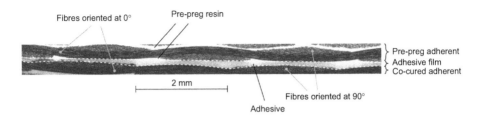

Figure 5.19 Transverse section of the specimen at the adhesive layer.

panels are woven fabrics. The resin of the pre-cured adherent is a lighter colour than the adhesive or the resin layer of the co-cured adherent. The surface of the pre-cured adherent was essentially flat resulting from the use of a peel-ply fabric. However, the bond line thickness varies with a wavy pattern that is repeated every 2 mm. This variation is caused by the waviness of the fabric used for the repair patch. During the co-curing process the adhesive flows and adapts to the fabric surfaces.

5.6 Design and simulation approaches

The design methodologies of bonded composite joints are strongly related to the available fatigue crack growth experimental data. As has been described in the previous sections, the experimental characterization of fatigue crack growth can be divided into two different tests types (Bak, Sarrado, Turon, & Costa, submitted for publication): crack-onset and crack-propagation tests. Crack-onset tests in FRP attempt to determine the number of cycles needed for a pre-existing flaw to begin to grow, whereas in crack propagation tests the crack length is monitored for different fatigue loads, so that a crack growth rate curve as a function of the applied load can be obtained. Based on the previous categorization of fatigue experimental tests, the design of bonded composite joints subjected to fatigue loads can be carried out by following two different approaches: a no-growth strategy, for which no flaw in the structure is allowed to grow; and a controlled crack growth strategy, for which crack growth might occur as long as it can be detected in the established inspection intervals of the structure and never reaches a critical length that would impair the load carrying capability of the structure.

The no-growth design strategy is based on limiting the load that is withstood by a structure to prevent any detectable flaw from growing. The no-growth methodologies available in the literature are summarized in the three following steps described in Composites Material Handbook 17 (CMH-17, 2012):

- Assumption of an existing initial flaw and estimation of the maximum energy release rate G_{max} to which the crack is submitted. The maximum energy release rate can be replaced by other fracture mechanics parameters, e.g. ΔG, K_{max}, ΔK, J_{max} or ΔJ, depending on the data reduction method behind the experimental results. To obtain this fracture mechanics parameter, some simulation technique is needed, as detailed later in this section.

- Determination of the maximum allowed energy release rate (or equivalent fracture mechanics parameter) for a given number of cycles. The number of cycles is estimated during the design process and the maximum allowed energy release rate is determined from the experimental crack-onset curves for this material.
- The maximum energy release rate obtained from the first step must be below the maximum energy release rate allowed in the second step. In that case, crack no-growth is assured for the given number of cycles.

For the first step of this methodology, some kind of technique must be used to determine the fracture mechanics parameter. LEFM-based techniques, such as the virtual crack closure technique (VCCT) (Krueger, 2002), are usually chosen because of their simplicity, ease of implementation and their similarity to the experimental data reduction methods, which are based on LEFM. However, although LEFM-based techniques have been shown to be accurate for delamination problems with small fracture process zones (FPZs), their accuracy has never been challenged in the presence of a large FPZ; as is the case for bonded joints where large plasticity and damage zones are usually present. In such cases, non-linear fracture mechanics parameters, such as the J integral, might be more suitable.

In the second step of the no-growth methodology, the crack-onset curves of the material need to be obtained. Crack-onset curves are equivalent to Whöler curves in metallic materials, also known as $S-N$ curves. In fact, the only standard available for fatigue delamination in FRP, the ASTM D6115-97 (2011), addresses the procedure to obtain the crack-onset curves. In the procedure described in the standard, a certain load level is applied and specimen compliance is monitored. The number of cycles required to ensure a significant change in the compliance of the structure is taken as the number of cycles to crack onset at the given load level. The crack-onset curves of Figure 5.13 can be generated by repeating the previous procedure for different load levels. It is worth highlighting that, although the onset curves are supposed to be a material property, the fact that they are obtained by monitoring the structure compliance makes them dependent on the geometry of the component. Besides that, the crack-onset curves also depend on the load ratio or the mixed-mode ratio (Hojo et al. 1987; Martin & Murri, 1990; O'brien, 1984), so the required experimental campaign is usually extensive and costly. Finally, it is also worth taking into account that the data reduction method proposed in the standard is based on LEFM, whose expected validity in cases with a large FPZ in bonded joints has not been challenged in the literature so far.

When whole-crack-onset curves of a material are not available, a conservative approach is to design the structure to be below the fatigue threshold load, so that crack growth is totally avoided. However, fatigue thresholds in FRP are low in most cases (Asp, Sjögren, & Greenhalgh, 2001), so this strategy becomes too conservative and leads to large structure oversizing.

The controlled crack growth design strategy allows a certain amount of subcritical crack growth provided that it can be detected and fixed within the established inspection intervals, so it does not endanger the performance of the structure. This design strategy relies on the experimental fatigue crack growth rate curve and needs a simulation model to estimate the crack growth rate for a given structure. The central region of the crack growth rate curve in Figure 5.14 is modelled either with Paris' law (Paris & Erdogan, 1963; Paris et al., 1961) or some of its modifications as proposed in the literature (Bak et al., submitted for publication).

The simulation methods available in the literature are commonly implemented in a finite element framework, and can be divided into two types: (1) LEFM-based methods, such as VCCT (Krueger, 2002) and (2) cohesive zone models (Alfano & Crisfield, 2001; Park & Paulino, 2013; Turon, Camanho, Costa, & Dávila, 2006; Tvergaard, 2001).

In LEFM-based methods, an existing crack is assumed, its energy release rate or equivalent fracture mechanics parameter is obtained and it is directly introduced into Paris' law to obtain an estimation of the crack growth rate (Krueger, 2011). In fatigue cohesive zone models, Paris' law is implicitly implemented into the formulation of the cohesive elements, which are placed at the interface along which crack growth might occur. With this simulation method, no existing crack is needed and therefore this methodology is also suitable in simulating crack initiation (May & Hallett, 2010; Serebrinsky & Ortiz, 2005). However, the models currently available have been calibrated using propagation data. Given that fatigue crack propagation is in nature different to fatigue crack initiation, some new experimental data reduction method to feed the models would be needed to properly account for fatigue crack initiation. The application range for cohesive zone models is also wider than LEFM-based techniques, given that they can account for a large FPZ. Although this particular LEFM limitation might not be relevant in delamination problems, it becomes critical in bonded-joint analysis, where plasticity and damage regions are usually large. Cohesive zone modelling also allows the simulation of the thickness of the adhesive (Huespe, Needleman, Oliver, & Sanchez, 2009) and of other dissipative phenomena. At the same time, the simulation of large fracture process zones implies some challenges under mixed-mode conditions that are still an open research topic (Turon, Camanho, Costa, & Renart, 2010; Sarrado, Turon, Renart, & Urresti, 2012), even under quasi-static loads.

From the two design strategies presented in the current section, the no-growth strategy has been traditionally adopted thanks to its simplicity and to the lack of alternative models and methods able to simulate fatigue crack growth. The models currently available have not yet been extensively verified and design best practice guidelines using these models are still missing. However, the no-growth strategy is very conservative and the recent developments in the framework of fatigue cohesive zone modelling are raising the interest of both research institutions and industry, as these could lead to the development of more reliable models that would, by applying controlled crack growth design strategies, eventually lead to more optimized structures.

5.7 Conclusions

The study of the fatigue behaviour of composite bonded joints often makes use of the methodologies established for the study of delamination but without any additional refinement or caution. However, in this chapter, it has been shown that the experimental determination and the simulation of the fatigue behaviour of bonded joints between composites have particularities that distinguish them from fatigue delamination.

On the one hand, the type of manufacturing process used to produce the bonded joint results in different potential locations where failure could occur. In addition, it has been shown that the fracture energy of the joint must be considered in relationship to the failure mechanism that caused it, in order to have a proper understanding of the phenomena involved. The procedures and methods to identify the failure modes have been presented.

On the other hand, delamination growth usually involves a failure-process zone (FPZ) that is small enough to be neglected in most cases (data reduction methods of experimental data and simulation approaches). However, the crack tip area in bonded joints includes zones with plastic deformation and/or damage, thus leading to a larger FPZ that cannot be ignored. This underlines the need to shift the framework where simulation models are developed from LEFM in delamination to fracture mechanics with plasticity and damage for bonded joints.

Test procedures for fatigue testing in bonded joints have not yet been standardized. The only available related standard concerns the measurement of fatigue-onset curves for delamination. There are several experimental aspects that influence the reliability of the results and these have been highlighted in this chapter. The determination of the crack length (needed to calculate crack growth rate) by continuously monitoring the dynamic compliance ensures more robust and less scattered results than making use of visual observation or methods based on the traditional compliance.

In any case, the microstructural complexity of the bonded joints, as compared to delamination (more interfaces, more materials with different properties involved), causes the fatigue phenomenology to be more sensitive to environmental factors and more difficult to interpret. The case presented with its undulations in the crack growth rate caused by the periodicity of the bond line thickness due to the geometry of the fabric in the co-cured adherent is one such example.

Acknowledgements

This work has been carried out within the framework of project MAT2012-37552-C03-03, funded by the Spanish Ministry of Economy and Competitiveness (MINECO).

References

Alfano, G., & Crisfield, M. A. (2001). Finite element interface models for the delamination analysis of laminated composites: mechanical and computational issues. *International Journal for Numerical Methods in Engineering, 50*(7), 1701−1736.

Argüelles, A., Viña, J., Canteli, A. F., & Bonhomme, J. (2010). Fatigue delamination, initiation, and growth, under mode I and II of fracture in a carbon-fiber epoxy composite. *Polymer Composites*, 700−706.

Argüelles, A., Viña, J., Canteli, A. F., Castrillo, M. A., & Bonhomme, J. (2008). Interlaminar crack initiation and growth rate in a carbon-fibre epoxy composite under mode-I fatigue loading. *Composites Science and Technology, 68*, 2325–2331.

Argüelles, A., Viña, J., Canteli, A. F., Viña, I., & Bonhomme, J. (2011). Influence of the matrix constituent on mode I and mode II delamination toughness in fiber-reinforced polymer composites under cyclic fatigue. *Mechanics of Materials, 43*, 62–67.

Ashcroft, I. A., Hughes, D. J., & Shaw, S. J. (2001). Mode I fracture of epoxy bonded composite joints: 1. Quasi-static loading. *International Journal of Adhesion and Adhesives, 21*(2), 87–99.

Asp, L. E., Sjögren, A., & Greenhalgh, E. S. (2001). Delamination growth and thresholds in a carbon/epoxy composite under fatigue loading. *Journal of Composites Technology and Research, 23*, 55–68.

ASTM D6115-97. (2011). *Standard test method for mode I fatigue delamination growth onset of unidirectional fiber-reinforced polymer matrix composites, annual book of ASTM standards*. West Conshohocken, PA: ASTM International. Ch. 15.03.

ASTM D5528-13. (2013). *Standard test method for mode I interlaminar fracture toughness of unidirectional fiber reinforced polymer matrix composites, annual book of ASTM Standards*. West Conshohocken, PA: ASTM International. Ch. 15.03.

Bak, B. L. V., Sarrado, C., Turon, A., & Costa, J. (submitted for publication). Delamination under fatigue loads in composite laminates: a review on the observed phenomenology and computational design procedures, *Applied Mechanics Reviews, AMR-13-1097*, http://dx.doi.org/10.1115/1.4027647.

Blackman, B. R. K., & Kinloch, A. J. (2001). *Fracture tests for structural adhesive joints, fracture mechanics testing methods for polymers, adhesives and composites*. Elsevier Science.

Blackman, B. R. K., Kinloch, A. J., Paraschi, M., & Teo, W. S. (2003). Measuring the mode I adhesive fracture energy, G_{IC}, of structural adhesive joints: the results of an international round-robin. *International Journal of Adhesion and Adhesives, 23*, 293–305.

Brunner, A. J., Murphy, N., & Pinter, G. (2009). Development of a standardized procedure for the characterization of interlaminar delamination propagation in advanced composites under fatigue mode I loading conditions. *Engineering Fracture Mechanics, 76*, 2678–2689.

BS 7991. (2001). *Determination of the mode I adhesive fracture energy, G_{IC}, of structural adhesives using the double cantilever beam (DCB) and tapered double cantilever beam (TDCB) specimens*. British Standards.

Caminero, M. A., Pavlopoulou, S., Lopez-Pedrosa, M., Nicolaisson, B. G., Pinna, C., & Soutis, C. (2013). Analysis of adhesively bonded repairs in composites: damage detection and prognosis. *Composite Structures, 95*, 500–517.

CMH-17. (2012). *Polymer matrix composites: materials usage, design, and analysis*. In *Composite materials handbook* (Vol. 3). SAE International.

Davies, P. (1992). *Protocols for interlaminar fracture testing of composites*. Plouzane, France: Marine Materials Lab.

Hart-Smith, L. J. (1999). A peel-type durability test coupon to assess interfaces in bonded, co-bonded, and co-cured composite structures. *International Journal of Adhesion and Adhesives, 19*(2–3), 181–191.

Hojo, M., Matsuda, S., Tanaka, M., Ochiai, S., & Murakami, A. (2006). Mode I delamination fatigue properties of interlayer-toughened CF/epoxy laminates. *Composites Science and Technology, 66*(5), 665–675.

Hojo, M., Tanaka, K., Gustafson, C. G., & Hayashi, R. (1987). Effect of stress ratio on near-threshold propagation of delamination fatigue cracks in unidirectional CFRP. *Composites Science and Technology, 29,* 273−292.
Huespe, A. E., Needleman, A., Oliver, J., & Sanchez, P. J. (2009). A finite thickness band method for ductile fracture analysis. *International Journal of Plasticity, 25,* 2349−2365.
ISO 15024. (2001). *Fibre-reinforced plastic composites − Determination of mode I interlaminar fracture toughness, G_{IC}, for unidirectionally reinforced materials.* ISO International Standards.
ISO 25217. (2009). *Adhesives − determination of the mode 1 adhesive fracture energy of structural adhesive joints using double cantilever beam and tapered double cantilever beam specimens.* ISO International Standards.
Kanerva, M., & Saarela, O. (2013). The peel ply surface treatment for adhesive bonding of composites: a review. *International Journal of Adhesion and Adhesives, 43,* 60−69.
Kenane, M., Azari, Z., Benmedakhene, S., & Benzeggagh, M. L. (2011). Experimental development of fatigue delamination threshold criterion. *Composites: Part B Engineering, 42,* 367−375.
Kim, K. S., Yoo, J. S., Yi, Y. M., & Kim, C. G. (2006). Failure mode and strength of unidirectional composite single lap bonded joints different bonding methods. *Composite Structures, 72,* 477−485.
Krueger, R. (2002). *The virtual approach crack closure technique: History, approach and applications.* NASA/CR-2002−211628.
Krueger, R. (2011). *Development of benchmark examples for static delamination propagation and fatigue growth predictions.* Technical Report, Langley Research Center.
Mall, S., & Ramamurthy, G. (1989). Effect of bond thickness on fracture and fatigue strength of adhesively bonded joints. *International Journal of Adhesion and Adhesives, 9,* 33−37.
Martin, R. H., & Murri, G. B. (1990). Characterization of mode I and mode II delamination growth and thresholds in AS4/PEEK composites, composite materials: testing and design, Ninth volume. *ASTM STP, 1059,* 251−270.
Marzi, S., Biel, A., & Stigh, U. (2011). On experimental methods to investigate the effect of layer thickness on the fracture behavior of adhesively bonded joints. *International Journal of Adhesion and Adhesives, 31,* 840−850.
Matthews, F. L., & Tester, T. T. (1985). The influence of stacking sequence on the strength of bonded CFRP single lap joints. *International Journal of Adhesion and Adhesives, 5,* 13−18.
May, M., & Hallett, S. R. (2010). A combined model for initiation and propagation of damage under fatigue loading for cohesive interface elements. *Composites Part A: Applied Science and Manufacturing, 41*(12), 1787−1796.
O'Brien, T. K. (1984). Mixed-mode strain energy release rate effects on edge delamination of composites. *Effects of defects on composite materials* (pp. 125−142). ASTM STP 836.
Paris, P., & Erdogan, F. (1963). A critical analysis of crack propagation laws. *Journal of Basic Engineering, 85*(4), 528−533.
Paris, P., Gomez, M., & Anderson, W. (1961). A rational analytical theory of fatigue. *The Trend in Engineering, 13,* 9−14.
Park, K., & Paulino, G. H. (2013). Cohesive zone models: a critical review of traction-separation relationships across fracture surfaces. *Applied Mechanics Reviews, 64*(6), 061002.
Parker, B. M. (1983). The effect of composite pre-bond moisture on adhesive-bonded CFRP-CFRP joints. *Composites, 14,* 226−232.
Renart, J., Vicens, P., Budhe, S., Costa, J., Mayugo, J. A. (submitted for publication). Compliance real time monitoring in delamination fatigue tests.

Sans, D., Stutz, S., Renart, J., Mayugo, J. A., & Botsis, J. (2012). Crack tip identification with long FBG sensors in mixed-mode delamination. *Composite Structures*, *94*(9), 2879−2887.

Sarrado, C., Turon, A., Renart, J., & Urresti, I. (2012). Assessment of energy dissipation during mixed-mode delamination growth using cohesive zone models. *Composites Part A: Applied Science and Manufacturing*, *43*(11), 2128−2136.

Serebrinsky, S., & Ortiz, M. (2005). A hysteretic cohesive-law model of fatigue-crack nucleation. *Scripta Materialia*, *53*(10), 1193−1196.

Song, M. G., Kweon, J. H., Choi, J. H., Byun, J. H., Song, M. H., Shin, S. J., et al. (2010). Effects of manufacturing methods on the shear strength of composite single−lap bonded joints. *Composite Structures*, *92*, 2194−2202.

Stelzer, S., Brunner, A. J., Argüelles, A., Murphy, N., Cano, G. M., & Pinter, G. (2014). Mode I delamination fatigue crack growth in unidirectional fiber reinforced composites: results from ESIS TC4 round-robins. *Engineering Fracture Mechanics*, *116*, 92−107.

Stelzer, S., Brunner, A. J., Argüelles, A., Murphy, N., & Pinter, G. (2012). Mode I delamination fatigue crack growth in unidirectional fiber reinforced composites: development of a standardized test procedure. *Composites Science and Technology*, *72*(10), 1102−1107.

Turon, A., Camanho, P. P., Costa, J., & Dávila, C. G. (2006). A damage model for the simulation of delamination in advanced composites under variable-mode loading. *Mechanics of Materials*, *38*(11), 1072−1089.

Turon, A., Camanho, P. P., Costa, J., & Renart, J. (2010). Accurate simulation of delamination growth under mixed-mode loading using cohesive elements: definition of interlaminar strengths and elastic stiffness. *Composite Structures*, *92*(8), 1857−1864.

Tvergaard, V. (2001). Crack growth predictions by cohesive zone model for ductile fracture. *Journal of the Mechanics and Physics of Solids*, *49*, 2191−2207.

Whittingham, B., Baker, A. A., Harman, A., & Bitton, D. (2009). Micrographic studies on adhesively bonded scarf repairs to thick composite aircraft structure. *Composite Part A: Applied Science and Manufacturing*, *40*(9), 1419−1432.

Mode I fatigue and fracture behavior of adhesively-bonded pultruded glass fiber-reinforced polymer (GFRP) composite joints

6

A.P. Vassilopoulos, M. Shahverdi, T. Keller
École Polytechnique Fédérale de Lausanne (EPFL), Lausanne, Switzerland

6.1 Introduction

The fatigue behavior and life time of composite materials and structural components depend on the loading patterns applied. For materials used for structures that function in the open air, these loading patterns are usually of a stochastic nature and can be simulated by a time series of variable amplitude and mean values. Nevertheless, although this is what happens in reality, it is impossible to experimentally investigate the fatigue behavior of any material of interest under all possible loading conditions. Therefore, standard experiments are performed in laboratories, and appropriate models are established to simulate the fatigue behavior of the examined materials and structural components.

The most common way of representing the fatigue data of composite materials and structural components, and for design based on phenomenological modeling concepts, is the $S-N$ curve. When the design is based on micromechanics modeling and a crack or cracks develop inside the material, the matrix crack density or delamination are normally conceived as being an acceptable damage metric.

During the loading of structural joints, a crack or cracks initiate naturally and propagate along the weakest path within the component. This uncontrollable phenomenon led scientists to focus on the investigation of fracture mechanics joints, i.e., precracked joints to produce pure Mode I stable crack propagation, e.g., the double cantilever beam (DCB), or Mode II fracture, e.g., the end-notched flexure (ENF) beam (Ashcroft, Hughes, & Shaw, 2001; Blackman, Hadavinia, Kinloch, Paraschi, & Williams, 2003; Hadavinia, Kinloch, Little, & Taylor, 2003; Shahverdi, Vassilopoulos, & Keller, 2012a). These tests resulted in the determination of fracture mechanics data, i.e., crack propagation rate da/dN and strain energy release rate G_{max} in a fatigue cycle under specific applied load F_{max}. When structural joints are used instead of fracture joints, the situation becomes far more complicated (Sarfaraz, Vassilopoulos, & Keller, 2011; Sarfaraz, Vassilopoulos, & Keller, 2013a, 2013b; Zhang, Vassilopoulos, & Keller, 2010). In this case, the failure mode is not pure Mode I or Mode II but a mixed mode, Mode I/II. The proportion of each failure mode in this mixed-mode failure depends on the material, joint geometry, type of loading, and environmental conditions. The strain energy release rate calculated for this mixed-mode failure is designated total strain energy release rate (Mall, Ramamurthy, & Rezaizdeh, 1987) and is assumed to be equal to the sum of G_I and G_{II}, $G_{tot} = G_I + G_{II}$. Other studies present mixed-mode failure criteria for adhesively-bonded composite—composite joints based on measurements of Mode I (DCB), Mode II (ENF), and mixed Mode I/II fracture parameters (Ducept, Davies, & Gamby, 2000; Shahverdi, Vassilopoulos, & Keller, 2013a). Previous studies have revealed that it is possible to establish a sigmoid relationship between da/dN and ΔG or G_{max}. Crack propagation rates decrease to very low values as the strain energy release rate approaches a limit, G_{th}, and increase significantly as the strain energy release rate approaches a critical value, G_c, which is designated critical strain energy release rate or fracture energy.

Fracture mechanics theory has been used by a number of researchers to study the fatigue behavior of structural joints. Analytical solutions and, more frequently, finite element models were used to correlate experimentally determined fracture mechanics data with the fatigue life of the examined material configurations. A review of previous works led to the conclusion that fracture mechanics data should be directly obtained by using structural joints rather than the standardized specimens. For example, adhesively-bonded composite-metal double-lap joints (Cheuk, Tong, Wang, Baker, & Chalkley, 2002), or carbon/epoxy laminates bonded with epoxy adhesive to form single-lap joints (Quaresimin & Ricotta, 2006), or pultruded adhesively-bonded glass fiber—reinforced polymer (GFRP) joints (Sarfaraz et al., 2011; Sarfaraz, Vassilopoulos, & Keller, 2012; Zhang, Vassilopoulos, & Keller, 2008; Zhang et al., 2010) were directly tested to obtain the necessary fracture mechanics data. The basic idea behind such an approach was to obtain fracture mechanics data that are more representative of the real structure rather than determining fracture mechanics data from DCB and/or ENF specimens. Furthermore, during the testing of these structural joints, crack initiation may continue even up to 70% of the fatigue life (Quaresimin & Ricotta, 2006), and in some cases even further (Zhang et al., 2008), imposing the need to incorporate this phase into the development of predicting models.

Together with crack development, other measures such as cyclic stress, cyclic strain, remaining stiffness, and remaining strength were used by scientists as damage metrics to quantify the phenomenon of the fatigue damage to FRP materials and structures and to predict fatigue life. Among the proposed damage metrics, the two that attracted the most attention from researchers for the prediction of adhesively-bonded joint fatigue life were cyclic stress and the corresponding stress-life theory and fatigue crack growth (FCG) with the related fracture mechanics theory. The first is based on the establishment of reliable $S-N$ curves, and the modeling of fatigue life is performed using a number of macroscopic fatigue data, whereas the second is based on accurate measurement of the developed fatigue crack and its correlation with a fracture mechanics parameter such as strain energy release rate.

6.1.1 Experimental characterization

The literature comprises a vast number of publications (Brunner, Murphy, & Pinter, 2009; Kawai & Kato, 2006; Philippidis & Vassilopoulos, 2002) on the aforementioned topics. For composite laminates, the $S-N$ curves are usually derived under given loading conditions to model the constant amplitude fatigue behavior of the examined materials. A significant effect of the stress ratio, $R = \sigma_{min}/\sigma_{max}$, on fatigue life has been reported in the literature and extensively investigated to establish theoretical models (Beheshty & Harris, 1998; Kawai & Koizumi, 2007; Vassilopoulos, Manshadi, & Keller, 2010b) for the subsequent prediction of fatigue life under more complicated loading conditions such as block and variable amplitude loading. A literature review revealed that the behavior of composite laminates is strongly affected by the stress ratio, in a consistent way, however, which allows the derivation of universal formulations that can simulate this effect independently of the examined material (Vassilopoulos, Manshadi, & Keller, 2010a).

Extensive research efforts were also devoted to the investigation of the FCG under Mode I (Costa, Mahdi, Vicens, Blanco, & Rodríguez-Bellido, 2009; Hojo, Matsuda, Tanaka, Ochiai, & Murakami, 2006; Shivakumar, Chen, Abali, Le, & Davis, 2006; Walls, Bao, & Zok, 1993), Mode II (Argüelles, Viña, Canteli, & Bonhomme, 2010; Tanaka & Tanaka, 1997), and mixed-mode fracture (Naghipour, Bartscha, & Voggenreitera, 2011; Shahverdi et al., 2013a; Zhang, Vassilopoulos, & Keller, 2009) of composite joints under fatigue loading. Instead of the $S-N$ curves, FCG curves are used to represent the fatigue behavior and also to provide information concerning the developed damage in terms of crack or cracks that propagate during fatigue loading. The FCG curves are plots of the stress intensity factor, SIF or K (Andersons, Hojo, & Ochiai, 2004; Paris, Gomez, & Anderson, 1961; Zheng & Powell, 1999) or the strain energy release rate, SERR or G (Mall, Yun, & Kochhar, 1989; Martin & Murri, 1990, pp. 251−270; Russell & Street, 1988; Wilkins, Eisenmann, Camin, Margolis, & Benson, 1982) versus the crack propagation rate, da/dN, usually on logarithmic axes. Although the stress intensity factor is used mainly for the derivation of FCG curves for metals/alloys, the strain energy release rate is preferred for composite materials, because for the calculation of G there is no need to directly calculate the local stress field close to the crack tip, which is difficult if not impossible for composite laminates (Rans, Alderliesten, & Benedictus, 2011). The strain energy release rate has been used by many researchers (Argüelles et al., 2010; Brunner et al., 2009; Costa et al., 2009; Hojo et al., 2006; Naghipour et al., 2011; Shivakumar et al., 2006; Tanaka & Tanaka, 1997; Zhang et al., 2009). A schematic representation of FCG curves in shown in Figure 6.1. Any strain energy release rate cyclic parameter can be used for representation of the results, with the maximum cyclic, G_{max}, and cyclic range, $\Delta G = G_{max} - G_{min}$, being the most common. The fracture behavior exhibited usually dictates which of the two parameters is the more appropriate for an accurate description of the behavior of the examined joints and the investigation of the R-ratio effect on fatigue life, which, in this kind of experiment that is normally performed under displacement control, represents the displacement and not the stress ratio, i.e., $R = \delta_{min}/\delta_{max}$.

FCG curves show three different regions (Figure 6.1). The first region, corresponding to lower SERR values, designated subcritical, is located close to the fatigue threshold where cracks propagate very slowly. The middle part, which is linear, covers most of the FCG curve, and a constant crack growth rate gradient is observed in this region. Finally, a rapid crack growth rate region is observed where the cyclic

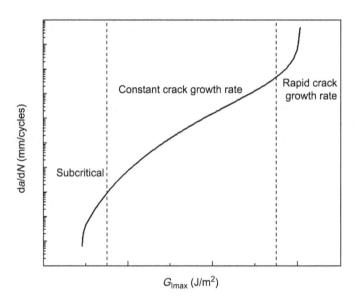

Figure 6.1 Schematic representation of a FCG curve.

SERR approaches the critical SERR of the examined material with fast crack propagation.

Although the shape of the derived FCG curves was similar for several different material and structural systems, contradictory results were found in the literature concerning the effect of the R-ratio on the derived curves. Moreover, in a lot of published works, only the middle part of the FCG curves is examined, since it is difficult to derive experimentally the fast part of the FCG curve, difficult to capture rapid crack propagation, and time consuming to estimate the fatigue threshold.

A thorough investigation of the fatigue fracture behavior of DCBs and cracked lap shear joints composed of graphite/epoxy adherends bonded with a toughened epoxy adhesive at different R-ratios (Mall et al., 1987) showed that the FCG curves, at least their middle parts, are independent of the applied R-ratios when plotted on the ΔG versus da/dN plane. These results were supported in another publication (Turon, Costa, Camanho, & Davila, 2007) where it was also shown that FCG curves based on ΔG are independent of the applied R-ratio along the middle part, although they resulted in different threshold values, ΔG_{th}, since loading under lower R-ratios led to higher ΔG_{th} values. Erpolat, Ashcroft, Crocombe, and Abdel-Wahab (2004) investigated the effect of the R-ratio on the FCG of delaminated carbon fiber–reinforced polymer (CFRP) laminates and observed crack growth through the adhesive, cohesive failure, and through the composite, interlaminar failure. The crack growth rate was found to be almost independent of the R-ratio for cohesive failure but quite sensitive to the R-ratio for interlaminar failure when ΔG is used as the SERR fatigue parameter. Similar results were reported for pultruded GFRP DCB joints (Shahverdi et al., 2012a).

Contrary to the aforementioned results, a study of the fracture behavior of adhesively-bonded steel joints showed an effect of the R-ratio on the FCG curve only when ΔG is used as the SERR parameter; the effect is diminished when G_{max} is used instead (Knox, Tan, Cowling, & Hashim, 1996). Nevertheless, most of the published experimental data are limited to the second region of the FCG curve (Allegri, Jones, Wisnom, & Hallett, 2011; Johnson & Mall, 1985; Mall & Johnson, 1986; Mall et al., 1987; Sarfaraz et al., 2011; Wilkins et al., 1982; Zhang et al., 2010), since it is very time consuming to estimate the fatigue threshold, and it is difficult to monitor the fast region close to G_c. The majority of publications on the subject (Allegri et al., 2011; Bathias & Laksimi, 1985; Martin & Murri, 1990, pp. 251−270; O'Brian 1990; Shahverdi et al., 2012a) report that higher R-ratios result in steeper FCG curves, independent of the strain energy release rate parameter (G_{max}, or ΔG) that is used for derivation of the curves. However, there are other experimental results showing FCG curves for different R-ratios that are parallel to each other (Jia & Davalos, 2004a; Mall et al., 1987).

A certain amount of experimental results concerning the effect of the R-ratio on the fracture behavior of composite joints under constant amplitude fatigue loading patterns exist in the literature. In addition, the results relate to a wide range of composite materials and adhesives and are not consistent, as in some cases they contradict each other. All of the hitherto examined materials are typical composites used in the aerospace and automotive industries, and only limited information can be found in the literature regarding GFRP joints (Sarfaraz et al., 2011, 2012; Shahverdi, Vassilopoulos, & Keller, 2012b; Zhang et al., 2008, 2009, 2010) a material typically used in civil engineering structures, which undergo different loading patterns during their operational life.

6.1.2 Characterization of the fatigue/fracture behavior

Numerous methods have been introduced to characterize the examined fatigue behavior of engineering materials and structures and to develop procedures to accurately model and predict their fatigue life. These methods can be classified into two main categories: those based on the phenomenological representation of the material/structural behavior (mainly expressed by the $S-N$ curves and the modeling/prediction based on this concept), and those based on a damage metric representative of the structure's durability. This damage metric is monitored (or measured, or calculated) during fatigue loading and can indicate how close to failure the material is. Strength, stiffness, crack density, and crack length are some of the damage metrics that have been used in the past. The methods based on each damage metric present certain advantages and disadvantages, which have been discussed in detail elsewhere (Sendeckyj, 1991; Vassilopoulos, 2010).

Use of the crack that is initiated and propagated in the material during fatigue loading as the damage metric has been proved very valuable for metals, where one crack is created and its propagation controls the material's behavior. However, it is questionable whether the same concept can be used for composite materials, in which failure is a result of the interaction of different phenomena (matrix cracking, delamination, fiber cracking, etc.). Nevertheless, the failure of adhesively-bonded composite joints is the result of a dominant crack that, if monitored during fatigue life, can be an acceptable damage metric (Bloyer, Rao, & Ritchie, 1998). In such cases, FCG curves like the one shown in Figure 6.1 are established to represent the fatigue behavior of the examined joints.

Paris et al. (1961) observed that in, the second region, the relationship between the crack propagation rate, da/dN, and the stress intensity factor range, ΔK, or the maximum stress intensity factor, K_{max}, follows a power law equation. This observation was validated by Mode I experimental data from three independent sources on two aluminum alloys, 2024-T3 and 7075-T6. The Paris law has also been extensively applied for modeling fatigue crack propagation in composite materials under pure Mode I (Bathias & Laksimi, 1985; Hojo, Ochiai, Gustafson, & Tanaka, 1994; Mall & Johnson, 1986; Mall et al., 1989; O'Brien, 1990, pp. 7−33; Shivakumar et al., 2006; Wilkins et al., 1982), mixed Mode I/II (Gustafson & Hojo, 1987; Johnson & Mall, 1985; Kenane, Azari, Benmedakhene, & Benzeggagh, 2011; Mall & Johnson, 1986; Mall et al., 1989; Wilkins et al., 1982), and pure Mode II (Allegri et al., 2011), and for adhesively-bonded structural joints (Curley, Hadavinia, & Kinloch, 2000; Jia & Davalos, 2004a, 2004b; Mall & Johnson, 1986; Sarfaraz et al., 2011; Zhang et al., 2010).

Martin and Murri (1990, pp. 251−270) introduced a phenomenological formulation that is able to model the FCG behavior over the entire range of applied G, from the first to the third region. The derived model, designated the "total fatigue life model," expresses the crack growth rate as a function of the maximum cyclic strain energy release rate, G_{max}, the strain energy release rate threshold, G_{th}, and the critical strain energy release rate, G_c. Shivakumar et al. (2006) used the total fatigue life model for characterizing the crack growth rate in glass/vinylester delaminated composite panels subjected to Mode I cyclic loading. None of the above-mentioned models takes into account the effect of the R-ratio on the derived FCG curves, however.

Experimental evidence (see previous section) proved that, independent of the examined material or joint, FCG curves show similar trends. The crack propagation under constant amplitude (CA) loading at any R-ratio was modeled using the Paris law. Several models can be found in literature for the simulation of the crack propagation only along the second region of the FCG curve (Andersons et al., 2004; Hojo et al., 1994; Jia & Davalos, 2004a, 2004b; Walker, 1970), incorporating the effect of the R-ratio so that all data could be condensed into a single master curve. Despite the fact that much research work has been devoted to the FCG rate in composite materials, there is only one study regarding the modeling of the fatigue fracture behavior of pultruded GFRP joints used in civil engineering structures (Shahverdi et al., 2012b). In addition, the aforementioned study presents a method that is able to model the effects of the R-ratio on the total fatigue life and to predict the fatigue/fracture behavior of the examined joints under different loading conditions. The development of such methods was hindered by the lack of complete databases containing fatigue/fracture data regarding the total fatigue life, from the subcritical to the unstable region.

6.1.3 Content of the chapter

The objective of this chapter is to present the fatigue/fracture behavior of adhesively-bonded pultruded GFRP DCB joints under Mode I loading of different displacement ratios. A complete fatigue/fracture database, derived during the last years (Shahverdi et al., 2012a) is used for the demonstration. The effect of the applied R-ratio on the resulting FCG curves is thoroughly analyzed and correlated with the exhibited failure processes of the specimens. The effect of the different locations of the crack path on the fracture behavior of the examined joints is also investigated, as was done in a previous publication (Shahverdi, Vassilopoulos, & Keller, 2011) by the authors for quasi-static loading. A new phenomenological FCG formulation that has been previously introduced by the authors (Shahverdi et al., 2012b) is used in this chapter for the modeling of the exhibited fatigue behavior. The model parameters, calibrated by fitting the experimental FCG curves under $R = 0.1$, 0.5, and 0.8, are subsequently used for predicting FCG curves of the examined joints under $R = 0.3$ and 0.65. It is shown in this chapter that if the model parameters are estimated accurately, the model can be used for the prediction of reliable FCG curves for several unknown loading conditions, and can therefore assist the development of methodologies for the fatigue life prediction of joints under realistic loading conditions.

6.2 Experimental investigation of adhesively-bonded pultruded glass fiber-reinforced polymer (GFRP) joints

6.2.1 Material

Adhesively-bonded pultruded GFRP DCB joints were examined under constant amplitude fatigue loads. The laminates, supplied by Fiberline A/S, Denmark, consisted of E-glass fibers embedded in isophthalic polyester resin and had a width of 40 mm and thickness of 6.0 mm. The laminates comprised two outer combined mat layers and a roving layer in the symmetry plane. One combined mat consisted of two outer chopped strand mats (CSM), and an inner woven 0°/90° fabric, all three stitched together. On the outside, a 40 g/m^2 polyester surface veil was added to protect against environmental attack. The fiber architecture of the laminates is shown in Figure 6.2. An estimation of the nominal thickness of each layer derived by optical microscopy is also given in Figure 6.2. The fiber content, determined by burn-off tests in accordance with ASTM D3171-99, was 43.2 vol.% based on the fiber density of 2560 kg/m^3 specified by the manufacturer and the assumption that no voids were present; the fiber fractions are listed in Table 6.1. The weight of the second, inner mat was almost double that of the first, outer mat, and the proportion of woven fabrics was much higher. The longitudinal strength and Young's modulus of the GFRP laminate were obtained from tensile experiments, according to ASTM D3039-08, as being 307.5 ± 4.7 MPa and 25.0 ± 0.5 GPa, respectively.

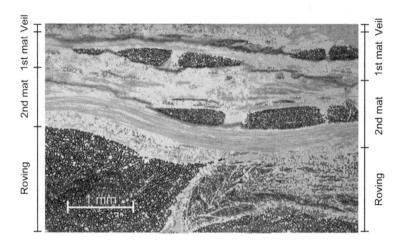

Figure 6.2 Fiber architecture of upper half of laminate cross-section, transverse to pultrusion direction.

A two-component epoxy adhesive system was used, Sikadur 330 (supplied by Sika AG Switzerland), as the bonding material. The tensile strength of the adhesive was 39 MPa and the stiffness 4.6 GPa. The epoxy exhibited an almost elastic behavior and a brittle failure under quasi-static tensile loading (de Castro & Keller, 2008).

Table 6.1 Fiber architecture and fractions by volume and weight of the pultruded laminates

Layers	Average thickness (mm)	% by volume	% by weight
Veil	2×0.05*		
First combined mat	2×0.63		
CSM		2×1.7	2×2.5
Woven 0°/90°		2×1.6	2×2.3
Second combined mat	2×1.07		
CSM		2×2.6	2×3.8
Woven 0°/90°		2×4.1	2×6.0
Roving (UD)	1×2.5	1×23.3	1×34.2
Total	6.0	43.3	64.4

* "2×" means on each side of the symmetry axis.

6.2.2 Specimen geometry and fabrication

The geometry of the DCB specimens used is shown in Figure 6.3. The specimen length was 250 mm including a precrack length of 50 mm. All surfaces subjected to bonding were mechanically abraded by approximately 0.3 mm to increase roughness and then chemically degreased using acetone. An additional 0.5 mm was abraded from the upper arm only along the precrack to ensure that the crack would propagate between the two mat layers of the upper pultruded laminate (Shahverdi et al., 2011). An aluminum frame was used to assist the alignment of the two pultruded laminates. The 2-mm thickness of the adhesive was controlled by using spacers embedded in the bonding area. In-house–developed piano hinges were glued, using the same epoxy adhesive, at the end of both specimen arms to allow load application. A Teflon film of 0.05-mm thickness was placed between the upper arm and the adhesive layer to introduce the precrack. After preparation of the configuration, the specimens were kept under laboratory conditions for 24 h and then placed for 24 h in a conditioning chamber at 35 °C and 50 ± 10% RH to ensure full curing of the adhesive. The resulting thickness of the DCB specimens was 13.4 mm and the crack was located 1.5 mm above the center axis of the joints due to the presence of the adhesive layer. However, as proved in an earlier work by the same group (Zhang et al., 2010) for similar joints with cracks propagating at 0.9 mm above the center axis, and also verified for the present case (Shahverdi, Vassilopoulos, & Keller, 2013b), this does not affect the strain energy release rate calculations. Only a limited Mode II SERR component, in the range of 1% of the corresponding Mode I SERR component was introduced due to the geometric asymmetry of the joints.

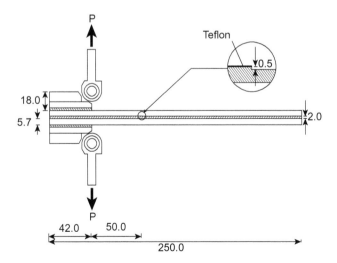

Figure 6.3 Specimen configuration, dimensions in millimeters.

6.2.3 Experimental set-up and loading

A 25-kN MTS Landmark servo-hydraulic testing rig, calibrated to 20% of its maximum capacity, was used for all of the fatigue experiments. All experiments were conducted in laboratory conditions (23 ± 5 °C and 50 ± 10% RH), under displacement control at a frequency of 5 Hz. Three different R-ratios were selected to cover a wide range of possible loading conditions: $R = 0.1$ representing fatigue loading with high amplitude values, an intermediate case, $R = 0.5$, and $R = 0.8$ to represent fatigue loading with low amplitude and high mean cyclic stresses.

The specimens were labeled according to the applied loading conditions; e.g., DCB0.1-02 represents the second specimen loaded under $R = 0.1$. In some of the examined cases, since only a short crack <20−25 mm was created, the same joint was available for a new experiment under the same or different loading conditions. Additional indices, a, b, and c, were used to indicate whether the result had been derived from a virgin specimen, a, or an already-tested specimen, loaded for a second, b, or a third, c, time.

The experimental process followed three steps:

- Step 1: The specimen was inserted into the grips of the machine. The specimen was aligned, and first the lower grip and then the upper grip were tightened. The load that was introduced due to this clamping process, less than 5 N, was manually set to zero by adjusting the position of the moving rod of the testing rig.
- Step 2: After specimen installation a loading ramp was applied to initiate and propagate the crack up to around 15−30 mm corresponding to the crack length required to reach the plateau in the R-curve of the examined specimen (Shahverdi et al., 2011). For any subsequent fatigue loading on an already used specimen, an initial loading ramp was applied only in order to reach the desired displacement values.
- Step 3: The fatigue loading was started at a maximum displacement equal to the maximum displacement reached during the quasi-static loading in order to record the initial fast crack propagation values corresponding to the fast crack growth rate region.

A schematic representation of the loading procedure, steps 2 and 3, is shown in Figure 6.4. The loading ramp and corresponding crack initiated and propagated up to 18 mm for the selected specimen, DCB0.5-06, are presented in Figure 6.4(a), whereas the fatigue displacement controlled loading, together with the crack development during this phase, are shown in Figure 6.4(b).

6.3 Interpretation of the fatigue/fracture experimental results and discussion

6.3.1 Failure modes analysis

In all the examined specimens the observed failure mode, according to ASTM D 5573-99, was a fiber-tear failure or light−fiber-tear failure. The crack paths were located between the two lower mat layers of the upper laminate as planned, therefore corresponding to Path II, according to the nomenclature given in Shahverdi et al. (2011). However, two different failure modes were observed: one when the crack

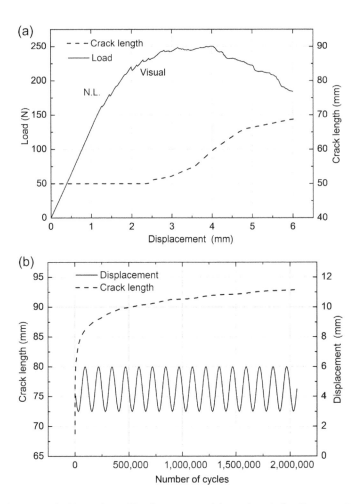

Figure 6.4 Schematic illustration of loading process: (a) quasi-static loading, step 2; (b) cyclic loading, step 3.

was propagating in the upper CSM of the first mat layer, Path II-A, and another when the crack was propagating in the lower CSM of the second mat layer, Path II-B. In some of the examined specimens, the crack tip stopped propagating between the mat layers, and a new crack appeared between the second mat layer and the roving layer. When this occurred, rovings of the woven fabric of the second mat layer bridged the two crack faces and significantly increased the applied load required for crack opening and the corresponding SERR. The term "roving bridging" was used in Shahverdi et al. (2011) to describe this phenomenon. According to the nomenclature used in Shahverdi et al. (2011), the crack path observed in this case was designated Path III with roving bridging.

Examples of typical failure surfaces with crack growth along Path II-A, Path II-B, and Path III with roving bridging are shown in Figure 6.5. Regarding the morphology of the fracture surfaces of the specimens under Mode I fatigue loading, a clear difference between the generated fracture surfaces is observed when the crack propagation

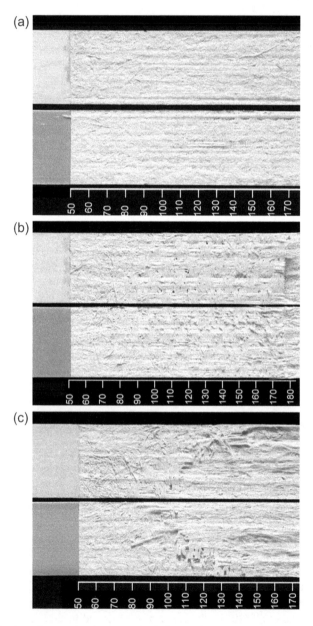

Figure 6.5 Crack surface comparisons: (a) Path II-A, DCB0.1-02a, (b) Path II-B, DCB0.1-01, and (c) roving bridging, DCB0.5-02b.

location changes (Shahverdi et al., 2011). The Path II-B surface shows much more fiber tear than Path II-A, while the results of roving bridging are visible in Figure 6.5(c) between crack lengths of 90 and 120 mm. The failure modes observed for all examined specimens are shown in Table 6.2.

6.3.2 Compliance and crack length measurements

The relationship between specimen compliance and crack length is required in order to estimate the SERR. The compliance of the specimen, C, defined as the ratio of the maximum displacement over the maximum load (δ_{max}/P_{max}), can be calculated at each number of cycles directly from the recorded data. However, derivation of the crack length during fatigue loading is not an easy task. Direct methods, crack gages, and visual observation, and also indirect methods, e.g., the dynamic compliance calibration introduced in Costa et al. (2009), exist for this purpose. The applicability of four methods for determination of the crack length—compliance relationship was investigated during this work. The first was based on visual observation; the second relied on measurements recorded by appropriate crack gages; and the third was a dynamic compliance calibration method (Costa et al., 2009). The fourth method was based on the average crack length—compliance relationship of the same type of joints with the same configuration under quasi-static loading.

According to the first method, the fatigue experiment was interrupted at a predetermined number of cycles, and the joint was opened until a displacement equal to the maximum cyclic displacement was reached. An optical microscope was used to record the crack length. The process was repeated several times to establish a relationship between joint compliance, C, and crack length, a. This is the simplest method for determining the relationship between crack length and compliance. The disadvantage of this method was the scatter caused by the interruptions of the experiment, as has also been reported elsewhere (Shivakumar et al., 2006). Preliminary experiments showed increased sensitivity of the obtained results to the load interruption, and this method was therefore not used in the present work.

The relationship between crack length and compliance can also be determined during the fatigue loading, without any interruption, by using crack gages. The crack gages, the HBM RDS20 model, included 20 parallel wires with a pitch of 1.15 mm placed perpendicular to the adhesive layer (Figure 6.6). As the crack propagated, the wires were progressively cut, and the electrical resistance of the gage increased. A Labview application and a multichannel electronic measurement unit, HBM-Spider8, were used for data acquisition. The high cost was the main drawback of this method, especially if it had to cover the total length of the specimen, i.e., to use several gages along its length. Therefore, crack gages were used for 13 specimens in order to compare the results obtained with those derived by other methods.

Another method for determining the relationship between crack length and compliance was based on a dynamic compliance calibration (Costa et al., 2009). According to this method, the fatigue experiment was terminated and the specimen was then clamped by a mechanical fixture (Figure 6.7). The mechanical fixture could be moved along the specimen to simulate different crack lengths reached

Table 6.2 Summary of all results with values of N in compliance fitting, $C = ka^n$

Specimen code	δ_{max} (mm)	Failure mode	Number of cycles	a_1 (mm)	a_2 (mm)	F_{max1} (N)	F_{max2} (N)	G_{Imax1} (J/m^2)	G_{Imax2} (J/m^2)	Crack gage	n
DCB0.1-01	3	Path II-B	53,000	65.7	78.3	240	153	473	253	Yes	2.58
DCB0.1-02a	3	Path II-A	30,000	65.0	77.0	223	143	344	186	Yes	2.65
DCB0.1-02b	7	Path II-A	45,000	95.0	114.4	136	85	324	167	Yes	2.56
DCB0.1-02c	9	Path II-A	52,000	114.5	133.5	108	73	280	160	Yes	2.61
DCB0.1-03	6	Path II-A	2,100,000	74.2	119.6	207	60	549	98	No	2.60
DCB0.1-04	6	Path II-A	3,000,000	78.1	125.1	183	59	427	85	Yes	2.41
DCB0.1-05	7	Path II-A	310,000	89.1	122.4	153	69	379	125	Yes	2.50
DCB0.1-06	7	Path II-A	3,200,000	64.3	99.8	205	67	407	86	No	2.54
DCB0.1-07	5	Path III + roving	340,000	73.6	85.3	229	141	646	345	Yes	3.30
DCB0.1-08	4	Path III + roving	1,180,000	65.4	81.7	284	143	672	272	Yes	3.07
DCB0.1-09a	6	Path III + roving	1,230,000	64.4	88.9	322	131	1055	310	Yes	2.79
DCB0.1-09b	11	Path III + roving	1,410,000	90.3	140.0	197	66	759	164	No	2.51
DCB0.1-10	4	Path II-A	1,340,000	65.7	100.5	210	67	430	90	No	2.69
DCB0.1-11	6	Path II-B	3,000,000	66.7	109.9	302	124	607	151	No	1.78

Specimen	n	Path	Cycles								
DCB0.1-12	5	Path II-A	1,690,000	75.0	117.2	214	64	488	93	Yes	2.71
DCB0.5-01	5	Path II-B	3,000,000	65.0	84.7	259	132	632	249	No	2.53
DCB0.5-02a	5	Path II-B	3,300,000	64.3	85.7	227	97	659	210	Yes	2.96
DCB0.5-02b	12	Path III + roving	3,000,000	89.7	121.0	224	98	1033	335	No	2.74
DCB0.5-03a	5	Path II-A	2,930,000	70.1	93.3	156	68	407	133	No	2.92
DCB0.5-03b	10	Path II-A	2,970,000	109.7	137.2	97	45	379	140	No	3.41
DCB0.5-04	4	Path II-B	2,420,000	65.1	80.5	227	118	538	227	Yes	3.07
DCB0.5-05	6	Path II-B	2,100,000	68.5	92.9	252	107	780	244	Yes	2.80
DCB0.5-06*	9	Path II-A	2,170,000	98.3	128.4	132	66	399	152	No	2.63
DCB0.8-01a	5	Path II-A	2,600,000	71.9	83.0	186	117	512	280	Yes	3.15
DCB0.8-01b	8	Path II-A	3,100,000	95.0	109.2	150	101	443	262	No	2.80
DCB0.8-02	5	Path II-A	2,090,000	67.8	80.1	194	120	521	271	No	2.89
DCB0.8-03*	5	Path II-A	3,300,000	75.5	91.6	192	114	512	252	Yes	2.68
DCB0.8-04a	10	Path II-B	3,420,000	80.8	98.5	266	172	908	482	No	2.19
DCB0.8-04b	14	Path III + roving	2,400,000	99.4	104.2	245	121	1111	526	No	2.57
DCB0.8-05	5	Path II-B	3,000,000	64.1	72.7	292	201	854	518	No	2.98

* Same specimen. DCB0.5-06 tested after DCB0.8-03.

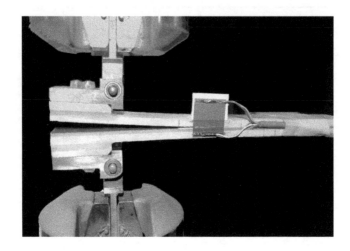

Figure 6.6 Crack gage at the crack tip, DCB0.1-04.

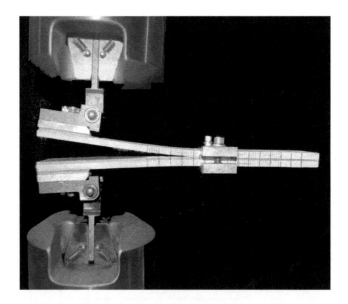

Figure 6.7 Clamping at crack tip, dynamic compliance method, DCB0.1-09b.

during fatigue loading. For each selected crack length, a fatigue block of around 1000 cycles with maximum and minimum displacements the same as those in the real fatigue experiment was applied, and the displacement and load were recorded to estimate the compliance of the joint. The drawback of this method was that it could not capture the effect of the fiber bridging on the compliance, since the measurements were performed on the cracked specimen, when the bridging fibers had already broken.

The specimens examined under fatigue loading exhibited the same failure modes as the same type of specimens examined under quasi-static loading, reported in Shahverdi et al. (2011). Therefore, the compliance of the examined specimens can be assumed as being equal to the average compliance of the same type of specimens examined under quasi-static loading. In this method, the effect of fiber bridging on the compliance of the examined joints is included but is, however, only partially correlated with the applied fatigue loading conditions, as different fiber bridging results under different R-ratios. Another disadvantage is the scatter of the compliance measurements of the quasi-static experimental results, which is considerable and makes the adoption of a single compliance for all cases very difficult.

The comparison between the results obtained using the three methods, i.e., crack gages, dynamic compliance calibration, and average static compliance, is shown in Figure 6.8, where the average compliance (of all examined specimens) versus average crack length is given. All examined specimens were used for the dynamic compliance calibration; the average static compliance was calculated from 16 joints of the same type examined under quasi-static loading and exhibiting the same failure modes, while crack gages were used for 13 of the examined specimens (Table 6.2). As expected, the compliance calculated using crack gages was lower than that calculated by the dynamic compliance method that does not consider the fiber bridging occurring during crack opening. The difference between the average static compliance and the dynamic compliance is due to the scatter of the experimental data in both methods and also the

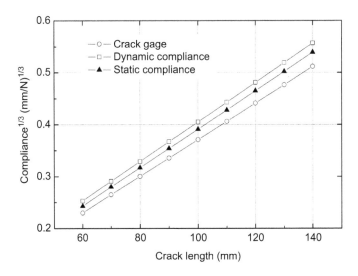

Figure 6.8 Comparison of the average compliance versus crack length relationship for different methods.

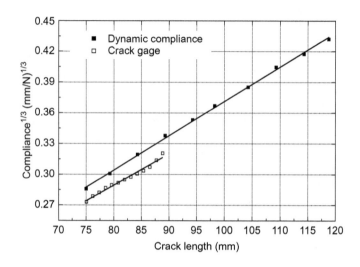

Figure 6.9 Comparison of dynamic compliance and crack gage measurements, DCB0.1-12.

effect of the fiber bridging. As a result of this comparison, the method based on the average static compliance was also abandoned to avoid mixing quasi-static and fatigue experimental results and averaging compliance values, which mask the exhibited scatter of each examined joint.

As shown in Figure 6.9, the remaining two methods, the crack gages and dynamic compliance method, resulted in similar slopes, although crack gages measured 5−10 mm longer cracks for the same compliance. However, the dynamic compliance calibration method could be applied to all specimens and for cracks running along the total length of the joints. Therefore it was used for derivation of the compliance versus crack growth rate relationship and calculation of the SERR values in the present work.

6.3.3 Load and crack length measurements

Figure 6.10(a) shows the variation of the maximum applied load during a fatigue test under different R-ratios and displacement control. The maximum load decreased rapidly at the beginning and then followed a smoother decreasing trend until it reached a plateau value, corresponding to the fatigue threshold. The decrease of the maximum load during fatigue loading was more significant for lower R-ratios because of the higher displacement range that was applied for the same maximum displacement. Consequently, crack length increased rapidly at the beginning of the experiment, when the maximum load was high, and at a reduced rate as the load decreased, reaching a plateau at around 3 million cycles, as shown in Figure 6.10(b). In correlation with the load decrease, longer cracks were observed in specimens examined under lower

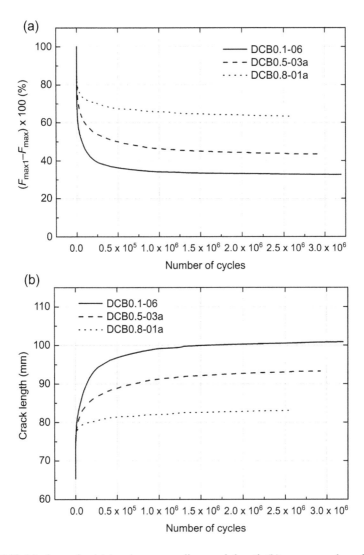

Figure 6.10 Maximum load (a) and corresponding crack length (b) versus number of cycles under different R-ratios, Path II-A.

R-ratios until the fatigue threshold was reached, since these specimens were loaded by higher displacement ranges.

Detailed information about the experiment conducted on each joint, including maximum cyclic displacement, δ_{max}, observed failure mode, number of cycles for each specimen, crack length at the starting point of cyclic loading, a_1, crack length attained at the end of cyclic loading, a_2, and maximum loads, F_{max}, corresponding to a_1 and a_2 are presented in Table 6.2.

6.4 Fracture mechanics data analysis

6.4.1 Crack growth rate calculation

The secant method and incremental polynomial fitting, according to ASTM E647-08, can be used to calculate the crack growth rate. According to the secant or point-to-point method, the crack propagation rate can be determined by calculating the slope of a straight line connecting two contiguous data points on the $a-N$ curve. The incremental polynomial method fits a second-order polynomial to sets of a specified number of successive data points, usually 3, 5, 7, or 9. The slope of the determined equation at any point corresponds to the crack propagation rate.

Both methods were applied as shown in Figure 6.11 for specimens representing all loading conditions. The secant method showed high sensitivity to scatter, while increasing the number of points in the polynomial method effectively decreased this sensitivity without changing the actual trend of experimental data. The 7-point polynomial method was selected in this work for calculation of the crack growth rate. As shown in Figure 6.11, the crack propagation rate continuously decreases with increasing crack length. However the rate of this decrease is not constant, being higher at the beginning and at the end of the experiment, close to the fatigue threshold, and more moderate for the major part of the fatigue loading. There is also a strong effect of the examined R-ratio. The crack growth rate decreases from c. 0.5 mm/cycle at the beginning to c. 10^{-6} mm/cycle at the end after only 10 mm of crack length for $R = 0.8$, while the same happened after 20 and 45 mm for $R = 0.5$ and $R = 0.1$, respectively.

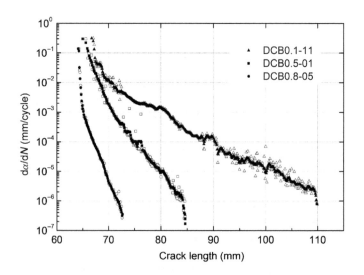

Figure 6.11 Comparison of fatigue crack growth rate versus crack length under different R-ratios, Path II-B, solid symbols: 7-points method; open symbols: secant method.

6.4.2 Strain energy release rate calculation

The strain energy release rate of the DCB joints can be calculated based on linear elastic fracture mechanics. According to this theory, for a DCB joint with width B and an existing crack length, a, the SERR is a function of the maximum cyclic load, F_{max}, and the rate of the compliance change, dC/da:

$$G_{Imax} = \frac{F_{max}^2}{2B} \frac{dC}{da} \quad (6.1)$$

Standard methods for the SERR calculation are based on this equation, the difference between them basically being the way in which the derivative dC/da is obtained. A thorough analysis of the applicability of several methods for the calculation of the SERR to similar composite joints is presented in Zhang et al. (2010). As shown, for similar pultruded GFRP DCB joints, all methods give similar results with the exception of simple beam theory. Therefore, in the present work, the experimental compliance method is used according to which the measured compliance is fitted to the measured crack length by a power law equation of the form: $C = ka^n$. The maximum cyclic SERR can then be calculated as:

$$G_{Imax} = \frac{nF_{max}\delta_{max}}{2Ba} \quad (6.2)$$

Correction factors for the loading blocks and moments resulting from large displacements were applied according to ASTM 5528-01 (2007).

Typical curves showing the variation of the G_{Imax} throughout fatigue life are presented in Figure 6.12 for specimens where a Path II-A crack path was observed. G_{Imax} decreases with an increasing number of cycles depending on the R-ratio. As observed in Figure 6.12, this reduction is more pronounced for lower R-ratio values.

6.4.3 Fatigue crack growth (FCG) curves

The FCG curves for all the examined R-ratios are shown in Figure 6.13 for Path II-A, Figure 6.14 for Path II-B, and Figure 6.15 for Path III with roving bridging. A total of 30 experimental results are presented in these figures as well as in Table 6.2. The results prove that the relationship between G_{Imax} and da/dN is highly dependent on the R-ratio. Regardless of the crack path location, a steeper curve reaching a higher fatigue threshold corresponds to higher R-ratios. All the plotted FCG curves converge to a value of around 600 J/m^2 for Path II-A, 1100 J/m^2 for Path II-B, and 1200 J/m^2 for cases when roving bridging is present. The G_{Imax} value obtained for Path II-B is the same as the critical SERR (G_{Ic}) for Path II and no Path II-A failure mode was observed in Shahverdi et al. (2011). Path III with roving bridging during the fatigue loading corresponds to a partially activated roving bridging due to the short crack length derived during the fatigue experiment.

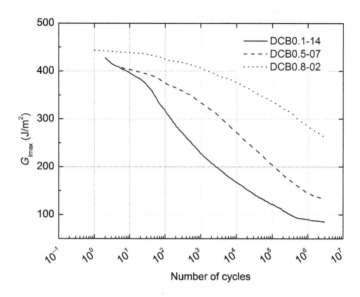

Figure 6.12 Comparison of G_{Imax} versus number of cycles under different R-ratios, Path II-A.

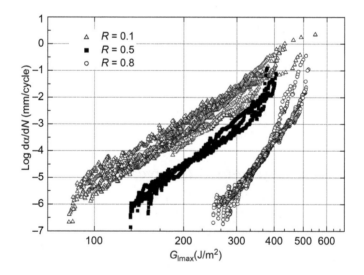

Figure 6.13 Crack growth rate versus G_{Imax}, Path II-A.

Therefore, the obtained G_{Imax} value, 1200 J/m², was less than the 1750 J/m² as mentioned in Shahverdi et al. (2011). The exhibited behavior is in agreement with the observed failure process of the joints and is related to the short fiber bridging that was present behind the crack tip.

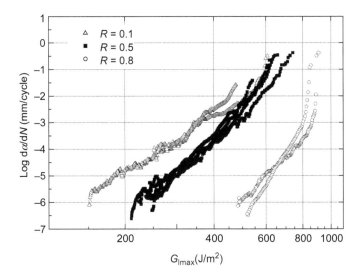

Figure 6.14 Crack growth rate versus G_{Imax}, Path II-B.

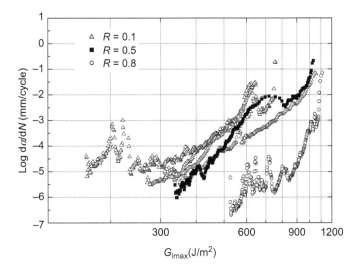

Figure 6.15 Crack growth rate versus G_{Imax}, Path III with roving bridging.

Initially fiber bridging occurs for all specimens independent of the R-ratio, therefore, for all experiments where the crack propagates along the same path, a similar SERR value is calculated. As the experiment developed, however, for lower R-ratios the crack closure was breaking the fibers that bridged the crack faces, and consequently

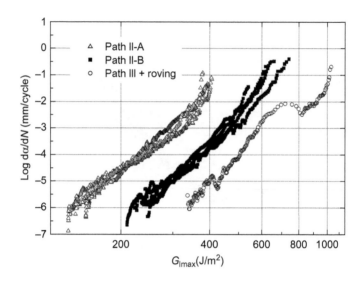

Figure 6.16 Crack growth rate versus G_{Imax} under $R = 0.5$.

the energy required for subsequent crack propagation was reduced. Therefore, the differences between FCG curves under different R-ratios were more pronounced close to the fatigue threshold.

The effect of the crack path location is also visible in these graphs. The corresponding FCG curves were shifted to higher SERR values when the path changed from Path II-A to Path II-B (a shift of c. 250–300 J/m^2) and from Path II-B to Path III with roving bridging (a shift of c. 300–400 J/m^2) (Figure 6.16). When roving bridging was no longer active, the fatigue/fracture behavior of the specimen was described by an FCG curve shifted back to lower SERR values, as shown in Figure 6.15 for two FCG curves obtained under $R = 0.1$, DCB0.1-09a and DCB0.1-09b, and one FCG curve obtained under $R = 0.8$, DCB0.8-0.4b.

Similar research findings have also been reported elsewhere in the literature for the fracture of joints composed of other types of composite materials (Erpolat et al., 2004; Hojo et al., 2006; Mall et al., 1987). FCG curves for DCB joints made of carbon epoxy laminates follow the same trend, exhibiting lower G_{Imax} values for the same rate of crack propagation under lower R-ratios.

The FCG curves can also be derived using the SERR cyclic range instead of its maximum value, as shown in Figure 6.17 for the specimens that exhibited Path II-A crack propagation. However, as has also been emphasized by other researchers (Jethwa & Kinloch, 1997), the use of ΔG_I instead of G_{Imax} may produce falsified FCG curves, especially for low R-ratios. In cases like this, during the unloading part of the fatigue cycle, the failed surfaces typically come into contact, producing an artificially higher value of G_{Imin}, and hence lower value for ΔG_I. Regardless of this, the difference among the three classes of data is obvious in Figure 6.17, with the FCG curves

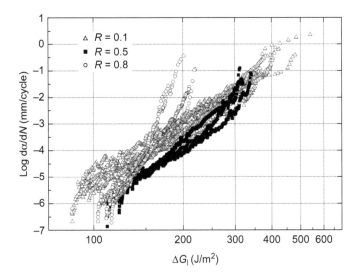

Figure 6.17 Crack growth rate versus ΔG_I, Path II-A.

corresponding to higher R-ratios being steeper. However, in contrast to the case shown in Figure 6.13 that shows FCGs based on G_{Imax}, the FCGs in Figure 6.17 do not converge to a single SERR value, as was the case for carbon epoxy DCB joints in Mall et al. (1987), either at the threshold region or at the critical region, making the development of a theoretical model for the description of the fatigue/fracture behavior of the examined specimens very difficult.

6.5 Fracture mechanics modeling

6.5.1 Total life fatigue model including the R-ratio effects

A phenomenological equation for the calculation of the crack growth rate as a function of the maximum cyclic strain energy release rate, the strain energy release rate threshold, and the static fracture toughness was first proposed by Martin and Murri (1990, pp. 251−270). This model covers all three FCG regions: the subcritical region around the fatigue threshold, the G_{Imax}−controlled region, and the critical region close to G_{Ic}.

The model was based on the experimental FCG curves for each R-ratio. Each FCG curve was divided into the three regions by visual inspection. Then, parameters D and m were estimated by a linear regression analysis after fitting Eqn (6.3) to the middle region:

$$\frac{da}{dN} = D(G_{Imax})^m \qquad (6.3)$$

In the subcritical region, the da/dN varies between 0 and a value corresponding to the lowest value in region 2. The da/dN equation can be written as follows:

$$\frac{da}{dN} = D(G_{\text{Imax}})^m \left(1 - \left(\frac{G_{\text{Ith}}}{G_{\text{Imax}}}\right)^{Q_1}\right) \tag{6.4}$$

The exponent Q_1 was determined using the already determined D and m by fitting the equation to the data in regions 1 and 2. A trial-and-error approach was found to work well.

In the unstable region, da/dN varies between infinite, when G_{Imax} is equal to G_c, and the transition value corresponding to the maximum value in region 2. The da/dN equation can be written as follows:

$$\frac{da}{dN} = D(G_{\text{Imax}})^m \frac{1}{1 - \left(\frac{G_{\text{Imax}}}{G_{\text{Ic}}}\right)^{Q_2}} \tag{6.5}$$

The exponent Q_2 was determined by fitting Eqn (6.5) to the experimental data in regions 2 and 3. Finally, the combined da/dN equation that covers all three regions is given by:

$$\frac{da}{dN} = D(G_{\text{Imax}})^m \frac{\left(1 - \left(\frac{G_{\text{Ith}}}{G_{\text{Imax}}}\right)^{Q_1}\right)}{\left(1 - \left(\frac{G_{\text{Imax}}}{G_c}\right)^{Q_2}\right)}, \tag{6.6}$$

where m, Q_1, and Q_2 are the empirical model parameters dependent on material and loading conditions. Equation (6.6) can be applied between the limits $G_{\text{Ith}} \leq G_{\text{Imax}} \leq G_{\text{Ic}}$. Therefore, as G_{Imax} approaches G_{Ith}, da/dN tends to become minimum. Also, as G_{Imax} approaches G_{Ic}, da/dN tends asymptotically to infinity. The values of the different parameters are given in Table 6.3. As shown in Figures 6.18 and 6.19, the model, Eqn (6.6) is capable of modeling successfully the entire FCG curve from the subcritical to the rapid crack growth region. However, the model in this form is not able to take into account the R-ratio effect on the FCG curve.

A new model, including the R-ratio effect has been introduced by Shahverdi et al. (2012b) as:

$$\frac{da}{dN}(R) = D(R)(G_{\text{Imax}})^{m(R)} \frac{\left(1 - \left(\frac{G_{\text{Ith}}(R)}{G_{\text{Imax}}}\right)^{25}\right)}{\left(1 - \left(\frac{G_{\text{Imax}}}{G_{\text{Ic}}}\right)^3\right)} \tag{6.7}$$

Table 6.3 Comparison of total fatigue life model parameters for paths II-A and II-B from experimental investigation and proposed model

Fracture mode	R-ratio		D	m	G_{Ith} (J/m²)	Q_1	Q_2
Path II-A	0.1	Exp	3.56E-20	7.072	90.00	25	3
		Pred	8.00E-19	6.678	83.25	25	3
		Error (%)	6.95	5.57	7.50		
	0.5	Exp	6.05E-26	8.987	140.00	25	3
		Pred	2.26E-28	9.624	150.91	25	3
		Error (%)	9.63	7.09	7.79		
	0.8	Exp	1.10E-44	14.654	265.00	25	3
		Pred	6.85E-44	14.412	260.85	25	3
		Error (%)	1.81	1.65	1.57		
Path II-B	0.1	Exp	1.80E-22	7.367	135.00	25	3
		Pred	1.67E-22	7.475	122.00	25	3
		Error (%)	0.14	1.46	9.63		
	0.5	Exp	8.74E-32	10.545	230.00	25	3
		Pred	1.37E-31	10.371	251.00	25	3
		Error (%)	0.63	1.65	9.13		
	0.8	Exp	2.34E-46	15.012	470.00	25	3
		Pred	2.35E-46	15.078	462.00	25	3
		Error (%)	0.00	0.44	1.70		

Exp, experimental; Pred, predicted.

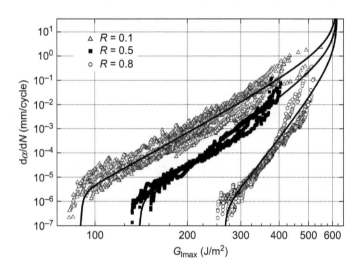

Figure 6.18 Crack growth rate versus G_{Imax} for Path II-A; solid lines are plots of model results.

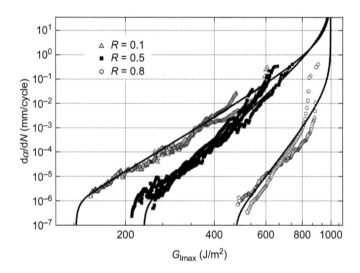

Figure 6.19 Crack growth rate versus G_{Imax} for Path II-B, solid lines are plots of model results.

The new model, Eqn (6.7), resembles Eqn (6.6), with parameters D, m, and G_{Ith} being functions of the R-ratio. For the examined DCB joints, the D, m, G_{Ith}, Q_1 and Q_2 model parameters were estimated by fitting Eqn (6.6) to the available experimental data for three different R-ratios, i.e., $R = 0.1$, 0.5, and 0.8, for both Path II-A and Path II-B.

The values, given in Table 6.3, indicate a strong dependence of D, m, and G_{Ith} on the R-ratio; however, exponents Q_1 and Q_2 remained constant for all of the examined loading conditions. The model parameters were assumed as being functions of the R-ratio squared (R^2), since, as proved (Bathias & Laksimi, 1985; Hojo et al., 1994), the ratio between G_{Imin} and G_{Imax} is approximately R^2. Parameter D was found to be an exponential function of R^2, whereas the exponent m and the strain energy release rate threshold, G_{Ith}, were shown to depend linearly on R^2.

The following functional forms were derived from the experimental data, as shown in Figures 6.20–6.22:

$$D(R) = A_1 e^{B_1 R^2} \tag{6.8}$$

$$m(R) = A_2 R^2 + B_2 \tag{6.9}$$

$$G_{\text{Ith}}(R) = A_3 R^2 + B_3 \tag{6.10}$$

Parameters A_i and B_i in Eqns (6.8)–(6.10) are derived by linear regression analysis after plotting the derived values of D, m, and G_{Ith} against the square of the R-ratio (Figures 6.18–6.20). For the DCB joints examined in this chapter, the specific form of functions $D(R)$, $m(R)$, and $G_{\text{Ith}}(R)$ are given in Eqns (6.11)–(6.13) for Path II-A:

$$D(R) = 2.10^{-18} e^{-91.62 R^2} \tag{6.11}$$

$$m(R) = 12.276 R^2 + 6.555 \tag{6.12}$$

$$G_{\text{Ith}}(R) = 281.9 R^2 + 80.43 \tag{6.13}$$

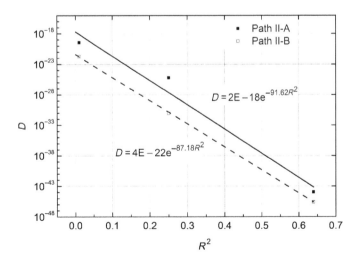

Figure 6.20 Effect of R-ratio on constant D in Paths II-A and II-B.

and in Eqns (6.14)–(6.16) for Path II-B:

$$D(R) = 4 \cdot 10^{-22} e^{-87.18 R^2} \tag{6.14}$$

$$m(R) = 12.068 R^2 + 7.354 \tag{6.15}$$

$$G_{\text{Ith}}(R) = 569.98 R^2 + 120.67 \tag{6.16}$$

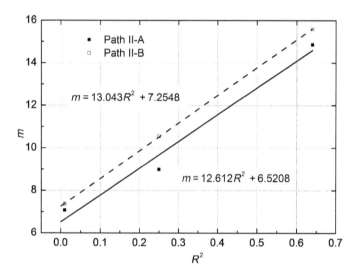

Figure 6.21 Effect of R-ratio on exponent m in Paths II-A and II-B.

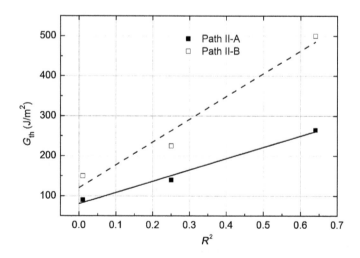

Figure 6.22 Effect of R-ratio on threshold strain energy release rate in Paths II-A and II-B.

The FCG curve for each desired R-ratio can be derived by Eqn (6.7) after substituting the functions from Eqns (6.11)–(6.16). Estimated parameters for the three R-ratios used are given in Table 6.3 for comparison with the corresponding values obtained after fitting each set of experimental data. As shown, the values derived by the introduced model approximate well those derived after fitting each experimental data set with deviations of the order of 10% maximum. The relative error which is used in Table 6.3, is calculated as:

$$\text{Error} = \frac{|\text{Exp} - \text{Pred}|}{\text{Exp}} \times 100\%, \qquad (6.17)$$

for m and G_{Ith}, which depend linearly on R^2. Since D was found to be an exponential function of R^2, logarithms of the D values were used.

6.5.2 Prediction of fatigue crack growth (FCG) curves for other R-ratios

Although the proposed model can be used for the prediction of FCG curves under any R-ratio, in this chapter two intermediate R-ratios have been chosen. The predicted FCG curves for R-ratios equal to 0.3 and 0.65, corresponding to Path II-A crack propagation based on Eqn (6.7) after substituting the functions of the model parameters given by Eqns (6.11)–(6.13), are plotted in Figure 6.23 by solid lines. The values of the model parameters estimated according to Eqns (6.11)–(6.13) are given in Table 6.4. Two experimentally derived FCG curves per loading case, and the fitted lines (dashed lines, according

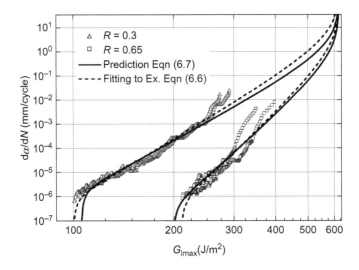

Figure 6.23 Modeling validation by comparison of experimental results and modeling prediction for Path II-A at $R = 0.3$ and $R = 0.65$.

Table 6.4 **Comparison of total fatigue life model parameters at R-ratios 0.3 and 0.65 for Path II-A from experimental investigation and proposed model**

Fracture mode	R-ratio		D	M	G_{Ith} (J/m^2)	Q_1	Q_2
Path II-A	0.3	Exp	1.50E-23	8.344	100.0	25	3
		Pred	5.25E-22	7.65	105.8	25	3
		Error (%)	6.77	8.32	5.80		
	0.65	Exp	2.18E-37	13.085	210.0	25	3
		Pred	3.09E-35	11.85	199.5	25	3
		Error (%)	5.87	9.44	4.99		

Exp, experimental; Pred, predicted.

to Eqn (6.6)) are also given for comparison. As shown, the predicted FCG curves are validated by the experimental data and fitted lines. The parameter values estimated according to the introduced model are compared to the fitted ones, based on Eqn (6.6), in Table 6.4. The maximum error (see also Eqn (6.17)), was less than 10% in all cases.

The total life fatigue model can be used for the derivation of other FCG curves, under different R-ratios, and for different crack paths, e.g., Path II-A and Path II-B, as long as the base line fatigue data exist and allow derivation of the relationships between the model parameters and R-ratio used.

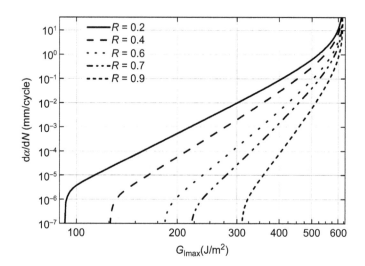

Figure 6.24 Predicted FCG curves at various R-ratios for Path II-A based on proposed model.

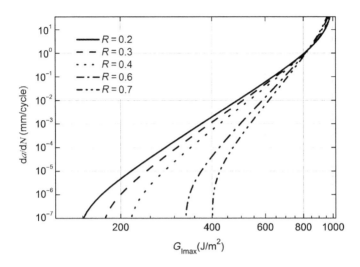

Figure 6.25 Predicted FCG curves at various R-ratios for Path II-B based on proposed model.

Theoretical FCG curves for different R-ratios corresponding to Path II-A and Path II-B crack propagation are shown in Figures 6.24 and 6.25. Such curves can be used for crack length estimation under block and variable loading conditions. In Figures 6.24 and 6.25, steeper FCG curves with higher values of G_{Ith} correspond to higher R-ratios. In addition, FCG curves corresponding to Path II-B crack propagation were steeper than those corresponding to Path II-A, simulating well the experimentally observed trends presented in Figures 6.18, 6.19, and 6.23.

6.6 Conclusions

The fatigue fracture behavior of composite materials can be investigated by using the fracture mechanics theory. A basic experiment used for the characterization of the fatigue/fracture behavior of several composite material systems is the Mode I experiment of double cantilever beam specimens. This experimental technique can lead to the derivation of the FCG curves, which can be used for the estimation of the life time of the examined joints (see Chapter 13).

The fatigue behavior of composite joints is affected by the loading type, in this case, the R-ratio. The fatigue and fracture behavior of adhesively-bonded double cantilever beam specimens was experimentally examined under the displacement ratios $R = 0.1$, 0.5, and 0.8 in order to investigate the effect of the different R-ratios on the derived FCG curves. The results, obtained from 30 experiments, showed that the change in R-ratio significantly affected the fracture behavior of the examined DCB joints.

Mainly two different crack path locations along the upper laminate were exhibited: one in the upper CSM of the first mat layer, and the other in the lower CSM of the second mat layer. In some of the examined cases, the crack shifted deeper, between the second mat layer and the roving layer accompanied by roving bridging. Short fiber bridging and roving bridging encountered during crack propagation increased the estimated strain energy release rate values.

FCG curves, derived for different crack paths, are parallel to each other for the same R-ratio. Crack growth in the upper CSM of the first mat layer attained the lowest G_{Imax}; however, crack growth between the second mat layer and the roving layer accompanied by roving bridging attained the highest G_{Imax}.

FCG curves exhibited high dependence on the R-ratio independent of the SERR fatigue parameter used for the representation. The FCG curves corresponding to higher R-ratios exhibited higher slopes in the middle part, resulting in higher fatigue threshold values. When plotted on the $G_{Imax}-da/dN$-plane, they all reach the same maximum value, similar to that of the critical strain energy release rate derived from quasistatic experiments. However, when plotted on the ΔG_I-da/dN-plane, FCG curves corresponding to higher R-ratios show higher slopes and tend to exhibit the same fatigue threshold value, compared to those corresponding to $R = 0.1$.

A total fatigue life model that takes into account the effect of the R-ratio on the FCG curve has been used in order to demonstrate the use of such models for the estimation of FCG curves under R-ratios for which experimental data do not exist. The derivation of the new model is phenomenological, relying on the fitting of the existing total life fatigue model to experimental data under different R-ratios in order to estimate the necessary model parameters as functions of the R-ratio.

The total fatigue life model was used for the modeling of the fatigue and fracture behavior of the examined joints under three different R-ratios. It was also demonstrated through comparisons to additional experimental results under two more R-ratios that the model is able to predict the material behavior exhibited under unseen loading conditions, i.e., different from the R-ratios used for estimating the model parameters.

This model and similar ones can be used as tools for the interpolation between FCG curves under known R-ratios to derive theoretical FCG curves under new R-ratios for which no experimental data exist. This process can be used for life estimation under block and variable amplitude loading conditions that normally develop in engineering structures.

References

Allegri, G., Jones, M. I., Wisnom, M. R., & Hallett, S. R. (2011). A new semi-empirical model for stress ratio effect on mode II fatigue delamination growth. *Composites, Part A: Applied Science and Manufacturing, 42*(7), 733−740.

Andersons, J., Hojo, M., & Ochiai, S. (2004). Empirical model for stress ratio effect on fatigue delamination growth rate in composite laminates. *International Journal of Fatigue, 26*(6), 597−604.

Argüelles, A., Viña, J., Canteli, A. F., & Bonhomme, J. (2010). Fatigue delamination, initiation, and growth, under mode I and II of fracture in a carbon-fiber epoxy composite. *Polymer Composites, 31*(4), 700−706.
Ashcroft, A. I., Hughes, D. J., & Shaw, S. J. (2001). Mode I fracture of epoxy bonded composite joints: 1. Quasi-static loading. *International Journal of Adhesion and Adhesives, 21*(2), 87−99.
Bathias, C., & Laksimi, A. (1985). Delamination threshold and loading effect in fiber glass epoxy composite. In W. S. Johnson (Ed.), *Delamination and debonding of materials* (pp. 217−237). Philadelphia: ASTM STP 876, American Society for Testing; and Materials.
Beheshty, M. H., & Harris, B. (1998). A constant life model of fatigue behavior for carbon fiber composites: the effect of impact damage. *Composites Science and Technology, 58*(1), 9−18.
Blackman, B. R. K., Hadavinia, H., Kinloch, A. J., Paraschi, M., & Williams, J. G. (2003). The calculation of adhesive fracture energies in mode I: revisiting the tapered double cantilever beam (TDCB) test. *Engineering Fracture Mechanics, 70*(2), 233−248.
Bloyer, D. R., Rao, K. T. V., & Ritchie, R. O. (1998). Fracture toughness and R-curve behavior of laminated brittle-matrix composites. *Metallurgical and Materials Transactions A: Physical Metallurgy and Materials Science, 29*(10), 2483−2496.
Brunner, A. J., Murphy, N., & Pinter, G. (2009). Development of a standardized procedure for the characterization of interlaminar delamination propagation in advanced composites under fatigue mode I loading conditions. *Engineering Fracture Mechanics, 76*(18), 2678−2689.
Cheuk, P. T., Tong, L., Wang, C. H., Baker, A., & Chalkley, P. (2002). Fatigue crack growth in adhesively bonded composite-metal double-lap joints. *Composite Structures, 57*(1−4), 109−115.
Costa, J., Mahdi, S., Vicens, J., Blanco, N., & Rodríguez-Bellido, A. (2009). Fatigue behavior of adhesive films and laminating resins in bonded joints for repair purposes. In *Proceedings of the ICCM 17 conference, July 27−31, 2009*, Edinburgh, UK.
Curley, A. J., Hadavinia, H., & Kinloch, A. J. (2000). Predicting the service life of adhesively-bonded joints. *International Journal of Fracture, 103*(1), 41−69.
de Castro, J., & Keller, T. (2008). Ductile double-lap joints from brittle GFRP laminates and ductile adhesives. Part-I: experimental investigation. *Composites, Part B: Engineering, 39*(2), 271−281.
Ducept, F., Davies, P., & Gamby, D. (2000). Mixed mode failure criteria for a glass/epoxy composite and an adhesively bonded composite/composite joint. *International Journal of Adhesion and Adhesives, 20*(3), 233−244.
Erpolat, S., Ashcroft, I. A., Crocombe, A. D., & Abdel-Wahab, M. M. (2004). Fatigue crack growth acceleration due to intermittent overstressing in adhesively bonded CFRP joints. *Composites, Part A: Applied Science and Manufacturing, 35*(10), 1175−1183.
Gustafson, C.-G., & Hojo, M. (1987). Delamination fatigue crack growth in unidirectional graphite/epoxy laminates. *Journal of Reinforced Plastics and Composites, 6*(1), 36−52.
Hadavinia, H., Kinloch, A. J., Little, M. S. G., & Taylor, A. C. (2003). The prediction of crack growth in bonded joints under cyclic-fatigue loading. I. Experimental studies. *International Journal of Adhesion and Adhesives, 23*(6), 449−461.
Hojo, M., Matsuda, S., Tanaka, M., Ochiai, S., & Murakami, A. (2006). Mode I delamination fatigue properties of interlayer-toughened CF/epoxy laminates. *Composites Science and Technology, 66*(5), 665−675.

Hojo, M., Ochiai, S., Gustafson, C. G., & Tanaka, K. (1994). Effect of matrix resin on delamination fatigue crack growth in CFRP laminates. *Engineering Fracture Mechanics*, *49*(1), 35–47.
Jethwa, J. K., & Kinloch, A. J. (1997). The fatigue and durability behaviour of automotive adhesives. Part I: fracture mechanics tests. *Journal of Adhesion*, *61*(1–4), 71–95.
Jia, J., & Davalos, J. F. (2004a). Loading variable effects on mode-I fatigue of wood–FRP composite bonded interface. *Composites Science and Technology*, *64*(1), 99–107.
Jia, J., & Davalos, J. F. (2004b). Study of load ratio for mode-I fatigue fracture of wood–FRP bonded interfaces. *Journal of Composite Materials*, *38*(14), 1211–1230.
Johnson, W. S., & Mall, S. (1985). A fracture mechanics approach for designing adhesively bonded joints. In W. S. Johnson (Ed.), *Delamination and debonding of materials* (pp. 189–199). Philadelphia: ASTM STP 876, American Society for Testing Materials.
Kawai, M., & Kato, K. (2006). Effects of R-ratio on the off-axis fatigue behavior of unidirectional hybrid GFRP/Al laminates at room temperature. *International Journal of Fatigue*, *28*(10 Special Issue 3), 1226–1238.
Kawai, M., & Koizumi, M. (2007). Nonlinear constant fatigue life diagrams for carbon/epoxy laminates at room temperature. *Composites, Part A: Applied Science and Manufacturing*, *38*(11), 2342–2353.
Kenane, M., Azari, Z., Benmedakhene, S., & Benzeggagh, M. L. (2011). Experimental development of fatigue delamination threshold criterion. *Composites, Part B: Engineering*, *42*(3), 367–375.
Knox, E. M., Tan, K. T. T., Cowling, M. J., & Hashim, S. A. (1996). *The fatigue performance of adhesively bonded thick adherend steel joints.* In *Proceedings of the European adhesion conference*, Cambridge, UK (Vol. 1) (pp. 319–324).
Mall, S., & Johnson, W. S. (1986). Characterization of mode I and mixed-mode failure of adhesive bonds between composite adherends. In J. M. Whitney (Ed.), *Composite materials: Testing and design (seventh conference)* (pp. 322–334). Philadelphia: ASTM STP 893, American Society for Testing Materials.
Mall, S., Ramamurthy, G., & Rezaizdeh, M. A. (1987). Stress ratio effect on cyclic debonding in adhesively bonded composite joints. *Composite Structures*, *8*(1), 31–45.
Mall, S., Yun, K. T., & Kochhar, N. K. (1989). *Characterization of matrix toughness effect on cyclic delamination growth in graphite fiber composites.* In P. A. Lagace (Ed.), *Composite materials: Fatigue and fracture* (Vol. 2) (pp. 296–310). Philadelphia: ASTM STP 1012, American Society for Testing and Materials.
Martin, R. H., & Murri, G. B. (1990). *Characterization of mode I and mode II delamination growth and thresholds is AS4/PEEK composites.* Philadelphia: ASTM STP 1059, American Society for Testing and Materials.
Naghipour, P., Bartscha, M., & Voggenreitera, H. (2011). Simulation and experimental validation of mixed mode delamination in multidirectional CF/PEEK laminates under fatigue loading. *International Journal of Solids and Structures*, *48*(6), 1070–1081.
O'Brien, T. K. (1990). *Towards a damage tolerance philosophy for composite materials and structures.* ASTM Special Technical Publication, No. 1059.
Paris, P. C., Gomez, M. P., & Anderson, W. E. (1961). A rational analytic theory of fatigue. *Trend in Engineering*, *13*, 9–14.
Philippidis, T. P., & Vassilopoulos, A. P. (2002). Complex stress state effect on fatigue life of GRP laminates. Part I, experimental. *International Journal of Fatigue*, *24*(8), 813–823.
Quaresimin, M., & Ricotta, M. (2006). Fatigue behaviour and damage evolution of single lap bonded joints in composite material. *Composites Science and Technology*, *66*(2), 176–187.

Rans, C. D., Alderliesten, R. C., & Benedictus, R. (2011). Misinterpreting the results: how similitude can improve our understanding of fatigue delamination growth. *Composites Science and Technology, 71*(2), 230−238.
Russell, A. J., & Street, K. N. (1988). A constant ΔG test for measuring mode I interlaminar fatigue crack growth rates. In J. D. Whitcomb (Ed.), *Composite materials testing and design (eighth conference)* (pp. 259−277). Philadelphia: ASTM STP 972, American Society for Testing and Materials.
Sarfaraz, R., Vassilopoulos, A. P., & Keller, T. (2011). Experimental investigation of the fatigue behavior of adhesively-bonded pultruded GFRP joints under different load ratios. *International Journal of Fatigue, 33*(11), 1451−1460.
Sarfaraz, R., Vassilopoulos, A. P., & Keller, T. (2012). Experimental investigation and modeling of mean load effect on fatigue behavior of adhesively-bonded pultruded GFRP joints. *International Journal of Fatigue, 44*, 245−252.
Sarfaraz, R., Vassilopoulos, A. P., & Keller, T. (2013a). Block loading fatigue of adhesively-bonded pultruded GFRP joints. *International Journal of Fatigue, 49*, 40−49.
Sarfaraz, R., Vassilopoulos, A. P., & Keller, T. (2013b). Variable amplitude fatigue of adhesively-bonded pultruded GFRP joints. *International Journal of Fatigue, 55*, 22−32.
Sendeckyj, G. P. (1991). *Life prediction for resin-matrix composite materials, fatigue of composite materials*. In K. L. Reifsneider (Ed.), *Composite materials series* (Vol. 4). Elsevier.
Shahverdi, M., Vassilopoulos, A. P., & Keller, T. (2011). A phenomenological analysis of mode I fracture of adhesively-bonded pultruded GFRP joints. *Engineering Fracture Mechanics, 78*(10), 2161−2173.
Shahverdi, M., Vassilopoulos, A. P., & Keller, T. (2012a). Experimental investigation of R-ratio effects on fatigue crack growth of adhesively-bonded pultruded GFRP DCB joints under CA loading. *Composites, Part A: Applied Science and Manufacturing, 43*, 1689−1697.
Shahverdi, M., Vassilopoulos, A. P., & Keller, T. (2012b). A total fatigue life model for the prediction of the R-ratio effects on fatigue crack growth of adhesively-bonded pultruded GFRP DCB joints. *Composites, Part A: Applied Science and Manufacturing, 43*(10), 1783−1790.
Shahverdi, M., Vassilopoulos, A. P., & Keller, T. (2013a). Mixed-mode fatigue failure criteria for adhesively bonded pultruded GFRP joints. *Composites, Part A: Applied Science and Manufacturing, 54*, 46−55.
Shahverdi, M., Vassilopoulos, A. P., & Keller, T. (2013b). Modeling modeling effects of asymmetry and fiber bridging on mode I fracture behavior of bonded pultruded composite joints. *Engineering Fracture Mechanics, 99*, 335−348.
Shivakumar, K., Chen, H., Abali, F., Le, D., & Davis, C. (2006). A total fatigue life model for mode I delaminated composite laminates. *International Journal of Fatigue, 28*(1), 33−42.
Tanaka, K., & Tanaka, H. (1997). Stress-ratio effect on mode II propagation of interlaminar fatigue cracks in graphite/epoxy composites. *Composite Materials: Fatigue Fracture ASTM Spec Tech Publ 1285*.
Turon, A., Costa, J., Camanho, P. P., & Davila, C. G. (2007). Simulation of delamination in composites under high-cycle fatigue. *Composites, Part A: Applied Science and Manufacturing, 38*(11), 2270−2282.
Vassilopoulos, A. P. (2010). Introduction to the fatigue life prediction of composite materials and structures: past, present and future prospects. In A. P. Vassilopoulos (Ed.), *Fatigue life prediction of composites and composite structures*. Woodhead Publishing Ltd.
Vassilopoulos, A. P., Manshadi, B. D., & Keller, T. (2010a). Influence of constant life diagram formulation on the fatigue life prediction of composite materials. *International Journal of Fatigue, 32*(4), 659−669.

Vassilopoulos, A. P., Manshadi, B. D., & Keller, T. (2010b). Piecewise non-linear constant life diagram formulation for FRP composite materials. *International Journal of Fatigue, 32*(10), 1731–1738.

Walker, K. (1970). The effect of stress ratio during crack propagation and fatigue for 2024-T3 and 7075-T6 aluminium. In M. S. Rosenfeld (Ed.), *Effects of environment and complex load history on fatigue life* (pp. 1–14). Philadelphia: ASTM STP 462, American Society for Testing Materials.

Walls, D. P., Bao, G., & Zok, F. W. (1993). Mode I fatigue cracking in a fiber reinforced metal matrix composite. *Acta Metallurgica et Materialia, 41*(7), 2061–2071.

Wilkins, D. J., Eisenmann, J. R., Camin, R. A., Margolis, W. S., & Benson, R. A. (1982). Characterizing delamination growth in graphite/epoxy. In K. L. Reifsnider (Ed.), *Damage in composite materials* (pp. 168–182). Philadelphia: ASTM STP 775, American Society for Testing and Materials.

Zhang, Y., Vassilopoulos, A. P., & Keller, T. (2008). Stiffness degradation and fatigue life prediction of adhesively-bonded joints for fiber-reinforced polymer composites. *International Journal of Fatigue, 30*(10–11), 1813–1820.

Zhang, Y., Vassilopoulos, A. P., & Keller, T. (2009). Environmental effects on fatigue behavior of adhesively-bonded pultruded structure joints. *Composites Science and Technology, 69*(7–8), 1022–1028.

Zhang, Y., Vassilopoulos, A. P., & Keller, T. (2010). Fracture of adhesively-bonded pultruded GFRP joints under constant amplitude fatigue loading. *International Journal of Fatigue, 32*(7), 979–987.

Zheng, J., & Powell, B. E. (1999). Effect of stress ratio and test methods on fatigue crack growth rate for nickel based superalloy Udimet720. *International Journal of Fatigue, 21*(5), 507–513.

Mixed-mode fatigue and fracture behavior of adhesively-bonded composite joints

M. Shahverdi, A.P. Vassilopoulos
École Polytechnique Fédérale de Lausanne (EPFL), Lausanne, Switzerland

7.1 Introduction

Adhesively-bonded joints are increasingly used as a joining technique for composite materials, thanks to their better performance as a permanent connection compared to other joining techniques such as bolted joints (Vallee, 2004). Nonetheless, failure in such joints can occur due to crack propagation in the adherend along paths outside the symmetry plane (Shahverdi, Vassilopoulos, & Keller, 2011). In order to increase design reliability, the fracture behavior of such joints under mixed-Mode I/II loading must be carefully studied. The most commonly used specimen for the characterization of the mixed-Mode I/II fracture behavior of composite materials is the mixed-mode bending (MMB) specimen (ASTM D6671, 2001). The MMB specimen proposed by Reeder and Crews (1990) is a combination of the double cantilever beam (DCB) and the end notched flexure (ENF) specimen, both standardized specimens for measuring pure Mode I and Mode II interlaminar fracture.

Almost any combination of Mode I and Mode II loading can be experimentally investigated by the MMB configuration. The MMB specimen has been extensively used for the characterization of the mixed-mode fracture behavior of different composite materials such as thermoplastic and thermosetting carbon fiber composites (Kinloch, Wang, Williams, & Yayla, 1993), carbon/epoxy laminates (Kim & Mayer, 2003), glass/epoxy composites (Benzeggagh & Kenane, 1996; Ozdil & Carlsson, 2000), glass/vinylester composites (Dharmawan, Simpson, Herszberg, & John, 2006), adhesively-bonded metal joints (Liu, Gibson, & Newaz, 2002), and adhesively-bonded composite/composite joints (Ducept, Davies, & Gamby, 2000).

The partitioning of the experimentally obtained total strain energy release rate, G_{tot}, into the G_I and G_{II} when mixed-mode conditions exist is challenging (Bhashyan & Davidson, 1997; Ozdil & Carlsson, 1999), especially in asymmetric specimens (Ducept, Gamby, & Davies, 1999; Harvey & Wang, 2012). There are two analytical methods in the literature that can be used for mode partitioning: the "global method" based on beam theory (Williams, 1988), and the "local method" based on the stress intensity factor calculation around the crack tip (Hutchinson & Suo, 1992). According to the "global method," the mode partition is performed globally by including the whole region around the crack faces, whereas the "local method" considers only the "near-the-crack-tip" region as explained by Harvey and Wang (2012) and de Morais and Pereira (2006). Therefore, the "global method" is more appropriate for fracture analysis of materials that exhibit crack propagation accompanied by a large process zone, while the "local method" is more appropriate for materials with small fracture process zones, e.g., ceramics (Harvey & Wang, 2012; de Morais & Pereira, 2006).

7.1.1 Mode partitioning in asymmetric specimens

The mode partition can also be performed using finite element (FE) models by means of the virtual crack closure technique (VCCT) (Krueger, 2004; Rybicki & Kanninen, 1977). This method is quite accurate for calculation of the fracture energy at the crack tip, especially when homogenous materials are analyzed. However, when the crack path lies in a bi-material interface, VCCT results concerning mode partition become sensitive to the mesh size around the crack tip (Agrawal & Karlsson, 2006; Raju, Crews, & Aminpour, 1988). In order to overcome this problem, a method has been proposed by Atkinson (1977) for analyzing isotropic fracture problems with a bi-material crack interface. This involves inserting a thin layer between the layers forming the interface and placing the crack within it. Because the crack tip is fully embedded in the resin layer, mode-mixity is not sensitive to mesh size.

Generally, the total fracture energy of a composite material comprises a fiber bridging component, G_{br}, and a tip component, G_{tip} (Sorensen, Botsis, Gmür, & Humbert, 2008). The VCCT is able to calculate the fracture energy at the crack tip (G_{tip}). The fiber bridging zone can be considered as part of the fracture process zone where the fracture energy is released. Many efforts have been made to model the effects of fiber bridging (de Morais, 2011; Tamuzs, Tarasovs, & Vilks, 2001), and to separate the two G components, mainly by FE modeling, with the cohesive zone model approach being the most commonly used for determination of the G_{br} (Shahverdi, Vassilopoulos, & Keller, 2013; Sorensen et al., 2008). The behavior of the cohesive element is based on a traction-separation law that defines the stresses at a particular location in a prescribed cohesive zone as a function of the opening displacement of the zone at that location. Cohesive laws in FE modeling have been used extensively during recent years (Shahverdi et al., 2013; Sørensen & Jacobsen, 2009). For example, the applicability of the CZM technique for modeling fiber bridging using a single layer of zero-thickness cohesive elements (COH2D4 in ABAQUS) along the delamination plane has been demonstrated by Sorensen et al. (2008).

7.1.2 Fatigue behavior characterization

Numerous approaches exist to characterize the fatigue behavior of engineering materials and structures and to develop procedures to accurately model and predict their fatigue life. These methods can be classified into two main approaches: those based on the phenomenological representation of the material/structural behavior (mainly expressed by the S—N relationship, e.g., Philippidis & Vassilopoulos, 2004; Zhang, Vassilopoulos, & Keller, 2010), and those based on fracture mechanics approaches, mainly expressed by the fatigue crack growth (FCG) curves (Martin & Murri, 1990; Shahverdi, Vassilopoulos, & Keller, 2012a). S—N curves, which can be easily derived based on the number of cycles to failure for a given cyclic stress, are simple, but they do not provide any information during fatigue life (Shahverdi et al., 2012a). On the other hand, the FCG curves represent the fatigue behavior by providing information concerning the damage developed in terms of crack or cracks that propagate during fatigue life (Shahverdi et al., 2012a).

The FCG curves are plots of the strain energy release rate, G (Martin & Murri, 1990; Shahverdi et al., 2012a) versus the crack propagation rate, da/dN, usually on logarithmic axes. Each FCG curve has a sigmoidal shape comprising three regions: the subcritical region, the linear region, and the unstable region. The Paris law has been extensively applied for modeling fatigue crack propagation in the stable region of FCG curves in composite materials under pure Mode I (Mall & Johnson, 1986; Mall, Yun, & Kochhar, 1989), mixed-Mode I/II (Asp, Sjögren, & Greenhalgh, 2001; Zhang et al., 2010), and pure Mode II (Bathias & Laksimi, 1985; Mall et al., 1989), and for adhesively-bonded structural joints (Shahverdi et al., 2012a; Shahverdi, Vassilopoulos, & Keller, 2012b). In Martin and Murri (1990) a phenomenological equation that is able to model the FCG behavior over the entire range of applied G, from the first to the third region is introduced. The derived model, designated the "total fatigue life model," expresses the crack growth rate as a function of the total maximum cyclic strain energy release rate, G_{tot}, the strain energy release rate threshold, $G_{tot,th}$, and the critical strain energy release rate, $G_{tot,c}$. The proposed model by Martin and Murri was used by Shahverdi et al. (2012a), and Shivakumar, Chen, Abali, Le, and Davis (2006) for characterizing the crack growth rate in pultruded glass fiber-reinforced polymer (GFRP) joints and glass/vinylester delaminated composite panels subjected to Mode I cyclic loading. In addition it was extended by Shahverdi et al. (2012a) to take into account the effect of the R-ratio on the FCG curves.

In the case of mixed-mode loading, it was shown by Mall et al. (1986) that the crack propagation rate is a function of the combined effect of G_I and G_{II}. Therefore it was suggested that the total strain energy release rate, G_{tot}, is more appropriate than G_I or G_{II} for derivation of the FCG curves. Additionally, the G_{tot} can be calculated much more easily than the individual G_I and G_{II} components under a mixed-mode loading condition. Experimental evidence in the literature shows that mode-mixity significantly affects the FCG curves obtained for different examined materials or joints (Fernández, De Moura, Da Silva, & Marques, 2013; Zhang, Peng, Zhao, & Fei, 2012). However, most of the published experimental data are limited to the second region of the FCG curve, since it is very time-consuming to derive a complete FCG curve including the region of very slow crack propagation, close to the fatigue threshold values, and difficult to monitor the rapid region close to $G_{tot,c}$. The majority of publications on the subject (Fernández et al., 2013; Mall et al., 1989) report that the slope of the FCG curves decreases as the G_{II}/G_{tot} increases. However, there are other experimental results (Kenane, Azari, Benmedakhene, & Benzeggagh, 2011; Kenane, Benmedakhene, & Azari, 2010) showing that the slope of the FCG curves decreases as the G_{II}/G_{tot} increases. On the other hand, it is reported by Mall et al. (1989) that the slope of the FCG curves initially decreases as the G_{II}/G_{tot} increases but increases again in the region close to pure Mode II, whereas Zhang et al. (2012) reported an initial increase and eventual decrease as G_{II}/G_{tot} increases.

An explanation for the above-mentioned different behaviors is still lacking, although some researchers have attempted to correlate them with the nature of the materials. For example, in O'Brien (1990), it is shown that, in general, the slope of the FCG curves decreases with increasing matrix toughness for any mode-mixity. However, this assumption is contradicted by the results presented by others (Mall et al., 1989; Zhang et al., 2012), who attribute changes of the FCG behavior to other factors such as fiber bridging and multiple cracking.

For each composite material/joint, a fatigue failure criterion can be established to describe and predict its behavior under any mode-mixity. In the literature, very few studies on the determination of mixed-Mode I/II fatigue failure criteria have been reported (Fernández et al., 2013; Kenane et al., 2011; Mall et al., 1989), and no attempt has been made to model the total fatigue life of bonded joints under mixed-Mode I/II loading conditions. The development of such methods is hindered by the lack of complete databases containing fatigue/fracture data regarding the total fatigue life of bonded joints.

7.1.3 Content of the chapter

One of the objectives of this chapter is to demonstrate the techniques for the characterization of the mixed-mode fracture behavior of adhesively-bonded pultruded GFRP joints. The partitioning of the fracture mode components and the modeling of the fiber bridging that affects the total fracture energy are presented as well. In such joints, the crack propagates along paths outside the symmetry plane where the materials on the two faces of the crack are different. In order to present the techniques and methodologies, the crack propagation under different mixed-mode loading conditions in asymmetric MMB adhesively-bonded pultruded GFRP specimens is experimentally investigated. A new analytical method, based on the existing "global method" and designated the "extended global method," (EGM) initially presented by Shahverdi, Vassilopoulos, and Keller (2014a) can be used to analyze the experimental results and thus take the asymmetry effect into account. The VCCT can also be used for calculation of the fracture components at the crack tip, and a CZM was established for the simulation and quantification of the fiber bridging.

The comprehensive database presented by Shahverdi et al. (2014a) and Shahverdi, Vassilopoulos, and Keller (2014b) has been used in this chapter for the demonstration of the presented approach. The database contains data from fatigue/fracture experiments that were performed up to 5 million cycles to obtain complete FCG curves for all examined specimens. A total fatigue life model previously introduced by Shahverdi et al. (2012a) was used to simulate the exhibited behavior. The model parameters were determined by developing the experimental FCG curves under different mode-mixity, and consequently have been correlated with the mode-mixity.

7.2 Mixed-mode fatigue and fracture experimental investigation

7.2.1 Material description

Adhesively-bonded pultruded GFRP asymmetric MMB joints were examined under quasi-static loading. The laminates, supplied by Fiberline A/S, Denmark, consisted of E-glass fibers embedded in isophthalic polyester resin and had a width of 40 mm and a thickness of 6.0 mm. The laminates, cut from I-beam profiles used for civil engineering applications, comprised two outer combined mat layers and a roving layer in the symmetry plane. One combined mat consisted of two outer chopped strand mats (CSM), and an inner woven 0°/90° fabric, all three stitched together. A 40-g/m^2 polyester surface veil was added on the outside. The fiber architecture of the laminates and corresponding thickness of each layer derived by optical microscopy are shown in Figure 7.1. The fiber content, determined by burn-off tests in accordance with ASTM D3171, 2011, was 43.2 vol.% based on the fiber density of 2560.0 kg/m^3 specified by the manufacturer and the assumption that no voids were present; the fiber fractions are listed in Table 7.1. The weight of the second, inner combined mat was almost double that of the first, outer mat, and the proportion of woven fabrics was much higher. The longitudinal strength and Young's modulus of the GFRP laminate were obtained from tensile experiments, according to ASTM D3039/D3039M, 2008, as being 307.5 MPa and 25.0 GPa, respectively (average values from Shahverdi et al. (2011)).

A two-component epoxy adhesive system, Sikadur 330 (supplied by Sika, A. G, Switzerland), was used as the bonding material. The tensile strength of the adhesive was 39.0 MPa and the longitudinal Young's modulus 4.6 GPa (average value from de Castro and Keller (2008)). The epoxy exhibited an almost elastic behavior and a brittle failure under quasi-static tensile loading.

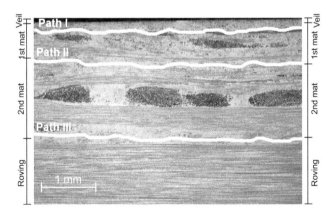

Figure 7.1 Fiber architecture of upper half of a laminate (section parallel to pultrusion direction) and observed crack propagation paths.

Table 7.1 **Fiber architecture and fractions by volume and weight of pultruded laminates**

Layer	Average thickness (mm)	Vol.%	Wt%
Veil	2 × 0.05*		
First combined mat	2 × 0.63		
−2 CSM		2 × 1.7	2 × 2.5
−Woven 0°/90°		2 × 1.6	2 × 2.3
Second combined mat	2 × 1.07		
−2 CSM		2 × 2.6	2 × 3.8
−Woven 0°/90°		2 × 4.1	2 × 6.0
Roving (UD)	1 × 2.5	1 × 23.3	1 × 34.2
Total	6.0	43.3	64.4

CSM, chopped strand mats; UD, unidirectional.
* 2× = on each side of the symmetry axis.

7.2.2 Specimen description, set-up and procedure

A drawing and a photograph of the MMB configuration are shown in Figures 7.2 and 7.3. The specimen length is 400 mm, and the half span length, L, is 170 mm (see Figure 7.4). All surfaces subjected to bonding were mechanically abraded by approximately 0.3 mm to increase roughness and then chemically degreased using acetone. A Teflon film of 0.05-mm thickness, sufficiently thin for the given joints,

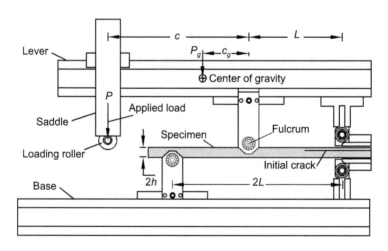

Figure 7.2 Schematic of mixed-mode bending apparatus.

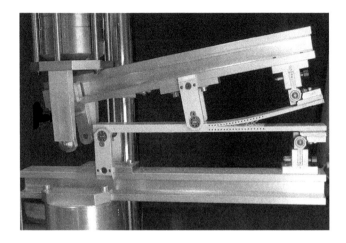

Figure 7.3 Photograph of mixed-mode bending apparatus.

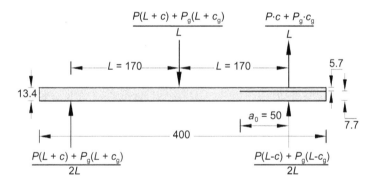

Figure 7.4 Mixed-mode bending specimen with applied loads and dimensions, in millimeters.

was placed between the upper arm and the adhesive layer to introduce the precrack. The length of the precrack was $a_0 = 50$ mm measured from the loading line. The joint and the precrack lengths were selected with the aim to have a long enough precrack length to achieve stable crack propagation, and enough overall joint length in order to reach a plateau in the R-curve. An aluminum frame was used to assist the alignment of the two pultruded laminates during the fabrication. The 2-mm thickness of the adhesive was maintained by using spacers embedded in the bonding area. The resulting joints are representative for civil engineering structures in which dimensions are significantly larger compared to aerospace or automotive applications. Typical adhesive thicknesses in such structures can vary significantly in order to compensate for tolerances.

In-house-developed piano hinges were glued, using the same epoxy adhesive, at the end of both specimen arms to allow load application. After preparation of the configuration, the specimens were kept under laboratory conditions for 24 h and then placed for 24 h in a conditioning chamber at 35 °C and 50% ± 10% RH to ensure full curing of the adhesive. The resulting thickness of the MMB specimens was 13.4 mm, and the precrack was located 1.0 mm above the center axis of the joints (Figure 7.4).

A 25-kN MTS Landmark servo-hydraulic testing rig, calibrated to 20% of its maximum capacity, was used for all the experiments conducted under laboratory conditions, 23 °C ± 5 °C and 50% ± 10% RH. The specimens were loaded under displacement control at a constant rate of 1 mm/min. The load was applied by means of a lever at a distance c from the fulcrum. The loading lever was an aluminum I-beam weighing 28.6 N, P_g, had a bending stiffness of around 170 times that of the MMB specimen and assumed to be rigid. The applied load, the mid-span load, and the left support reaction are applied via bearing-mounted rollers to reduce the frictional force. The right end of the specimen is loaded using in-house—developed piano hinges. The length of the loading lever, denoted c in Figure 7.2, determines the mixed-mode ratio. The applied loads and displacements were continuously recorded. A total of 21 specimens with four different lever lengths, $c = 227, 150, 100$, and 60 mm, were examined (Table 7.2).

The crack length was measured by means of a video extensometer. For this purpose, pairs of black dots were marked at equal intervals of approximately 5 mm on the upper and lower lateral surfaces of the specimen (Figures 7.3 and 7.5). The X—Y coordinates of these dots were monitored by a video-extensometer camera and recorded continuously with a precision of 10^{-5} m by a Labview application. The crack length was determined based on the change in relative distance between each pair of upper and lower black dots with a simple data processing module. According to this module, when the relative distance between each pair of black dots changed by more than 0.1 mm, it was considered that the crack reached the position of these dots.

7.2.3 Experimental fatigue procedure

A 25-kN MTS Landmark servo-hydraulic testing rig, calibrated to 20% of its maximum capacity, 5-kN, was used for all the fatigue experiments conducted under laboratory conditions, 23 °C ± 5 °C and 50% ± 10% RH. All constant amplitude fatigue experiments were performed under displacement control with R-ratios (the ratio of minimum/maximum displacement) equal to 0.5 at a frequency of 5 Hz.

The experimental procedure followed three steps:

- Step 1: The specimen was installed into the test set-up and aligned.
- Step 2: A loading ramp was applied to initiate and propagate the crack up to around 25—45 mm for ELS and MMB specimens, corresponding to the crack length required to reach the plateau in the R-curve and equal to the fiber bridging length.
- Step 3: The displacement-controlled fatigue experiment was performed with 100% of the crack opening displacement reached in Step 2, in order to record the initial rapid crack propagation values corresponding to the rapid crack growth rate region.

Table 7.2 **MMB specimen geometrical parameters**

Specimen code	c (mm)	c_g (mm)
MMB-01	227	54
MMB-02	227	54
MMB-03	227	54
MMB-04	227	54
MMB-05	227	54
MMB-06	150	38
MMB-07	150	38
MMB-08	150	38
MMB-09	150	38
MMB-10	150	38
MMB-11	150	38
MMB-12	100	27
MMB-13	100	27
MMB-14	100	27
MMB-15	100	27
MMB-16	100	27
MMB-17	100	27
MMB-18	60	18
MMB-19	60	18
MMB-20	60	18
MMB-21	60	18
MMB0.5-01	227	54
MMB0.5-02	227	54
MMB0.5-03	227	54
MMB0.5-04	150	38
MMB0.5-05	150	38
MMB0.5-06	150	38
MMB0.5-07	100	27
MMB0.5-08	100	27
MMB0.5-09	100	27

MMB, mixed-mode bending.

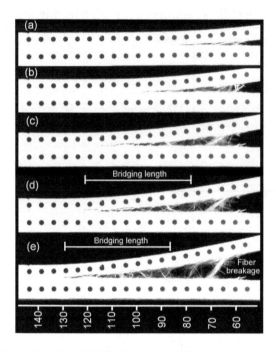

Figure 7.5 Side view of specimen MMB-02 showing different stages of crack development, dimensions in millimeters. MMB, mixed-mode bending.

The load was applied, via in-house—developed piano hinges, at the lever at distance c from the fulcrum (Figure 7.2). The position of the loading lever determines the mode-mixity (Table 7.2). The MMB experiments were performed for three different lever lengths, $c = 227$, 150, and 100 mm. Relatively high G_I/G_{II} ratio experiments require a very long lever which is not practical, and it was in fact not possible to conduct low G_I/G_{II} ratio fatigue experiments due to instability of crack propagation, according to the preliminary experiments performed with a $c = 60$ mm.

Determination of the crack length during fatigue loading is not an easy task. Direct methods, crack gages, visual observation and video extensometer, and also indirect methods, e.g., the dynamic compliance calibration, exist for this purpose. The crack length determination by video extensometer is a method that allows the direct measurement without interruption of the loading. For this purpose, pairs of black dots are marked at equal intervals (of ~2.5 and 5 mm for the examined joints) respectively on the upper and lower lateral surfaces of the specimen. The X—Y coordinates of these dots are monitored by a video-extensometer camera and recorded at each 1000 cycles at the maximum applied displacement with a precision of 10^{-2} mm by a Labview application. Similar to the quasi-static tests, the crack length is determined based on the change in relative distance between each pair of upper and lower black dots with a simple data processing module.

According to this module, when the relative distance between each pair of black dots is changed by more than 0.1 mm, it is considered that the crack reached the position of these dots.

7.2.4 Experimental results

7.2.4.1 Observed failure modes

In all of the examined specimens, the observed failure mode, according to ASTM D5573 (1999), was a fiber-tear failure or light-fiber tear failure (Figure 7.5). Fiber bridging started to develop with increasing crack opening displacement. Fibers from both arms of the specimen bridged the crack, transferring the load from one side to the other. At a certain crack opening displacement, fibers far from the crack tip were broken or pulled out (see crack length of up to ~85 mm in Figure 7.5(e)). The length along which fibers were not broken or pulled out is designated the "fiber bridging length", l_{br} (crack length of ca. 75−120 mm in Figure 7.5(d)), and remained almost constant, following the crack tip for the rest of the fracture process (see Figure 7.5(e) crack length of ~85−130 mm).

The precrack was located between the adhesive and first mat layer of the upper laminate; therefore the crack initiated and propagated along Path I, according to the nomenclature given by Shahverdi et al. (2011) (Figure 7.1). However, in all of the examined specimens, after a crack propagation of 5−20 mm, the crack penetrated the first mat layer and propagated between the first and second mat layers corresponding to Path II. In some cases, the crack tip stopped propagating between the mat layers, and a new crack appeared between the second mat layer and the roving layer, Path III. When this occurred, rovings of the woven fabric of the second mat layer bridged the two crack faces and significantly increased the applied load required for crack propagation. The term "roving bridging" was used by Shahverdi et al. (2011) to describe this phenomenon. According to the nomenclature used by Shahverdi et al. (2011), the crack path observed in this case was designated Path III with roving bridging. This phenomenon was observed in 11 of the 21 examined specimens.

7.2.4.2 Load−displacement responses

The load and crack length responses versus load-point displacement, δ_P, of representative specimens for a Path II crack under different lever lengths are shown in Figures 7.6−7.9. Linear responses until crack initiation were observed in all cases. The slopes of the load−displacement curves in the linear part were 27.7, 48.5, 79.0, and 106.6 N/mm for $c = 227$, 150, 100, and 60 mm, respectively. The load increased until a maximum value was reached and then gradually decreased. The maximum load obtained varied significantly when the lever length changed and increased as lever length decreased. The maximum load obtained from the representative specimen for $c = 60$ mm was around 1375 N, and was almost six times higher than that of the representative specimen for $c = 227$ mm, around 225 N.

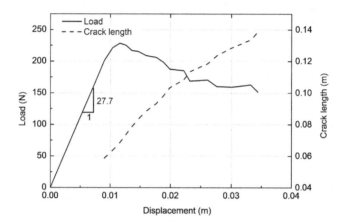

Figure 7.6 Load and crack length versus load-point displacement, MMB-04, $c = 227$ mm. MMB, mixed-mode bending.

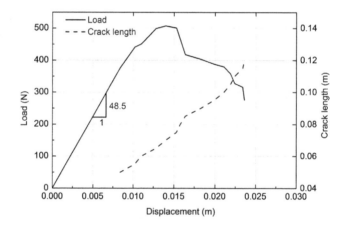

Figure 7.7 Load and crack length versus load-point displacement, MMB-09, $c = 150$ mm. MMB, mixed-mode bending.

As illustrated in Figures 7.6–7.9, crack propagation was faster for lower lever lengths. For example, the crack length increment from 60 to 120 mm corresponded to around 17.0, 15.0, 6.0, and 2.5 mm displacement increment for $c = 227$, 150, 100, and 60 mm, respectively. In addition, with longer lever lengths, the crack propagated steadily by displacement increment. However, with shorter lever lengths, unsteady crack propagation was observed (see Figure 7.9), with a sudden crack length increase at around 17 mm.

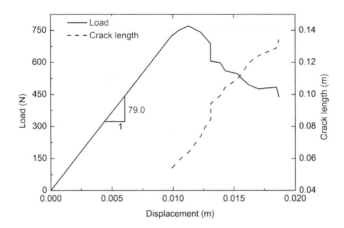

Figure 7.8 Load and crack length versus load-point displacement, MMB-16, $c = 100$ mm. MMB, mixed-mode bending.

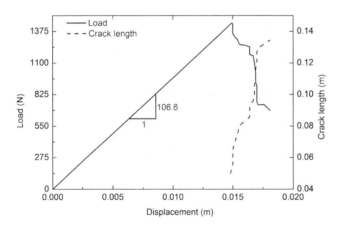

Figure 7.9 Load and crack length versus load-point displacement, MMB-19, $c = 60$ mm. MMB, mixed-mode bending.

7.2.4.3 Load and crack length versus number of cycles

Figure 7.10 shows the variation of the maximum applied load and the corresponding crack length during the fatigue test for a representative MMB specimen examined under $G_I/G_{II} = 1.08$ loading condition. The maximum load decreased and the crack length increased rapidly at the beginning and then followed a smoother decreasing and increasing trend in load and displacement, respectively, before approaching a

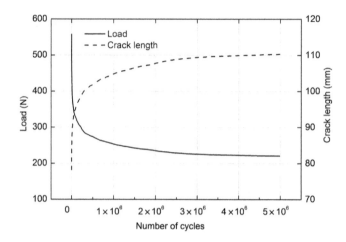

Figure 7.10 Load and crack length versus number of cycles, for MMB0.5-09 specimen with $G_I/G_{II} = 1.08$. MMB, mixed-mode bending.

plateau value, corresponding to the fatigue threshold values. Detailed information about the experiment conducted on each joint, including maximum cyclic displacement, δ, number of cycles for each specimen, crack length at the starting point of cyclic loading, a_1, crack length attained at the end of cyclic loading, a_2, and maximum loads, P, corresponding to a_1 and a_2, are presented in Table 7.3.

Table 7.3 **Summary of fatigue mixed-mode bending (MMB) results**

Specimen code	δ (mm)	Number of cycles	a_1 (mm)	a_2	P_1 (N)	P_2	G_{tot1} (J/m^2)	G_{tot2}
MMB0.5-01	15	3,225,000	79	114	232	75	993	241
MMB0.5-02	15	4,216,000	70	107	271	86	1011	267
MMB0.5-03	15	4,165,000	92	120	212	81	1012	292
MMB0.5-04	12	4,187,000	67	115	460	128	1274	309
MMB0.5-05	12	2,896,000	66	112	485	160	1335	429
MMB0.5-06	12	3,952,000	75	122	422	117	1411	298
MMB0.5-07	10	3,500,000	75	138	577	119	1167	470
MMB0.5-08	10	4,620,000	74	131	572	188	1130	389
MMB0.5-09	10	4,986,000	78	111	558	220	1185	375

MMB, mixed-mode bending.

The load decrease with number of cycles for MMB specimens depended on mode-mixity and was around 580 to 200 N, 480 to 130 N, and 270 to 85 N for mode-mixities of 1.08, 2.20, and 3.70, respectively. The corresponding crack length increments were 28−37 mm, 44−48 mm, and 33−62 mm for MMB specimens under mode-mixities of 1.08, 2.20, and 3.70, respectively.

7.3 Fatigue and fracture data analysis

7.3.1 Experimental compliance method

The total strain energy release rate can be calculated by the experimental compliance method (ECM), based on experimentally derived values of loads, displacements, and crack lengths, as follows:

$$G = \frac{P^2}{2B} \frac{dC}{da} \tag{7.1}$$

where P is the applied load, C is the compliance of the specimen, a is the crack length, and B is the specimen width. The MMB specimen compliance is defined as:

$$C = \frac{\delta_P}{P} \tag{7.2}$$

where δ_P is the load-point displacement. From among different models for fitting compliance−crack length curves, Eqn (7.3) was selected because it better fits the experimental results according to Bhashyan and Davidson (1997):

$$C = C_0 + ma^3 \tag{7.3}$$

The ECM has also been used for mode partition in symmetric MMB specimens (Bhashyan & Davidson, 1997; Reeder & Crews, 1990), in which it is possible to determine the Modes I and II components of displacement along with the Modes I and II components of load (see Reeder and Crews (1990), for details).

However, ECM cannot be used for the mode partitioning of mixed-mode results as is the case for the asymmetric and layered joint configurations used in this chapter. A new method designated the EGM, based on the "global method" (Williams, 1988), is therefore proposed in the next section for the mode partitioning of the examined specimens.

7.3.2 Extended global method

Williams (1988) developed beam theory-based equations for calculating the energy release rate from the values of bending moments and loads in a cracked laminate.

In this chapter, the equations have been modified in order to solve a crack propagation problem in which the crack is asymmetric and lies between two different orthotropic layers under the bending moments M_1 and M_2, as shown in Figure 7.11. According to linear-elastic analysis, the total strain energy release rate is (de Morais & Pereira, 2006):

$$G = \frac{6}{B^2}\left(\frac{M_1^2}{E_1 h_1^3} + \frac{M_2^2}{E_2 h_2^3} - \frac{(M_1 + M_2)^2}{E(h_1 + h_2)^3}\right) \tag{7.4}$$

where the bending moments (assumed to be positive when counterclockwise) are evaluated at a section of the specimen surrounding the crack tip. According to the "global method" (Williams, 1988), pure Mode I exists when symmetric moments act on the joint arms, i.e., $M_1 = M_I$ and $M_2 = -M_I$, and pure Mode II requires equal curvature of both arms, i.e., $M_1 = M_{II}$ and $M_2 = \psi M_{II}$. Furthermore, ψ is defined based on the ratio of the joint arm thicknesses h_1 and h_2. However, because the curvature of the orthotropic-layered arms depends on the bending stiffness rather than just the thickness, in the present chapter this definition is replaced by the equivalent bending stiffness ratio:

$$\psi = \frac{(EI)_{\text{eq}2}}{(EI)_{\text{eq}1}} \tag{7.5}$$

Therefore, under mixed-Mode I/II, loadings are

$$M_1 = M_I + M_{II} \tag{7.6}$$

$$M_2 = -M_I + \psi M_{II} \tag{7.7}$$

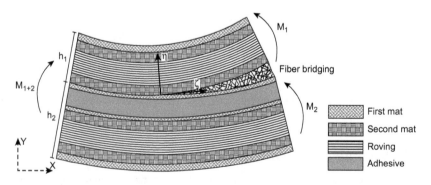

Figure 7.11 Schematic illustration of Path II crack in joint subject to bending moments.

Substitution of Eqns (7.6) and (7.7) into Eqn (7.4) leads to the partition of G into G_I and G_II as:

$$G_\text{I} = \frac{M_\text{I}^2}{2B(EI)_\text{eq1}} \left(\frac{1+\psi}{\psi}\right) \tag{7.8}$$

$$G_\text{II} = \frac{M_\text{II}^2}{2B(EI)_\text{eq1}} \left(1+\psi - \xi(1+\psi)^2\right) \tag{7.9}$$

with

$$\xi = \frac{(EI)_\text{eq1}}{(EI)_\text{eq}} \tag{7.10}$$

The ξ parameter, the ratio of the equivalent upper arm bending stiffness over the equivalent joint stiffness, similar to the ψ parameter, is defined as bending stiffness rather than thickness ratio. In the case of the asymmetric MMB specimen, as shown in Figure 7.4, the bending moments at the section surrounding the crack tip are:

$$M_1 = \frac{Pc + P_g c_g}{L} a \tag{7.11}$$

$$M_2 = \frac{P(L-c) + P_g(L-c_g)}{2L} a \tag{7.12}$$

Substitution of Eqns (7.11) and (7.12) into Eqns (7.6) and (7.7) leads to:

$$M_\text{I} = \frac{(2\psi + 1)(Pc + P_g c_g) - (P + P_g)L}{2L(1+\psi)} a \tag{7.13}$$

$$M_\text{II} = \frac{P(L+c) + P_g(L+c_g)}{2L(1+\psi)} a \tag{7.14}$$

Substitution of Eqns (7.13) and (7.14) into Eqns (7.8) and (7.9) leads to:

$$G_\text{I} = \frac{[(2\psi + 1)(Pc + P_g c_g) - (P + P_g)L]^2}{8\psi\xi(1+\psi)BL^2(EI)_\text{eq}} a^2 \tag{7.15}$$

$$G_\text{II} = \frac{[P(L+c) + P_g(L+c_g)]^2}{8\xi(1+\psi)^2 BL^2(EI)_\text{eq}} \left(1+\psi - \xi(1+\psi)^2\right) a^2 \tag{7.16}$$

Equations (7.15) and (7.16) are closed form equations for the calculation of G_I and G_{II} in which all the parameters are obtained directly from the experiments.

7.3.3 Finite element method

Two-dimensional (2D) plane–strain models were developed in ANSYS (academic version 13.0) to calculate the Modes I and II fracture components for different lever lengths (Table 7.2). All layers of the laminates were modeled according to the thicknesses estimated by optical microscopy (Table 7.1). The material properties are given in Table 7.4 and were adopted from Shahverdi et al. (2013). The element PLANE182, a 4-node structural solid, was used to model different layers. A manual mesh with controlled mesh size was used. The aspect ratio of the elements in the vicinity of the crack tip was kept at 1/1 (Krueger, 2004). Elements of the same size were used on both sides of the crack tip to avoid any non-convergence that may be caused by different sizes. Nonlinear-elastic analysis was performed, allowing calculation of the specimen deformation, nodal forces, and nodal displacements.

Fiber bridging along the crack faces was modeled by using a single layer of zero-thickness cohesive elements, INTER202, along the crack plane. INTER202 is a 2D 4-node interface element with 2 degrees of freedom at each node. The cohesive element behavior is based on a traction-separation law that defines the stresses, σ_{br}, at a particular location as a function of the opening displacement, δ. The traction-separation relationship is such that, with increasing opening displacement, the traction across the interface reaches a maximum, at ($\bar{\delta}$, σ_{max}), then decreases and eventually reaches zero, presenting a complete separation at an opening

Table 7.4 Properties used for finite element modeling

Material data	First combined mat	Second combined mat	Roving	Veil	Adhesive
E_{11} (GPa)	12.8	15.1	38.9	3.2	4.6
E_{22} (GPa)	12.8	15.1	3.2	3.2	4.6
E_{33} (GPa)	3.2	3.2	3.2	3.2	4.6
G_{12} (GPa)	6.2	6.7	2.7	1.2	1.7
G_{23} (GPa)	1.4	1.4	1.4	1.2	1.7
G_{31} (GPa)	1.4	1.4	2.7	1.2	1.7
ν_{12}	0.27	0.27	0.32	0.38	0.37
ν_{23}	0.36	0.36	0.27	0.38	0.37
ν_{31}	0.36	0.36	0.35	0.38	0.37

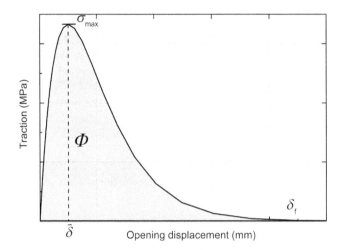

Figure 7.12 Schematic illustration of cohesive traction—separation law.

displacement of δ_f, at the end of the fiber bridging length, l_{br} (Figure 7.12). The length of the fiber bridging zone was obtained from the experimental observation. The area under the $\sigma-\delta$ curve represents the amount of energy dissipated during crack propagation in the cohesive zone, the cohesive energy. The three parameters cohesive energy, Φ, maximum traction, σ_{max}, and maximum opening displacement, δ_f, are interdependent, and therefore the CZM can be described by two of them (ASTM D3171, 2011), assuming an appropriate traction-separation cohesive law model, which can be linear, polynomial, exponential or user-defined (de Morais, 2011; Sorensen et al., 2008; Tamuzs et al., 2001). In this chapter, for modeling the fiber bridging, an exponential law was used, which, according to Sorensen et al. (2008), can model this effect better than the other laws. The applied exponential laws implemented in ANSYS software and used in this chapter are following the assumption that the work of separation under pure shear and pure normal conditions are assumed to be equal to each other (Xu & Needleman, 1994). The normal and tangential tractions, T_n and T_t, respectively, are:

$$T_n = e\sigma_{max}\frac{\delta_n}{\bar{\delta}_n}e^{-\frac{\delta_n}{\bar{\delta}_n}e^{-\left(\frac{\delta_t}{\bar{\delta}_t}\right)^2}} \tag{7.17}$$

and

$$T_t = 2e\sigma_{max}\frac{\bar{\delta}_n}{\bar{\delta}_t}\frac{\delta_t}{\bar{\delta}_t}\left(1+\frac{\delta_n}{\bar{\delta}_n}\right)e^{-\frac{\delta_n}{\bar{\delta}_n}e^{-\left(\frac{\delta_t}{\bar{\delta}_t}\right)^2}} \tag{7.18}$$

where δ_n and δ_t are the normal and tangential opening displacements along the cohesive zone, and $\bar{\delta}_n$ and $\bar{\delta}_t$ are the arbitrary normal and tangential opening displacements at maximum traction. The values of $\bar{\delta}$ and σ_{max} required by the CZM were estimated by an iterative procedure. The selected $\bar{\delta}$ is the one that allows the FE model to predict an opening displacement equal to δ_f, in which the traction reaches zero at the location of l_{br} behind the crack tip. Accordingly, selected σ_{max} values were those that resulted in the same loads computed by the FE models as those obtained from the experiments, for identical displacements and crack lengths. The estimated cohesive element model parameters for different lever lengths are listed in Table 7.5.

The following equation represents the amount of energy dissipated in the crack-bridging zone, G_{br}, according to the CZM approach (Sorensen et al., 2008).

$$G_{br} = \int_0^{\delta_f} \sigma_{br} d\delta \tag{7.19}$$

where σ_{br} is the bridging traction and δ is the relative opening displacement along the fiber bridging length of the upper and lower arms. The G_{br} can be partitioned into Mode I and Mode II components as:

$$(G_{br})_I = \int_0^{\delta_{f-n}} (\sigma_{br})_\eta d\delta_\eta \tag{7.20}$$

and

$$(G_{br})_{II} = \int_0^{\delta_{f-t}} (\sigma_{br})_\zeta d\delta_\zeta \tag{7.21}$$

Table 7.5 **Traction—separation cohesive model parameters for different lever lengths**

Specimen (lever length)	σ_{max} (MPa)	$\bar{\delta}_n$ (mm)	$\bar{\delta}_t$ (mm)
MMB-04 ($c = 227$ mm)	0.65	0.37	0.24
MMB-09 ($c = 150$ mm)	0.85	0.41	0.24
MMB-17 ($c = 100$ mm)	0.75	0.45	0.23
MMB-19 ($c = 060$ mm)	0.85	0.48	0.23

MMB, mixed-mode bending.

where δ_{f-n} is the maximum normal crack displacement, δ_{f-t} is the maximum tangential crack displacement, and ζ and η are the local axes (Figure 7.11). In Eqns (7.19)–(7.21), the bridging traction is obtained from the cohesive elements in the FE models along the bridging length.

The VCCT was used for calculation of the fracture parameters at the crack tip. Bi-material interfaces were present in all specimens. Therefore, the calculated $G_{\text{I-tip}}$ and $G_{\text{II-tip}}$ components and the calculated mode-mixity, $G_{\text{I}}/G_{\text{II}}$, depended on the FE mesh size around the crack tip, Δa, and did not represent the actual fracture development (see the dashed line in Figure 7.13). The approach proposed by Atkinson (1977) was applied in Shahverdi et al. (2013) and in the present chapter in order to diminish the effect of the Δa. A thin "resin" interlayer was inserted that had the average properties of the adjacent layers of the interface. The thickness of the resin interlayer was selected as being 0.10 mm as a compromise resulting in almost no changes in the stiffness of the model (less than 1%) and also introducing a reasonable number of elements into the FE model. The mesh size was gradually varied from 0.050 mm to 0.005 mm, representing two to 20 elements through the resin interlayer. As shown in Figure 7.13, the $G_{\text{I}}/G_{\text{II}}$ component obtained for a Path II crack with a resin interlayer is independent of the mesh size.

The total strain energy release rates, G_{tot}, calculated by FE analysis were the sum of the G_{tip} and G_{br} under both Mode I and Mode II, i.e., $G_{\text{tot}} = (G_{\text{tip}})_{\text{I}} + (G_{\text{tip}})_{\text{II}} + (G_{\text{br}})_{\text{I}} + (G_{\text{br}})_{\text{II}}$. These values are compared with the obtained experimental values in the next section.

7.3.4 Fatigue crack growth (FCG) curves

The FCG curves are plots of the strain energy release rate, G, versus the crack propagation rate, da/dN, usually on logarithmic axes. Each FCG curve has a

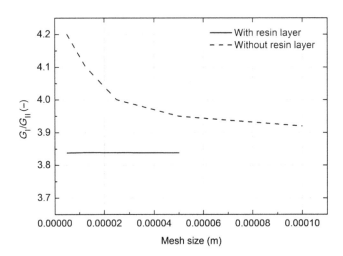

Figure 7.13 Mesh sensitivity analysis for a crack along Path II, $a = 100$ and $c = 227$ mm.

sigmoidal shape comprising three regions: the subcritical region, the linear region, and the unstable region. Phenomenological models (Martin & Murri, 1990; Shahverdi et al., 2012a) can be used to express the crack growth rate as a function of the maximum cyclic total strain energy release rate, G_{tot}, the strain energy release rate threshold, $G_{tot,th}$, and the quasi-static strain energy release rate values, $G_{tot,c}$.

Each FCG curve can be divided into three regions, namely, subcritical, linear, and unstable regions. Parameters D and m can then be estimated by a linear regression analysis after fitting Eqn (7.22) to the middle region:

$$\frac{da}{dN} = D(G_{tot})^m \tag{7.22}$$

In the subcritical region, the da/dN varies between zero and a value corresponding to the lowest value in region 2. The da/dN equation can then be written as follows:

$$\frac{da}{dN} = D(G_{tot})^m \left(1 - \left(\frac{G_{tot,th}}{G_{tot}}\right)^{Q_1}\right) \tag{7.23}$$

where $G_{tot,th}$ is the total strain energy release rate threshold and the exponent Q_1 is determined using D and m estimated in the previous step by fitting the equation to the data in regions 1 and 2.

In the unstable region, da/dN varies between infinite, when G_{tot} is equal to $G_{tot,c}$ and the transition value corresponding to the maximum value in region 2. The corresponding da/dN equation is as follows:

$$\frac{da}{dN} = D(G_{tot})^m \frac{1}{1 - \left(\frac{G_{tot}}{G_{tot,c}}\right)^{Q_2}} \tag{7.24}$$

The exponent Q_2 can be determined by fitting Eqn (7.24) to the experimental data in regions 2 and 3. Finally, the combined da/dN equation that covers all three regions is given by:

$$\frac{da}{dN} = D(G_{tot})^m \frac{\left(1 - \left(\frac{G_{tot,th}}{G_{tot}}\right)^{Q_1}\right)}{\left(1 - \left(\frac{G_{tot}}{G_{tot,c}}\right)^{Q_2}\right)} \tag{7.25}$$

where m, Q_1, and Q_2 are the empirical model parameters. For each mode-mixity ratio, these model parameters can be determined experimentally. Equation (7.25) is a total fatigue life model that covers all three FCG regions.

7.4 Results and discussion

The total strain energy release rate, G_{tot}, obtained by the ECM, the EGM, and FE modeling is illustrated in Figures 7.14–7.17 for representative specimens of the four different mode-mixity ratios. Similar results were obtained from the three different methods with almost ideal R-curves. The mean value of the visually determined plateau, taking the typical scatter of this type of material into account, was assumed to represent the G_{tot} for propagation. The values of G_{tot} for the initiation and propagation based on EGM for all of the specimens with the average values and standard deviations are presented in Table 7.6 according to the observed failure modes. In agreement with the results reported by Dharmawan et al. (2006) and Ducept et al. (2000), the total strain energy release rate corresponding to all paths increased as the lever length decreased, i.e., the Mode II contribution increased.

The mode partitioning results according to the EGM and the nonlinear FE analyses are also shown in Figures 7.14–7.17. Similar mode partition was achieved using these two methods. The EGM is more practical because it provides the fracture mode partition under mixed-mode loading conditions based on closed form equations with their parameters obtained directly from the experiments. In contrast, the use of the FE method requires the establishment of complex models containing parameters that must be estimated via iterative trial-and-error procedures. The mode-mixity obtained by the EGM and the FE models for different lever lengths versus crack length are presented in Figure 7.18. Slight variations of between 1% and 5% are observed between

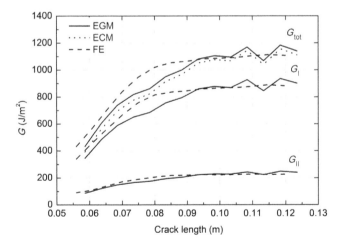

Figure 7.14 G versus crack length from MMB-04 determined by extended global and experimental compliance methods and finite element modeling. EGM, extended global method; ECM, experimental compliance method; FE, finite element; MMB, mixed-mode bending.

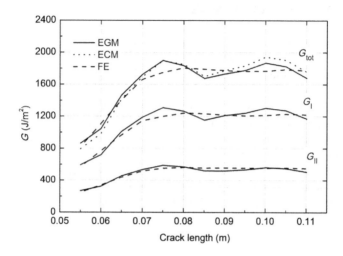

Figure 7.15 G versus crack length from MMB-09 determined by extended global and experimental compliance methods and finite element modeling. EGM, extended global method; ECM, experimental compliance method; FE, finite element; MMB, mixed-mode bending.

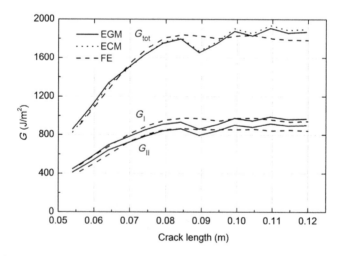

Figure 7.16 G versus crack length from MMB-16 determined by extended global and experimental compliance methods and finite element modeling. EGM, extended global method; ECM, experimental compliance method; FE, finite element; MMB, mixed-mode bending.

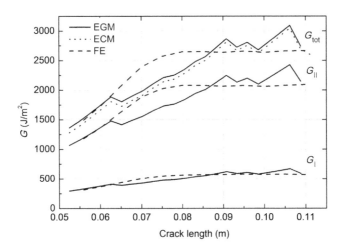

Figure 7.17 G versus crack length from MMB-19 determined by extended global and experimental compliance methods and finite element modeling. EGM, extended global method; ECM, experimental compliance method; FE, finite element; MMB, mixed-mode bending.

the two sets of results for lever lengths ranging from 227 to 60 mm. Nominal G_I/G_{II} ratios for Path II crack propagation corresponding to $c = 227$, 150, 100, and 60 mm were assumed to be 3.70, 2.20, 1.08, and 0.28, respectively, and are higher than the corresponding nominal values, 2.21, 1.02, 0.31, and 0.00, calculated according to ASTM D6671 (2001) by considering a symmetric joint configuration. Nominal G_I/G_{II} ratios for Paths I and III crack propagation are also presented in Table 7.6, with the values of G_I and G_{II} as a function of the associated crack paths for all specimens calculated by the EGM. The G_I/G_{II} ratios increased as the crack propagated in a deeper path. The average values with standard deviations for the four different configurations are also given in this table.

The G_I and G_{II} curves shown in Figures 7.14–7.17 for different mode-mixity ratios are expressed as the sum of the energy release rate at the crack tip, G_{tip}, and the contribution of the fiber bridging, G_{br}, in Figures 7.19–7.22. Fiber bridging results in an increase of G with increasing crack length. This phenomenon is expressed by the R-curve that follows an initially increasing trend before reaching a plateau as from which the bridging length remains constant. Using the presented FE models, it was possible to compute the G_{tip} and G_{br} for the representative specimens. The summation of these two values, shown as "$G_{I\text{-}FE}$ and $G_{II\text{-}FE}$" in Figures 7.19–7.22, was in good agreement with the experimentally derived values according to the EGM. The contribution of the fiber bridging was not constant, depended on the G_I/G_{II}, and decreased as the G_I/G_{II} decreased. The G_{br}/G ratio was around 60%, 55%, 52%, and 45% for MMB-04, MMB-09, MMB-17, and MMB-19, respectively.

Table 7.6 **Extended global method strain energy release rate at different path under different mixed-mode loading (J/m^2) with average values and standard deviations**

Specimen Code	Initiation				Path I			
	G_I	G_{II}	G_{tot}	G_I/G_{II}	G_I	G_{II}	G_{tot}	G_I/G_{II}
MMB-01	205	85	290	2.41				
MMB-02	166	69	235	2.41	410	151	561	2.71
MMB-03	133	55	188	4.41				
MMB-04	159	66	225	2.41				
MMB-05	163	67	230	2.41	497	183	680	2.71
Average	165 ± 26	68 ± 11	233 ± 36		454 ± 62	167 ± 23	620 ± 84	
MMB-06	162	128	290	1.27				
MMB-07	210	165	375	1.27	550	374	924	1.47
MMB-08	182	143	325	1.27	462	314	776	1.47
MMB-09	182	143	325	1.27				
MMB-10	181	143	324	1.27				
MMB-11	189	149	338	1.27	597	406	1003	1.47
Average	184 ± 16	145 ± 12	329 ± 28		536 ± 69	365 ± 47	901 ± 115	
MMB-12	166	329	495	0.50	504	805	1309	0.63
MMB-13	149	296	445	0.50				
MMB-14	118	235	353	0.50	351	560	911	0.63
MMB-15	167	331	498	0.50				
MMB-16	156	308	464	0.50	445	710	1155	0.63
MMB-17	170	337	507	0.50				
Average	154 ± 19	306 ± 38	460 ± 58		433 ± 77	692 ± 124	1125 ± 201	
MMB-18	49	932	981	0.05	104	1126	1230	0.09
MMB-19	45	852	897	0.05				
MMB-20	43	821	864	0.05	137	1484	1621	0.09
MMB-21	43	827	870	0.05				
Average	45 ± 3	858 ± 51	903 ± 54		121 ± 23	1305 ± 253	1426 ± 276	

MMB, mixed-mode bending.

Path II				Path III				Roving Bridging		
G_I	G_{II}	G_{tot}	G_I/G_{II}	G_I	G_{II}	G_{tot}	G_I/G_{II}	G_I	G_{II}	G_{tot}
1061	287	1348	3.70	1024	144	1168	7.11			
963	260	1223	3.70							
837	226	1063	3.70					559	97	656
890	241	1131	3.70							
924	250	1174	3.70					424	74	498
935 ± 84	253 ± 23	1188 ± 107		1024	144	1168		492 ± 95	86 ± 17	577 ± 112
851	387	1238	2.2	383	89	472	4.27	601	169	770
1247	567	1814	2.2					1156	326	1482
945	430	1375	2.2							
1190	541	1731	2.2							
1012	460	1472	2.2	700	164	864	4.27	800	225	1025
1302	592	1894	2.2					1201	338	1539
1091 ± 181	496 ± 82	1587 ± 263		542 ± 224	127 ± 52	668 ± 277		940 ± 288	265 ± 81	1204 ± 369
1061	982	2043	1.08	921	366	1287	2.52	859	445	1304
776	719	1495	1.08							
918	850	1768	1.08							
975	903	1878	1.08					1073	556	1629
1072	993	2065	1.08					331	172	503
921	853	1774	1.08							
954 ± 109	883 ± 101	1837 ± 211		921	366	1287		754 ± 382	391 ± 198	1145 ± 580
514	1836	2350	0.28							
616	2200	2816	0.28							
495	1768	2263	0.28	976	1041	2017	0.94	129	222	351
588	2100	2688	0.28					308	531	839
553 ± 58	1976 ± 207	2529 ± 265		976	1041	2017		219 ± 127	377 ± 218	595 ± 345

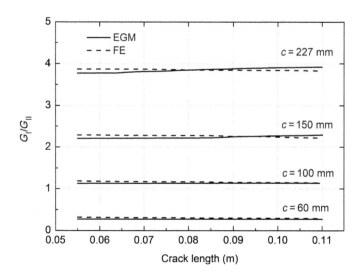

Figure 7.18 Mode ratio, G_I/G_{II}, versus crack length determined by extended global method and finite element modeling. EGM, extended global method; FE, finite element.

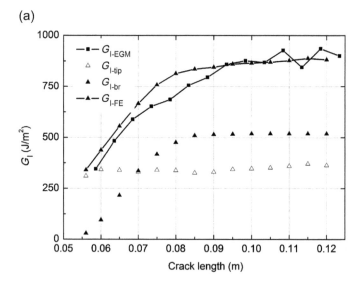

Figure 7.19 Separation of G into G_{tip} and G_{br}, MMB-04 ($c = 227$ mm). (a) G_I and (b) G_{II}. EGM, extended global method; FE, finite element; MMB, mixed-mode bending.

Continued

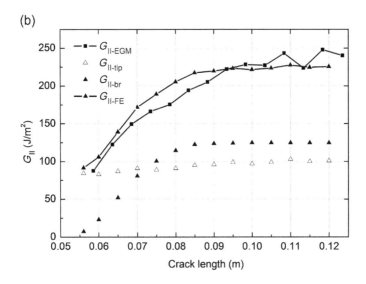

Figure 7.19 Continued

The FCG curves for all examined specimens, shown in Figures 7.23–7.25, clearly show the linear region corresponding to the Paris law and the threshold region. In some of the examined specimens, the upper limit was not determined experimentally due to the rapid crack propagation, and therefore it was considered equal to the corresponding quasi-static strain energy release rate values for crack propagation for each joint. The total fatigue life formulation, as in Eqn (7.10), was used to model the FCG behavior. The values of the estimated model parameters, given in Table 7.7, indicate a strong dependence of D, m, $G_{tot,th}$, and $G_{tot,c}$ on the mode-mixity. On the other hand, as proved in previous studies (Shahverdi et al., 2012a; Shivakumar et al., 2006), exponents Q_1 and Q_2 did not significantly affect the model results. The obtained values for Q_1 and Q_2 for all of the examined loading conditions were between 24.5–25.5 and 2.8–3.2, and therefore for simplicity considered constant.

7.5 Conclusions

The mixed-mode fatigue and fracture behaviors of adhesively-bonded joints can be established based on experimental investigations using asymmetric MMB specimens. For crack propagation in asymmetric specimens, the EGM in which the thickness ratios are extended to the bending stiffness ratios can be used for the analysis of the experimental data and the mode partitioning. The EGM is able to accurately

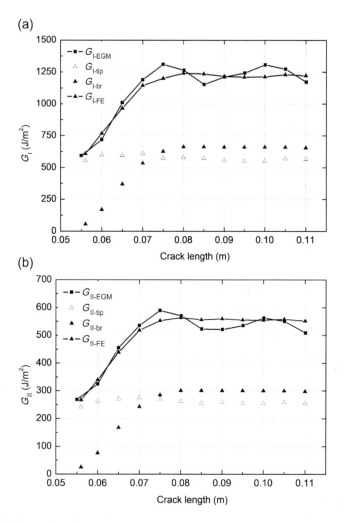

Figure 7.20 Separation of G into G_{tip} and G_{br}, MMB-09 ($c = 150$ mm). (a) G_I and (b) G_{II}. EGM, extended global method; FE, finite element; MMB, mixed-mode bending.

determine the energy release rate and mode-mixity ratios from a set of closed form equations.

FE models can also be developed in order to validate the approach. Zero-thickness cohesive elements can be used to model the fiber bridging zone. Comparison of the G_{tot} values estimated/calculated according to the EGM, the ECM, and FE modeling shows the agreement among the three approaches. FE models can be used for the modeling of the fracture process. The bridging zone can be modeled with cohesive elements and an exponential traction-separation law.

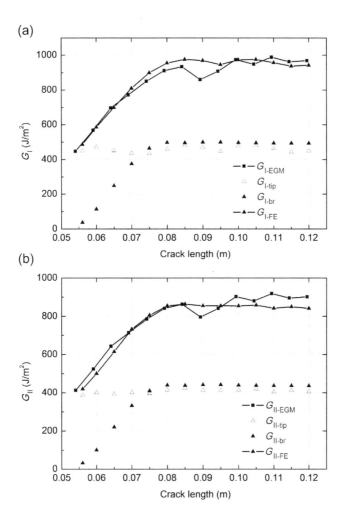

Figure 7.21 Separation of G into G_{tip} and G_{br}, MMB-16 ($c = 100$ mm). (a) G_I and (b) G_{II}. EGM, extended global method; FE, finite element; MMB, mixed-mode bending.

When the crack propagates in a bi-material interface, the mode-mixity ratios obtained from FE models are a function of the crack extension length. This problem can be solved by introducing a resin interlayer with the average properties of the adjacent layers of the interface.

For each mode-mixity, results can be presented in the form of G_{tot} versus da/dN plots, fitting a total fatigue life model to the results. The parameters of the total fatigue life model (D, m, G_{th}, and G_c) are depend on the mode-mixity ratio.

The approach presented in this chapter can be used for establishment of mixed-mode fatigue and fracture behavior of adhesively-bonded joints. The results

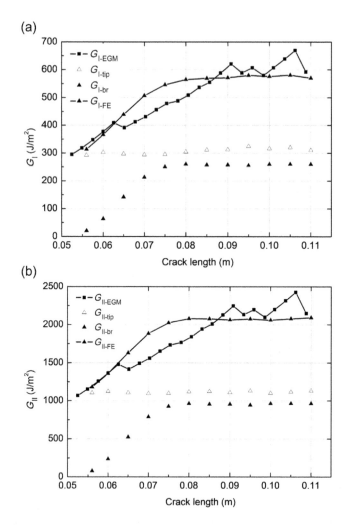

Figure 7.22 Separation of G into G_{tip} and G_{br}, MMB-19 ($c = 60$ mm). (a) G_I and (b) G_{II}. EGM, extended global method; FE, finite element; MMB, mixed-mode bending.

obtained from this approach, combined with the results obtained from pure Mode I and Mode II, can be used to establish initiation and propagation failure criteria for the examined joints, as presented in subsequent chapters of this volume. These failure criteria can be used for design structural joints with the same adherends and adhesive.

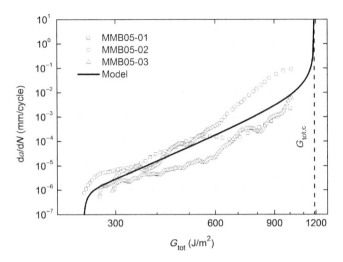

Figure 7.23 Crack growth rate versus G_{tot}, MMB specimens with $G_I/G_{II} = 3.70$. MMB, mixed-mode bending.

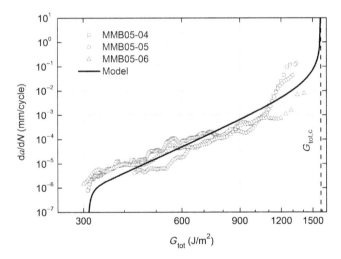

Figure 7.24 Crack growth rate versus G_{tot}, MMB specimens with $G_I/G_{II} = 2.20$. MMB, mixed-mode bending.

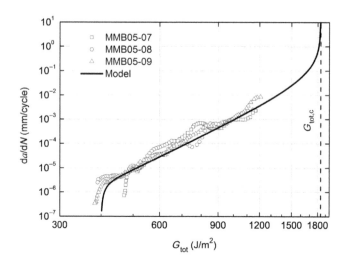

Figure 7.25 Crack growth rate versus G_{tot}, MMB specimens with $G_I/G_{II} = 1.08$. MMB, mixed-mode bending.

Table 7.7 Total fatigue life model parameters

Specimen	D	m	$G_{tot,th}$ (J/m^2)	$G_{tot,c}$ (J/m^2)	Q_1	Q_2
MMB ($c = 227$)	5.00e-21	5.90	240	1188	25	3
MMB ($c = 150$)	1.10e-21	6.00	310	1587	25	3
MMB ($c = 100$)	8.00e-24	6.68	400	1837	25	3

MMB, mixed-mode bending.

References

Agrawal, A., & Karlsson, A. M. (2006). Obtaining mode mixity for a biomaterial interface crack using the virtual crack closure technique. *International Journal of Fracture*, *141*(1−2), 75−98.

Asp, L. E., Sjögren, A., & Greenhalgh, E. S. (2001). Delamination growth and thresholds in a carbon/epoxy composite under fatigue loading. *Journal of Composites Technology and Research*, *23*(2), 55−68.

ASTM D3039/D3039M-08. Standard test method for tensile properties of polymer matrix composite materials, in annual book of ASTM standards. Adhesive section 15.06.

ASTM D3171−11. Standard test methods for constituent content of composite materials, in annual book of ASTM standards. Adhesive section 15.03.

ASTM D5573-99. (2005). Standard practice for classifying failure modes in fiber-reinforced-plastic (FRP) joints, in annual book of ASTM standards. 15.06.

ASTM D6671-01. Stand test method mixed mode I–mode interlaminar fracture toughness unidirectional fiber reinforced polymer matrix composites, Annual Book ASTM Standard, 15.03, 392–403.

Atkinson, C. (1977). On stress singularities and interfaces in linear elastic fracture mechanics. *International Journal of Fracture*, *13*(6), 807–820.

Bathias, C., & Laksimi, A. (1985). Delamination threshold and loading effect in fiber glass epoxy composite. In W. S. Johnson (Ed.), *Delamination and debonding of materials, ASTM STP 876* (pp. 217–237). Philadelphia: American Society for Testing; and Materials.

Benzeggagh, M. L., & Kenane, M. (1996). Measurement of mixed-mode delamination fracture toughness of unidirectional glass/epoxy composites with mixed-mode bending apparatus. *Composites Science and Technology*, *56*(4), 439–449.

Bhashyan, S., & Davidson, B. D. (1997). Evaluation of data reduction methods for the mixed mode bending test. *AIAA Journal*, *35*(3), 546–552.

de Castro, J., & Keller, T. (2008). Ductile double-lap joints from brittle GFRP laminates and ductile adhesives. Part I: experimental investigation. *Composites Part B: Engineering*, *39*(2), 271–281.

Dharmawan, F., Simpson, G., Herszberg, I., & John, S. (2006). Mixed mode fracture toughness of GFRP composites. *Composite Structures*, *75*(1–4), 328–338.

Ducept, F., Davies, P., & Gamby, D. (2000). Mixed mode failure criteria for a glass/epoxy composite and an adhesively bonded composite/composite joint. *International Journal of Adhesion and Adhesive*, *20*(3), 233–244.

Ducept, F., Gamby, D., & Davies, P. (1999). A mixed-mode failure criterion derived from tests on symmetric and asymmetric specimens. *Composites Science and Technology*, *59*(4), 609–619.

Fernández, M. V., De Moura, M. F. S. F., Da Silva, L. F. M., & Marques, A. T. (2013). Mixed-mode I + II fatigue/fracture characterization of composite bonded joints using the single-leg bending test. *Composites Part A: Applied Science and Manufacturing*, *44*(1), 63–69.

Harvey, C. M., & Wang, S. (2012). Experimental assessment of mixed-mode partition theories. *Composite Structures*, *94*(6), 2057–2067.

Hutchinson, J. W., & Suo, Z. (1992). Mixed mode cracking in layered materials. *Advances in Applied Mechanics*, *29*(c), 64–122.

Kenane, M., Azari, Z., Benmedakhene, S., & Benzeggagh, M. L. (2011). Experimental development of fatigue delamination threshold criterion. *Composites Part B: Engineering*, *42*(3), 367–375.

Kenane, M., Benmedakhene, S., & Azari, Z. (2010). Fracture and fatigue study of unidirectional glass/epoxy laminate under different mode of loading. *Fatigue and Fracture of Engineering Materials and Structures*, *33*(5), 284–293.

Kim, B. W., & Mayer, A. H. (2003). Influence of fiber direction and mixed-mode ration on delamination fracture toughness of carbon/epoxy laminates. *Composites Science and Technology*, *63*(5), 695–713.

Kinloch, A. J., Wang, Y., Williams, J. G., & Yayla, P. (1993). The mixed-mode delamination of fibre composite materials. *Composites Science and Technology*, *47*(3), 225–237.

Krueger, R. (2004). The virtual crack closure technique: history, approach and applications. *Applied Mechanics Reviews*, *57*(2), 109–143.

Liu, Z., Gibson, R. F., & Newaz, G. M. (2002). The use of a modified mixed mode bending test for characterization of mixed-mode fracture behavior of adhesively bonded metal joints. *The Journal of Adhesion*, *78*(3), 223–244.

Mall, S., & Johnson, W. S. (1986). Characterization of mode I and mixed-mode failure of adhesive bonds between composite adherends. In J. M. Whitney (Ed.), *Composite materials:*

Testing and design (seventh conference), ASTM STP 893 (pp. 322−334). Philadelphia: American Society for Testing Materials.

Mall, S., Yun, K. T., & Kochhar, N. K. (1989). Characterization of matrix toughness effect on cyclic delamination growth in graphite fiber composites. In P. A. Lagace (Ed.), *Composite materials: Fatigue and fracture, second volume, ASTM STP 1012* (pp. 296−310). Philadelphia: American Society for Testing and Materials.

Martin, R. H., & Murri, G. B. (1990). Characterization of mode I and mode II delamination growth and thresholds in AS4/PEEK composites. In S. P. Garbo (Ed.), *Composite materials: Testing and design, ASTM STP 1059* (pp. 251−270). Philadelphia: American Society for Testing and Materials.

de Morais, A. B. (2011). A new fibre bridging based analysis of the double cantilever beam (DCB) test. *Composites Part A: Applied Science and Manufacturing, 42*(10), 1361−1368.

de Morais, A. B., & Pereira, A. B. (2006). Mixed mode I + II interlaminar fracture of glass/epoxy multidirectional laminates − part 1: analysis. *Composites Science and Technology, 66*(13), 1889−1895.

O'Brien, T. K. (1990). Towards a damage tolerance philosophy for composite materials and structures. In S. P. Garbo (Ed.), *Composite materials: Testing and design, ASTM STP 1059* (pp. 7−33). Philadelphia: American Society for Testing and Materials.

Ozdil, F., & Carlsson, L. A. (1999). Beam analysis of angle-ply laminate mixed-mode bending specimens. *Composites Science and Technology, 59*(6), 937−945.

Ozdil, F., & Carlsson, L. A. (2000). Characterization of mixed mode delamination growth in glass/epoxy composite cylinders. *Journal of Composite Materials, 34*(5), 420−441.

Philippidis, T. P., & Vassilopoulos, A. P. (2004). Life prediction methodology for GFRP laminates under spectrum loading. *Composites Part A: Applied Science and Manufacturing, 35*(6), 657−666.

Raju, I. S., Crews, J. H., & Aminpour, M. A. (1988). Convergence of strain energy release rate components for edge delaminated composite materials. *Engineering Fracture Mechanics, 30*(3), 383−396.

Reeder, J. R., & Crews, J. R. (1990). Mixed-mode bending method for delamination testing. *AIAA Journal, 28*(7), 1270−1276.

Rybicki, E. F., & Kanninen, M. F. (1977). A finite element calculation of stress intensity factors by a modified crack closure integral. *Engineering Fracture Mechanics, 9*(4), 931−938.

Shahverdi, M., Vassilopoulos, A. P., & Keller, T. (2011). A phenomenological analysis of mode I fracture of adhesively-bonded pultruded GFRP joints. *Engineering Fracture Mechanics, 78*(10), 2161−2173.

Shahverdi, M., Vassilopoulos, A. P., & Keller, T. (2012a). A total fatigue life model for the prediction of the R-ratio effects on fatigue crack growth of adhesively-bonded pultruded GFRP DCB joints. *Composites Part A: Applied Science and Manufacturing, 43*(10), 1783−1790.

Shahverdi, M., Vassilopoulos, A. P., & Keller, T. (2012b). Experimental investigation of R-ratio effects on fatigue crack growth of adhesively-bonded pultruded GFRP DCB joints under CA loading. *Composites Part A: Applied Science and Manufacturing, 43*(10), 1689−1697.

Shahverdi, M., Vassilopoulos, A. P., & Keller, T. (2013). Modeling effects of asymmetry and fiber bridging on mode I fracture behavior of bonded pultruded composite joints. *Engineering Fracture Mechanics, 99*, 335−348.

Shahverdi, M., Vassilopoulos, A. P., & Keller, T. (2014a). Mixed-mode I/II fracture behavior of asymmetric adhesively-bonded pultruded composite joints. *Engineering Fracture Mechanics, 115*, 43−59.

Shahverdi, M., Vassilopoulos, A. P., & Keller, T. (2014b). Mixed-mode fatigue failure criteria for adhesively-bonded pultruded GFRP joints. *Composites Part A: Applied Science and Manufacturing*, *53*, 46−55.

Shivakumar, K., Chen, H., Abali, F., Le, D., & Davis, C. (2006). A total fatigue life model for mode I delaminated composite laminates. *International Journal of Fatigue*, *28*(1), 33−42.

Sorensen, L., Botsis, J., Gmür, Th, & Humbert, L. (2008). Bridging tractions in mode I delamination: measurements and simulations. *Composites Science and Technology*, *68*(12), 2350−2358.

Sørensen, B. F., & Jacobsen, T. K. (2009). Characterizing delamination of fibre composites by mixed mode cohesive laws. *Composites Science and Technology*, *69*(3−4), 445−456.

Tamuzs, V., Tarasovs, S., & Vilks, U. (2001). Progressive delamination and fiber bridging modeling in double cantilever beam composite specimens. *Engineering Fracture Mechanics*, *68*(5), 513−525.

Vallee, T. (2004). *Adhesively bonded lap joints of pultruded GFRP shapes* (Ph.D. thesis). Switzerland: EPFL.

Williams, J. G. (1988). On the calculation of energy release rates for cracked laminates. *International Journal of Fracture*, *36*(2), 101−119.

Xu, X. P., & Needleman, A. (1994). Numerical simulations of fast crack growth in brittle solids. *Journal of the Mechanics and Physics of Solids*, *42*(9), 1397−1434.

Zhang, J., Peng, L., Zhao, L., & Fei, B. (2012). Fatigue delamination growth rates and thresholds of composite laminates under mixed mode loading. *International Journal of Fatigue*, *40*, 7−15.

Zhang, Y., Vassilopoulos, A. P., & Keller, T. (2010). Fracture of adhesively-bonded pultruded GFRP joints under constant amplitude fatigue loading. *International Journal of Fatigue*, *32*(7), 979−987.

Fatigue and fracture behavior of adhesively-bonded composite structural joints

A.P. Vassilopoulos, T. Keller
École Polytechnique Fédérale de Lausanne (EPFL), Lausanne, Switzerland

8.1 Introduction

Engineering structures are subject to complex loading histories combining cyclic thermal/humidity and mechanical loading. These loading patterns are frequently of variable amplitude and cause damage in the material that eventually leads to functional and/or structural integrity problems. Adhesively-bonded fiber-reinforced polymer-matrix (FRP) joints currently represent critical elements in numerous engineering structures and they must therefore be able to transfer the developed stresses (of a complex nature) from one part of the structure to another. One of the key objectives of the scientific research community is thus to develop reliable methodologies for the fatigue life prediction of adhesively-bonded FRP joints under realistic loading patterns. For this purpose, the fatigue behavior of the joints must be extensively examined.

The behavior of structural FRP joints under fatigue loads has been examined, mainly in relation to bolted joints in automotive and aerospace applications. Although bolted and bonded joints have advantages and disadvantages, adhesively-bonded joints are preferred for permanent connections in engineering structures. This type of joint offers cost-effective structures with uniform geometrical shapes resulting in a material-adapted stress transfer. Also, it is more appropriate for corrosion-resistant lightweight structures because no metallic parts are used, and owing to the absence of moving parts, adhesively-bonded joints exhibit good fatigue behavior.

Adhesively-bonded glass-fiber-reinforced polymer-matrix (GFRP) joints have been used in engineering applications and their quasi-static and constant amplitude fatigue behavior have been investigated (e.g., Burgueno, Karbhari, Seible, & Kolozs, 2001; Keller & Gürtler, 2005; Keller & Tirelli, 2004; Keller & Zhou, 2006; Zhang, Vassilopoulos, & Keller, 2008). Full-scale structural adhesively-bonded joints composed of pultruded GFRP laminates and epoxy adhesive have been investigated under tensile constant amplitude fatigue loading (Keller & Tirelli, 2004; Zhang et al., 2008; Zhang, Vassilopoulos, & Keller, 2009; Zhang, Vassilopoulos, & Keller, 2010). Questions related to developed failure mechanisms, the fatigue limit, and environmental effects have been discussed in Keller and Tirelli (2004) and Zhang et al. (2008, 2009, 2010). The applicability of stiffness degradation and fracture mechanics models to these types of structures has also been examined in Sarfaraz, Vassilopoulos, and Keller (2011) and Zhang et al. (2008, 2010).

Single-lap joints and double-lap joints are two main types of structural joints found in many structural applications. During the loading of a structural joint, a crack or cracks initiate naturally and propagate along the weakest path within the joint. This uncontrollable phenomenon led scientists to focus on investigating fracture mechanics joints, i.e., pre-cracked joints, to produce pure Mode I crack propagation, e.g., double-cantilever beam (DCB) or Mode II fracture, e.g., end-notched flexure (ENF) or end-loaded split (ELS) beam (Ashcroft, Hughes, & Shaw, 2001; Blackman, Hadavinia, Kinloch, Paraschi, & Williams, 2003; Hadavinia, Kinloch, Little, & Taylor, 2003). Fracture mechanics data can thus be derived, i.e., crack length, a, crack propagation rate, da/dN, and strain energy release rate, G_{max}, in a fatigue cycle under a specific applied maximum cyclic displacement (δ_{max}) or maximum cyclic load (F_{max}).

When structural joints are used instead of fracture joints, the situation becomes more complicated because the failure mode is not pure Mode I or Mode II failure but a mixed-mode failure. The proportion of each failure mode depends on the material, joint geometry, type of loading, and environmental conditions. The strain energy release rate calculated for this mixed-mode failure is designated the total strain energy release rate (Mall, Ramamurthy, & Rezaizdeh, 1987) and is assumed to be equal to the sum of G_I and G_{II}, $G_{tot} = G_I + G_{II}$. Moreover, crack initiation in a structural joint depends on a number of uncontrollable parameters and is random rather than based on the length and shape of a pre-crack incorporated in the joint.

Common methods used for modeling the fatigue behavior of composite laminates have been successfully applied to model constant amplitude fatigue behavior exhibited by adhesively-bonded pultruded FRP joints (e.g., Sarfaraz, Vassilopoulos, & Keller, 2012a, 2012b, 2013b; Zhang et al., 2008, 2009, 2010). The applicability of Linear Elastic Fracture Mechanics (LEFM) theory to the fatigue life modeling of bonded joints has been also proved in previous studies (Ashcroft & Shaw, 2002; Hadavinia et al., 2003; Zhang et al., 2010).

Despite the vast number of scientific publications concerning the fatigue of FRP composite joints under constant amplitude loading, numerous aspects related to their behavior under realistic loading patterns require further examination. One of these aspects is the effect of the mean stress and the application of compressive loading components on the life of the examined joints. The mean stress effect is also referred to as the R-ratio (the ratio of the minimum over the maximum cyclic stress) effect. Although a considerable amount of information exists regarding the mean stress or R-ratio effect on the fatigue life of composite laminates, there is little literature regarding similar investigations for adhesively-bonded structural joints. Experimental studies on joints are based on tensile fatigue loads because they focus on joints exhibiting a cohesive or adhesive failure. Nevertheless, as shown in several studies (e.g., Quaresimin & Ricotta, 2006; Renton & Vinson, 1975), different failure modes can be observed depending on the adherend materials and joint geometry. There is no evidence that the fatigue behavior of the examined joints is the same under compressive loads, a loading pattern that is common during the operation of a structure. This phenomenon can be more pronounced for joints in which cracks in the adherend lead the failure process: for example, pultruded FRP joints (see Keller & Gürtler, 2005; Keller & Tirelli, 2004; Keller & Zhou, 2006; Zhang et al., 2008, 2009, 2010). Moreover, significant R-ratio effects were reported (e.g., Crocombe & Richardson, 1999), especially for pultruded FRP joints in which cracks in the adherend led the failure process and different failure modes were observed under tension and compression fatigue (Sarfaraz et al., 2011).

The fatigue behavior of adhesively-bonded pultruded FRP double-lap joints (DLJs) under different constant amplitude loading patterns, including tensile, compressive, and reversed (combination of tensile and compressive) loading, was experimentally investigated in Sarfaraz et al. (2011). These constant amplitude fatigue results can form the basis for further investigation of their behavior under realistic loading patterns (see, for example, Sarfaraz et al., 2013a, 2013c). The failure process of the examined joints was thoroughly examined and analyzed. The fatigue life was simulated using load-life curves (similar to the $S-N$ curves used for composite laminates) and a Goodman-like constant life diagram (CLD) was employed in an attempt to model the effect of the load ratio on fatigue life. The results showed that the examined adhesively-bonded joints exhibited significant creep—fatigue interaction under R-ratios close to 1 owing to the presence of low load amplitude and high mean values that characterize fatigue loading in this region. Stiffness fluctuations during loading were recorded and analytically presented. Stiffness can be used as a damage metric and stiffness-based phenomenological models can be established to derive fatigue design allowable (Degrieck & Paepegem, 2001; Philippidis & Vassilopoulos, 2000; Zhang et al., 2008) models. Linear Elastic Fracture Mechanics theory was used to interpret fracture mechanics data acquired during the experiments to describe the different failure modes observed under tensile and compressive loading. Fatigue life curves, corresponding to a predetermined (allowed) crack length, can be derived based on the fracture mechanics measurements, thus establishing a method to determine damage-tolerant design allowables (Zhang et al., 2010).

8.2 Experimental investigation of adhesively-bonded structural joints — experimental program description

8.2.1 Material

Symmetric adhesively-bonded double-lap joints composed of pultruded GFRP laminates bonded using an epoxy adhesive system were examined under axial tensile, compressive, and reversed fatigue loads. The pultruded GFRP laminates, supplied by Fiberline A/S, Denmark, consisted of E-glass fibers and isophthalic polyester resin. The fiber architecture of the laminates is shown in Figure 8.1. The laminate is composed of two mat layers on each side and a roving layer in the middle, with a thin layer of polyester veil on the outer surfaces of the laminate. Each mat layer is composed of a woven fabric stitched to a chopped strand mat. An estimation of the nominal thickness of each layer derived by optical microscopy is also given in Figure 8.1. The fiber content, determined by burn-off tests in accordance with

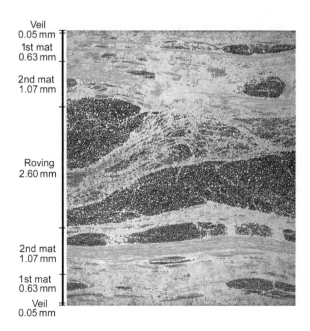

Figure 8.1 Microscopic view of laminates (cross-section perpendicular to pultrusion direction).

ASTM D3171-99, was 43.2 vol.% based on the fiber density of 2560 kg/m^3 specified by the manufacturer. The burn-off tests were carried out on three specimens cut from the long laminates and kept for 5 h at 600 °C in an electric oven.

The longitudinal strength and Young's modulus of the GFRP laminate were obtained from tensile experiments, according to ASTM D3039-08, as being 307.5 ± 4.7 MPa and 25.1 ± 0.5 GPa, respectively. A two-component epoxy adhesive system was used (Sikadur 330, Sika AG Switzerland) as the bonding material. The tensile strength of the adhesive was 38.1 ± 2.1 MPa and the stiffness was 4.6 ± 0.1 GPa. The epoxy showed almost elastic behavior and brittle failure under quasi-static tensile loading (de Castro & Keller, 2008).

8.2.2 Specimen geometry and fabrication

All surfaces subjected to bonding were mechanically abraded (to a depth of approximately 0.3 mm) to increase roughness and then chemically degreased using acetone. An aluminum frame was employed to assist in aligning the laminates. The thickness of the adhesive was controlled by using 2-mm-thick spacers embedded in the bonding area. The specimens were kept for at least 10 h at 30 °C to ensure full curing of the adhesive.

Two different joint configurations were prepared: one with a total length of 410 mm (Figure 8.2) and used only for tensile loading, and another with a reduced total length of 350 mm, which was used when compressive loads were applied to avoid buckling of the joints. To achieve the latter configuration, the free length of the inner laminate was reduced from 100 to 40 mm without changing the bonding and gripping length. These

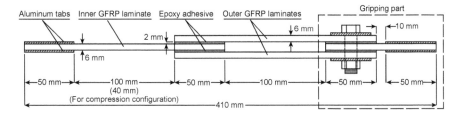

Figure 8.2 Double-lap joint geometry. GFRP, glass-fiber-reinforced polymer-matrix.

dimensions were selected after preliminary testing and modal analysis using the finite element software ANSYS, v.10, which indicated that this length reduction sufficed to prevent buckling of the laminates. Moreover, finite element stress analysis showed that there was no change in the stress field close to the bonded area owing to the decreased laminate length. The bond line was kept constant at 50 mm for both joint configurations. The gripping areas were also 50 mm long to allow load transfer through shear. Aluminum tabs were used to deter the wedges of the testing frame from crushing the laminate. The gripping part (shown on the right side of the specimen in Figure 8.2), which was supported by a bolted connection, was designed to adapt the thickness of the specimen to the opening of the jaw faces of the machine. No failure or crack initiation was observed in the gripping part of all specimens during the entire experimental program.

8.2.3 Experimental set-up

All experiments were carried out on an INSTRON 8800 servohydraulic machine under laboratory conditions (23 ± 5 °C and $50 \pm 10\%$ relative humidity (RH)). Quasi-static tensile and compressive experiments were performed under two different loading modes: a displacement-control mode with a ramp rate of 1 mm/min—designated the low loading rate (LLR)—and a load-control mode with a ramp rate of around 350 kN/s—designated the high loading rate (HLR). The loading rate selected for the load-control mode is similar to the highest loading rate applied during fatigue loading. Five specimens were examined under tensile loading and LLR and three samples per loading condition for the other cases, i.e., tension-HLR, compression-HLR, and compression-LLR.

Fatigue experiments were performed under load control, using a constant amplitude sinusoidal waveform, at a frequency of 10 Hz. It has already been shown (Zhang et al., 2009) that the fatigue performance of similar specimens is not affected by the frequency when it lies between 2 and 10 Hz. Nine R-ratios (denoting the ratio of the minimum over the maximum applied cyclic load) were selected to cover as many loading domains as possible; $R = 0.1$, $R = 0.5$, and $R = 0.9$ for the T–T domain, $R = 10$, $R = 2$, and $R = 1.1$ for the C–C domain, $R = -0.5$ for the TC–C domain, $R = -2$ for the C–T domain, and $R = -1$ for reversed loading. The fatigue experimental program was designed to derive experimental data that cover the entire lifetime between one cycle and several decades of life. At least six specimens were examined under each

R-ratio to cover the entire lifetime between low and high cycle fatigue. The behavior of the examined joints under representative R-ratios from each loading domain was investigated more intensively. Therefore, seven load levels were examined under $R = 0.1$ and five load levels under $R = -1$ and $R = 10$. At least three specimens were tested at each load level to obtain information regarding the scatter of the fatigue life. The specimens were labeled accordingly, e.g., R018503 represents the third specimen (R01850**3**) loaded at a level of 85% (R0**185**03) of the ultimate tensile load under $R = 0.1$ (**R01**8503). One specimen per load level at $R = 10$ and 0.1 was instrumented on one side by two crack gages (HBM crack gage-type RDS20) that cover the whole bonding length and monitor crack initiation and propagation during the fatigue life. Preliminary results proved that the cracks initiate and propagate in a similar way along the two sides of the bond line of each specimen (Zhang et al., 2010). The crack gages included 20 parallel wires with a pitch of 1.15 mm placed perpendicular to the adhesive layer. As the crack propagated, the wires were progressively broken and the electrical resistance of the gage increased. A Labview application and a multichannel electronic measurement unit (HBM-Spider8) were used for data acquisition. The fatigue crack was optically monitored at $R = -1$ using a 40× microscope because of the different possible locations for crack initiation and propagation.

8.3 Interpretation of quasi-static and fatigue/fracture experimental data

8.3.1 Quasi-static investigation

The examined DLJs showed an almost linear load-elongation behavior up to a brittle failure under both tension and compression independent of the applied loading rate. Similar behavior was reported in (de Castro & Keller, 2008; Zhang and Keller, 2008) for DLJs of the same material under tension loading. The ultimate tensile load (UTL), ultimate compressive load (UCL), and stiffness for all examined cases are given in Table 8.1. Joint stiffness was defined as the slope of the load—displacement curve in the range of 0—10 kN where no crack formed in the bond line. The higher joint stiffness under compression resulted from the reduced length of the inner laminate. The difference between the results obtained from both

Table 8.1 Quasi-static data

	UTL or UCL (kN)	Stiffness (kN/mm)
Tension (LLR)	25.5 ± 0.97	23.1 ± 0.20
Tension (HLR)	27.7 ± 2.17	24.6 ± 0.28
Compression (LLR)	−29.0 ± 1.07	30.5 ± 0.39
Compression (HLR)	−27.1 ± 1.92	30.2 ± 0.46

UTL, ultimate tensile load; UCL, ultimate compressive load; LLR, low loading rate; HLR, high loading rate.

applied loading rates was insignificant, comparable to the scatter of the results. Therefore, it was concluded that the loading rate effect is not a significant parameter for the quasi-static strength of the examined DLJs.

The observed failure mode was a fiber-tear failure, as presented in Figure 8.3 for a specimen tested under tensile loads. A dominant crack initiated from the joint corner of one of the bond lines (the upper in the figure) between the adhesive and the inner laminate and then shifted deeper, between the first and the second mat layers of the inner laminate and propagated along this path up to failure. The cracks observed along the lower bond line and at the right side of the inner laminate of the specimen shown in Figure 8.3 are secondary cracks that occurred after the failure of the specimen. The same failure mode was observed for double-lap joints composed of similar materials as documented in Zhang and Keller (2008).

A different failure mode was observed for the specimens examined under compression loading as shown in Figure 8.4. The dominant crack initiated and propagated from the right side of the inner laminate (Figure 8.4) inside the roving layer where the highest peeling stresses developed owing to the applied loading.

8.3.2 Fatigue loading: failure modes analysis

Different failure modes were observed for the three applied load ratios. Under tensile fatigue loading (T−T), i.e., a load ratio of 0.1, the failure mode was similar to that

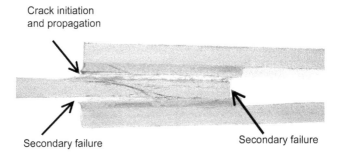

Figure 8.3 Double-lap joint failure mode under tension loading.

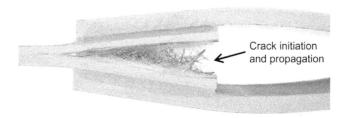

Figure 8.4 Double-lap joint failure mode under compression loading.

observed under quasi-static tension failure (Figure 8.3); a crack occurred only in one of the joint bond lines for all of the applied load levels. Similarly, for compression—compression (C—C) fatigue, i.e., $R = 10$, the crack, as for the compression quasi-static failure, occurred within the roving layer of the inner laminate (Figure 8.4). Under tension-compression (T—C) fatigue ($R = -1$), different failure modes were observed. In most of the examined cases, the failure process was similar to T—T mode. For some of the examined specimens, in addition to the dominant crack along one of the bond lines, a smaller crack of approximately 1 mm was observed in the middle of the inner laminate at a similar location as for C—C loading. However, during the fatigue life the dominant crack was propagating and leading the failure process, whereas the crack created by the compressive component of the applied cyclic load reached a maximum length of 5 mm before failure of the joint.

Under T—T ($R = 0.5$ and $R = 0.9$) and T—C ($R = -0.5$) fatigue, the failure mode was similar to that observed under T—T ($R = 0.1$). Under C—C loading ($R = 2$ and $R = 1.1$), as under $R = 10$, failure occurred within the roving layer of the inner laminate. Under C—T loading ($R = -2$), the failure mode was similar to $R = -1$. These observations showed that the failure transition occurred beyond the load ratio $R = -1$ in the C—T region, which is consistent with the higher fatigue strength of joints under compressive loading.

8.3.3 Fatigue life: R-ratio effect

The experimental program was designed and maximum load levels were applied to obtain representative experimental data in the range between 1 and 10^8 cycles. The experimental results are presented in Tables 8.2—8.5. The fatigue life of the examined joints under the applied R-ratios is plotted against the cyclic load amplitude in Figures 8.5 and 8.6 for experiments under positive and negative mean loads, respectively. Although several formulations can be used for the interpretation of the fatigue data (Sarfaraz et al., 2012, 2013b) the classic power law relationship was employed to simulate fatigue behavior:

$$\sigma_a = \sigma_o N^{-k} \tag{8.1}$$

where σ_a corresponds to the load amplitude, N to the number of cycles, and σ_o and k are the model parameters that can be obtained by fitting Eqn (8.1) to the experimental data. A more detailed discussion about the modeling of bonded-joints fatigue life is presented in Chapter 16 of this volume. The values of parameters σ_o and k as estimated by a linear regression analysis are shown in Table 8.6. Although the designation "$S-N$" curves was preserved for convenience, the fatigue data refer to applied load and not to stress levels because the use of a stress value is not meaningful for joints.

The $S-N$ curve for $R = -1$ exhibits the highest slope ($k = 0.1038$), demonstrating the sensitivity of the examined joints to reversed loading. As shown in Figures 8.5 and 8.6, the fatigue strength of joints decreases under higher R-ratios at tension and tension-dominated loading. In contrast, it decreases under lower R-ratios at compression and compression-dominated loading. Under high mean loads (e.g., $R = 0.9$ and

Table 8.2 Fatigue data at $R = 0.1$

Specimen ID	Nominal load level (% of UTL)	Applied maximum cyclic load (F_{max}) (kN)	Cycles to failure, N_f	No. of cycles to crack initiation, N_i	N_i/N_f (%)
R018501	85	22.8	217	–	–
R018502			415	–	–
R018503			864	–	–
R018504*			261	1	0.38
R018001	80	21.6	1083	–	–
R018002			799	–	–
R018003			720	–	–
R018004*			1229	1	0.08
R017001	70	19.2	9493	–	–
R017002			5773	–	–
R017003			5043	–	–
R017004*			16,624	1	0.01
R016001	60	16.8	7132	–	–
R016002			11,873	–	–
R016003			82,646	–	–
R016004*			85,025	261	0.31
R015001	50	14.4	154,191	–	–
R015002			131,493	–	–
R015003			169,674	–	–
R015004			231,260	–	–
R015005*			124,195	1967	1.58
R014501	45	13.2	187,063	–	–
R014502			299,261	–	–
R014503			183,874	–	–
R014504*			215,718	1680	0.78
R014001	40	12.0	1,317,105	–	–
R014002			2,173,519	–	–
R014003			1,309,163	–	–
R014004*			1,581,478	18,603	1.18

UTL, ultimate tensile load.
* Specimens instrumented by crack gages.

Table 8.3 Fatigue data at $R = 10$

Specimen ID	Nominal load level (% of UCL)	Applied maximum cyclic load (F_{max}) (kN) (absolute value)	Cycles to failure, N_f	No. of cycles to crack initiation, N_i	N_i/N_f (%)
R109001	90	26.1	32	–	–
R109002			235	–	–
R109003*			385	1	0.26
R108001	80	23.2	1144	–	–
R108002			13,623	–	–
R108003			6222	–	–
R108004			2766	–	–
R108005*			944	45	4.77
R107001	70	20.3	116,281	–	–
R107002			57,472	–	–
R107003			121,639	–	–
R107004*			44,821	2370	5.29
R106501	65	18.9	308,732	–	–
R106502*			278,641	38,170	13.70
R106503			58,755	–	–
R106001	60	17.4	2,490,433	–	–
R106002			4,363,735	–	–
R106003*			2,240,724	22,047	0.98

UCL, ultimate compressive load.
* Specimens instrumented by crack gages.

$R = 1.1$), close to the static strength of the joints, the fatigue life is sensitive to the change of load amplitude and $S-N$ curves become flatter with lower slopes ($k_{(R=0.9)} = 0.0314$ and $k_{(R=1.1)} = 0.0154$).

However, one exception to this rule is observed when the mean load is decreased from zero, $R = -1$, to a negative level, i.e., $R = -2$. The $S-N$ data for $R = -2$ in Figure 8.6 is located slightly higher than the fatigue data for $R = -1$. As already explained, although the compressive part of the cyclic load was dominant at this R-ratio, the observed failure mode, similar to the reverse loading ($R = -1$), was tensile failure. This behavior occurs as a result of the higher fatigue strength of the examined joints under compression fatigue. Compared with $R = -1$, a higher load amplitude is required to reach the same maximum load level under $R = -2$. Therefore the highest

Table 8.4 Fatigue data at $R = -1$

Specimen ID	Nominal load level (% of UTL)	Applied maximum cyclic load (F_{max}) (kN)	Cycles to failure, N_f	No. of cycles to crack initiation, N_i	N_i/N_f (%)
R-17501	75	20.4	180	–	–
R-17502			352	–	–
R-17503			22	1	4.55
R-17001	70	19.2	4628	–	–
R-17002			13,173	–	–
R-17003			3632	1	0.03
R-16001	60	16.8	19,087	–	–
R-16002			1370	–	–
R-16003			8372	1	0.01
R-14001	40	12	41,726	–	–
R-14002			46,269	–	–
R-14003			140,176	700	0.50
R-13501	35	9.6	354,021	–	–
R-13502			2,920,391	–	–
R-13503			874,667	5000	0.57

UTL, ultimate tensile load.

load amplitude occurs at a load ratio other than $R = -1$ i.e., under a C–T loading condition where a transition in failure occurs from tensile to compressive mode.

The effect of load ratio on the fatigue life of the examined joints can also be visualized by using constant life diagrams (CLDs) (see Vassilopoulos, Manshadi, & Keller, 2010a). For the derivation of a CLD, the fatigue data are normally, although not necessarily (see Vassilopoulos, Manshadi, & Keller, 2010b) plotted on the "mean-amplitude" (σ_m–σ_a) plane, as radial lines emanating from the origin of the coordinate system. Each radial line represents a single S–N curve under a given R-ratio and can be reproduced using the following equation:

$$\sigma_a = \left(\frac{1-R}{1+R}\right)\sigma_m \tag{8.2}$$

Constant life diagrams are formed by joining in a linear or nonlinear way the points (creating iso-life curves) corresponding to the same number of cycles on consecutive radial lines.

Table 8.5 **Constant amplitude fatigue data**

Load ratio (R)	Specimen ID*	Nominal load level (% of UTL or UCL)	Applied maximum cyclic load (kN) (absolute value)	Cycles to failure, N_f
0.9	R098901	89	24.0	658
	R098601	86	23.3	666,356
	R098401	84	22.8	637,004
	R098201	82	22.0	4063
	R098001	80	21.6	257,184
	R097801	78	21.0	425,510
	R097201	72	19.3	22,867,961
0.5	R059001	90	24.0	832
	R058501	85	22.8	4929
	R058001	80	21.6	30,229
	R057001	70	19.2	71,410
	R056001	60	16.8	225,253
	R055001	50	14.4	1,679,838
−0.5	R-058001	80	21.6	533
	R-057001	70	19.2	1660
	R-056001	60	16.8	1924
	R-055001	50	14.4	21,472
	R-054001	40	12.0	82,510
	R-053801	38	10.8	1,141,900
	R-053501	35	9.6	1,897,288
−2	R-29001	90	26.1	1554
	R-28001	80	23.2	5861
	R-27001	70	20.3	8153
	R-27002	70	20.3	25,390
	R-26001	60	17.4	155,025
	R-25001	50	14.5	1,118,434
2	R29001	90	26.1	3
	R28501	85	24.6	44,743
	R28001	80	23.2	5607
	R27501	75	21.8	22,720

Continued

Fatigue and fracture behavior of adhesively-bonded composite structural joints 237

Table 8.5 **Continued**

Load ratio (R)	Specimen ID*	Nominal load level (% of UTL or UCL)	Applied maximum cyclic load (kN) (absolute value)	Cycles to failure, N_f
	R27001	70	20.3	304,338
	R26501	65	18.9	1,273,000
1.1	R1.19301	93	27.0	1
	R1.19302	93	27.0	1453
	R1.19101	91	26.4	229,071
	R1.19102	91	26.4	313,705
	R1.19001	90	26.0	82,056
	R1.18301	83	24.0	341,995
	R1.18001	80	23.0	41,524,855

UTL, ultimate tensile load; UCL, ultimate compressive load.
* R098001 designates the first specimen (R098**0**01) loaded at a level of 80% (R0**98**001) of the ultimate tensile load under $R = 0.9$ (**R09**8001).

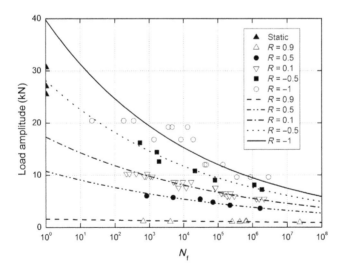

Figure 8.5 $S-N$ data for tension and tension-dominant fatigue loading.

The constant life diagram for the examined joint configuration is shown in Figure 8.7. It is obvious that the CLD is not symmetric with respect to the zero mean cyclic load axis and shifted somewhat toward the compression-dominated domain with the apex corresponding to the $S-N$ curve under $R = -2$. This behavior

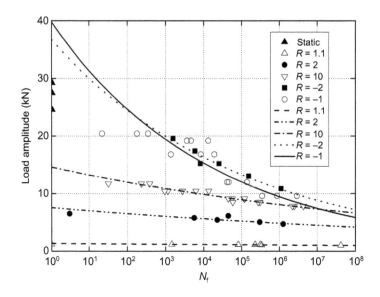

Figure 8.6 $S-N$ data for compression and compression-dominant fatigue loading.

Table 8.6 Material constants of Eqn (8.1) for all load ratios

	R-ratio								
	0.9	0.5	0.1	−0.5	−1	−2	10	2	1.1
σ_o	1.60	10.78	17.32	28.15	39.77	36.74	14.60	7.59	1.37
k	0.0314	0.0752	0.0828	0.0949	0.1038	0.0883	0.0426	0.0320	0.0154

can be attributed to the difference in fatigue strength under tension and compression loading, as discussed earlier. An inflection in the curvature of the iso-life curves is observed. The iso-life curves change from concave to convex when the loading condition shifts from T−T or C−C to combined tension-compression fatigue loading. Moreover, the ultimate tensile and compressive load (UTL $= 27.7 \pm 2.17$ kN and UCL $= -27.1 \pm 1.92$ kN) values are not appropriate for description of the fatigue behavior under zero load amplitude because, as shown in Figure 8.7, a fatigue−creep interaction occurs under R-ratios close to 1 owing to the presence of very low amplitude and high mean values that characterize the cyclic loading in this region.

8.3.4 Stiffness degradation

Stiffness degradation was also recorded during fatigue life for the joints loaded under $R = 0.1$, 10, and −1. Stiffness can be used as a nondestructive damage metric to evaluate the structural integrity of constructions permitting the establishment of fatigue

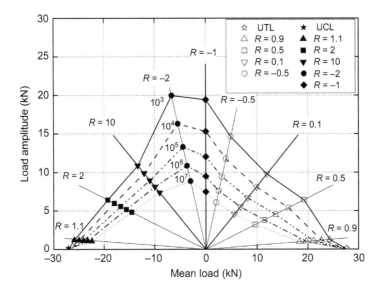

Figure 8.7 Variation of alternating load versus mean load at different fatigue lives. UTL, ultimate tensile load; UCL, ultimate compressive load.

design allowables, which can be easily implemented in design codes, as described in Zhang et al. (2008). The slope of the load—displacement loops, calculated by fitting a linear equation to a set of 1 to 10 consecutive hysteretic loops depending on load level at approximately every 1/40 of the fatigue life, was used to describe the structural stiffness of the examined joints.

In theory, stiffness degradation results from crack propagation and degradation of laminate stiffness. However, previous studies on similar adherends (Keller, Tirelli, & Zhou, 2005) showed that degradation of laminate stiffness is almost insignificant at the level of the loads applied here and the degradation can therefore be attributed solely to crack propagation.

The stiffness degradation results for different load levels of the examined load ratios are shown in Figures 8.8—8.10. Average stiffness degradation values obtained from load—displacement measurements of the three to five specimens tested per load level are presented.

The differences in the initial values can be attributed to the small geometry tolerances and the scatter of the experimental results. Independent of load ratio and load level, a similar trend was observed for all the curves; a rapid stiffness degradation was recorded at the beginning and up to 10% of the total life, an almost linear and less steep stiffness degradation in the range of 10—90% of life, and a very rapid drop after 90% of total life, related to the final failure of the joint. The degradation of the joints that were loaded under $R = 0.1$ and -1 was around 8% up to failure (see Figures 8.8 and 8.9), whereas the value was less than 2% for joints tested under pure compressive loading. This behavior is in agreement with the observed failure modes of the specimens.

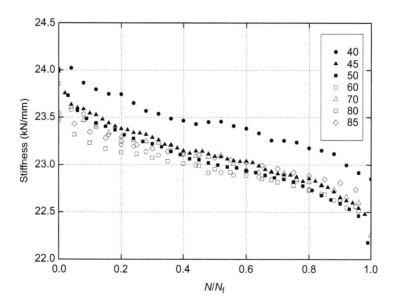

Figure 8.8 Stiffness degradation curves at $R = 0.1$ for loads between 40 and 85% of UTL.

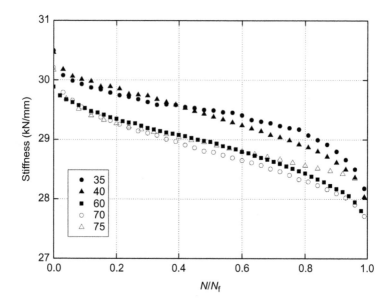

Figure 8.9 Stiffness degradation curves at $R = -1$ for loads between 35 and 75% of UTL.

Fatigue and fracture behavior of adhesively-bonded composite structural joints   241

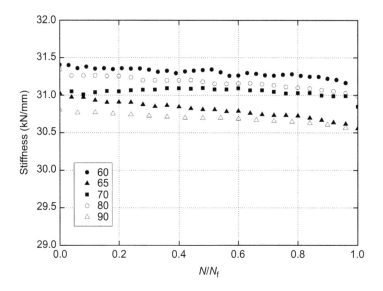

Figure 8.10 Stiffness degradation curves at $R = 10$ for loads between 60 and 90% of UCL.

A linear model was used in Zhang et al. (2008) to simulate the stiffness degradation of adhesively-bonded double-lap joints, similar to the ones examined in the frame of this chapter. The model had the form:

$$\frac{E(N)}{E(0)} = 1 - k_1 \left(\frac{F}{F_u}\right)^{k_2} N \tag{8.3}$$

where F denotes the applied load level and could correspond to its amplitude, maximum value, or a normalized value of it. Model parameters k_1 and k_2 depend on available experimental data for stiffness degradation, and Zhang et al. (2008) assumed that they depend on the number of stress cycles and level of the applied load.

Equation (8.3) also establishes a stiffness-based design criterion because for a preset value of stiffness degradation, $E(N)/E(0) = p$, N can be solved for to obtain an alternative form of the $F-N$ curve, corresponding not to material failure but to a specific stiffness degradation percentage:

$$N = \frac{E(N) - E(0)}{E(0) k_1 \left(\frac{F}{F_u}\right)^{k_2}} \tag{8.4}$$

The term "stiffness-controlled curves" or "stiffness-based curves" was initially introduced in Philippidis and Vassilopoulos (2000). A heuristic procedure was established to define the so-called Sc–N curves, where Sc denotes "stiffness-controlled."

Compared with conventional $S-N$ curves, they offer a significant advantage because they provide information concerning both allowable stiffness degradation and probability of survival.

For a given specimen, the residual stiffness is assumed to follow Eqn (8.3) with the term $k_1(F/F_u)^{k_2}$ representing the rate of stiffness degradation and is assumed to depend on the applied load level. Model parameters were estimated by plotting the stiffness degradation rate against the relevant load levels for all available experimental data, i.e., 12 specimens for the DLJs examined in Zhang et al. (2008), as presented in Figure 8.11. The resulting estimations of parameters k_1 and k_2 were $k_1 = -0.00126$ and $k_2 = 14.176$.

After deriving the model parameters k_1 and k_2 the expected $F-N$ behavior can be extrapolated using Eqn (8.4). The results are presented in Figure 8.12 and are compared with the experimentally determined $F-N$ data. As shown in Figure 8.12, the linear model was able to produce theoretical predictions that compare well with experimental data. The slight overestimation of fatigue life is attributed to ignorance regarding the initial and final periods of stiffness degradation. However, for the DLJs, the effect of these two periods on the entire life is almost negligible. In addition to the $F-N$ curves, Sc$-N$ curves corresponding to predetermined stiffness reduction and not to failure data can be plotted and used as design allowables. For DLJs in which total stiffness degradation at failure was less than 7%, the Sc$-N$ curve for a 2% decrease in stiffness can be plotted based on the linear model as shown in Figure 8.12.

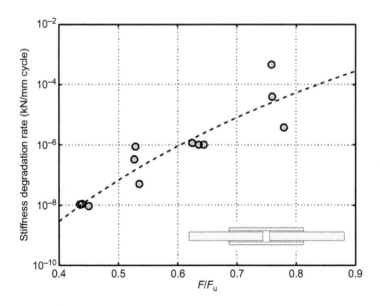

Figure 8.11 Stiffness degradation rate (absolute value) of double-lap joints at different load levels.

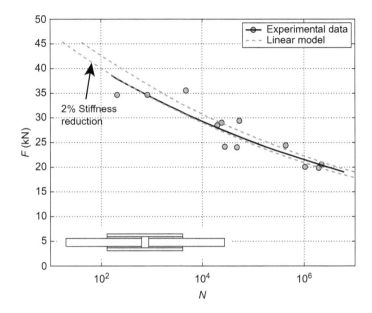

Figure 8.12 Comparison between predicted $F-N$ curve of double-lap joints and experimental results, design allowable corresponding to 2% stiffness reduction.

Although this type of modeling requires more effort than the simpler stress-based approach (in terms of equipment, complicated recording setup, and calculations), it has the merit of also being able to specify allowable stiffness reduction levels. Moreover, because these methods are based on stiffness measurements that can be performed during the operational life of structures without interruptions and in a nondestructive manner, they can be adapted by design codes as on-line health monitoring tools.

8.4 Analysis of the fracture mechanics measurements

As shown in Tables 8.2–8.4 the crack initiated early in the fatigue life of all examined cases independent of load level. Nevertheless, a trend was apparent in the average of (N_i/N_f) for different R-ratios because initiation was observed later in the lifetime in the case of C–C fatigue loading (0.62 ± 0.54 for $R = 0.1$, 5.00 ± 4.78 for $R = 10$, and 1.13 ± 1.72 for $R = -1$). The crack lengths, measured by crack gages or the visual method as previously described for $R = -1$, as a function of the normalized number of cycles are given in Figures 8.13–8.15 for the three types of loading. A common trend was observed: rapid crack propagation at the beginning and at the end of fatigue life, with a significantly lower rate between 10% and 90% of the fatigue life. This behavior concurs with the joint stiffness degradation trends shown in

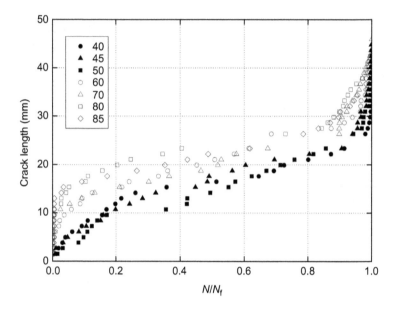

Figure 8.13 Crack length versus normalized number of cycles at $R = 0.1$ for loads between 40 and 85% of UTL.

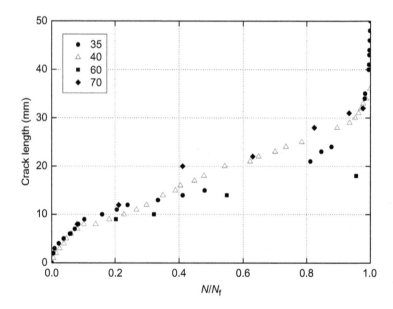

Figure 8.14 Crack length versus normalized number of cycles at $R = -1$ for loads between 35 and 70% of UTL.

Fatigue and fracture behavior of adhesively-bonded composite structural joints   245

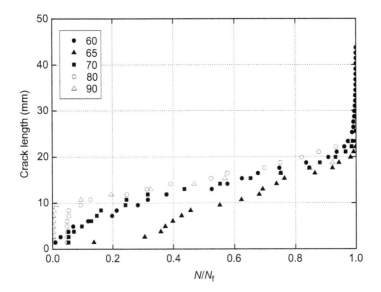

Figure 8.15 Crack length versus normalized number of cycles at $R = 10$. for loads between 60 and 90% of UCL.

Figures 8.8–8.10. Under load conditions of $R = 0.1$ and 10, another trend is apparent in Figures 8.13 and 8.15. The specimens tested at higher load levels exhibited faster crack propagation up to around 10% of the fatigue life.

The strain energy release rate (SERR), G, can be calculated based on linear elastic fracture mechanics. According to this theory, for a double-lap joint with width B, and an existing crack of length α, the strain energy release rate is a function of the applied load F, and the rate of the compliance change $dC/d\alpha$:

$$G = \frac{F^2}{2B} \frac{dC}{d\alpha} \tag{8.5}$$

For cyclic loading, the maximum value of the strain energy release rate during one fatigue cycle can be deduced accordingly:

$$G_{max} = \frac{F_{max}^2}{2B} \frac{dC}{d\alpha} \tag{8.6}$$

where F_{max} is the maximum cyclic load applied during the fatigue cycle. The joint compliance as a function of crack length is given in Figures 8.16–8.18. As shown, in the crack length range between 10 and 25 mm, corresponding to approximately 10–90% of fatigue life, the relationship between compliance and crack length is linear, with a slope that is almost independent of the applied load level under each load ratio. Thus, for each loading case, the average values of $dC/d\alpha$, as shown in Table 8.7, were used to derive the maximum SERR by means of Eqn (8.6).

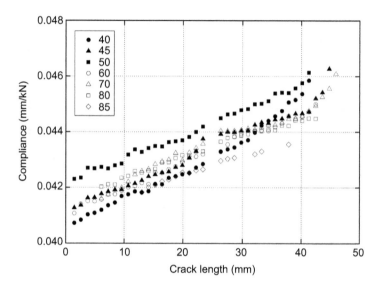

Figure 8.16 Compliance versus crack length at $R = 0.1$ for loads between 40 and 85% of UTL.

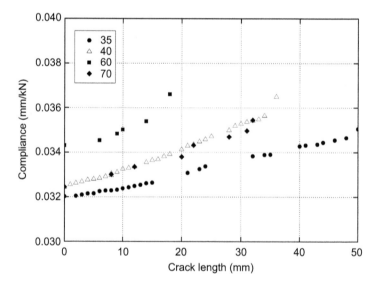

Figure 8.17 Compliance versus crack length at $R = -1$ for loads between 35 and 70% of UTL.

If G_{max} is plotted against the crack propagation rate, $d\alpha/dN$, on logarithmic axes to derive the fatigue crack growth (FCG) graphs, the major part of the relationship is linear and can be simulated by the following equation:

$$\frac{d\alpha}{dN} = D(G_{max})^m \tag{8.7}$$

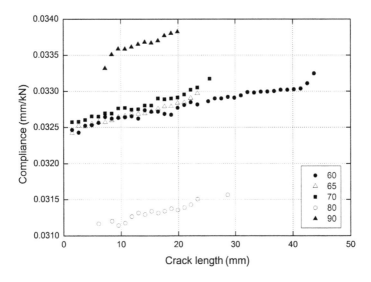

Figure 8.18 Compliance versus crack length at $R = 10$ for loads between 60 and 90% of UCL.

Table 8.7 Compliance change rate under different load ratios (average of all load levels)

	$R = -1$	$R = 0.1$	$R = 10$
$dC/d\alpha$ (1/N)	9.04e-8	8.83e-8	1.73e-8
Standard deviation	0.14e-9	7.32e-9	4.79e-9

where D and m are the fitting parameters, depending on the loading conditions. In previous studies (e.g., Abdel-Wahab, Ashcroft, Crocombe, & Smith, 2004; Hadavinia et al., 2003), D and m were considered to be material parameters with values independent of joint configuration and loading conditions. However, as was proved earlier (Zhang et al., 2010), this argument fails to produce reliable life prediction results.

A phenomenological equation for the calculation of the crack growth rate as a function of the maximum cyclic strain energy release rate, the strain energy release rate threshold, and the static fracture toughness was first proposed by Martin and Murri (1990). This model covers all three regions of the FCG curve: the subcritical around the fatigue threshold, the G_{Imax}-controlled region (the linear one, as described in Eqn (8.7)), and the critical region, close to G_{Ic}.

The total fatigue life model resembles the following equation (Shahverdi, Vassilopoulos, & Keller, 2012):

$$\frac{da}{dN} = D(G_{\text{Imax}})^m \frac{\left(1 - \left(\frac{G_{\text{Ith}}}{G_{\text{Imax}}}\right)^{Q_1}\right)}{\left(1 - \left(\frac{G_{\text{Imax}}}{G_c}\right)^{Q_2}\right)}, \tag{8.8}$$

where m, Q_1, and Q_2 are the empirical model parameters dependent on material and loading conditions. Equation (8.8) can be applied between the limits $G_{\text{Ith}} \leq G_{\text{Imax}} \leq G_{\text{Ic}}$. Therefore, as G_{Imax} approaches G_{th}, da/dN tends to become minimal. Also, as G_{Imax} approaches G_{Ic}, da/dN tends asymptotically to infinity. However, in cases such as the one examined in this chapter, in which most of the fatigue life corresponds to the linear part of the FCG curve, the simplified Eqn (8.7) can be used instead.

The secant method and incremental polynomial fitting (according to ASTM E647-99) were used to calculate the crack propagation rate. According to the secant or point-to-point technique, the crack propagation rate can be determined by calculating the slope of a straight line connecting two contiguous data points on the $\alpha-N$ curve. The incremental polynomial method fits a second-order polynomial to sets of a specified number of successive data points, usually 3, 5, 7, or 9. The slope of the determined equation at any point corresponds to the crack propagation rate. The secant method is simple and accurately represents experimental data, but is sensitive to scatter in the latter (Ashcroft & Shaw, 2002). The incremental polynomial method can reduce scatter but involves the risk of masking real effects, especially when only small data sets are available (Ashcroft & Shaw, 2002). The curve is expected to become smoother when more points are used for the calculations but there is a risk of inadequate modeling, usually at the start or end of the lifetime.

Both methods were applied, as seen in Figures 8.19–8.21, for representative specimens for all loading conditions. The secant method showed high sensitivity to the scatter whereas increasing the number of points in the polynomial method effectively decreased this sensitivity without changing the actual trend of experimental data. The crack growth rate was almost constant in the range of 10–90% for all load ratios and load levels. Hence, in this range, the slope of the fitted line to the experimental data on the $\alpha-N$ curve was considered to be the crack propagation rate.

The FCG curves for all the examined load ratios are presented in Figure 8.22 and the parameters of the plotted FCG curves are given in Table 8.8. The results prove that the relationship between G_{max} and (da/dN) is highly dependent on the load ratio. A steeper curve with lower strain energy release rate value corresponds to $R = 10$, under which specimens failed owing to a crack propagating through the roving layer of the inner laminate without significant fiber bridging. The dominant cracks for $R = 0.1$ and -1 propagated through the mat layers with a considerable amount of fiber bridging. The

Fatigue and fracture behavior of adhesively-bonded composite structural joints   249

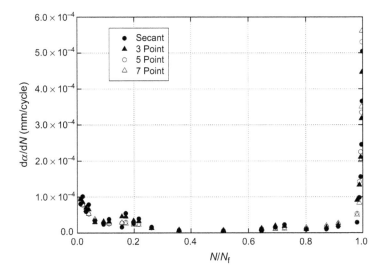

Figure 8.19 Comparison of secant method and incremental polynomial method for estimation of crack growth rate at $R = 0.1$.

lower slope of the derived FCG curve for $R = -1$ compared with the $R = 0.1$ results from the closing of the crack during the compressive part of the cycles under reversed loading. The crack closure breaks the fibers that cause the fiber bridging and consequently reduces the energy required for subsequent crack propagation.

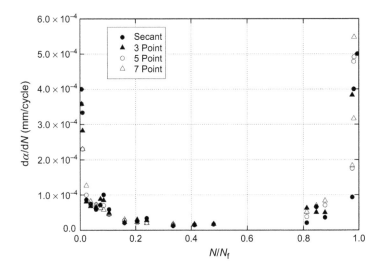

Figure 8.20 Comparison of secant method and incremental polynomial method for estimation of crack growth rate at $R = -1$.

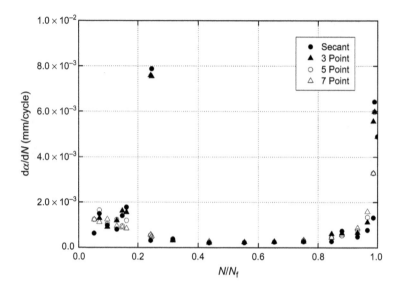

Figure 8.21 Comparison of secant method and incremental polynomial method for estimation of crack growth rate at $R = 10$.

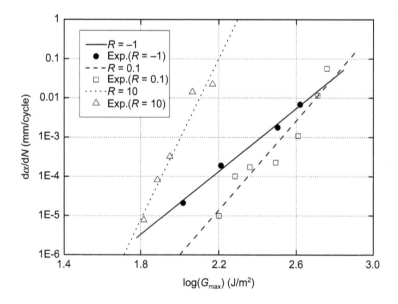

Figure 8.22 Comparison of developed FCG curves for all load ratios.

Table 8.8 **Estimated constant parameters of Eqn (8.7) and corresponding R^2**

	$R = -1$	$R = 0.1$	$R = 10$
D	1.95e-13	3.92e-18	3.35e-24
m	4.012	5.700	10.239
R^2	0.99	0.92	0.95

The life of the examined structure can be modeled directly by integration of the crack propagation rate between two different crack lengths.

$$N - N_\mathrm{i} = \int_{N_\mathrm{i}}^{N} \mathrm{d}N = \int_{a_\mathrm{i}}^{a} \frac{1}{(\mathrm{d}a/\mathrm{d}N)} \mathrm{d}a \qquad (8.9)$$

where N_i denotes the number of cycles for crack initiation and a_i the initial crack length. $N - N_\mathrm{i}$ corresponds to the number of cycles for crack propagation between crack lengths a_i and a. The application of Eqn (8.9) is straightforward, because only a method for the calculation of the crack propagation rate, da/dN, has to be selected.

This application is directly linked to the specific experimental data and does not take into account the materials and/or geometry of the structural element. Based on fracture mechanics data, however, predictive methods can be developed that combine the experimental evidence obtained from one type of structural element and analytical or numerical solutions for other types of structural elements, made from the same materials, to predict the strength or fatigue life of the latter (Curley, Jethwa, Kinloch, & Taylor, 1998; Shahverdi, Vassilopoulos, & Keller, 2013).

By substituting da/dN with its equivalent from Eqns (8.7) and (8.9) becomes:

$$N - N_\mathrm{i} = \int_{a_\mathrm{i}}^{a} \frac{1}{D(G_{\max})^m} \mathrm{d}a \qquad (8.10)$$

Depending on the values of $G_{\max}$ and the corresponding limits of the integration, Eqn (8.10) allows the calculation of conservative or nonconservative design allowables in line with a damage tolerance design philosophy (e.g., estimation of the number of cycles required to attain a specific crack length under a specific applied load).

When D and m are known, the fatigue life for crack propagation can be calculated based on Eqn (8.8). Fatigue life ($F-N$) curves that correspond to failure or a predetermined crack length can then be easily calculated, thus establishing a method for the determination of damage-tolerant design allowables. Corresponding curves are presented in Figure 8.23 for the DLJs examined in Zhang et al. (2010). Curves corresponding to joint failure agree well with the experimental data (data from specimens

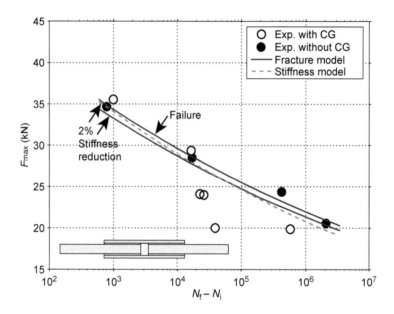

Figure 8.23 $F-N$ curves obtained from fracture and stiffness models (CG = crack gage).

with crack gages are designated CG). In addition, design allowables corresponding to predetermined crack lengths were derived and compared with design allowables derived from stiffness degradation measurements, as shown earlier where a stiffness degradation of 2% was considered for the DLJs. As shown in (Zhang et al., 2010), each crack length can be attributed to a specific compliance increase and consequently to a specific stiffness degradation of the joint. The crack length corresponding to a 2% DLJ stiffness degradation was estimated as being 20 mm. The resulting $F-N$ curves derived from the fracture model, corresponding to failure and 2% stiffness degradation, are shown in Figure 8.23 (solid lines) together with the corresponding curves obtained from the stiffness degradation model (dashed lines). The stiffness models show systematically steeper $F-N$ curves that tend to give more conservative results, especially toward the high cycle fatigue region. However, results from both models seem reasonable and accurate, proving their potential use for the derivation of reliable design allowables.

8.5 Conclusions

The fatigue behavior of adhesively-bonded pultruded GFRP double-lap joints was experimentally examined under nine different load ratios to investigate the effect of the mean stress on fatigue life, stiffness degradation, and crack propagation. The results showed that the change in load ratio significantly affected the fatigue behavior of the examined adhesively-bonded joints.

The examined joints exhibited different behavior under quasi-static tension and compression loading. Fiber-tear failure was observed under tensile loading, with the failure of the specimens dominated by cracks in the mat layers of the inner laminate. Under compression, failure occurred in the roving layer in the middle of the inner laminate. The failure modes of joints under different loading conditions can be classified according to the loading type, i.e., T−T, C−C, and T−C. A transition of the failure mode from tensile to compressive was observed when the mean load was decreased from zero to negative values.

The fatigue failure mode of DLJs under load ratios in the T−T region, e.g., 0.9, 0.5, 0.1, and −0.5, as well as for reversed loading, $R = -1$, were similar to the failure mode observed under quasi-static tension loads. The failure of specimens under R-ratios located in the compression-dominated region of the constant life diagram, e.g., under $R = 1.1, 2, 10$, and -2, was similar to the compressive quasi-static failure mode. The constant life diagram derived for the examined bonded joints was asymmetric and shifted toward the compressive domain. This shift was consistent with the higher fatigue strength of joints under compressive loading.

Similar stiffness degradation trends were recorded under different load ratios. However, joints examined under $R = 10$ exhibited lower stiffness degradation at failure than those loaded under $R = 0.1$ and -1. A general trend regarding crack propagation rate was observed for all specimens independent of load ratio and load level. A higher rate at the beginning, up to around 10%, and at the end, between 90% and 100%, of the fatigue life were exhibited, with a linear crack propagation rate observed during the remaining 80% of the fatigue life. A linear relationship between compliance and crack length was established for each load ratio in the stable crack propagation phase, independent of load level.

Linear elastic fracture mechanics was applied to derive the FCG curves. A steeper curve was derived for $R = 10$ because the crack propagated in the roving layer without significant fiber bridging. On the other hand, significant fiber bridging was associated with cracks that propagated between the mat layers of the inner laminates of the joints examined under $R = 0.1$ and -1. However, the bridged fibers broke during the closing of the crack during the compressive component of each cycle under $R = -1$ and the corresponding FCG curve therefore has a lower slope than the FCG curve for $R = 0.1$.

Fracture mechanics measurements and stiffness degradation measurements can be used to derive reliable design allowables. Stiffness and actual damage (in terms of crack length in this study) can be eventually be used as valuable damage metrics and can assist in the development of damage-tolerant design processes.

References

Abdel-Wahab, M. M., Ashcroft, I. A., Crocombe, A. D., & Smith, P. A. (2004). Finite element prediction of fatigue crack propagation lifetime in composite bonded joints. *Composites, Part A: Applied Science and Manufacturing, 35*(2), 213−222.

Ashcroft, A. I., Hughes, D. J., & Shaw, S. J. (2001). Mode I fracture of epoxy bonded composite joints: 1. Quasi-static loading. *International Journal of Adhesion and Adhesives, 21*(2), 87−99.

Ashcroft, A. I., & Shaw, S. J. (2002). Mode I fracture of epoxy bonded composite joints: 2. Fatigue loading. *International Journal of Adhesion and Adhesives, 22*(2), 151–167.
Blackman, B. R. K., Hadavinia, H., Kinloch, A. J., Paraschi, M., & Williams, J. G. (2003). The calculation of adhesive fracture energies in mode I: revisiting the tapered double cantilever beam (TDCB) test. *Engineering Fracture Mechanics, 70*(2), 233–248.
Burgueno, R., Karbhari, V. M., Seible, F., & Kolozs, R. (2001). Experimental dynamic characterization of an FRP composite bridge superstructure assembly. *Composite Structures, 54*(4), 427–444.
de Castro, J., & Keller, T. (2008). Ductile double-lap joints from brittle GFRP laminates and ductile adhesives. Part I: experimental investigation. *Composites, Part B: Engineering, 39*(2), 271–281.
Crocombe, A. D., & Richardson, G. (1999). Assessing stress and mean load effects on the fatigue response of adhesively-bonded joints. *International Journal of Adhesion and Adhesives, 19*(1), 19–27.
Curley, A. J., Jethwa, J. K., Kinloch, A. J., & Taylor, A. C. (1998). The fatigue and durability behavior of automotive adhesives. Part III: predicting the service life. *Journal of Adhesion, 66*(11), 39–59.
Degrieck, J., & Paepegem, W. M. (2001). Fatigue damage modeling of fiber reinforced composite materials: a review. *Applied Mechanics Reviews, 54*(4), 279–300.
Hadavinia, H., Kinloch, A. J., Little, M. S. G., & Taylor, A. C. (2003). The prediction of crack growth in bonded joints under cyclic-fatigue loading. I. Experimental studies. *International Journal of Adhesion and Adhesives, 23*(6), 449–461.
Keller, T., & Gürtler, H. (2005). Quasi-static and fatigue performance of a cellular FRP bridge deck adhesively-bonded to steel girders. *Composite Structures, 70*(4), 484–496.
Keller, T., & Tirelli, T. (2004). Fatigue behavior of adhesively connected pultruded GFRP profiles. *Composite Structures, 65*(1), 55–64.
Keller, T., Tirelli, T., & Zhou, A. (2005). Tensile fatigue performance of pultruded glass fiber reinforced polymer profiles. *Composite Structures, 68*(2), 235–245.
Keller, T., & Zhou, A. (2006). Fatigue behavior of adhesively-bonded joints composed of pultruded GFRP adherends for civil infrastructure applications. *Composites, Part A: Applied Science and Manufacturing, 37*(8), 1119–1130.
Mall, S., Ramamurthy, G., & Rezaizdeh, M. A. (1987). Stress ratio effect on cyclic debonding in adhesively-bonded composite joints. *Composite Structures, 8*(1), 31–45.
Martin, R. H., & Murri, G. B. (1990). *Characterization of mode I and mode II delamination growth and thresholds is AS4/PEEK composites*. Philadelphia: ASTM STP 1059, American Society for Testing and Materials, 251–270.
Philippidis, T. P., & Vassilopoulos, A. P. (2000). Fatigue design allowables for GFRP laminates based on stiffness degradation measurements. *Composites Science and Technology, 60*(15), 2819–2828.
Quaresimin, M., & Ricotta, M. (2006). Fatigue behaviour and damage evolution of single lap bonded joints in composite material. *Composites Science and Technology, 66*(2), 176–187.
Renton, W. J., & Vinson, J. R. (1975). Fatigue behavior of bonded joints in composite material structures. *Journal of Aircraft, 12*(5), 442–447.
Sarfaraz, R., Vassilopoulos, A. P., & Keller, T. (2011). Experimental investigation of the fatigue behavior of adhesively-bonded pultruded GFRP joints under different load ratios. *International Journal of Fatigue, 33*(11), 1451–1460.
Sarfaraz, R., Vassilopoulos, A. P., & Keller, T. (2012a). Experimental investigation and modeling of mean load effect on fatigue behavior of adhesively-bonded pultruded GFRP joints. *International Journal of Fatigue, 44*, 245–252.

Sarfaraz, R., Vassilopoulos, A. P., & Keller, T. (2012b). A hybrid S-N formulation for fatigue life modeling of composite materials and structures. *Composites, Part A: Applied Science and Manufacturing, 43*, 445−453.
Sarfaraz, R., Vassilopoulos, A. P., & Keller, T. (2013a). Block loading fatigue of adhesively-bonded pultruded GFRP joints. *International Journal of Fatigue, 49*, 40−49.
Sarfaraz, R., Vassilopoulos, A. P., & Keller, T. (2013b). Modeling the constant amplitude fatigue behavior of adhesively-bonded pultruded GFRP joints. *Journal of Adhesion Science and Technology, 27*(8), 855−878.
Sarfaraz, R., Vassilopoulos, A. P., & Keller, T. (2013c). Variable amplitude fatigue of adhesively-bonded pultruded GFRP joints. *International Journal of Fatigue, 55*, 22−32.
Shahverdi, M., Vassilopoulos, A. P., & Keller, T. (2012). A total fatigue life model for the prediction of the R-ratio effects on fatigue crack growth of adhesively-bonded pultruded GFRP DCB joints. *Composites, Part A: Applied Science and Manufacturing, 43*(10), 1783−1790.
Shahverdi, M., Vassilopoulos, A. P., & Keller, T. (2013). Modeling effects of asymmetry and fiber bridging on mode I fracture behavior of bonded pultruded composite joints. *Engineering Fracture Mechanics, 99*, 335−348.
Vassilopoulos, A. P., Manshadi, B. D., & Keller, T. (2010a). Influence of constant life diagram formulation on the fatigue life prediction of composite materials. *International Journal of Fatigue, 32*(4), 659−669.
Vassilopoulos, A. P., Manshadi, B. D., & Keller, T. (2010b). Piecewise non-linear constant life diagram formulation for FRP composite materials. *International Journal of Fatigue, 32*(10), 1731−1738.
Zhang, Y., Vassilopoulos, A. P., & Keller, T. (2008). Stiffness degradation and fatigue life prediction of adhesively-bonded joints for fiber-reinforced polymer composites. *International Journal of Fatigue, 30*(10−11), 1813−1820.
Zhang, Y., Vassilopoulos, A. P., & Keller, T. (2009). Environmental effects on fatigue behavior of adhesively-bonded pultruded structure joints. *Composites Science and Technology, 69*(7−8), 1022−1028.
Zhang, Y., Vassilopoulos, A. P., & Keller, T. (2010). Fracture of adhesively-bonded pultruded GFRP joints under constant amplitude fatigue loading. *International Journal of Fatigue, 32*(7), 979−987.

Block and variable amplitude fatigue and fracture behavior of adhesively-bonded composite structural joints

A.P. Vassilopoulos
École Polytechnique Fédérale de Lausanne (EPFL), Lausanne, Switzerland

9.1 Introduction

Several experimental investigations of the fatigue behavior of composite laminates show their sensitivity to the loading sequence. Experiments composed of two blocks of constant amplitude (CA) loading passing from a low stress level to a higher stress level (L-H sequence) or vice versa (H-L sequence) usually are used to study the sequence effect on composite materials. However, the results obtained from these experiments are not consistent and show a greater damaging effect caused by the L-H sequence (e.g., Bartley-Cho, Lim, Hahn, & Shyprykevich, 1998; Broutman & Sahu, 1972; Found & Kanyanga, 1996; Hosoi, Kawada, & Yoshino, 2006; Jen, Kau, & Wu, 1994; Otani & Song, 1997; Van Paepegem & Degrieck, 2002; Wahl, Mandell, & Samborsky, 2001; Yang & Jones, 1980, 1983) or the opposite (less damaging) behavior (e.g., Gamstedt & Sjögren, 2002; Han & Abdelmohsen, 1986; Hwang & Han, 1987, 1989; Lee & Jen, 2000a, 2000b), depending on the material and loading parameters. These conclusions are mainly drawn based on comparisons between the fatigue life under block and CA loading, athough little information concerning the failure mechanisms that cause sequence effects is available (Gamstedt & Sjögren, 2002; Plumtree, Melo, & Dahl, 2010).

Several different loading parameters such as the R ratio (the ratio of the minimum to the maximum applied cyclic load) and the cyclic load levels govern the sequence effects (Adam, Gathercole, Reiter, & Harris, 1994; Bonnee, 1996; Harris, Gathercole, Reiter, & Adam, 1997; Lee & Liu, 1994). Tension or compression loading blocks can produce different damage compared to mixed tension–compression blocks. The difference between the applied load levels in a two-stage block loading (BL) sequence also can be an important parameter because the sequence effect can be magnified when the difference between two load levels is increased (see, e.g., Lee & Liu, 1994).

Although numerous publications are dedicated to the study of the sequence effect, explicit explanations regarding the contributing failure mechanisms are limited (Bonnee, 1996; Found & Quaresimin, 2003; Gamstedt & Sjögren, 2002; Plumtree et al., 2010). The activation of competing failure mechanisms, such as initiation mechanisms versus progressive failure mechanisms or resin cracking versus fiber breakage or delamination under different stress levels, was considered for the explanation of the sequence effects observed for different types of composite materials (Found & Quaresimin, 2003; Gamstedt & Sjögren, 2002). For instance, transverse cracking dominates the failure of cross-ply laminates under high stress levels, whereas delamination is activated under lower stress levels (see, e.g., Gamstedt & Sjögren, 2002). Therefore the H-L sequence results in shorter fatigue durations than the L-H sequence since the transverse cracks, created under a high stress level, are potential places for the initiation of delamination. A reverse effect was observed after an experimental investigation of multidirectional carbon/epoxy laminates and explained by the assumption that since most of the applied load is borne by the matrix under low stress levels and by the fibers under high stress levels, the damage mainly involves the growth of microcracks in the matrix throughout the specimen under lower stress levels, which can induce rapid failure in the following high-stress stage (Adam et al., 1994).

The balance between the damage state and the stress levels and its effect on stress intensity also was proposed as a way of explaining sequence effects in angle ply laminates. The longer life of $[\pm 45]_{2s}$ carbon/epoxy laminates under both L-H and H-L sequences compared to the expected life, characterized by the Palmgren-Miner sum, was thus attributed to the decrease in local stress intensity because of a large number of well-distributed matrix cracks when the stress level decreased (Plumtree et al., 2010). In addition to the loading sequence effect, the significant influence of loading transition and its frequency of occurrence on the duration of fatigue in composite materials has been discussed in several investigations (Filis, Farrow, & Bond, 2004; Schaff & Davidson, 1997a, 1997b; Van Paepegem & Degrieck, 2002). The effect of the frequent transition of cyclic load level on the duration of fatigue was found to be more significant than the loading sequence effect (Van Paepegem & Degrieck, 2002). The transition effect, as defined by the term "cycle mix," was introduced to model the damage accumulated under block and variable amplitude (VA) loading by Schaff & Davidson (1997a, 1997b) and Filis et al. (2004).

The load sequence also affects the fatigue behavior of fiber-reinforced polymer (FRP) composite joints, although only a limited number of works exist concerning this phenomenon. Similar to composite laminates, several parameters were found to contribute to the effect of loading sequence on the duration of fatigue. A significant load interaction effect (overloads and loading sequence effect) was identified (Erpolat, Ashcroft, Crocombe, & Abdel-Wahab, 2004a) for adhesively-bonded double-lap joints composed of carbon/epoxy laminates and a single-part epoxy adhesive and (Sarfaraz, Vassilopoulos, & Keller, 2013a) for pultruded bonded laminates with a two-part epoxy adhesive resin. The acceleration of cracking caused by the load interaction was put forward as the main reason for the variability in fatigue duration exhibited by the joints under investigation. The cycle mix effect and the variation in mean stress also were investigated; it has been proved that they both caused crack growth to accelerate, whereas overloads were shown to increase the likelihood of fatigue to initiate cracking.

In addition to BL, a limited number of investigations of the VA fatigue behavior of bonded joints have been performed, and they are mainly related to bonded joints with metallic adherends (Ashcroft, 2004; Bond and Ansell, 1998; Erpolat et al., 2004a; Erpolat, Ashcroft, Crocombe, & Abdel-Wahab, 2004b; Jeans, Grimes, & Kan,

1983; Nolting, Underhill, & DuQuesnay, 2008; Sarkani, Michaelov, Kihl, & Beach, 1999; Shenoy, Ashcroft, Critchlow, & Crocombe, 2010; Smith and Hardy, 1977). Regardless of the adherend material, load interaction effects such as load transition, load sequence, and overload have been reported in different investigations. Overloads can accelerate the initiation of fatigue cracking (Erpolat et al, 2004b) or increase the damaging effect of the following cycles of lower amplitude, although their effect is reduced when the number of low-amplitude cycles following the overloads is increased (Nolting et al., 2008). The change of mean load can also accelerate the growth of cracks in bonded composite joints (Erpolat et al, 2004a, 2004b). The significant damaging effect of introducing a small number of cycles at a higher mean load also was addressed (Shenoy et al., 2010). The load interaction effect was observed for pultruded glass FRP (GFRP) joints in which cracks in the multidirectional laminate (adherend) lead to failure of the process. The investigation of two-stage BL sequences under tension loading demonstrated a retardation effect under H-L and an acceleration effect under L-H loading sequences. However, the damaging effect of frequent load transitions in a spectrum dominated the load sequence effect (Sarfaraz et al., 2013a; Sarfaraz, Vassilopoulos, & Keller, 2013b).

The aforementioned retardation or acceleration of the fatigue crack growth rate due to load interactions is common for metals, where one dominant crack mainly governs the fracture behavior. In composite materials, which exhibit several contributing fatigue-induced failure mechanisms, identifying a single dominant crack for this investigation is difficult. The situation is less complicated for adhesively-bonded lap joints under cyclic loading, however, since experimental observation in previous investigations (e.g., Sarfaraz, Vassilopoulos, & Keller, 2011; Zhang, Vassilopoulos, & Keller, 2010) showed that in several cases only one dominant crack led to final failure, even for joints composed of composite materials. Therefore, for adhesively-bonded joints, the load interaction effects can be correlated with the acceleration or retardation of the propagation rate of the dominant crack, consequently explaining the fatigue behavior under VA loading.

The earlier review highlights the significant influence of load interaction on the fatigue behavior of materials and structures under realistic loading patterns. It also shows that this interaction strongly depends on the materials as well as the applied loading spectrum. It is the aim of this chapter to investigate the effect of load sequence under both block and VA loading conditions on the fatigue behavior of adhesively-bonded, pultruded GFRP joints. The loading sequence effect, L-H versus H-L sequences, and the effect of load transition frequency on the duration of fatigue on the examined joints are experimentally investigated under both tension and compression loading patterns. The examined bonded joints exhibited complex failure modes, which were different under tension and compression (Sarfaraz et al., 2013a). The failure process of the examined joints is thoroughly investigated and the acquired data concerning the crack initiation and propagation are analyzed to explore the load interaction effects. The comparison of the data concerning crack propagation under CA (Sarfaraz et al., 2011), BL (Sarfaraz et al., 2013a), and VA loading (Sarfaraz et al., 2013b) acquired during experiments provides a clear insight into the effect of load history and interactions on the duration of fatigue of the bonded joints.

9.2 Experimental investigation of the block and variable amplitude fatigue behavior of adhesively-bonded joints

A complete database comprising block and VA fatigue experimental data from adhesively-bonded, pultruded, double-lap joints (Sarfaraz et al., 2013a, 2013b) is used in this chapter to demonstrate the load sequence and the load transition effects on the duration of fatigue.

9.2.1 Materials and specimens

Symmetric adhesively-bonded double-lap joints, shown in Figure 9.1, composed of 40-mm-wide pultruded GFRP laminates bonded by an epoxy adhesive system, were examined. The pultruded GFRP laminates, supplied by Fiberline A/S, Denmark, consisted of E-glass fibers and isophthalic polyester resin. The laminate comprises two mat layers on each side and a roving layer in the middle, with a thin layer of polyester veil on the outer surfaces of the laminates. Each mat layer comprises a 0°/90° woven fabric stitched to a chopped strand mat. A two-component epoxy adhesive system (Sikadur 330; Sika AG, Switzerland) was used as the bonding material. The resulting joints are representative of civil engineering structures, in which dimensions are significantly larger compared to aerospace or automotive applications. More details about the constituent materials and joint configurations are given in Sarfaraz et al. (2011) and in Chapter 8 of this volume.

9.2.2 Experimental program

All experiments were carried out on an INSTRON 8800 servohydraulic machine under load control, using a sinusoidal waveform, at a frequency of 10 Hz under laboratory conditions (23 ± 5 °C and 50 ± 10% relative humidity). The loading sequence effect in a two-stage BL and the effect of load level transitions in a multi-BL were investigated. As previously explained, since the load ratio ($R = F_{min}/F_{max}$) can also affect the results, it was kept constant for tension ($R = 0.1$) and compression ($R = 10$) loading blocks, and only the load levels were altered.

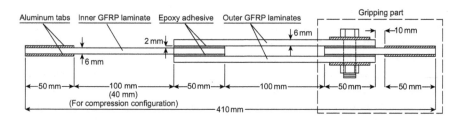

Figure 9.1 Double-lap joint geometry. GFRP, glass fiber-reinforced polymer.

Table 9.1 Experimental program for investigating loading sequence effect

R	Sequence	load level under first block, up to 25−35% of fatigue duration (nominal % of UTL or UCL)	load level under second block, up to failure (nominal % of UTL or UCL)
0.1	L-H	50*	80
		40	70
	H-L	80	50
		70	40
10	L-H	60†	70
		65	80
	H-L	70	60
		80	65

* Nominal % of ultimate tensile load (UTL = 27.7 ± 2.2 kN).
†Nominal % of ultimate compressive load (UCL = −27.1 ± 1.9 kN).

The experimental program shown in Table 9.1 was used to investigate the loading sequence effect. The program consisted of two-stage BL sequences with transitions from L-H and H-L load levels under R ratios of 0.1 and 10, representing tensile and compressive fatigue, respectively. A schematic representation of the applied loads is shown in Figure 9.2(a) for the L-H and in Figure 9.2(b) for the H-L loading sequences. Two types of loading blocks with different load levels were applied under each R ratio. The load levels in the BL sequences were chosen based on the availability of fracture data in the CA fatigue database. The length of the first loading block in terms of number of cycles (n_1) was predetermined as being equal to 25−35% of the CA fatigue duration of the joints (N_1), given in Tables 9.2 and 9.3. After completing the first step (n_1), the second loading block was applied up to failure, which resulted in n_2. The allowable number of cycles under the second load level, N_2, is given in Tables 9.2 and 9.3. Two specimens were examined under each loading sequence. For two more specimens under tension loading, annotated in Table 9.2, the length of the first block was equal to 50% of the CA fatigue duration.

The BL experimental matrix and detailed results concerning different loading sequences for tension ($R = 0.1$) and compression ($R = 10$) fatigue are presented in Tables 9.2 and 9.3, respectively. The applied number of cycles (n_1) and corresponding maximum absolute cyclic load level in the first block (F_{max1}) and the same parameters for the second loading block (n_2, F_{max2}) are shown in the same tables. The allowable numbers of cycles corresponding to the CA loading under each F_{max} were calculated by fitting a power law model to the experimental data and are indicated by N_1 and N_2 in Tables 9.2 and 9.3.

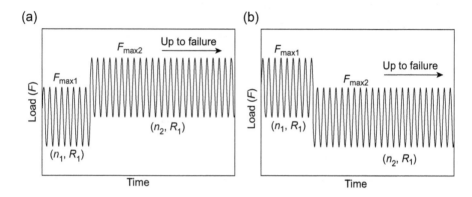

Figure 9.2 Schematic representation of applied two-stage block loading sequences: low to high sequence (a) and high to low sequence (b).

The damage index (D) was calculated according to the Palmgren-Miner rule, as in Eqn (9.1):

$$D = \sum_{i=1}^{k} \frac{n_i}{N_i} \quad (9.1)$$

with k denoting the number of applied loading blocks. According to this model, the specimen under a cyclic loading pattern fails when the damage index reaches 1. In the present study this index was used as a reference value to compare the effect of the parameters being investigated.

The specimens were labeled accordingly; for example, R01F30B704002 represents the second specimen (R01F30B704002) loaded at the nominal level of 70%, followed by the level of 40% (R01F30B704002) of the ultimate tensile load under $R = 0.1$ (R01F30B704002), and the length of the first loading block was equal to 25–35% of the CA fatigue duration in the joints under the first load level (R01F30B704002). The letter x is also used in labels to designate a group of specimens with partially similar conditions. For example, R01FxxB7040xx refers to all specimens loaded under $R = 0.1$ and the loading sequence 70–40% of the ultimate tensile load.

One specimen per loading sequence was instrumented on one side by two crack gauges (HBM crack gauge type RDS20), which cover the whole bonding length and monitor crack initiation and propagation throughout the duration of fatigue (see Figure 9.3). Preliminary results from a previous study of similar joint configurations (Zhang et al., 2010) proved that the cracks initiate and propagate in a similar way along the two sides of the bond line of each specimen. The crack gauges included 20 parallel wires, with a pitch of 1.15 mm, placed perpendicular to the adhesive layer. As the crack propagated, the wires were progressively broken and the electrical resistance of the gauge increased. A Labview application and a multichannel electronic measurement unit (HBM Spider8) were used to acquire data.

Table 9.2 Two-stage block loading results at $R = 0.1$

	Specimen ID	F_{max1} (kN)	n_1	N_1	F_{max2} (kN)	n_2	N_2	Damage index
L-H sequence	R01F30B508001	14.4	48,649	142,978	21.6	483	1070	0.792
	R01F30B508002*		48,649			3		0.343
	R01F30B407001	12.0	456,038	1,291,993	19.2	2592	4434	0.938
	R01F30B407002*		387,598			9815		2.514
	R01F50B407001§		645,996			3		0.501
H-L sequence	R01F30B805001	21.6	287	1070	14.4	207,559	142,978	1.720
	R01F30B805002*		287			686,108		5.067
	R01F30B704001	19.2	1330	4434	12.0	4,832,688	1,291,993	4.040
	R01F30B704002*		1330			5,496,607		4.554
	R01F50B704001§		2217			7,329,969		6.173

*Specimens instrumented by crack gauges.
§Length of the first block corresponds to 50% of the CA fatigue life.

Table 9.3 Two-stage block loading results at $R = 10$

	Specimen ID	F_{max1} (kN)[†]	n_1	N_1	F_{max2} (kN)[†]	n_2	N_2	Damage Index
L-H sequence	R10F30B607001	17.4	682,023	2,273,411	20.3	1,009,548	60,834	16.895
	R10F30B607002*		682,023			1,829,254		30.370
	R10F30B658001	18.9	104,054	346,848	23.2	5691	2642	2.454
	R10F30B658002*		104,054			824,363		312.322
H-L sequence	R10F30B706001	20.3	18,250	60,834	17.4	454,047	2,273,411	0.500
	R10F30B706002*		18,250			1,614,479		1.010
	R10F30B806501	23.2	793	2642	18.9	63,357	346,848	0.483
	R10F30B806502*		261			0		0.099
	R10F30B806503*		793			297,046		1.157

* Specimens instrumented by crack gauges.
† Absolute value.

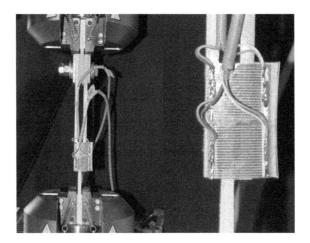

Figure 9.3 Double-lap joint instrumented with two crack gauges.

Table 9.4 **Experimental program for investigating the transition effect at $R = 0.1$**

	First load level (nominal % of UTL)	n_1	Second load level (nominal % of UTL)	n_2
High transition	70	10	40	2914
	40	2914	70	10
Low transition	70	100	40	29,140
	40	29,140	70	100

UTL, ultimate tensile load.

The experimental program used to investigate the transition effect on the lifetime of the examined bonded joints is summarized in Tables 9.4 and 9.5. A schematic representation of the applied loads is shown in Figure 9.4(a) and (b). Each loading sequence in Figure 9.4(a) and (b) is composed of two load levels with the same R ratio but a different length and was applied repeatedly until specimen failure. The length of the blocks was predetermined to provide the same amount of partial damage (n_i/N_i) based on the linear damage accumulation model, according to the CA fatigue data, that is, $n_1/N_1 = n_2/N_2$. Accordingly, when the number of cycles in the first block is increased (e.g., at $R = 0.1$ from 10 to 100 cycles), the length of the second block also increases by the same ratio (e.g., from 2914 to 29,140 cycles; see Table 9.4). Two specimens with different starting loading block were examined under each loading sequence. The letters L and H used in Tables 9.6 and 9.7 for specimen labeling (e.g.,

Table 9.5 **Experimental program for investigating the transition effect at $R = 10$**

	First load level (nominal % of UCL)	n_1	Second load level (nominal % of UCL)	n_2
High transition	70	10	60	374
	60	374	70	10
Low transition	70	1000	60	37,400
	60	37,400	70	1000

UCL, ultimate compression load.

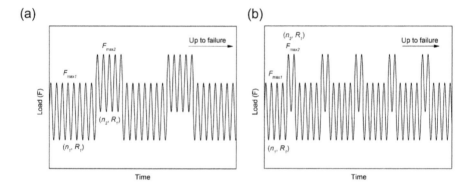

Figure 9.4 Schematic representation of applied multiple block loading sequences: low transition (a) and high transition (b).

R01B4070H01) denote the low and high number of transitions in the experiments, respectively.

The VA loading experiments were performed under load control, using a standard loading spectrum at a constant frequency of 10 Hz. It was already shown that the fatigue performance of similar specimens is not affected by the frequency when it lies in the range between 2 and 10 Hz (Zhang et al., 2010). Keeping the frequency constant during the whole spectrum imposed different loading rates depending on the amplitude of each cycle, and thus loading rates were similar to those in previously performed CA experiments (Sarfaraz et al., 2011).

Although any VA spectrum would be appropriate to investigate the behavior of the examined joints, the WISPERX time series that has been developed for wind turbine applications (ten Have, 1993) was used. The WISPERX spectrum is well documented,

Table 9.6 Multi-block loading results at $R = 0.1$

	Specimen ID	F_{max1} (kN)	n_1	F_{max2} (kN)	n_2	NB_1	NB_2	Damage index
High transition	R01B4070H01	12.0	2914	19.2	10	34	34	0.153
	R01B7040H01	19.2	10	12.0	2914	164.6	164	0.741
Low transition	R01B4070L01	12.0	29,140	19.2	100	19	18.1	0.837
	R01B7040L01	19.2	100	12.0	29,140	42.8	42	1.913

Table 9.7 **Multi-block loading results at $R = 10$**

	Specimen ID	F_{max1} (kN)*	n_1	F_{max2} (kN)*	n_2	NB_1	NB_2	Damage index
High transition	R10B6070H01	17.4	374	20.3	10	366	365.5	0.120
	R10B7060H01	20.3	10	17.4	374	730	729.5	0.240
Low transition	R10B6070L01	17.4	37,400	20.3	1000	8	7.1	0.248
	R10B7060L01	20.3	1000	17.4	37,400	10.9	10	0.345

* Absolute value.

optimized in terms of length, and includes a variety of load amplitudes, mean loads, and overloads (Sarfaraz et al., 2013b).

To obtain the desired maximum applied loads (F_{max}), representing four load levels at 22, 20, 18, and 16 kN, each of the spectrum integers was multiplied by appropriate factors. The scaled spectra were repeatedly applied to the specimens up to failure. At least three specimens were examined under each load level to obtain information regarding the scatter of fatigue life. The specimens were labeled accordingly; for example, WX1603 represents the third specimen loaded at the maximum load level of 16 kN under the WISPERX loading spectrum. As for BL, one specimen per load level was instrumented on one side with two crack gauges (HBM crack gauge type RDS20) that covered the whole bonding length and monitored crack initiation and propagation throughout the duration of fatigue.

9.3 Experimental results and discussion of the effect of loading

9.3.1 Failure modes

The observed failure modes under block and VA loading patterns were analogous to the failure modes exhibited by similar joints under tensile and compressive CA loading, as reported by Sarfaraz et al. (2011). Under loading sequences composed of tensile loading blocks ($R = 0.1$), a dominant crack initiated from the joint corner of one of the bond lines between the adhesive and the inner laminate. The crack then shifted deeper, between the first and second mat layers of the inner laminate, and propagated along this path up to failure, as shown in Figure 9.5(a) for BL and Figure 9.6 for a specimen loaded under the selected VA spectrum. The cracks observed along the lower bond line and at the right side of specimen between the outer laminate and the adhesive, shown in Figure 9.6, are secondary cracks that occurred only after the specimen failed.

Similar to the observed failure mode under CA compression loading at $R = 10$, under loading sequences composed of compressive loading blocks, the dominant crack initiated from the right side of bonded area, shown in Figure 9.5(b), and propagated in the middle of the inner laminate inside the roving layer. The crack observed in the outer laminates is a secondary crack that occurred after the failure of the specimen. Accordingly, no visual difference was observed in failure location because of the load interaction, which is a basic required condition for comparison of the BL and CA fatigue data.

9.3.2 Block loading results

Under tensile loading blocks, the calculated damage indices in Table 9.2 for the L-H sequences, independent of load level, are less than 1 except for one experiment (R01F30B407002), whereas they are higher than 1 for the H-L sequences. The results for two joints subjected to the longer first loading blocks, R01F50B407001 and R01F50B704001, also were consistent with the obtained results. The calculated damage indices for different loading sequences are presented in Figure 9.7 against the ratio

270 Fatigue and Fracture of Adhesively-bonded Composite Joints

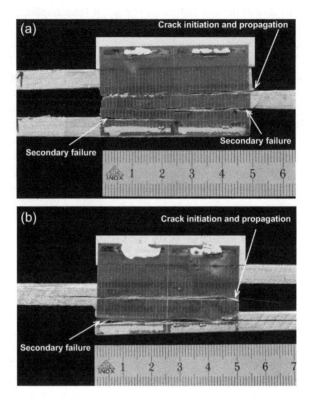

Figure 9.5 Double-lap joint failure modes under block loading tension (a) and compression (b).

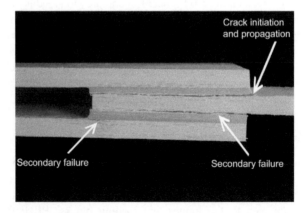

Figure 9.6 Failure mode under VA loading (tensile mode).

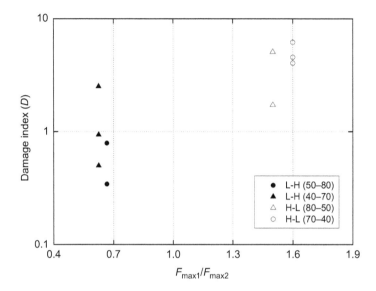

Figure 9.7 Comparison of calculated damage indices for low to high (L-H) and high to low (H-L) sequences composed of tensile loading blocks ($R = 0.1$).

of the maximum load of the first block to the second block (F_{max1}/F_{max2}). The damage indices corresponding to the L-H sequence are close to or less than 1, whereas for the H-L sequences the D values are higher than 1.

In contrast to tensile loading blocks, the loading sequence effect was more damaging under H-L than L-H compressive loading blocks. The damage index for experiments carried out under the L-H sequence was higher than 1, whereas the H-L sequence led to values of less than or close to 1 (see Figure 9.8). The fatigue failure occurred for one of the specimens, R10F30B806502, during the first block before the predetermined number of cycles had been completed.

The results of the experimental program designed to study the transition effect are presented in Tables 9.6 and 9.7 for tensile and compressive loading, respectively. The number of cycles in each block (n_i), number of blocks (NB_i) (number of occurrences of each block), and also the damage indices calculated using the Palmgren-Miner rule given in Eqn (9.1) are shown in these tables. The fatigue failure occurred during both loading blocks, blocks with decimal NB_i, independent of load level. The integer part of NB_i denotes the number of transitions (NT). The expected duration of CA fatigue for each load level can be found in Tables 9.2 and 9.3. The damage indices for all specimens except one, R01B7040L01, are less than 1.

9.3.2.1 Loading sequence effect

The data acquired from the crack gauges concerning the developing crack in the joints provide valuable information regarding the variation in the crack propagation rate during different loading blocks. Because of the scatter in the fracture mechanics data for

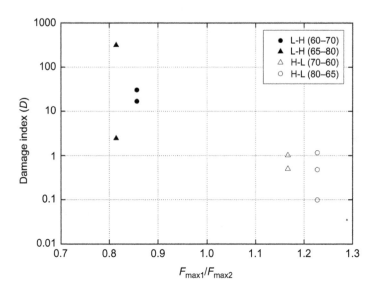

Figure 9.8 Comparison of calculated damage indices for low to high (L-H) and high to low (H-L) sequences composed of compressive loading blocks ($R = 10$).

composite materials, the incremental polynomial fitting method (according to ASTM E647) is usually preferred to the secant method for calculating da/dN. The calculated crack propagation rate (da/dN) based on the incremental polynomial fitting against the crack length, a, is shown in Figures 9.9–9.12 for tensile loading sequences and Figures 9.13–9.16 for compressive loading sequences. The data corresponding to each block are indicated by different symbols. In all figures the solid circles and triangles are related to the first and second blocks, respectively.

The fracture data obtained from the BL experiments can be compared with the CA data since similar failure modes were observed under both CA and BL conditions. The comparison showed that the crack propagation rate during the first block for all loading cases conformed well to the CA data, as expected. However, when the load level was altered and the second block started, a noticeable change compared to the corresponding CA behavior occurred. The analysis of da/dN during the second loading block explains the shorter or longer duration of fatigue compared to the expected duration based on CA data.

Under tensile loading, the acceleration of crack growth when the load level of the second block is higher than the first block can be seen in Figures 9.9 and 9.10. The da/dN during the second block is slightly higher than the CA results, and rapid, nonlinear growth of the crack occurs earlier, around when the crack is 25 mm long, under BL, as shown in Figure 9.9. Under the second loading sequence, R01F30B508002, the crack propagated very rapidly (Figure 9.10), and a sudden rupture occurred when the second load level was applied, similar to the R01F30B407002 sequence.

A retardation effect was identified under the H-L sequences during the second loading block. The crack growth rate decelerated to levels lower than the recorded

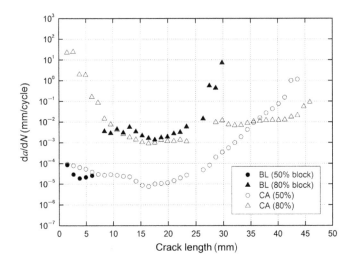

Figure 9.9 Comparison of crack propagation rate under two-stage block loading (BL) and corresponding constant amplitude (CA) loading for R01F30B407002.

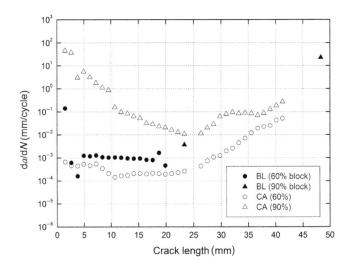

Figure 9.10 Comparison of crack propagation rate under two-stage block loading (BL) and corresponding constant amplitude (CA) loading for R01F30B508002.

rate under CA loading, as presented in Figures 9.11 and 9.12. The retardation effect remains for a long period of time and affects the growth rate until the crack length reaches the rapid nonlinear propagation phase. This behavior leads to fatigue durations longer than that expected based on the CA data and damage indices higher than 1.

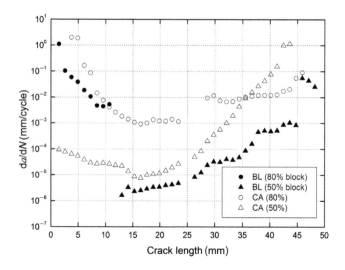

Figure 9.11 Comparison of crack propagation rate under two-stage block loading (BL) and corresponding constant amplitude (CA) loading for R01F30B704002.

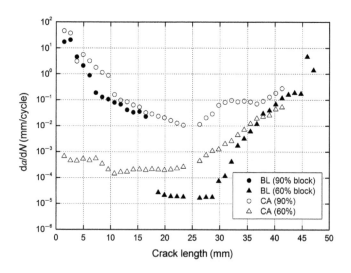

Figure 9.12 Comparison of crack propagation rate under two-stage block loading (BL) and corresponding constant amplitude (CA) loading for R01F30B805002.

The retardation effect caused by overloads is known for metals and can be attributed to several mechanisms (Suresh, 1998). However, in contrast to what was observed for the examined joints, a progressive reduction in crack growth rate continues in metals over a certain distance, known as the delay distance, up to a minimum and then starts to increase until it eventually attains the propagation rate before the overload (Suresh, 1998).

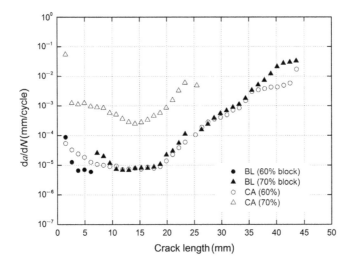

Figure 9.13 Comparison of crack propagation rate under two-stage block loading (BL) and corresponding constant amplitude (CA) loading for R10F30B607002.

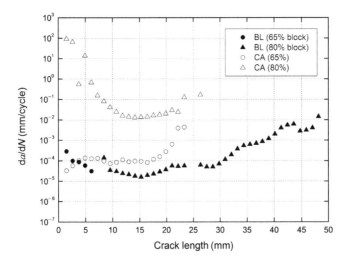

Figure 9.14 Comparison of crack propagation rate under two-stage block loading (BL) and corresponding constant amplitude (CA) loading for R10F30B658002.

Under the L-H compressive sequences (Figures 9.13 and 9.14), when the load level was increased, the da/dN was accelerated over a short crack distance. However, this rate never reached the rate of CA crack propagation under the second load level and slowed down to a rate similar to the CA rate corresponding to the first load level.

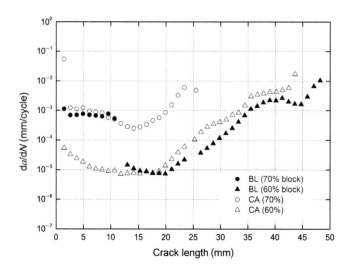

Figure 9.15 Comparison of crack propagation rate under two-stage block loading (BL) and corresponding constant amplitude (CA) loading for R10F30B706002.

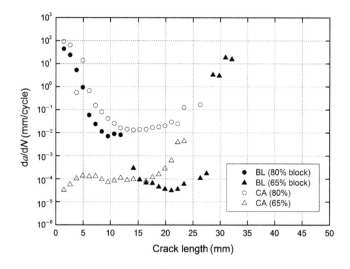

Figure 9.16 Comparison of crack propagation rate under two-stage block loading (BL) and corresponding constant amplitude (CA) loading for R10F30B806503.

This behavior clearly explains the extension of the fatigue duration under the L-H compressive loading sequence and the calculated damage indices that are significantly higher than 1. Under the H-L loading blocks, a minor sequence effect was observed, consistent with the fracture data presented in Figures 9.15 and 9.16. The da/dN for CA and BL was found to be similar under both load levels, obviously resulting in the damage index close to 1.

The acceleration and retardation of the crack growth rate is due to the damage mechanisms activated under different load levels. Under tensile loading, the dominant crack is located between the first and second mat layers of the laminate, whereas under compression loading the dominant crack initiates and propagates through the roving layer of the laminates. Between the mat layers, the chopped strands randomly stitched to the 0°/90° fabrics make the material prone to the initiation of several microcracks and the formation of considerable fiber bridging.

The detailed fracture analysis of double cantilever beam joints composed of pultruded laminates similar to those presented by Shahverdi, Vassilopoulos, and Keller (2011) showed that the amount of fiber bridging strongly depends on the location of the crack in the laminate. Based on this research, maximum fiber bridging occurs between the two mat layers with the highest strain energy release rate, whereas the bridging between the mat and roving layers is less significant. Higher fiber bridging for 0°/90° interfaces than the 0°/0° interfaces also has been reported by Pereira and de Morais (2004) for carbon/epoxy multidirectional laminates. Therefore, for the examined joints, under tension loading the bridging greatly contributes to the fracture behavior, whereas under compression loading the strain energy release rate mainly results from the contribution of the matrix.

Under the H-L tensile loading sequences, when the load level is decreased, the input energy for crack propagation also is decreased, while the fiber bridging developed during the first loading block remains constant. Therefore, the crack propagation rate significantly decreases (see Figures 9.11 and 9.12) and leads to a longer fatigue duration compared to that based on the CA fatigue data. An inverse process occurs under the L-H sequences: the increase in load levels in the second loading blocks provide the energy required to break the fiber bridging developed during the first block and consequently accelerate the crack growth rate (Figures 9.9 and 9.10).

Under compressive loading sequences no significant fiber bridging was observed and the failure was dominated by other mechanisms causing microdamage. Since the first block is short (applied for less than 5% of the total life), there is no sufficient time for the development of microdamage mechanisms when the H-L loading sequences are applied. Therefore, the crack propagation rate decreases to the corresponding levels of the CA experiments, as seen in Figures 9.15 and 9.16, when the load is decreased. Under the L-H sequences, the specimens were loaded for a long period under the initial low loading block, allowing the time required for the multiple microdamage that governs the joint failure to develop. This damage develops further during the second step (under higher loads), absorbing energy and thus preventing the crack propagation rate from increasing, as presented in Figures 9.13 and 9.14, and therefore extending the lifetime of the joint.

Although there is no model for bonded FRP joints that takes into account the load sequence effect, several attempts to formulate such effects in composite materials have been made. A comprehensive review of damage accumulation models has been provided by Post, Case, and Lesko (2008). These models are mainly expressed by nonlinear damage accumulation models such as

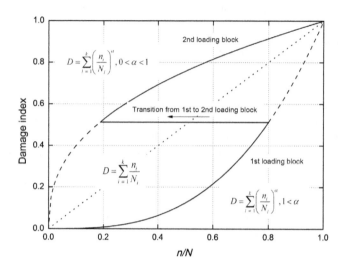

Figure 9.17 Comparison of linear (dotted line) and nonlinear (dashed lines) damage accumulation models and simulation of two-block loading sequence (solid lines).

$$D = \sum_{i=1}^{k} \left(\frac{n_i}{N_i}\right)^{\alpha} \tag{9.2}$$

which provides a nonlinear form of the Palmgren-Miner damage summation rule, depending on the exponent α, which can be a function of the applied load level and R ratio. The nonlinear form of the Palmgren-Miner rule is schematically shown in Figure 9.17 (dashed lines) compared to the linear form (dotted line).

According to these models, the damage accumulation trend (dashed lines in Figure 9.17) is a function of the applied stress level. Figure 9.17 is used as an example to quantitatively describe the nonlinear damage accumulation process in the following text. In this figure, when the load level is altered, for example, from the first to the second loading block, the damage accumulation path also changes, but the amount of damage remains constant (see solid lines in Figure 9.17). In this process, although approximately 80% of the total fatigue duration occurs during the first block ($n_1/N_1 = 0.8$), when the load changes, the fatigue that has already occurred is equivalent to 20% of the total duration of fatigue under the second loading block, and the partial damage required to cause failure is $n_2/N_2 = 0.8$. Therefore the sum of partial damage (n_i/N_i) due to both loading blocks is equal to $D = 1.6$. Based on these models, the calculated damage in a two-stage BL experiment never reaches $D = 2$ because the partial damage under each loading block independent of the model exponent is less than 1. Although these models have been successfully used for some composite material systems, they are not able to accurately model the accumulated damage under two-stage BL sequences for the examined bonded joints. As shown in Tables 9.2 and 9.3, the damage indices for several cases are greater than 2.

Moreover, the fracture mechanics data acquired under different CA load levels for similar joints shows a linear correlation between crack length and the number of cycles in the range of 10—90% of fatigue duration independent of the applied load level (Sarfaraz et al., 2011). Hence, if the length of the dominant crack in the joint that leads to the final failure is considered as the total damage, the damage accumulation for the examined joints occurs in a linear manner, and only a model that takes into account the retardation and acceleration phenomena is able to accurately predict the remaining fatigue life. Therefore the application of nonlinear models to the examined joints not only provides inaccurate results but also is physically meaningless. A detailed testing program including all the parameters involved, such as the load level ratio (F_{max1}/F_{max2}), length of the first loading block, and applied load ratio (R) should be investigated to develop a damage accumulation model that takes into account load sequence effects.

9.3.2.2 Load transition effect

The consistency of the damage indices in Table 9.6 for the tensile loading block shows the strong damaging effect of the number of transitions on the fatigue duration in spite of the small amount of experimental data and the few loading cases available. The damage indices decreased for the sequences starting with the high load level (R01B7040xxx), from 1.913 to 0.741, or for sequence R01B4070xxx, from 0.837 to 0.153, when the number of transitions was increased. In addition, for the R01B7040xxx sequence, both of these values are lower than the calculated damage index for similar loading conditions but with only one transition (R01FxxB7040xx), that is, 4.040, 4.554, and 6.173 (given in Table 9.2). Although the applied sequences are repeated several times in these experiments, it seems that the starting loading block still has a minor effect consistent with the loading sequence effect discussed earlier. For both high and low transition cases the damage indices corresponding to the experiments starting with higher load levels (0.741 and 1.913) are greater than the sequences that started with low load levels, that is, 0.153 and 0.837.

The number of load transitions has a greater damaging effect under compressive fatigue loading. The damage indices for the high transition cases (R10BxxxxH01; given in Table 9.7), independent of the applied load levels, are lower than those of the low transition experiments (R10BxxxxL01). Furthermore, the Palmgren-Miner indices for all experiments are less than 0.4, which indicates a stronger effect of the number of loading transitions than the loading sequence effect. In contrast to the tensile fatigue loading, the effect of loading sequence is completely eliminated for compressive loading when the number of transitions is increased.

9.3.3 Variable amplitude results

The VA fatigue durations obtained for the examined joints under the WISPERX spectrum with different load levels are presented in Table 9.8. The fatigue duration is given both in terms of number of spectrum passes (N_p), and total number of cycles (N_f) until failure. The crack lengths versus the normalized number of cycles are shown in Figure 9.18 for one specimen per load level that was equipped with crack gauges. A

Table 9.8 **Fatigue data under WISPERX spectrum**

Specimen ID	Maximum load level, F_{max} (kN)	No. of spectrum passes to failure, N_p	No. of cycles to failure, N_f
WX2201	22.0	5.21	66,805
WX2202		10.26	131,641
WX2203*		3.05	39,091
WX2001	20.0	23.04	295,647
WX2002		3.21	41,142
WX2003*		14.74	189,102
WX1801	18.0	304.74	3,910,140
WX1802		19.12	245,336
WX1803*		9.71	124,628
WX1804		170.92	2,193,086
WX1601	16.0	201.60	2,586,685
WX1602		80.16	1,028,587
WX1603*		98.23	1,260,387

* Specimens instrumented with crack gauges.

common trend, independent of load level, was observed; it showed rapid crack propagation at the beginning (region A), followed by a moderate crack propagation rate represented by a constant slope between c. 10% and 90% of fatigue duration (region B) and another rapid crack propagation at the end (region C) of the fatigue life. This behavior is similar to the crack propagation observed under CA loading reported by Sarfaraz et al. (2011), and the similarity shows that this behavior is independent of the applied loading patterns.

The incremental polynomial method has been used to analyze the BL fatigue/fracture results described in the previous paragraph. Nevertheless, the secant method was used to calculate the crack propagation rate during VA loading to avoid obscuring the possible load interaction effects of the VA spectrum on the crack propagation rate. According to the secant or point-to-point method, the crack propagation rate is determined by calculating the slope of a straight line connecting two contiguous data points on the $a-N$ curve. Characteristic fracture mechanics data and their relationship with the applied loading spectrum for the lowest applied load of 16 kN are presented in Figures 9.19 and 9.20 and for the highest applied load of 22 kN in Figures 9.21 and 9.22.

The plots of the crack propagation rate, da/dN, against the normalized number of cycles, N/N_f, are presented in Figures 9.19 and 9.21. They showed a trend consistent with the $a-N$ curves presented in Figure 9.8. The crack propagation rate decreased at

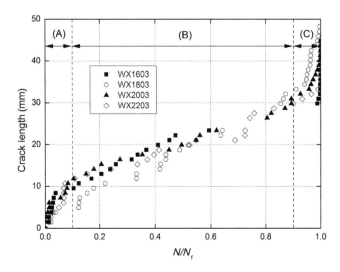

Figure 9.18 Crack length versus normalized number of cycles under different load levels.

the beginning of fatigue, followed by a stable phase with an almost constant rate, and finally accelerated at the end of fatigue. Sharp variations of da/dN during the stable phase are likely for composite materials, even those under CA loading, because of the complex nature of the progressing crack, as reported by Sarfaraz et al. (2011). However, the changes observed in the stable phase of crack propagation (region B in Figure 9.18) were thoroughly analyzed to explore any possible correlation between them and the corresponding periods of the applied spectrum. In Figures 9.19 and 9.21, the points that indicate an unusual change in da/dN, designated "event," are marked with capital letters in ascending order based on the corresponding spectrum passes and the number of cycles they represent. The numbers in brackets denote the spectrum passes of each event. The location of each event in the applied spectrum is shown in Figures 9.20 and 9.22.

Figure 9.19 represents the crack propagation rate under the lowest applied maximum load (16 kN). Four points marked with letters A–D indicate the locations where a significant increase in da/dN was observed. Comparisons of the loading cycles corresponding to these events, shown in Figure 9.20(a) and (b), do not indicate any specific correlation between loading and increase of da/dN, except at point C, where several load transitions occurred before this increase was recorded. It should be noted, however, that there are several cycles between two adjacent points in Figure 9.19 (e.g., 27,450 cycles between point C and the previous data point), and the observed increase cannot be directly attributed to the load transitions just before point C. Therefore the variations identified in da/dN are attributed to the usual scatter of the fracture data.

Eight events were identified under the highest maximum load level (22 kN), as shown in Figure 9.21. Sudden changes in da/dN observed at the beginning of the spectrum coincided with the overload cycles indicated by points A, C, E, and F in

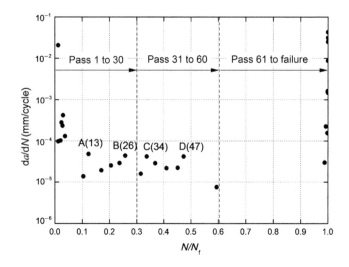

Figure 9.19 Crack propagation rate versus a normalized number of cycles for WX1603.

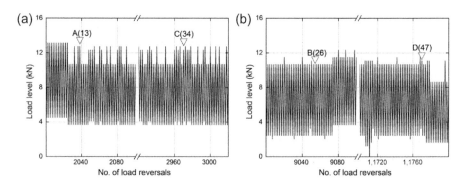

Figure 9.20 Detailed locations in WISPERX spectrum corresponding to specific events for WX1603, as defined in Figure 9.19.

Figure 9.22(a). Two subsequent events after points C and F, that is, D and G, occurred almost at the same location in the spectrum (Figure 9.22(b)), showing that both events D and G still were a consequence of the overload.

According to these observations, a sudden increase in the crack propagation rate can be correlated to spectrum overloads. Nevertheless, the crack propagation rate returned to the preceding levels after applying a certain number of additional loading cycles, without any observed retardation. In addition, an increase in da/dN after several consecutive load transitions was observed at different locations of the applied spectra. These interpretations are in agreement with the discussion by Sarfaraz et al. (2013a), who reported that frequent load transitions in a BL spectrum introduced a more

Block and variable amplitude fatigue and fracture behavior 283

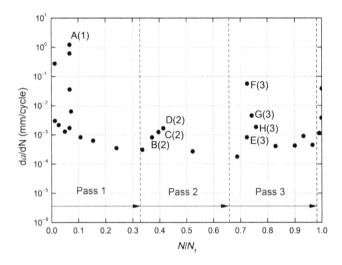

Figure 9.21 Crack propagation rate versus normalized number of cycles for WX2203.

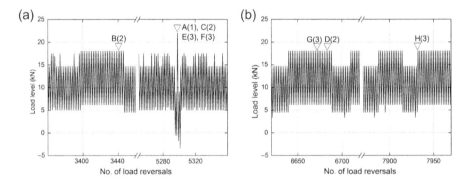

Figure 9.22 Detailed locations in WISPERX spectrum corresponding to specific events for WX2203, as defined in Figure 9.21.

damaging effect and obscured the load sequence effects. Therefore, in spite of the various changes in load amplitude, the crack growth rate remained almost constant (excluding the sudden jumps caused by the overloads) during the stable crack propagation phase (region B in Figure 9.18), showing a trend similar to that measured under CA loading.

9.4 Conclusions

The fatigue behavior of adhesively-bonded, pultruded, double-lap GFRP joints under block and VA loading conditions was experimentally investigated. The crack propagation data recorded during cyclic loading were used to analyze the effects of loading sequence and number of load transitions on the fatigue behavior of the examined joints under the different applied loading patterns. The results showed a strong sequence effect for a small number of transitions and its dependence on the type of loading and, consequently, the failure mode.

The failure modes of bonded joints under different BL and VA loading patterns were consistent with those exhibited by similar joints loaded under tensile and compressive CA loading. The similarity of the failure modes under different loadings allows the CA loading experimental results to be used to predict the duration of fatigue in the examined joints under block and VA loading.

Analysis of the experimental results revealed a significant loading sequence effect and showed that it is a function of the type of loading and the applied load levels. The L-H sequences were found to be more damaging than the H-L sequences under tensile loading, whereas under compressive loading this trend was reversed. The effect of loading sequence on the duration of fatigue in the examined joints was associated with the crack growth rate during the applied loading blocks. The H-L tensile loading blocks led to retardation of crack growth rate, whereas acceleration was observed under L-H sequences. In contrast, the crack growth rate under L-H compressive loading blocks did not increase significantly when the load level was increased and led to longer duration of fatigue. However, under the H-L sequences, the first loading block did not affect the expected crack propagation rate under the second loading block. The difference in sequence effects under tension and compression was attributed to the difference in failure modes.

Three stages in the crack propagation behavior of the examined bonded joints under VA loading were identified: rapid crack propagation at the beginning, a stable phase in the middle, and another accelerated phase at the end of the duration of fatigue. The crack propagation rate was correlated to the applied loading spectrum. Under higher loads, where the fatigue duration was limited to a small number of spectrum passes, acceleration in crack propagation caused by the overloads was observed. However, the crack propagation rate decreased to the previous rate after a short time.

The frequent change of load levels showed a very strong damaging effect compared to the sequence effect. The sequence effect is considerably reduced when the number of transitions is increased. Thus this parameter is more critical than the sequence effect in predicting duration of fatigue under complex loading patterns. In spite of the presence of loading sequence effects under two-stage BL for the examined bonded joints, the crack propagation rate was almost stable during 80% of the fatigue duration. This behavior was attributed to the frequent change of the load level under VA loading, which accelerates the crack propagation rate, and, as was proved in this work, has more of an effect on the duration of fatigue in the examined joints than does the loading sequence effect.

The Palmgren-Miner rule was found to be inadequate for estimating the accumulated damage under BL and VA sequences. However, applying the developed nonlinear models for composite materials is also questionable for the examined bonded joints because the fracture mechanics data show no sign of nonlinear damage accumulating under fatigue loading. Therefore, a detailed study including all parameters involved is required to develop a reliable damage accumulation model.

References

Adam, T., Gathercole, N., Reiter, H., & Harris, B. (1994). Life prediction for fatigue of T800/5425 carbon-fibre composites: II. Variable amplitude loading. *International Journal of Fatigue, 16*(8), 533–547.
Ashcroft, I. A. (2004). A simple model to predict crack growth in bonded joints and laminates under variable-amplitude fatigue. *Journal of Strain Analysis for Engineering Design, 39*(6), 707–716.
Bartley-Cho, J., Lim, S. G., Hahn, H. T., & Shyprykevich, P. (1998). Damage accumulation in quasi isotropic graphite/epoxy laminates under constant-amplitude fatigue and block loading. *Composites Science and Technology, 58*(9), 1535–1547.
Bond, I. P., & Ansell, M. P. (1998). Fatigue properties of jointed wood composites: part II life prediction analysis for variable amplitude loading. *Journal of Materials Science, 33*(16), 4121–4129.
Bonnee, W. J. A. (1996). NLR investigation of polyester composite materials. In C. W. Kensche (Ed.), *Fatigue of materials and components for wind turbine rotor blades. German Aerospace Establishment, European Commission-EUR 16684* (pp. 39–70).
Broutman, L. J., & Sahu, S. (1972). *A new theory to predict cumulative fatigue damage in fibreglass reinforced plastics*. Philadelphia: ASTM STP 497, American Society for Testing and Materials, 170–188.
Erpolat, S., Ashcroft, I. A., Crocombe, A. D., & Abdel-Wahab, M. M. (2004a). A study of adhesively bonded joints subjected to constant and variable amplitude fatigue. *International Journal of Fatigue, 26*(11), 1189–1196.
Erpolat, S., Ashcroft, I. A., Crocombe, A. D., & Abdel-Wahab, M. M. (2004b). Fatigue crack growth acceleration due to intermittent overstressing in adhesively bonded CFRP joints. *Composites Part A: Applied Science and Manufacturing, 35*(10), 1175–1183.
Filis, P. A., Farrow, I. R., & Bond, I. P. (2004). Classical fatigue analysis and load cycle mix-event damage accumulation in fibre reinforced laminates. *International Journal of Fatigue, 26*(6), 565–573.
Found, M. S., & Kanyanga, S. B. (1996). The influence of two-stage loading on the longitudinal splitting of unidirectional carbon-epoxy laminates. *Fatigue & Fracture of Engineering Materials & Structures, 19*(1), 65–74.
Found, M. S., & Quaresimin, M. (2003). Two-stage fatigue loading of woven carbon fibre reinforced laminates. *Fatigue & Fracture of Engineering Materials & Structures, 26*(1), 17–26.
Gamstedt, E. K., & Sjögren, B. A. (2002). An experimental investigation of the sequence effect in block amplitude loading of cross-ply laminates. *International Journal of Fatigue, 24*(2–4), 437–446.
Han, K. S., & Abdelmohsen, M. H. (1986). Fatigue life scattering of RP/C, Int J of Vehicle Design, Technological advances in Vehicle Design Series, SP6. *Designing with Plastics and Advanced Plastic Composites*, 218–227.
Harris, B., Gathercole, N., Reiter, H., & Adam, T. (1997). Fatigue of carbon-fibre reinforced plastics under block-loading conditions. *Composites Part A: Applied Science and Manufacturing, 28*(4), 327–337.

ten Have, A. A. (1993). WISPER and WISPERX: summary paper describing their backgrounds, derivation and statistics. *American Society of Mechanical Engineers, Solar Energy Division, 14*, 169–178.

Hosoi, A., Kawada, H., & Yoshino, H. (2006). Fatigue characteristic of quasi-isotropic CFRP laminates subjected to variable amplitude cyclic loading of two-stage. *International Journal of Fatigue, 28*(10 Special Issue), 1284–1289.

Hwang, W., & Han, K. S. (1987). Statistical study of strength and fatigue life of composite materials. *Composites, 18*(1), 47–53.

Hwang, W., & Han, K. S. (1989). *Fatigue of composite materials − Damage model and life prediction*. Philadelphia: ASTM STP 1012, American Society for Testing and Materials, 87–102.

Jeans, L. L., Grimes, G. C., & Kan, H. P. (1983). Fatigue sensitivity of composite structure for fighter aircraft. *Journal of Aircraft, 20*(2), 102–110.

Jen, M. H. R., Kau, Y. S., & Wu, I. C. (1994). Fatigue damage in a centrally notched composite laminate due to two-step spectrum loading. *International Journal of Fatigue, 16*(3), 193–201.

Lee, C. H., & Jen, M. H. R. (2000a). Fatigue response and modeling of variable stress amplitude and frequency in AS-4/PEEK composite laminates, part 1: experiments. *Journal of Composite Materials, 34*(11), 906–929.

Lee, C. H., & Jen, M. H. R. (2000b). Fatigue response and modelling of variable stress amplitude and frequency in AS-4/PEEK composite laminates, part 2: analysis and formulation. *Journal of Composite Materials, 34*(11), 930–953.

Lee, B. L., & Liu, D. S. (1994). Cumulative damage of fiber-reinforced elastomer composites under fatigue loading. *Journal of Composite Materials, 28*(13), 1261–1286.

Nolting, A. E., Underhill, P. R., & DuQuesnay, D. L. (2008). Variable amplitude fatigue of bonded aluminum joints. *International Journal of Fatigue, 30*(1), 178–187.

Otani, N., & Song, D. Y. (1997). Fatigue life prediction of composite under two-step loading. *Journal of Materials Science, 32*(3), 755–760.

Pereira, A. B., & de Morais, A. B. (2004). Mode I interlaminar fracture of carbon/epoxy multidirectional laminates. *Composites Science and Technology, 64*(13–14), 2261–2270.

Plumtree, A., Melo, M., & Dahl, J. (2010). Damage evolution in a $[\pm 45]_{2S}$ CFRP laminate under block loading conditions. *International Journal of Fatigue, 32*(1), 139–145.

Post, N. L., Case, S. W., & Lesko, J. J. (2008). Modeling the variable amplitude fatigue of composite materials: a review and evaluation of the state of the art for spectrum loading. *International Journal of Fatigue, 30*(12), 2064–2086.

Sarfaraz, R., Vassilopoulos, A. P., & Keller, T. (2011). Experimental investigation of the fatigue behavior of adhesively-bonded pultruded GFRP joints under different load ratios. *International Journal of Fatigue, 33*(11), 1451–1460.

Sarfaraz, R., Vassilopoulos, A. P., & Keller, T. (2013a). Block loading fatigue of adhesively-bonded pultruded GFRP joints. *International Journal of Fatigue, 49*, 40–49.

Sarfaraz, R., Vassilopoulos, A. P., & Keller, T. (2013b). Variable amplitude fatigue of adhesively-bonded pultruded GFRP joints. *International Journal of Fatigue, 55*, 22–32.

Sarkani, S., Michaelov, G., Kihl, D. P., & Beach, J. E. (1999). Stochastic fatigue damage accumulation of FRP laminates and joints. *Journal of Structural Engineering, 125*(12), 1423–1431.

Schaff, J. R., & Davidson, B. D. (1997a). Life prediction methodology for composite structures. Part I. Constant amplitude and two-step level fatigue. *Journal of Composite Materials, 31*(2), 128–157.

Schaff, J. R., & Davidson, B. D. (1997b). Life prediction methodology for composite structures. Part II — spectrum fatigue. *Journal of Composite Materials, 31*(2), 158—181.

Shahverdi, M., Vassilopoulos, A. P., & Keller, T. (2011). A phenomenological analysis of mode I fracture of adhesively-bonded pultruded GFRP joints. *Engineering Fracture Mechanics, 78*(10), 2161—2173.

Shenoy, V., Ashcroft, I. A., Critchlow, G. W., & Crocombe, A. D. (2010). Fracture mechanics and damage mechanics based fatigue lifetime prediction of adhesively bonded joints subjected to variable amplitude fatigue. *Engineering Fracture Mechanics, 77*(7), 1073—1090.

Smith, M. A., & Hardy, R. (1977). Fatigue research on bonded carbon fibre composite/metal joints. *Composites, 8*(4), 255—261.

Suresh, S. (1998). *Fatigue of materials*. Cambridge University Press.

Van Paepegem, W., & Degrieck, J. (2002). Effects of load sequence and block loading on the fatigue response of fibre-reinforced composites. *Mechanics of Advanced Materials and Structures, 9*(1), 19—35.

Wahl, N. W., Mandell, J. F., & Samborsky, D. D. (2001). Spectrum fatigue lifetime and residual strength for fiberglass laminates in tension. In *ASME wind energy symposium, AIAA-2001-0025, ASME/AIAA*.

Yang, J. N., & Jones, D. L. (1980). Effect of load sequence on the statistical fatigue of composite. *AIAA Journal, 18*(12), 1525—1531.

Yang, J. N., & Jones, D. L. (1983). Load sequence effects on graphite/epoxy [±35]$_{2s}$ laminates. In *Long-term behavior of composites* (pp. 246—262). Philadelphia: ASTM STP 813, American Society for Testing and Materials.

Zhang, Y., Vassilopoulos, A. P., & Keller, T. (2010). Fracture of adhesively-bonded pultruded GFRP joints under constant amplitude fatigue loading. *International Journal of Fatigue, 32*(7), 979—987.

Durability and residual strength of adhesively-bonded composite joints: the case of F/A-18 A–D wing root stepped-lap joint

W. Seneviratne[1], J. Tomblin[1], M. Kittur[2]
[1] Wichita State University, Wichita, KS, USA; [2] Naval Air System Command, Patuxent River, MD, USA

10.1 Introduction

The first generations of primary structural composite components are currently entering the twilight of their certified service lives. Life-extension efforts underway for many aircraft are primarily geared to the continued safe operation of the metallic components of the airframe, for which a fatigue life can be quantified. However, many of the composite components do not have a measurable fatigue life, so the extension of their lives cannot be accomplished using the same methodology. The McDonnell Douglas (now Boeing) F/A-18 Hornet was first introduced in the 1970s for use by the U.S. Navy and Marine Corps. F/A-18s currently are used by several foreign air services including the Royal Australian Air Force, Canadian Forces, Finnish Air Force, and Swiss Air Force. The F/A-18 models A through D (Hornets) and E and F (Super Hornets) use about 10% and 19% of composites, respectively, by their structural weights. In addition, wing skins use bonded joints in primary and secondary structural details. The wing root stepped-lap joint, discussed in this chapter, is one of the key examples of a bonded primary structure that is certified and deployed on an air vehicle in the United States.

The primary goal of this case study was to evaluate the residual static strength and remaining life of this joint area after one lifetime of aircraft service and to assess the remaining service life based on its usage history. Furthermore, tests were designed to address one of the biggest fears in managing aging aircraft fleet—unknown failure mechanisms that emerge with little or no warning, raising concerns that an unexpected phenomenon may suddenly jeopardize an entire fleet's flight safety, mission readiness, and/or support costs.

The results of these tests are useful in developing a quantifiable, risk-based assessment methodology for determining the capability to extend the life of a composite structure. This research provides useful information for the following:

- Evaluating assumptions used in establishing the service life of the joint
- Determining the remaining (or unused) life of the structure
- Developing structural modifications to mitigate risks with regard to extending service life.

Fatigue and Fracture of Adhesively-bonded Composite Joints. http://dx.doi.org/10.1016/B978-0-85709-806-1.00010-0
Copyright © 2015 Elsevier Ltd. All rights reserved.

Life-extension methodologies developed in this research use the test data to tie together both original certification and operational usage as outlined by Seneviratne and Tomblin (2010), and they succeed in making repetition of initial certification testing unnecessary. In addition, this research program was designed to investigate the possible damage growth (i.e., disbonds in stepped-lap joint areas, delamination, and microcrack growth) under further fatigue tests (Seneviratne & Tomblin, 2011). All fatigue tests were closely monitored with periodic inspections and continuous data acquired from strain gages.

10.2 Bonded joint applications in F/A-18

The F/A-18 has bonded primary structures that have successfully performed since its inception into an active fleet in the early 1980s. Despite the added manufacturing cost, the development and life-cycle costs of bonded joints are significantly lower than those for mechanically fastened joints. The wing skin is primarily constructed of an AS4/3501-6 350°-cure carbon/epoxy composite. Each wing skin is attached to the center fuselage bulkheads using three sets of lugs at the end of a 6Al-4V titanium splice fitting. As shown in Figure 10.1, both top and bottom composite wing skins are bonded to the titanium alloy splice fittings through double-stepped-lap joints using FM-300 film adhesive. The double-overlap configuration has better mechanical performance than the

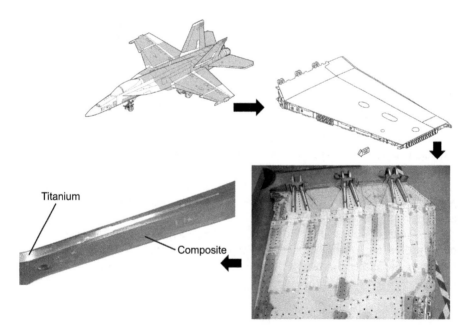

Figure 10.1 Inner wing composite-to-titanium stepped-lap bonded joint.

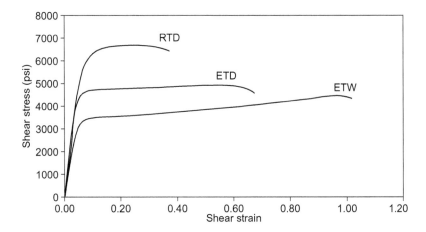

Figure 10.2 Characteristic stress–strain curves for FM-300. ETD, elevated temperature dry; ETW, elevated temperature wet; RTD, room temperature dry.

single-overlap configuration and has the ability to carry significant compressive loads. These hybrid bonded joints are designed to carry a significant load and have complex stress distributions. In addition to the wing root joint, both inboard and outboard titanium pylon attachment fittings are bonded stepped-lap joints.

As shown in Figure 10.2 (Tomblin & Seneviratne, 2002), FM-300 film adhesive can sustain large shear strains, has high toughness, and ultimate shear strength. It compared well against other adhesive systems available during the initial design and production of F/A-18 aircraft (Krieger, 1988). These characteristics are instrumental in redistributing the high concentrations of shear stress on composite-to-metal bonded joints and accommodate the low interlaminar shear strength of composites. This adhesive also has demonstrated excellent moisture resistance in highly humid environments and fatigue resistance with no significant reduction in mechanical properties.

10.3 Stress analysis of stepped-lap joints

Compared to aluminum, titanium has a low coefficient of thermal expansion and does not cause galvanic corrosion when in contact with carbon composites. Therefore, titanium is a good candidate for composite–metal hybrid joints. However, mismatches of the coefficient of thermal expansion and stiffness between the titanium and the composite significantly affect joint efficiency and must be accounted for during joint design. On the other hand, the alternative bolted configuration requires about 1000 fasteners (Krieger, 1993). In addition to cost and weight penalties, this alternative would have an adverse effect because of stress concentrations caused by fastener holes in the composite structure. Thus, the use of a stepped-lap bonded-joint configuration is well suited for the wing root joint.

Bonded stepped-lap joints have the characteristics of both scarf joints and double-lap joints. The stepped-lap joint analysis derived by Hart-Smith (1973c) includes features from both uniform lap joints (Hart-Smith, 1973a, 1973b) as well as scarf joints (Hart-Smith, 1973c). Their analysis is a continuum mechanics-based closed-form solution for adhesive stress. It accounts for the stiffness and thermal mismatch in the adherends (Figure 10.3). It includes adhesive plasticity and is used effectively for optimizing stepped-lap joint designs. This analysis identifies three key design features for optimization: (1) end step, (2) outermost step, and (3) overlap lengths. The stepped-lap joint consists of a series of single-lap joints and has a nonuniform shear stress distribution with high stresses at the ends of each step. The end step (step 9 in Figure 10.3) is designed to prevent failure at the end of the overlap, where titanium terminates, as a result of higher-than-average load transfer and overloading. The outermost step is designed to be sufficiently thin to reduce significant peel stresses and to prevent interlaminar tension failure of the adherend. The analysis also shows that the load-carrying capability of a stepped-lap joint cannot be increased indefinitely by increasing the overlaps.

Stiffness imbalance in the F/A-18 wing root stepped-lap joint has resulted in adhesive stress distribution, as shown in Figure 10.3. Compared to load transfer in the middle of a uniform lap joint, the middle steps in a stepped-lap joint significantly contribute to load transfer, thereby increasing joint efficiency.

Orientation of the ply adjacent to the composite—adhesive interface with respect to the primary load path can affect joint performance and failure mode. Transverse matrix (intraply) cracking of off-axis plies adjacent to the composite—adhesive interface, especially 90° plies, cause delamination or adherend failure. Conversely, fibers of 0° plies adjacent to the composite—adhesive interface are capable of carrying high loads and prolonging the adherend failure caused by transverse cracking of interfacial plies. This also causes significantly high shear strains at the interface. Although the fracture toughness of typical aerospace-grade adhesives are significantly higher than that of the matrix material, at high strain levels an interface with 0° plies develops microcracks in the adhesive closer to the interface, which results in cohesive failure very close to the interface (which may even be mistaken as adhesive failure) or in the first ply. In a bonded stepped-lap joint, different steps may have different ply

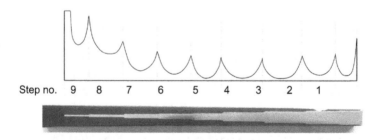

Figure 10.3 Representative adhesive shear stress distribution of stiffness-unbalanced stepped-lap joint.

orientations at the interface; therefore, stress analysis must account for this observation during sizing and optimize the joint to withstand design stresses throughout the life of the structure, accounting for environmental effects.

10.4 End-of-life residual strength evaluation of wing root stepped-lap joint

This effort was designed to evaluate the residual static strength of a wing root stepped-lap joint (WR-SLJ) after one lifetime, or design service goal, of aircraft service and to evaluate the remaining service life based on the usage history. Data generated through this research can be useful for assessing the conservatism associated with data scatter used for the original approach for bolted composite structures (Sanger, 1986; Whitehead, Kan, Cordero, & Saether, 1986) and for determining retirement life, with the potential for extending life (Seneviratne & Tomblin, 2010).

10.4.1 Test specimen geometry

Figure 10.4 shows the test specimen geometry used by Chrissos and Coffey (1996). Similar specimen geometry was used for the end-of-life residual strength evaluation (Tillman, Tsai, & Peek, 2009), which assessed the residual strength of specimens extracted from several decommissioned wing skins. All test specimens were extracted from wing skins from decommissioned F/A-18 aircraft wings (models A—D) supplied by the U.S. Navy; the wings had expended wing root fatigue cycles ranging from a half lifetime to one lifetime of fleet service. Before extracting the specimens, nondestructive inspection (NDI) using a pulse-echo ultrasonic technique was carried out to locate the titanium step termination (outboard edge of the overlap region), as shown in Figure 10.5. Then, the test specimens were laid out so that the loading axis of each specimen was parallel to the third intermediate spar. The specimens were cut using a waterjet and labeled sequentially starting from the specimen that was closest to the leading edge. Since the waterjet starting point exerts a significant amount of pressure through the thickness of the composite material, the starting point must be several inches away from the specimens.

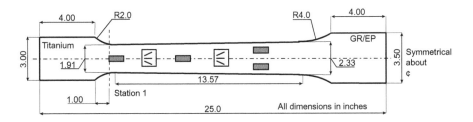

Figure 10.4 Wing root stepped-lap joint specimen with approximate strain gauge locations (one of many strain gauge configurations used in the test program). EP, epoxy; GR, graphite.

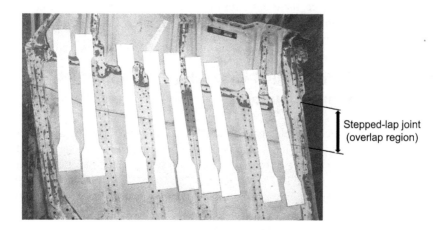

Figure 10.5 Test specimen cut locations of lower wing skin.

The particular wing skin from which each specimen was extracted was tracked. Specimens extracted from lower skins were somewhat straight, so they required minor shimming or smaller tabs during mechanical loading. Specimens extracted from upper skins, however, required additional machining and tabbing because of the kink toward the wing root, as shown in Figure 10.1, to ensure that the specimen was parallel to the loading axis. As can be seen in Figure 10.5, some of the specimens were extracted from areas where there were fastener holes. Thus, three specimen categories were tested:

- No hole (NH)—specimen did not have fastener holes in the gage section.
- Open hole (OH)—specimen had fastener holes in the gage section.
- Filled Hole (FH)—specimen had a fastener installed through the hole in the gage section using a torque of 70 in-lbf.

10.4.2 Test matrices

The static test matrix used to evaluate the end-of-life residual strength of the WR-SLJ specimens is shown in Table 10.1. Test specimens were extracted—14 specimens from lower wing skins and 10 specimens from upper wing skins—and tested in tension and compression loading. Test specimens were labeled by a wing identification number followed by L or U for lower or upper skin, respectively, and NH, OH, or FH, corresponding to the specimen configuration and specimen number; for example, 1L-OH-7 corresponds to the seventh specimen from the leading-edge side extracted from the lower skin of wing number 1 that has an open hole in the gage section. Thickness at the step from the root end that has the first composite—metal bond overlap closer to the metal side (station 1, shown in Figure 10.4) of the specimen is included in Table 10.1.

Table 10.1 Static test specimens

Loading mode	Skin	Specimen configuration	Specimen ID	Thickness at station 1 (in)	Width at station 1 (in)
Tension	Lower	Open hole	1L-OH-1	0.4824 was assumed since measurements were not available	1.983 was assumed since measurements were not available
			1L-OH-2		
			1L-OH-4		
			1L-OH-7		
			2L-OH-1		
			2L-OH-5		
			2L-OH-7		
Compression			1L-OH-3		
			1L-OH-5		
			1L-OH-6		
			2L-OH-2		
			2L-OH-3		
			2L-OH-4		
			2L-OH-6		
Tension	Upper	No hole	4U-NH-3	0.4908	1.994
			4U-NH-4	0.4953	2.003
			3U-NH-3	0.5198	1.957
			3U-NH-4	0.5192	1.958
			1U-NH-3	0.4832	1.994
Compression			1U-NH-10	0.5115	1.996
			2U-NH-4	0.4827	1.980
			2U-NH-5	0.4819	1.976
			2U-NH-6	0.4804	1.974
			3U-NH-2	0.5166	1.950

TDFS, tension-dominant fatigue spectrum; CDFS, compression-dominant fatigue spectrum; FH, filled hole; LSF, load severity factor; OH, open hole; NH, no hole.

The pool of test specimens extracted from eight different wing skins had the following variables:

- Wings with different service history
- Upper or lower wing skins
- NH/OH/FH
- Location of hole in OH/FH specimens
- Step angles and locations

Therefore, the final statistics may have been influenced by some of these variables; however, because of the limited availability of military assets allocated for this program, these variables were unavoidable. Since the primary goal of this research was to investigate the end-of-life performance of the WR-SLJ, the data pool included in this chapter represents different parts of the wing root structure and provides valuable insight into end-of-life residual strength assessment and the estimate of unused life of the WR-SLJ.

10.4.3 Experimental procedure and instrumentation

All mechanical tests at the laboratory were conducted in an ambient room temperature environment. The quasi-static residual strength was tested with a displacement control rate of 0.05 in/min using computer-controlled servo-hydraulic test systems. All compression strength tests used antibuckling fixtures to stabilize the specimen in the lateral (out-of-plane) direction. To accommodate thickness variations in the test specimens, either hardened rubber rollers or aluminum rollers with thin polycarbonate sleeves were positioned as shown in Figure 10.6. Aluminum rollers were used only for the static compression specimens from the upper wing and for the residual strength tests. Hardened rubber rollers were used for fatigue tests to prevent excessive side loads and to minimize abrasion of the specimen surfaces in contact with the rollers. Tension static strength specimens were tested without the antibuckling fixture.

Before testing, all specimens were inspected for potential defects that may have occurred during service, including environmental degradation; no evidence of such cases was found. All test specimens were instrumented with strain gauges strategically placed to detect both increases in progressive damage during static/residual strength tests and any compliance change and/or damage growth during fatigue tests. In addition, full-field strain measurements of selected specimens were acquired using ARAMIS photogrammetry.

10.4.4 Baseline static and fatigue evaluation of pristine specimens

Hurd and Coffey (1995) described the results of static and fatigue tests they conducted using similar types of specimens obtained from scrapped production articles and test panels. These data were generated to evaluate the effects of ply migration on the structural adequacy of the F/A-18 inner wing stepped-lap joint. Compared to specimens included in the present study, these samples did not have drilled fastener holes and

Figure 10.6 Antibuckling fixtures with (a) aluminum and (b) 180A hardness rollers as lateral supports.

were not subjected to effects from any operational environment. Fatigue tests of baseline specimens were conducted using both the tension-dominant fatigue spectrum (TDFS) and the compression-dominant fatigue spectrum (CDFS) based on the normalized fatigue spectrum shown in Figure 10.7. Upon completion of two test lifetimes at ambient laboratory conditions, the residual strength of both tension and compression loading modes was determined. The average ultimate static tension and compression loads were 75,740 lbf and −80,253 lbf, respectively. Following TDFS loading for two tests lives, the residual tension and compression strengths of specimens extracted from the scrapped lower wing panel were noted as 79,396 lbf and −74,653 lbf, respectively, whereas for a specimen extracted from the test panel, tension residual strength was 107,250 lbf. Following CDFS loading for two test lives, the residual compression strength of specimens extracted from the test panel was −79,470 lbf. The ultimate strain levels for all baseline specimens ranged from 5000 to 7000 microstrains.

10.4.5 Residual strength after fleet service

Table 10.2 includes a summary of all static tests. Tension specimens indicated that the upper-skin failure loads were more than 10% higher than those for the lower skin. Upper-skin specimens were approximately 2.5% thicker than lower-skin specimens. Compression data indicated approximately 5% higher ultimate loads for upper-skin specimens (excluding the upper-skin specimen that indicated buckling) than that for

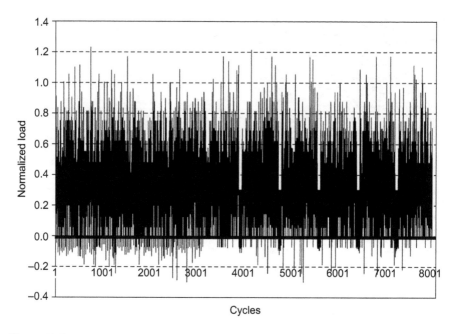

Figure 10.7 Normalized fatigue spectrum.

lower-skin specimens. Considering data scatter and thickness variations, compressive strength data from lower and upper skins were considered comparable. Back-to-back strains of tensile specimens were comparable, and the strain at failure ranged from approximately 6500 to 7500 microstrains. However, compression specimens indicated significant variations between back-to-back strain gauges, and the strain at failure ranged from approximately 4500 to 9500 microstrains. Specimens indicating extremely high strain on one side had extremely low strain on the opposite side, thus indicating buckling. Although the antibuckling fixture was used for these tests, the variation in thickness across both length and width caused localized buckling that, in turn, created such back-to-back strain variations. Overall, the static test data are comparable or higher than the static test data reported by Hurd and Coffey (1995) for pristine specimens (Figure 10.8). These data indicate that the service history, including environmental exposure, did not degrade the structural integrity of the bonded stepped-lap joint.

10.4.6 Failure modes for residual strength tests

To gain a full understanding of the failure mechanism of the joint being investigated, the mode of failure must be characterized. Three typical characterizations for the failure mode of an adhesively-bonded joint are the following:

- *Cohesive failure* is characterized by failure of the adhesive itself because of an inability to resist internal separation. The observation that adhesive stuck to both surfaces is an indication of this failure mode.

Table 10.2 Summary of static test results

Loading mode	Skin	Specimen ID	Ultimate load (lbf)	Ultimate strength (psi)	Strength		
					Mean (psi)	Standard deviation (psi)	Coefficient of variation (%)
Tension	Lower	1L-OH-1	92,832	97,048	96,013	2668	2.8
		1L-OH-2	95,695	100,041			
		1L-OH-4	87,025	90,977			
		1L-OH-7	91,967	96,144			
		2L-OH-1	91,719	95,885			
		2L-OH-5	91,898	96,072			
		2L-OH-7	91,756	95,923			
Compression	Lower	1L-OH-3	−84,690	−88,536	−84,695	3143	3.7
		1L-OH-5	−79,437	−83,044			
		1L-OH-6	−84,051	−87,868			
		2L-OH-2	−76,756	−80,242			
		2L-OH-3	−80,066	−83,702			
		2L-OH-4	−83,287	−87,069			
		2L-OH-6	−78,822	−82,402			

Continued

Table 10.2 Continued

Loading mode	Skin	Specimen ID	Ultimate load (lbf)	Ultimate strength (psi)	Strength Mean (psi)	Standard deviation (psi)	Coefficient of variation (%)
Tension	Upper	4U-NH-3	99,799	102,001	102,328	3654	3.6
		4U-NH-4	95,563	96,349			
		3U-NH-3	104,647	102,876			
		3U-NH-4	106,365	104,629			
		1U-NH-3	101,931	105,785			
Compression	Upper	1U-NH-10	−78,753	−77,139	−83,376	10,395	12.5
		2U-NH-4	−67,833*	−70,969			
		2U-NH-5	−77,317	−81,193			
		2U-NH-6	−86,114	−90,811			
		3U-NH-2	−97,483	−96,767			

TDFS, tension-dominant fatigue spectrum; CDFS, compression-dominant fatigue spectrum; FH, filled hole; LSF, load severity factor; OH, open hole; NH, no hole.
* Some lateral supports moved, and the test specimen indicated buckling before failure; thus, failure load was lower than other specimens in this category.

Durability and residual strength of adhesively-bonded composite joints 301

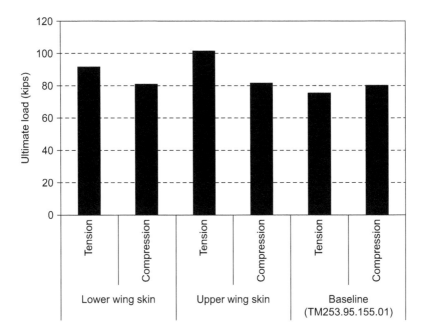

Figure 10.8 Comparison of average residual strength after one lifetime in service to static strength.

- *Adhesive failure* is characterized by a failure of the joint at the adhesive—adherend interface and is typically caused by inadequate chemical and/or mechanical preparation of the surface. Specimens that fail adhesively tend to have excessive peel stresses, which lead to failure and often do not yield a strength value for the adhesive joint but rather indicate unsuitable surface qualities of the adherend.
- *Substrate (adherend) failure* is characterized by failure of the adherend rather than the adhesive. In metals, this occurs when the adherend yields. In composites, the laminate typically fails by way of interlaminar failure, that is, when the matrix between plies fails. In substrates, failure occurs when the adhesive is stronger than the adherend in the joint being tested.

Typical failure modes for static specimens are shown in Figure 10.9. Tensile specimens primarily experience adherend failure. Since the joint included several lap joints, detailed inspection after failure was required to determine the failure mechanism. Inspection of the bonded surface of the joint after failure indicated adherend failure (e.g., intralaminar/interlaminar cracks in composite and tensile fracture in titanium) and cohesive failure. As a result of the catastrophic nature of the failure, the primary or initial failure mode was inconclusive. Because of the robust antibuckling fixture used, most of the compression specimens indicated compressive failure. However, several compression test specimens experienced composite failure of the unsupported region, whereas some indicated titanium buckling. The latter two failure modes were possibly due to load eccentricity caused by uneven specimen surfaces. Those specimens later had two failure modes, resulting in lower ultimate loads

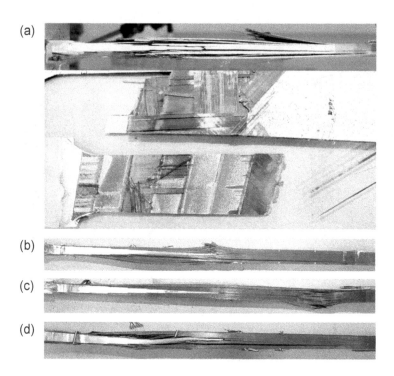

Figure 10.9 Typical failure modes for static specimens. (a) Adherend tensile (1U-NH-3), (b) compressive (2U-NH-6), (c) composite (2U-NH-5), and (d) titanium failure (1U-NH-10).

than the compression specimens that failed in the compression failure mode, thus indicating premature failure.

10.5 Remaining life after fleet service

Upon evaluation of the residual strength after one service life, a set of specimens was subjected to fatigue loading to determine the remaining life of the wing root joint. Extracted and dog-bone specimens were prepared in a manner similar to that used for residual strength test specimens from several wing skins and included NH, OH, and FH specimens from both upper and lower skins. All fatigue tests were carried out with antibuckling fixtures to stabilize the specimen in the lateral (out-of-plane) direction.

10.5.1 Fatigue test spectrum

Fatigue tests were conducted using a load spectrum generated based on the flight-by-flight loading of the wing root, which is the same as the spectrum used by Hurd and Coffey (1995). Because of the variation in thickness of the wing skins across the bonded region and the differences in construction of the upper and lower wings, the spectrum used by Hurd and Coffey (1995) was normalized by the designed limit loads. The normalized fatigue spectrum that represents 300 spectrum fatigue hours (SFHs) or one test profile is shown in Figure 10.7. It was multiplied by either a tension or compression load limit (reference) along with a thickness-correction factor to obtain the tension-dominant or compression-dominant fatigue spectrum, respectively. When generating compression-dominant fatigue spectra, the negative sign of the compression load limit is maintained so that a mostly positive normalized spectrum is changed to a mostly negative (compressive) load spectrum. As can be seen in Figure 10.7, the aircraft is designed to withstand loads above the load limit that occur during its maneuvers (e.g., normalized loads >1.0). To reduce the test duration, the test spectrum was multiplied by a load severity factor (LSF) while maintaining the original stress ratio of each segment. The test frequency was 3 Hz. One lifetime is represented by 20 test profiles, which is equivalent to 6000 SFHs.

10.5.2 Fatigue test matrix

The fatigue test matrix included 21 specimens that were tested using the tension-dominant fatigue spectrum, and 13 specimens were tested using the compression-dominant fatigue spectrum, as shown in Table 10.3. All fatigue tests were conducted in a laboratory environment. Except for the limited number of specimens that were conditioned in a salt fog chamber, specimens were tested in the as-machined condition. The LSF equal to 1.15 corresponds to a 15% increase in applied load in the original fatigue spectrum. Note that the tension and compression reference loads used to generate the test spectrum by multiplying the normalized spectrum in Figure 10.7 are different in magnitude. Therefore, the tension and compression stress levels are uniquely identified as LT and LC, respectively, in Table 10.3.

Based on preliminary tests conducted on TDFS specimens, 10 lifetimes of laboratory fatigue cycles were considered runout, while 30 lifetimes of laboratory fatigue cycles were considered for CDFS specimens. Test specimens that survived the number of targeted repetitions (runouts) were tested for residual strength. TDFS and CDFS specimens were loaded in tension and compression, respectively, during residual strength tests.

10.5.3 Fatigue data

Table 10.4 shows the summary of all fatigue tests and the ultimate loads from residual strength tests. For comparison purposes, the ultimate load is normalized by the thickness and width at station 1 to calculate the residual strength. Note that the fatigue lifetimes indicated in Table 10.4 are only the laboratory test duration, excluding the service history. Figure 10.10 shows a summary of all stepped-lap joint test data in Tables 10.2 and 10.4. Six specimens survived the tension-dominant fatigue spectrum for 10 lifetimes. Furthermore, 10 specimens survived the compression-dominant fatigue spectrum for 30 lifetimes, while one specimen survived the same for 60 lifetimes! Figure 10.11 shows a summary of fatigue and residual strength results. Residual strength of the runout specimen did not indicate a significant strength reduction due to additional fatigue loading.

Table 10.3 **Fatigue test specimens**

Fatigue spectrum	Skin	Specimen configuration	Stress level	LSF	Specimen ID	Thickness at station 1 (in)	Width at station 1 (in)
TDFS	Lower	Open hole	LT-1	1.15	4L-OH-6	0.4960	2.000
					4L-OH-7	0.4870	2.000
					4L-OH-9	0.5030	1.984
	Lower	Filled hole			3L-FH-2	0.4765	1.983
					4L-FH-3	0.4800	1.996
					3L-FH-8	0.4776	1.982
					3L-FH-6	0.4713	1.971
					3L-FH-4	0.4718	1.973
					3L-FH-5	0.4818	1.974
					3L-FH-7	0.4835	1.977
					3L-FH-3	0.4737	1.995
	Lower	Filled hole			4U-NH-10	0.5018	1.996
	Upper	No hole			3U-NH-5	0.5054	1.951
					3U-NH-6	0.5045	1.967
	Lower	Filled hole	LT-2	1.30	3L-FH-1	0.3722	1.976
	Upper	No hole	LT-3	1.45	2U-NH-9	0.4937	1.984
					1U-NH-5	0.4863	1.987
					1U-NH-2	0.4839	1.987

Durability and residual strength of adhesively-bonded composite joints

CDFS	Upper (salt-fog*)			1U-NH-9	0.4886	1.995
	Lower	Open hole	LC-1	1.15	3U-NH-7	0.5019	1.970
					3U-NH-8	0.4918	1.960
	Upper	No hole	LC-3	1.45	4L-OH-4	0.4635	1.989
					4L-OH-8	0.4920	2.008
					4L-OH-5	0.4620	1.989
	Upper	No hole			4U-NH-9	0.4830	1.990
					3U-NH-1	0.4643	1.953
					1U-NH-6	0.4859	1.998
					1U-NH-7	0.4947	1.996
					4U-NH-2	0.4820	2.002
					4U-NH-5	0.4978	1.993
					2U-NH-8	0.4874	1.990
					2U-NH-10	0.4993	1.976
	Upper (salt fog*)	No hole	LC-4	1.60	1U-NH-8	0.4898	1.999
	Upper				2U-NH-7	0.4985	1.985

TDFS, tension-dominant fatigue spectrum; CDFS, compression-dominant fatigue spectrum; FH, filled hole; LSF, load severity factor; OH, open hole; NH, no hole.
* Conditioned in salt fog environment for 60 days at 120 °F and tested at ambient room temperature.

Table 10.4 Summary of fatigue test results

Fatigue spectrum	Stress level	Specimen ID	Fatigue segments	Test lifetimes	Ultimate load (lbf)	Residual strength (psi)
TDFS	LT-1	4L-OH-6	839,637	5.2		
		4L-OH-7	843,570	5.3		
		4L-OH-9	983,239	6.1		
		3L-FH-2	1176,553	7.3	Fatigue failures	
		4L-FH-3	1474,892	9.2		
		3L-FH-8	948,298	5.9		
		3L-FH-6	978,445	6.1		
		3L-FH-4	1334,296	8.3		
		3L-FH-5	1470,248	9.2		
		3L-FH-7	1,606,000	10.0	85,821	89,776
		3L-FH-3	1,606,000	10.0	86,820	91,870
		4U-NH-10	1,606,000	10.0	97,101	96,947
		3U-NH-5	1,606,000	10.0	89,043	90,304
		3U-NH-6	1,606,000	10.0	91,434	92,148
	LT-2	3L-FH-1	1,606,000	10.0	55,383*	75,303
	LT-3	2U-NH-9	775,782	4.8	−80,236†	−81,912
		1U-NH-5	898,458	5.6	Fatigue failures	
		1U-NH-2	1,777,288	11.1		
		1U-NH-9	771,408	4.8		
		3U-NH-7	768,254	4.8	Fatigue failures	
		3U-NH-8	787,710	4.9		

CDFS					
LC-1	4L-OH-4	4,818,000	30.0	−44,379[‡]	−48,139
	4L-OH-8	4,818,000	30.0	−73,238	−74,132
	4L-OH-5	4,818,000	30.0	−81,492	−88,683
	4U-NH-9	4,818,000	30.0	−78,325	−81,489
	3U-NH-1	9,636,000	60.0	−73,257	−80,788
LC-3	1U-NH-6	4,818,000	30.0	−95,421	−98,292
	1U-NH-7	4,818,000	30.0	−87,846	−88,962
	4U-NH-2	4,818,000	30.0	−75,048	−77,783
	4U-NH-5	4,818,000	30.0	−96,307	−97,085
	2U-NH-8	4,818,000	30.0	−92,167	−95,025
	2U-NH-10	4,441,347	27.7	Fatigue failure	
LC-4	1U-NH-8	4,818,000	30.0	−100618	−102758
	2U-NH-7	3,003,978	18.7	Fatigue failure	

TDFS, tension-dominant fatigue spectrum; CDFS, compression-dominant fatigue spectrum; FH, filled hole; LSF, load severity factor; OH, open hole; NH, no hole.
* Relatively thin specimen obtained closer to leading edge.
[†] Test specimen failed during setup; load cell value reported as residual strength.
[‡] Test fixture malfunction caused premature failure due to excessive buckling.

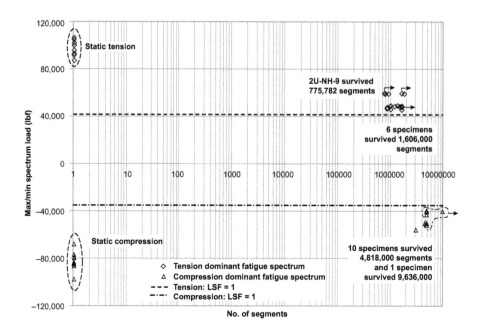

Figure 10.10 Summary of wing root stepped-lap joint tests. LSF, load severity factor.

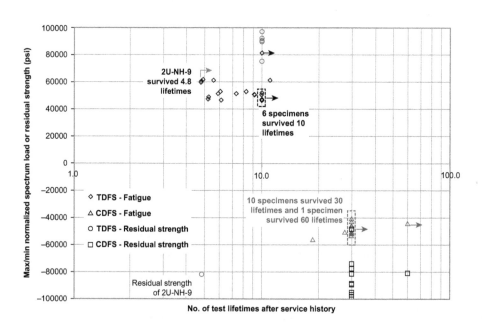

Figure 10.11 Summary of fatigue and residual strength tests. CDFS, compression-dominant fatigue spectrum; TDFS, tension-dominant fatigue spectrum.

10.5.4 Fatigue damage growth

All TDFS OH specimens failed during fatigue; the primary failure mode was a fracture across the fastener hole. All but three of the TDFS FH specimens had fatigue failures. The full-field strain from a digital image correlation indicated a significant concentration of strain at the hole, which led to growth of the crack in the titanium under fatigue. The three TDFS FH specimens that survived 10 lifetimes indicated compliance changes or a global stiffness change during fatigue cycling, similar to the TDFS FH specimens that had fatigue failures. Failure analysis and inspections of these specimens after residual strength testing revealed the presence of fatigue-induced cracks near the fastener hole (Figure 10.12), which seem to have contributed to the compliance change. The residual strength of TDFS FH specimens, except for the relatively thin 3L-FH-1, which had a higher LSF, indicated no significant degradation in the load-carrying capacity of the joint.

Similarly, all TDFS NH specimens that were tested using an LSF of 1.15 (LT-1) survived 10 lifetimes with no significant degradation. TDFS NH specimens that were tested with an LSF of 1.45 (LT-3) had large delaminations that initiated around the areas where fatigue-induced cracks formed in the titanium, as shown in Figure 10.13. Periodic microscopic and visual inspections showed that the fracture initiated in the metal and then propagated into the composite as a delamination across the remaining length of the specimen. Furthermore, it was noted that the crack began as a corner crack on one side of the titanium and propagated across the width of the specimen. Even after the titanium failed across the fillet region between steps 7 and 8 and large delaminations were present, these specimens were able to transfer loads across the remainder of the stepped-lap joint for a significant number of fatigue cycles. Based on the strain gauge data, the load redistribution was noticeably greater over the remainder of the joint, resulting in final failure.

Although the presence of microcracks in the matrix was noted before failure, the global compliance did not change until cracks were fully developed in the titanium,

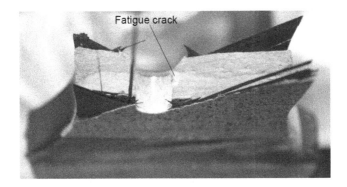

Figure 10.12 Fatigue damage from a corner crack at a fastener hole in sample 3L-FH-7 (inspected after a residual strength test).

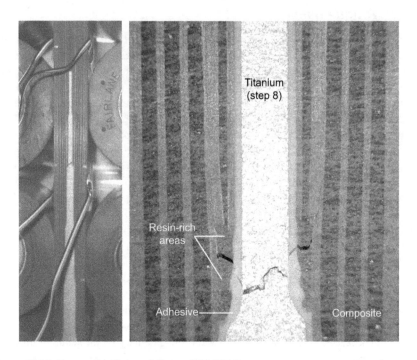

Figure 10.13 Progressive fatigue failure of 1U-NH-5.

as shown in Figure 10.14(a). Once the crack in the titanium started to propagate, the compliance changed rapidly. Figure 10.14(b) shows crack growth around the fillet region between steps 7 and 8 of 1U-NH-2, which indicates the same progressive failure mechanism as 1U-NH-5. Figure 10.15(a) and (b) shows the failure surfaces of 1U-NH-2 closer to the outer mold line (OML) and inner mold line (IML), respectively. A mostly light-color area on the OML side indicates cohesive failure, whereas the IML side shows a combination of cohesive failure and adherend (composite) failure—both indicating satisfactory adhesion.

Except for two specimens, all CDFS specimens, including the OH specimens, survived 30 lifetimes (the 3U-NH-1 specimen survived 60 lifetimes) and did not indicate significant degradation of load-carrying capacity with respect to static strength data, as shown in Table 10.2. Of the two that failed, 2U-NH-10, with an LSF of 1.45, survived more than 27 lifetimes, whereas 2U-NH-7, with an LSF of 1.60, survived more than 18 lifetimes. Both of these specimens indicated progressive delamination through multiple layers of the composite. Visual inspection and compliance monitoring (Figure 10.16(a)) at the end of each profile indicated that the progressive damage occurred in two stages. During the first stage, microcracks formed because of fatigue loading. At the end of this stage, these microcracks coalesced and caused delamination a few plies below the IML of the specimen, as shown in Figure 10.16(b). The initial decrease in slope of the load−displacement curve at the beginning of stage 2 also confirmed the visual inspections. The formation of a

Durability and residual strength of adhesively-bonded composite joints 311

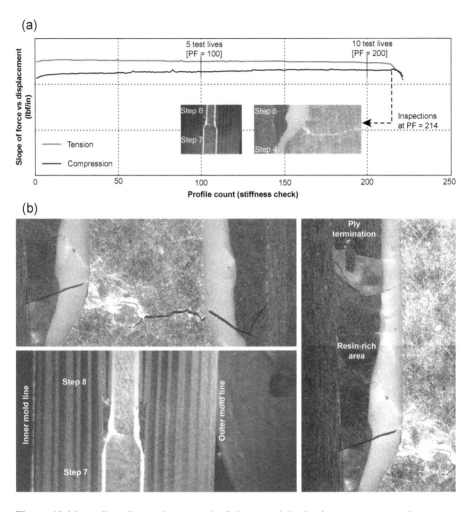

Figure 10.14 (a) Compliance change as the fatigue crack in titanium progresses and (b) inspections of 1U-NH-2 after 214 test profiles (PF) indicating propogation of the crack from titanium to the composite through the adhesive layer and resin-rich areas.

large delamination resulted in a significant reduction in specimen stiffness, which is depicted in the change in slope of the load—displacement curve at the end of profile 237. This slope continued to decrease during stage 2, indicating the softening of the specimen due to fatigue damage, especially for compression loading. As the first delamination propagated toward the grips and completely separated the outer plies from the load path, the second delamination formed several plies below the initial delamination. As stage 2 fatigue loading continued, the second delamination propagated toward the grip section and completely separated the next set of plies from the load path. This caused a majority of the load to be redistributed to the opposite side of

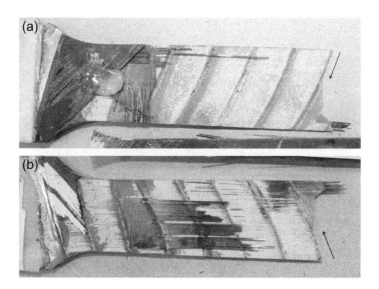

Figure 10.15 (a) Failed surfaces of the outer mold line and (b) inner mold line of 1U-NH-2 (arrows indicate the path of the crack on titanium).

the specimen. Finally, the load eccentricity and increase in localized stresses resulted in compression buckling and failure of the composite, as shown in Figure 10.16(b). Back-to-back strain gauges on the compression specimens, especially those with high failure loads, indicated a significant amount of localized buckling, regardless of the use of the antibuckling fixture.

10.5.5 Inspections

Load, displacement, and strain data were continuously monitored and saved; the specimen compliance during tension and compression was monitored quasi-statically each time after repeating the 300 SFH profile. Specimen compliance was monitored periodically to detect potential stiffness losses because of an increase in damage. In addition to compliance checks and visual inspections, periodic nondestructive examinations were carried out using through-transmission ultrasonic C-scans, an optical microscope, and pulse thermography. Since compliance checks do not require the test specimen to be removed from the fixture, they are effective compared to other detailed NDI inspections. However, this technique may not be able to detect small stiffness changes caused by microcrack formation or minor wearout caused by fatigue loading. NDIs around the joint area were instrumental in understanding the formation of microcracks in composite details near the joint area so that field inspectors could interpret similar findings during aircraft maintenance. Fatigue data indicated that despite the formation of microcracks, the joints were able to carry fatigue loads and showed no change in stiffness.

Through-transmission ultrasonic C-scans of stepped-lap joint specimens indicated anomalies (darkened lines compared to scans before fatigue cycling), as shown in

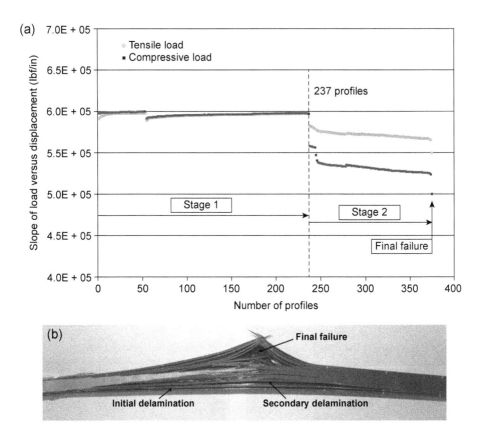

Figure 10.16 (a) Specimen stiffness change caused by progressive damage and (b) failure mode of 2U-NH-7 indicating delaminations in multiple composite layers.

Figure 10.17(a). Microcracks were noted on both the bond line and in resin-rich areas during early fatigue loading. Some of these microcracks resulted in intralaminar cracks, as shown in Figure 10.17(b). Since the toughness of the adhesive against fracture is significantly higher than that of the composite resin system, it was noted that the majority of microcracks nucleated from the composite or from the resin-rich areas. Although the microscopic inspections of the sides of the specimens suggest that the darkened lines in Figure 10.17(a) represent microcracks in the fillet regions, further study is required to properly interpret the NDI data from field inspections of the aircraft wing, that is, anomalies representing microcracks in adhesive and/or composite.

10.5.6 Environmental effects

Mechanical properties of the adhesive degrade with the absorption of moisture from humid environments. Water diffuses to the adhesive and/or to the adhesive—adherend

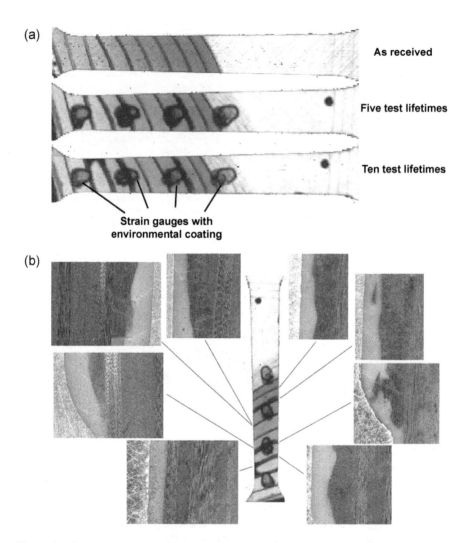

Figure 10.17 (a) Through-transmission ultrasonic inspections at major fatigue intervals and (b) microscopic inspection results of 1U-NH-2 after 10 test lifetimes.

interface. Some effects, such as plasticization, affect the mechanical properties as well as the glass transition temperature, but are reversible. Moisture ingression can also cause irreversible changes such as swelling-induced cracks, further cross-linking, or chemical degradation when mixed with other fluids. In addition, the operating temperature has significant effects on adhesive behavior, that is, adhesives become weak and ductile at high temperatures and more brittle at low temperatures. In general, the yield stress of all adhesives increases with decreasing temperature, as does the modulus (Figure 10.2).

Approximately 34% of the F/A-18 external surfaces are constructed of graphite/epoxy composites (Vodicka, Nelson, van den Berg, & Chester, 1999). An environmental protective layer prevents diffusion of moisture into the composites (and the adhesive layer). However, long-term exposure of the environmental protective coating to ultraviolet radiation can degrade its effectiveness. Seven traveler specimens were extracted from one of the decommissioned composite wing skins to determine their residual moisture content at the end of service life. These traveler specimens were dried in a chamber, and the weight loss was measured every week until equilibrium was reached; it was found that the average residual moisture content (as received) of wing skin was approximately 0.251% by weight.

U.S. Navy aircraft, including the F/A-18, are exposed to severe environments as salt water splashes on them while on carrier decks. To evaluate the effect of the operating environment, four specimens with geometry similar to that shown in Figure 10.4 were exposed to salt fog in a chamber at 120 °F for 60 days, in accordance with ASTM standards B117 and G85. Sides of the specimens were completely exposed to the chamber environment (to simulate complete failure of the environmental protective coating), subjecting the composite to accelerated moisture absorption through diffusion and subjecting titanium to a corrosive environment. Periodic inspections of the specimens conditioned in the salt fog environment indicated that the titanium showed excellent resistance to corrosion. When titanium is exposed to an environment containing oxygen, it forms an oxide layer that protects the metal from corrosive elements. In addition to protection from direct exposure, Mahoon (1982) showed that the chemical composition of the oxide during surface preparation before bonding has a strong influence on bond durability, and the oxide is unaffected by the exposure to humidity, thus providing stable interfacial characteristics.

Specimens were then fatigue tested in a laboratory environment—three in a tension-dominant fatigue spectrum and one in a compression-dominant fatigue spectrum. Salt fog-conditioned specimens subjected to the TDFS failed just before reaching five test lives in the laboratory environment. The salt fog-conditioned specimens subjected to CDFS survived 30 test lives, with $-100{,}618$ lbf failure load during the residual strength test. This is the highest compressive strength recorded of all the specimens in this investigation. Although composite, titanium, and adhesive were directly exposed on both sides, these fatigue results indicated that the salt fog conditioning did not have any significant effect on the residual strength or fatigue performance of the joint. This indicates the robustness of the material for this application and provides confidence that the failure of the environmental protective layer in the short term (until it is detected by a periodic inspection and repaired) will not have a significant effect on the joint performance.

10.6 Inner-wing full-scale fatigue test

Evaluations of the residual strength and remaining life of WR-SLJ described in previous sections were done using specimens extracted from wing skins and subjected only to in-plane loading. Bonded joints are sensitive to out-of-plane loading caused by bending and twisting of the structure. To investigate loads that are representative of what the bonded joint experiences during service, that is, in-plane and out-of-plane loading, a full-scale fatigue test of the F/A-18 inner wing was conducted (Seneviratne & Tomblin, 2013). The test article consisted of a center fuselage, inner wings, and trailing-edge flaps (Figure 10.18). A simulated inboard leading-edge flap and outboard wing were attached to each inner-wing box for fatigue load application. Both the left and right inner wings expended one fatigue lifetime in service and were exposed to various operational environmental conditions. The center barrel section of the fuselage was used as a part of the inner-wing test fixture to ensure proper load transfer at the wing root lugs. The simulated inboard leading-edge flap and the outboard wing were attached to each inner-wing box for fatigue load application. This structure provided an opportunity to look at the management of service life and to deliver useful insight into sustaining the active fleet.

Current methods of certification for a composite and bonded aircraft structure rely on the development of a safe-usage life through fatigue testing. Since the composite structure is conservatively designed, with considerable analytical reductions in strength to account for environmental effects, it is rare that the full-scale fatigue testing of aircraft components exercises the capabilities of the composite structural members. Thus, these factors combine to prevent composite structures from failing during fatigue testing. In addition, the expense of fatigue testing rarely permits continued assessment past the original design goals for the program. As a result, over the course of the aircraft's life there is little capability to relate in-service events to known fatigue limitations of the original certification test and no mechanism based on engineering principles for life extension. In many instances, the ability to reuse structural components from retired aircraft (even after a full-life service) has been shown to be beneficial in supporting existing fleet. This research provides the first step in such an approach to fleet maintenance.

To simulate the maneuver loads, the full-scale F/A-18 test article was loaded using 16 servo-hydraulic actuators. Each wing was divided into four zones, and distributed loads were applied using tension-compression whiffletree mechanisms (Figure 10.19). The trailing-edge flap included a one-zone whiffletree. The leading-edge flap loads and outer-wing loads were applied through dummy (simulated) structures. Loads applied to the wing were reacted through the fuselage, mainly at three bulkheads located in the center barrel. The weight of the test article was counterbalanced by tare loads before applying the fatigue loads.

The fatigue spectrum used for the full-scale test consisted of loads from four reference load cases defined at a particular altitude and a Mach number: (1) symmetric pull-up, (2) rolling pullout, (3) symmetric push-down, and (4) 1-g roll. Reference fatigue loads were adjusted for altitude, Mach number, and g-force used on each load segment. Each 300-SFH segment of the flight-by-flight fatigue loads included 9354 load segments (one test life $= 187,080$ segments). The test article was instrumented using both resistance and fiber optic strain gauges at strategic locations to monitor any degradation or load redistribution due to damage. Wing and flap deflections were

Durability and residual strength of adhesively-bonded composite joints 317

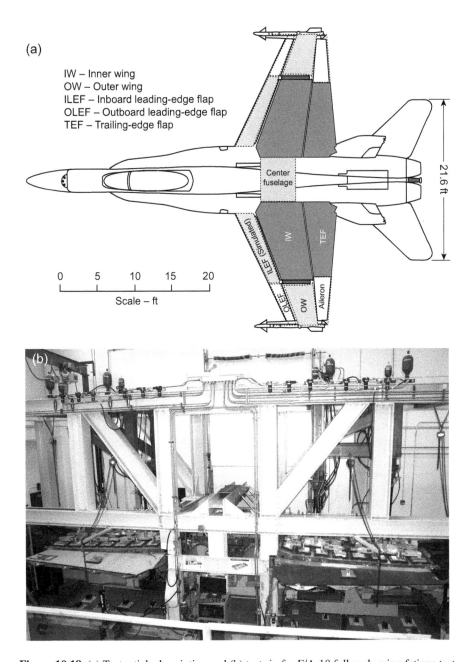

Figure 10.18 (a) Test article description and (b) test rig for F/A-18 full-scale wing fatigue test.

Figure 10.19 Four-actuator tension–compression whiffletree over wing.

measured using string-pot displacement gages. Periodic inspections were carried out using visual, ultrasonic, thermal, and photogrammetry full-field image correlation techniques.

The wing structure was cycled for one lifetime of test loads while monitoring for potential fatigue-induced damage using various types of health monitoring techniques. This full-scale test program also provided an opportunity to evaluate current field inspection techniques and structural health monitoring systems used to detect known or suspected damage threats found during teardown inspections for the period of the extended life. It also assessed novel inspection techniques to detect such damage threats in a controlled environment and implications of introducing them as field inspection techniques. Periodic NDI and strain data did not indicate degradation of the bonded region after an additional one lifetime of fatigue loading in the laboratory environment.

10.7 Conclusions

Static strength tests conducted on the specimens extracted from decommissioned F/A-18 inner wing root stepped-lap joints indicated that the load-carrying capacity of the joint is comparable to the test data generated for pristine specimens. This signifies that one lifetime of service history, including environmental exposure of the wings, had no significant effect on the residual strength of the joint. This also indicates that the environmental protective measures used on wing skins performed well throughout the service life. The salt fog-conditioned fatigue tests indicated the robustness of the material selection of the joint to perform effectively even with short-term exposure to salt fog and moisture after failure of environmental protective measures.

Furthermore, additional spectrum fatigue tests indicated that the remaining life of the joints was substantial and that the residual strength was unaffected by the additional fatigue cycles induced under laboratory environmental conditions.

The double stepped-lap joint configuration was well suited for a wing root that exhibits complex load distribution, including significant compressive fatigue loads. Microscopic inspection of the bond line indicated that its thickness was well controlled across the joint. Also, these inspections indicated minimal fiber kinking and waviness at ply termination at each step. The use of 0° plies at the interface also increased joint efficiency and prevented premature adherend failure due to transverse microcracking.

NDI findings indicated that there is a reasonable window of opportunity to detect microcracks that can cause detrimental localized load redistribution due to softening of the composite structure. The progressive failure mechanism of the wing root stepped-lap joint is complex and thus requires detailed failure analysis and investigation into the NDI findings during fatigue. The failure of TDFS specimens during fatigue initiated in titanium either from a hole (on OH specimens) or as a corner crack on the transition region of step 8 (on NH specimens). The long-term durability of an adhesive joint hinges on the interfacial properties and the adhesive resistance to environmental effects. Overall, the detailed failure analysis indicated that the surface preparation technique and the selection of moisture-resistant adhesive performed well, thus providing durable adhesion to the substrate.

The full-scale fatigue test of the inner wings after one lifetime of service indicated no fatigue damage caused by the additional lifetime of fatigue loading in the laboratory environment. Once the wing is assembled, only the bond line interface closer to the OML can be inspected using NDI equipment since the titanium interferes with the penetration of signals to the far side (closer to the interior of the wing) of the bond line interface. Thus, the damage to the bond line or the composite on the far side cannot be detected during service. The progressive fatigue-induced damage observed during specimen-level tests indicated a significant number of cycles from when the titanium fails until large delamination forms on both sides of the joint. This makes it feasible to consider using a field/depot inspection to detect delaminations of the stepped-lap joint on the outer/accessible side of the wing. Since the damage progresses slowly, a one-time maneuver-restricted flight to the depot for wing removal can be considered.

We recommend additional full-scale fatigue tests using wings from retired aircraft to establish both the durability of wing structures with extensive damage and the probability of detection curves for the field NDI techniques most suitable for the failure modes seen in the test.

Acknowledgments

This research program was funded by the Office of Naval Research and monitored by the Naval Air System Command (NAVAIR). The authors thank Travis Cravens, Caleb Saathoff, Brandon Saathoff, and Nathan Alexander for their assistance during experiments and data analysis.

References

Chrissos, P., & Coffey, F. (1996). *Process verification testing for the F/A-18C/D net resin inner wing root steplap joint*. Report No. TWD LMA03.12−006 (Suppl. 532).
Hart-Smith, L. J. (1973a). *Adhesive-bonded double-lap joints*. Douglas Aircraft Company. NASA Langley Contract NAS1−11234, NASA CR-112235.
Hart-Smith, L. J. (1973b). *Adhesive-bonded single-lap joints*. Douglas Aircraft Company. NASA Langley Contract NAS1−11234, NASA CR-112236.
Hart-Smith, L. J. (1973c). *Adhesive-bonded scarf and stepped-lap joints*. Douglas Aircraft Company. NASA Langley Contract NAS1−11234, NASA CR-112237.
Hurd, M., & Coffey, F. J. (1995). *F/A-18 C/D steplap joint overlap evaluation—axial pull specimens*. Technical Memorandum 253.95.0155.01. St. Louis, MO: The Boeing Company.
Krieger, R. B. (1988). Stress analysis concepts for adhesive bonding of aircraft primary structure. *ASTM STP, 981*. Philadelphia: American Society for Testing and Materials.
Krieger, R. B. (1993). Bonding structural composites for aircraft. In *Composites design: Vol. 4. Proceedings of ICCM/9*. Madrid.
Mahoon, A. (1982). Improved surface pretreatments for adhesive bonding of titanium alloys. In *Proceedings of 27th national SAMPE symposium*.
Sanger, K. B. (1986). *Certification testing methodology for composite structures*. Report No. NADC-86132−60.
Seneviratne, W., & Tomblin, J. (2010). Determination of retirement life based on service history and load-life combined approach. In *Proceedings of aircraft airworthiness & sustainment conference*. Austin, Texas.
Seneviratne, W., & Tomblin, J. (2011). Aging evaluation of advanced materials used for military aircraft. In *Proceedings of aircraft airworthiness & sustainment conference*. San Diego, California.
Seneviratne, W., & Tomblin, J. (2013). Full-scale fatigue test of F/A-18 A-D composite structure for aging evaluation. In *Proceedings of aircraft structural integrity program*. Bonita Springs, Florida.
Tillman, M. S., Tsai, H. C., & Peek, M. (2009). *An investigation of the end-of-life residual strength of the F/A-18A-d inner wing step lap joint*. NAWCADPAX/TR-2009/139.
Tomblin, J., & Seneviratne, W. (2002). *Shear stress-strain data for structural adhesives*. DOT/FAA/AR-02/97.
Vodicka, R., Nelson, B., van den Berg, J., & Chester, R. (1999). *Long-term environmental durability of F/A-18 composite material*. DSTO-TR-0826.
Whitehead, R. S., Kan, H. P., Cordero, R., & Saether, E. S. (1986). Report No. NADC-87042−60. *Certification testing methodology for composite structures* (Vols. I and II).

Part Three

Modelling fatigue and fracture behaviour

Simulating mode I fatigue crack propagation in adhesively-bonded composite joints

M.M. Abdel Wahab
Ghent University, Zwijnaarde, Belgium

11.1 Introduction

One of the main advantages of adhesive bonding is its high resistance to fatigue and its long fatigue lifetime when compared to classical mechanical joining techniques. As adhesives are used in many industries including aerospace, automotive and civil engineering, they are often subjected to fatigue loading conditions. Because a structure may fail under the action of a fatigue load equivalent to a small percentage of static strength, fatigue analysis and performance are desirable for failsafe and damage tolerance design. The analysis and prediction of fatigue behaviour are challenges due to many factors, e.g. the complicated nature of fatigue crack initiation and propagation, the complex geometry of bonded joints involving different materials, and the complex material behaviour of adhesives under loading and unloading regimes (Abdel Wahab, 2012).

The techniques used to simulation mode I fatigue crack propagation in adhesively-bonded joints using finite element analysis (FEA) are divided into two main categories: fracture mechanics (FM) and continuum damage mechanics (CDM). In case of a joint subjected only to mode I loading, i.e. only peel stresses, the FM approach requires the knowledge of mode I fatigue crack growth behaviour and fracture mechanics deriving parameter as a function of crack length (Pirondi & Nicoletto, 2006). Under mode I loading conditions, the fatigue crack growth is often experimentally characterized using double cantilever beam (DCB) or tapered double cantilever beam (TDCB), and its steady state stage is expressed by Paris' equation (Wahab, Ashcroft, Crocombe, & Smith, 2004).

The application of the CDM approach to crack propagation implies the use of the cohesive zone model (CZM) to account for the process zone ahead of the crack tip. CZM can be used to model both damage initiation and crack propagation under monotonic and fatigue loading regime. Recently, it has been used to analyse fatigue problems in adhesively-bonded joints, e.g. Khoramishad, Crocombe, Katnam, and Ashcroft (2010), Moroni and Pirondi (2011) and Pirondi and Moroni (2010). In such an approach, cohesive traction-separation is characterized by a non-linear failure law between the cohesive traction vector and the displacement separation vector acting across the cohesive surfaces. In case of high-cycle fatigue, where the number of cycles is usually larger than 10,000, a combination of both CZM and FM concepts is desirable in order to avoid a cycle-by-cycle analysis and to reduce the computational costs.

In the following sections, first the finite element modelling techniques used to model mode I crack propagation are reviewed. Techniques such as re-meshing, node release and the extended finite element method (XFEM) are considered. Secondly, the FM approach and the techniques used to extracted fracture mechanics parameters from FEA are reviewed. Third, the CZM approach, the cohesive traction−separation law and the cohesive fatigue model are presented. Next, a mixed FM and CZM approach is proposed. Finally, conclusions are presented.

11.2 Finite element (FE) modelling

11.2.1 Modelling of a double cantilever beam (DCB) joint

DCB is a standard test specimen used to characterize mode I fracture in adhesively-bonded joints. It is often used to measure the fracture toughness of a cracked adhesive layer bonded to two substrates, as shown in Figure 11.1. For fatigue crack propagation measurements, DCB is used to determine the mode I crack growth rate as a function of the range of a fracture parameter, e.g. stress intensity factor or strain energy release rate. DCB test specimen was originally described by ASTM (American standard) in 1990 and at a later stage further developed by BSI (British standard) in 2001 and ISO (international standard) in 2009. An initial crack of length a_0 is introduced during the fabrication of the specimen. The dimensions of the specimen, according to the ASTM standard, are: length $l = 300$ mm, width $w = 25$ mm and height $h = 12.7$ mm.

When modelling DCB using FEA, a two-dimensional simplification is often used. As the joint width is 25 mm, it may be considered to be large enough to assume a plane strain condition in the mid-plane ($z = -w/2$). Near the free surfaces, i.e. $z = 0$ or $z = -w$, a plane stress condition may be assumed. However, a two-dimensional plane strain analysis is usually enough to extract the mode I fracture parameter and to analysis the crack growth behaviour. In order to capture the high stress gradient near the

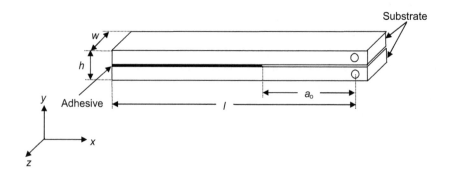

Figure 11.1 DCB mode I test specimen.

crack tip and the interface between the adhesive layer and the substrates, higher-order elements are recommended (Cook, Malkus, & Plesha, 1989; Taylor & Zienkiewicz, 2005; Zienkiewicz, Taylor, & Nithiarasu, 2005). Two-dimensional quadratic, eight-noded elements are usually used to model both adhesives and substrates. Because adhesives are often used to bond very thin substrates, large deformations are expected, and geometric non-linearity should be taken into account in the finite element (FE) simulations. Geometric non-linearity is characterized by large displacements and/or rotations. In case of large deformation, a higher-order term is considered in the strain–displacement relationship, e.g. for one-dimensional analysis the strain is expressed as $\varepsilon_x = \dfrac{\partial u_x}{\partial x} + \dfrac{1}{2}\left(\dfrac{\partial u_x}{\partial x}\right)^2$. Thus, the element stiffness matrix, $[K]$, is a function of the displacement vector and is given by:

$$[K] = \int_\Omega [B]^\mathrm{T}[D]\left(I + \frac{1}{2}[B][\delta \mathbf{d}]\right)[B]\mathrm{d}\Omega \tag{11.1}$$

where $[B]$ is the strain-displacement matrix, $[D]$ is the stress-strain matrix, $[\delta \mathbf{d}]$ is the nodal displacement vector and Ω is the element surface area (for a two-dimensional element). In case of small deformation, the second term in the parenthesis on the right-hand side in Eqn (11.1) disappears and the element stiffness matrix is reduced to its standard well-known form:

$$[K] = \int_\Omega [B]^\mathrm{T}[D][B]\mathrm{d}\Omega \tag{11.2}$$

Therefore, in order to solve a large deformation FE problem, a non-linear numerical procedure is required and the non-linear set of equations should be incrementally solved.

11.2.2 Material models

Although some adhesives may behave as linear elastic materials, many modern adhesives exhibit large plasticity and high non-linear behaviour. Therefore, material non-linearity is very often used in modelling adhesive layer in bonded joints. A universally acceptable non-linear material model, which fits well the experimental stress–strain ($\sigma - \varepsilon$) of adhesive materials and takes the elastic strain effect into consideration, is based on the Ramberg–Osgood equation (Ramberg & Osgood, 1943):

$$\varepsilon = \frac{\sigma}{E}\left[1 + \alpha\left(\frac{\sigma}{\sigma_y}\right)^{m-1}\right] \tag{11.3}$$

where E is Young's modulus, α and m are the constants, and σ_y is a nominal yield stress. In implementing non-linearity in FEA, the stiffness matrix in Eqn (11.1) or

(11.2) will be a function of stresses and strains because the stress–strain matrix $[D]$ varies according to Eqn (11.3). Therefore, the solution of the classical linear FEA equation $[K][\delta d] = [F]$, where $[F]$ is the external force vector, will not be satisfied in one step, and an iterative solution will be required. Therefore, in elastic–plastic problems, the material stiffness matrix is continually varying and the instantaneous incremental stress–strain relationship is calculated as $[\Delta\sigma] = [D_{ep}][\Delta\varepsilon]$, where $[D_{ep}]$ is the elastic–plastic material matrix. In order to evaluate the material matrix $[D_{ep}]$ at any stage during the solution, the incremental solution is in the following form:

$$[\Delta\psi] = [K_T][\delta d] - [\Delta F] \tag{11.4}$$

where $[K_T]$ is the material tangential stiffness, $[K_T] = \int_\Omega [B]^T [D_{ep}][B]\,d\Omega$, and $[\Delta\psi]$ is the residual vector, which approaches zero for a converged solution.

Composite substrates are usually modelled with linear elastic material behaviour. As the main interest in the DCB test is the mode I crack propagation in the adhesive layer, or along the adhesive–substrate interface, composites are usually modelled using apparent moduli of elasticity when performing two-dimensional plane strain analysis. The apparent moduli of elasticity and Poisson's ratios for a composite laminate are given by (Tsai, 1992):

$$E_1 = \frac{1}{a^*_{11}}; \quad E_2 = \frac{1}{a^*_{22}}; \quad E_6 = \frac{1}{a^*_{66}};$$

$$\nu_{21} = -\frac{a_{21}}{a_{11}}; \quad \nu_{61} = \frac{a_{61}}{a_{11}}; \quad \nu_{62} = \frac{a_{62}}{a_{11}}; \tag{11.5}$$

$$\nu_{12} = -\frac{a_{12}}{a_{22}}; \quad \nu_{16} = \frac{a_{16}}{a_{66}}; \quad \nu_{26} = \frac{a_{26}}{a_{66}}$$

where 1, 2 and 6 refer to laminate axes in one to two planes, and the constants a_{11}, a_{22}, ..., are the elements of the compliance matrix, $[a]$, which is defined as the inverse of the stiffness matrix $[A]$. The stiffness matrix is calculated from the off-axis ply stiffness, $[Q]$ as:

$$[A] = \int_{-h/2}^{h/2} [Q]\,dz; \quad [A^*] = \frac{1}{h}[A] \tag{11.6}$$

where h is the laminate thickness and z is the coordinate in the through-thickness direction. The apparent moduli of elasticity and Poisson's ratios can be determined using any laminate plate theory analysis software, which solves Eqns (11.5) and (11.6) for any given stacking sequence. If delamination of the first ply adjacent to the interface is expected, explicit modelling of the first ply will be required. In three-dimensional analysis, layered composite elements, in which each ply is explicitly defined, can be used to model the substrates.

11.2.3 Modelling of crack propagation

11.2.3.1 Re-meshing

As the model geometry changes during crack propagation, re-meshing of the structure is required to account for the new geometry and the new crack configuration. In much FEA software, automatic fatigue crack propagation using the re-meshing technique can be performed. The FE mesh is modified and a new analysis is carried out after each crack increment. In the automatic re-meshing technique, each crack propagation increment involves the following steps: (1) performing FEA for the initial mesh with an initial crack, (2) extracting fracture parameters, (3) determining the direction of crack propagation and the new crack tip location, (4) updating the crack geometry and (5) automatic re-meshing. A flow chart for the crack propagation re-meshing technique is shown in Figure 11.2. The re-meshing technique is suitable when the crack path is not well known in advance.

11.2.3.2 Node release

Unlike the re-meshing technique, the node release technique does not required the generation of new FE mesh. The crack propagation is simulated by separating the crack tip node into two nodes, one on each crack face. However, the crack path should be known in advance so that nodes behind the initial crack tip are released during the progress of crack propagation. Although avoiding re-meshing is an advantage, the node release technique requires mesh refinement along the expected crack path.

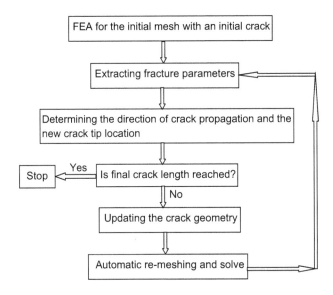

Figure 11.2 Flow chart for crack propagation re-meshing technique.

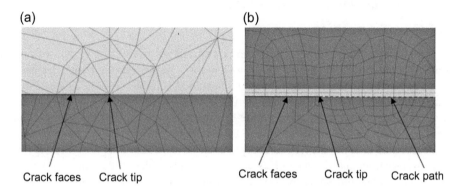

Figure 11.3 Crack propagation techniques: (a) re-meshing and (b) node release.

A comparison between the FE idealization used for both techniques is shown in Figure 11.3.

11.2.3.3 Extended finite element method (XFEM)

Another technique used for studying crack propagation without re-meshing makes use of the XFEM, which is an extension of the classical FEA or FEM (finite element method). XFEM is based on the concept of the partition of unity and improves the capabilities of FEM to model discontinuities (Belytschko & Black, 1999). In XFEM, enrichment functions for nodal displacements are used to simulate the separation of crack faces and the crack tip behaviour. The elements along the crack faces are enriched with a discontinuous function, while the elements at the crack tip are enriched with a near-tip asymptotic displacement function. In general, the enrichment functions are automatically created by the XFEM algorithm around the crack tip and crack faces, and continuously adjusted to follow the crack propagation direction. Therefore, re-meshing during crack growth is not required, which is one of the advantages of XFEM. In such a technique, a crack is modelled through discontinuity in displacements, i.e. the displacement approximation is given by:

$$u = \sum_{i=1}^{n} N_i \left(u_i + Ha_i + \sum_{j=1}^{4} Q_j b_{ij} \right)$$ (11.7)

where N is the shape function, n is the number of nodes, u is the classical degree of freedom, a and b are the additional degrees of freedom, H is the Heaviside function that represents discontinuity, and Q is a crack tip function that represents a square root displacement variation at a crack tip. The Heaviside function, H, is a discontinuous function that has a value of $+1$ above the crack face and -1 below the crack face. Figure 11.4 shows XFEM idealization for a crack.

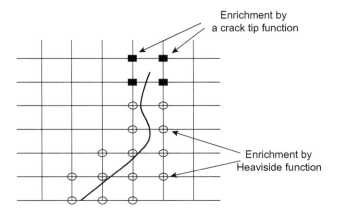

Figure 11.4 XFEM idealization for a crack.

11.3 Fracture mechanics (FM) approach

11.3.1 Crack growth laws

In order to study mode I fatigue crack propagation, the relationship between a fracture parameter and the crack growth rate should be identified using a DCB test specimen. Figure 11.5 shows a common mode I fatigue crack propagation curve for adhesively-bonded joints, in which crack growth rate (da/dN) is plotted against the mode I maximum strain energy release rate (G_{Imax}) using a logarithmic scale. The fatigue crack propagation curve has a sigmoidal shape and can be divided into three different regions: a threshold region, a linear region and a fast fracture region. The threshold

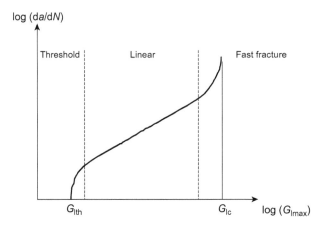

Figure 11.5 A typical mode I fatigue crack propagation.

region is defined by mode I fatigue threshold, G_{Ith}, below which no crack growth takes place. The linear or steady-state crack growth region can be well described by Paris' law, whereas the fast or unstable crack growth region is where catastrophic failure occurs, and takes place when mode I fracture toughness, G_{Ic}, is reached. Knowing the relationship between the crack length a and G_{Imax}, the integration of the fatigue crack propagation curve leads to an estimation of the crack propagation lifetime of an adhesively-bonded joint.

For adhesively-bonded joints, the strain energy release rate is often used as a fracture parameter in Paris's equation. The mode I fatigue crack growth rate in the steady crack (linear) propagation region using a DCB test specimen leads to a linear trend on Paris' law (Wahab et al., 2004):

$$\frac{da}{dN} = C_1 \times \left(\frac{\Delta G_I}{G_{Ic}}\right)^m \tag{11.8}$$

where a is the crack length, N is the number of cycles, G_I is the mode I strain energy release rate and G_{Ic} is the mode I fracture toughness. The constants C_1 and m can be obtained by curve fitting of Eqn (11.8) to DCB experimental data. The power constant m represents the sensitivity of the crack to its growth, and is higher for adhesives than for metals. A modified mode I crack growth law, which describes the full da/dN versus G_{Imax} curve, is given by (Martin & Murri, 1990, pp. 251–270; Wahab, Ashcroft, Crocombe, & Smith, 2002):

$$\frac{da}{dN} = C_1 G_{Imax}^n \left(\frac{1 - (G_{Ith}/G_{Imax})^{n_1}}{1 - (G_{Imax}/G_{Ic})^{n_2}}\right) \tag{11.9}$$

Again the constants n, n_1 and n_2 can be obtained by fitting Eqn (11.9) to experimental data of a DCB test specimen. The difference between Eqn (11.9) and the classical Paris' law Eqn (11.8) is that the former accounts for the threshold and fast crack regions.

11.3.2 Finite element (FE) meshing

Mesh refinement is applied near the crack tip and usually quarter-point singularity elements are used in the first row of elements. Quarter-point singularity elements are obtained by shifting the mid-side nodes to the quarter position at the crack tip in order to achieve the theoretical square root stress singularity for cracks in homogenous media (Barsoum, 1976). Although the stress singularity of cracks in bi-material structures, e.g. an adhesive/substrate system, is different from 0.5, quarter-point elements still can be used and provide good results. However, the FE mesh should be fine enough near the crack tip to capture the high stress variation. Figure 11.6 shows the FE mesh of a DCB specimen and details arrangement near an interfacial (adhesive/substrate) crack tip. It has been proven that good results can be obtained if the radius of the first row of elements around the crack tip is equal to or smaller than $a/8$, where a is the crack length, or $t_a/4$, where t_a is the thickness of the adhesive layer (Wahab et al., 2002). Furthermore, one element every 30° or 40° is recommended.

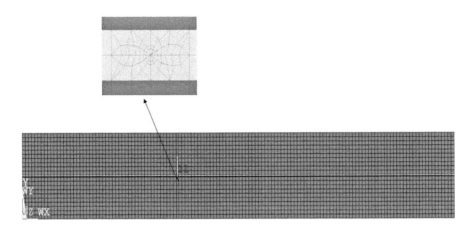

Figure 11.6 FE mesh of DCB and details near an interfacial crack tip.

It is worth mentioning that improper meshing near the crack tip may lead to unacceptable numerical errors.

11.3.3 Extracting fracture parameters from finite element analysis

11.3.3.1 Virtual crack closure technique

To compute the strain energy release rate from FEA results, the amount of energy released dU due to a crack extension da, can be approximated as:

$$G = \frac{\Delta U}{\Delta a} = \frac{U_2 - U_1}{\Delta a} \tag{11.10}$$

where U_1 and U_2 are the strain energies associated with crack lengths a and $a + \Delta a$, respectively. Therefore, two FEA models for two different crack lengths, differing by an incremental amount Δa, are required to evaluate the strain energy release rate. As this approach does not depend on the localized stress and displacement fields at the crack tip, it provides accurate results even with a coarse FE mesh. However, high computation time is required because of the need to perform two analyses. An alternative technique that makes use of one single analysis is based on Irwin's crack closure integral, also known as the virtual crack closure technique (VCCT), and can be used to extract the strain energy release rate from FEA results. For the four-noded linear elements shown in Figure 11.7(a), the strain energy release rate for mode I can be calculated using the nodal force at the crack tip F_{yi} and the opening displacements of the crack faces elements (u_{yk}, u'_{yk}) as follows (Rybicki & Kanninen, 1977):

$$G_{\mathrm{I}} = \frac{1}{2\Delta a}\left[F_{yi}\left(u_{yk} - u'_{yk}\right)\right] \tag{11.11}$$

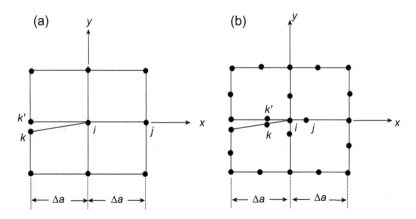

Figure 11.7 Virtual crack closure technique. (a) Linear elements. (b) Singularity elements.

where Δa is equivalent to the crack tip element size. If singularity elements are used around the crack tip, as shown in Figure 11.7(b), the strain energy release rate for mode I is expressed in terms of the nodal forces at and ahead of the crack tip (F_{yi}, F_{yj}) and the opening displacements behind it (u_{yk}, u'_{yk}) as (Sethuraman & Maiti, 1988):

$$G_I = \frac{(u_{yk} - u'_{yk})}{\Delta a}[F_{yj} + (1.5\pi - 4)F_{yi}] \tag{11.12}$$

The application of the Irwin VCCT is limited to linear elastic material and to very small plastic zone size at the crack tip. This is due to the fact that VCCT assumes that the work required to extend a crack by an infinitesimal amount is equal to that required to close it by the same amount. This assumption is not valid in case of a large plastic zone at the crack tip due to the large amount of energy dissipation. If a large amount of plasticity takes place, J-integral, which is explained in the next section, is more suitable to characterize crack behaviour.

11.3.3.2 J-integral

When large plastic deformation takes place at the crack tip, the elastic strain energy release rate, G, introduced in the previous section, cannot be used to characterize the crack behaviour. In elastic–plastic analysis when the plastic strain cannot be ignored, the path-independent J-integral (Rice, 1968a) is used to determine the energy release rate. For a two-dimensional case, J-integral is expressed as:

$$J = \int_\Gamma \left(W dy - \mathbf{T}\frac{\partial u}{\partial x} ds \right) \tag{11.13}$$

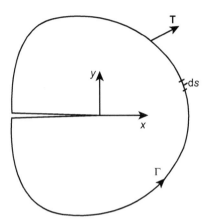

Figure 11.8 Path-independent J-integral.

where W is the strain energy per unit volume $\left(W = \int_0^\varepsilon \sigma d\varepsilon\right)$, Γ is a closed contour anti-clockwise as shown in Figure 11.8, **T** is the traction vector perpendicular to Γ, u is the displacement in the x-direction and ds is an element of Γ. If no crack exists along the contour, J will be equal to zero. J-integral is path independent, i.e. it has a unique value regardless the shape or the length of the contour Γ. For linear elastic material, J is identical to G.

The numerical implementation of Eqn (11.13) is based on extracting the displacements, stresses and strain energy from the results along a defined contour path in the post-processor, mathematically manipulated and numerically integrated within the FEA package.

11.3.4 Integration of crack growth laws

In order to obtain the number of cycles to failure (N_f), the crack growth law, Eqn (11.8) or (11.9), should be integrated from an initial crack size, a_o, to a final crack size, a_f, i.e.:

$$N_f = \int_{a_o}^{a_f} \frac{1}{da/dN} da = \int_{a_o}^{a_f} \frac{1}{C_1 G_{\text{Imax}}^n \left(\frac{1-(G_{\text{Ith}}/G_{\text{Imax}})^{n_1}}{1-(G_{\text{Imax}}/G_{\text{Ic}})^{n_2}}\right)} da \qquad (11.14)$$

To integrate Eqn (11.14), G_{Imax} should be expressed in terms of the crack length a. Although G_{Imax} may be analytically determined, its expression is too complicated and its predictions may show high discrepancies with the experimental data as it has been observed in the literature. A numerical procedure to integrate Eqn (11.14) was proposed by Wahab et al. (2002). Figure 11.9 summarizes the different steps involved in this numerical approach, which can be automated in any FEA package. The crack

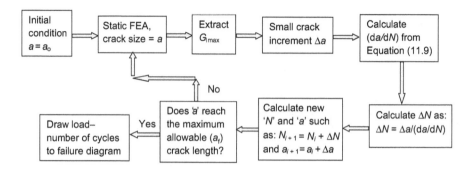

Figure 11.9 Numerical integration for fatigue crack growth lifetime.

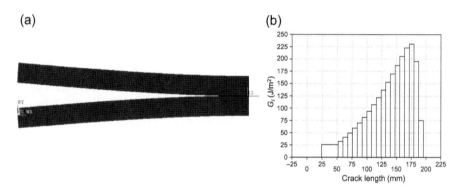

Figure 11.10 FE model of DCB fatigue crack growth. (a) Deformed shape at last crack increment. (b) G_I versus crack length.

length is defined as a parameter that varies between the initial crack length (a_o) and the final crack length (a_f). The final crack length can be defined in the FEA model as the crack length at which G_I reaches the mode I fracture toughness G_{Ic}. The crack increment (Δa) may be defined as:

$$\Delta a = \frac{(a_f - a_o)}{n_d} \quad (11.15)$$

where n_d is the number of crack increments or divisions used in the numerical integration.

The procedures presented in Figure 11.9 are applied to DCB and the results are shown in Figure 11.10, where the deformed shape at the last crack increment and the mode I strain energy release rate versus crack length are plotted.

11.4 Cohesive zone model (CZM) approach

The application of the cohesive zone model (CZM) to fatigue problems in adhesively-bonded joints has been intensively investigated in the last decade (Khoramishad et al., 2010; Moroni & Pirondi, 2011; Pirondi & Moroni, 2010). The CZM can be used to model fatigue damage initiation and crack growth. The fatigue CZM is formulated by adjusting the static CZM, taking into account the number of load cycles.

11.4.1 Cohesive traction–separation law

The cohesive traction–separation law is characterized by a relationship between a cohesive traction vector and displacement separation vector acting across the cohesive surfaces. For an isotropic material, it is represented by three parameters: the critical energy release rate, the critical tensile cohesive failure stress and the shape of the traction–separation law. The traction–separation law may have different shapes: (1) exponential form (Barenblatt, 1962), as shown in Figure 11.11(a); (2) polynomial form (Needleman, 1987), as shown in Figure 11.11(b); (3) constant form (Dugdale, 1960), as shown in Figure 11.11(c); (4) tri-linear form (Tvergaard & Hutchinson, 1992), as shown in Figure 11.11(d); (5) linear form (Camacho & Ortiz, 1996), as shown in Figure 11.11(e); and (6) bi-linear form (Geubelle & Baylor, 1998), as shown in Figure 11.11(f). The bi-linear form is commonly used in adhesively-bonded joints. For a bi-linear cohesive traction–separation law, the relationship between the cohesive normal traction, T_n, and the crack opening displacement, Δ_n, for fracture mode I, is shown in Figure 11.12 and can be mathematically expressed as (Maiti & Geubelle, 2005):

$$T_n = \frac{d}{1-d} \times \frac{\Delta_n}{\Delta_{nc}} \times \frac{\sigma_{max}}{d_{init}} \qquad (11.16)$$

where Δ_{nc} is the critical opening displacement jump and d is the damage parameter, which monotonically varies from an initial value, d_{init}, to zero. The numerical value of the initial damage parameter, d_{init}, is usually close to unity. σ_{max} is defined as the tensile cohesive failure strength. The damage parameter, d, quantifies the evolution of damage and is defined as follows:

$$d = \min\left(d_p, \left\langle \frac{1-\Delta_n}{\Delta_{nc}} \right\rangle \right) \qquad (11.17)$$

where d_p is the value of the damage parameter in the previous load step and the operator $\langle \rangle$ is defined as $\langle x \rangle = x$ if $x \geq 0$ otherwise. As the decrease in d monotonically takes place, according to Eqn (11.17), damage is not recovered upon unloading and the cohesive zone is not healed as shown in Figure 11.12. If reloading occurs, the stiffness of the cohesive zone keeps its most recent value and decreases according to the cohesive separation law up to failure.

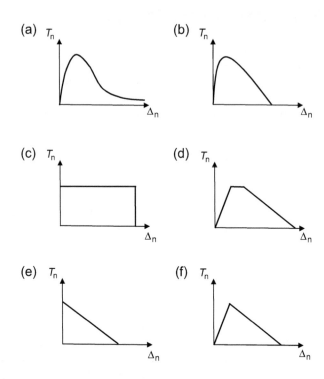

Figure 11.11 Different shapes of the cohesive traction−separation law: (a) exponential (Barenblatt, 1962), (b) polynomial (Needleman, 1987), (c) constant (Dugdale, 1960), (d) tri-linear (Tvergaard & Hutchinson, 1992), (e) linear (Camacho & Ortiz, 1996) and (f) bi-linear (Geubelle & Baylor, 1998).

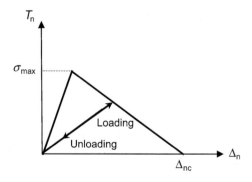

Figure 11.12 Cohesive traction−separation law.

Simulating mode I fatigue crack propagation in adhesively-bonded composite joints

The implementation of the cohesive traction–separation law in FEA is performed through the principle of virtual work, i.e.:

$$\int_\Omega [\sigma]:[\delta\varepsilon]\mathrm{d}\Omega - \int_{\Gamma_{\mathrm{ex}}} [T_{\mathrm{ex}}].[\delta u]\mathrm{d}\Gamma_{\mathrm{ex}} - \int_{\Gamma_{\mathrm{c}}} T_n \delta\Delta_n \mathrm{d}\Gamma_{\mathrm{c}} = 0 \quad (11.18)$$

where $[\varepsilon]$ and $[\sigma]$ are the internal strain and stress tensors, respectively, $[T_{\mathrm{ex}}]$ is the externally applied tractions, $[u]$ is the displacement vector, Ω is the volume, Γ_{ex} is the external boundary and Γ_{c} is the cohesive boundary. In Eqn (11.18), the first term represents the internal virtual work, the second term represents the virtual work done by the external traction, while the third term represents the virtual work done by the cohesive traction. For a mode I DCB specimen, if symmetry is considered, only half of the structure should be modelled, as shown in Figure 11.13, and the crack opening displacement is implemented as:

$$\Delta_n = 2[N][u_n] \quad (11.19)$$

where $[N]$ is the shape function vector and $[u_n]$ is the nodal displacement vector. The cohesive element stiffness matrix is then given by:

$$[k_{\mathrm{coh}}] = 2\int_0^l [N]^T k_c [N] \mathrm{d}\Gamma_{\mathrm{c}} \quad (11.20)$$

where l is the cohesive element length and k_c is the cohesive stiffness, which according to Eqn (11.16) for the bi-linear traction–separation law can be written as:

$$k_c = \frac{d}{1-d} \times \frac{1}{\Delta_{\mathrm{nc}}} \times \frac{\sigma_{\mathrm{max}}}{d_{\mathrm{init}}} \quad (11.21)$$

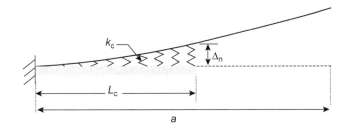

Figure 11.13 Mode I cohesive zone model.

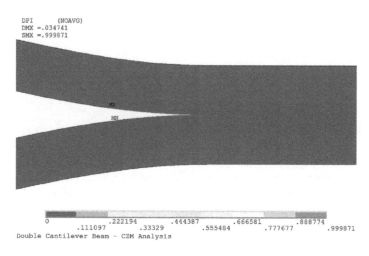

Figure 11.14 CZM model of DCB, de-bonding parameter using contact elements.

It has been demonstrated that the cohesive zone size, L_c, which should contain a sufficient number of cohesive elements in the active fracture process zone, is estimated as (Rice, 1968b):

$$L_c = \frac{\pi}{8} \times \frac{E}{1-\nu^2} \times \frac{G_{Ic}}{\sigma_{ave}^2} \qquad (11.22)$$

where E is the elastic modulus, ν is Poisson's ratio, G_{Ic} is the mode I fracture toughness, which is equal to $\dfrac{\sigma_{max} \times \Delta_{nc}}{2}$ for the bi-linear traction–separation law and σ_{ave} is the average stress in the cohesive zone, which is equal to $\dfrac{\sigma_{max}}{2}$ for the bi-linear traction–separation law.

An example of the application of the bi-linear traction–separation law to DCB using FEA package analysis system (ANSYS) contact elements to simulate the separation at the middle of the adhesive layer is shown in Figure 11.14, where the de-bonding parameter or damage variable, d, is plotted.

11.4.2 Cohesive fatigue model

In order to develop a cohesive fatigue model for mode I crack propagation, the evolution of fatigue damage law under cyclic loading, which incorporates changes in the cohesive strength, should be taken into account. Therefore, when the cracked structure is reloaded after unloading, the cohesive stiffness does not remain unchanged as in the case of static monotonic loading (Figure 11.12). The cohesive stiffness during fatigue loading will then be degraded upon reloading, as shown in Figure 11.15. The evolution of the cohesive stiffness, k_c, can be expressed as a function of the number of cycles,

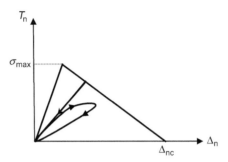

Figure 11.15 Fatigue cohesive zone model.

denoted by N_f, experienced by the material. The following evolution law was proposed by (Maiti & Geubelle, 2005):

$$k_c = \frac{dT_n}{d\Delta_n} = -\frac{1}{\alpha} \times N_f^{-\beta+1} \times T_n \qquad (11.23)$$

where α is a material parameter that has the dimension of length and β is a material parameter that describes the history of the cohesive failure process. β can be obtained from the slope of the Paris fatigue curve and α from curve fitting of Paris' equation to fatigue experimental data. Equation (11.23) represents an exponential decay of the cohesive strength.

In terms of the rate of change of cohesive stiffness $\dot{k}_c$, and the rate of change of the normal separation, $\dot{\Delta}_n$, the fatigue cohesive evolution law is written as:

$$\dot{k}_c = -\frac{1}{\alpha} \times N_f^{-\beta} \times k_c \times \dot{\Delta}_n \quad \text{if } \dot{\Delta}_n \geq 0$$
$$\dot{k}_c = 0 \quad \text{if } \dot{\Delta}_n \leq 0 \qquad (11.24)$$

This indicates that the cohesive stiffness changes during loading and remains constant during unloading. The FEA implementing of Eqn (11.24) can be performed using the following approximation:

$$k_c^{i+1} - k_c^i = -\frac{1}{\alpha} \times \left(N_f^i\right)^{-\beta} \times k_c^i \times \left(\Delta_n^{i+1} - \Delta_n^i\right) \qquad (11.25)$$

where the superscripts i and $i+1$ refer to two successive loading steps. Substituting Eqn (11.16) into Eqn (11.25) and using $k_c^i = \frac{T_n^i}{\Delta_n^i}$, the cohesive stiffness at loading step $i+1$ is given by:

$$k_c^{i+1} = k_c^i \left(1 - \frac{1}{\alpha} \times \left(N_f^i\right)^{-\beta} \times \left(\Delta_n^{i+1} - \Delta_n^i\right)\right) \quad \text{for } \Delta_n^{i+1} \geq \Delta_n^i \qquad (11.26)$$

11.5 Mixed CZM and FM approach

The CZM approach presented in the previous section requires the analysis of cycle per cycle in order to accumulate the crack propagations cycles up to failure. It is, therefore, not convenient in the case of a large number of cycles, especially in high cycle fatigue where the number of cycles is usually larger than 10,000. A more convenient approach makes use of both CZM and FM concepts. Consider the CZM model in Figure 11.16. The normal contact stress, σ_n, can be expressed in terms of the initial contact stiffness, k_n, and the crack opening displacement, Δ_n, as:

$$T_n = k_n \Delta_n (1 - d_n) \tag{11.27}$$

where d_n is as before the damage parameter, which represents the loss in stiffness during the separation process. From Figure 11.16, it can be easily shown that:

$$d_n = \frac{\Delta_{nc}(\Delta_n - \Delta_o)}{\Delta_n(\Delta_{nc} - \Delta_o)} \tag{11.28}$$

where Δ_o is the tripping opening displacement taking place at the maximum (tripping) normal stress σ_{max}. The area under the whole diagram $0-\sigma_{max}-\Delta_{nc}$ represents the fracture toughness G_{Ic}, whereas the area under the diagram $0-\sigma_{max}-\Delta_{max}$ represents the maximum strain energy release rate G_I^{max} (due to maximum fatigue load), which can be expressed as:

$$G_I^{max} = \frac{\sigma_{max}}{2}\left(\Delta_o + \frac{(\Delta_{max} - \Delta_o)^2}{\Delta_{nc} - \Delta_o}\right) \tag{11.29}$$

The load ratio, R, is defined, as:

$$R^2 = \frac{G_I^{min}}{G_I^{max}} \tag{11.30}$$

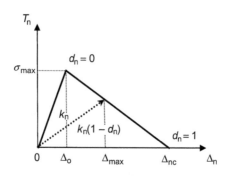

Figure 11.16 CZM model.

From Eqns (11.29) and (11.30), the range of mode I strain energy release rate can be written as:

$$\Delta G_I = G_I^{max} - G_I^{min} = \frac{\sigma_{max}}{2}\left(\Delta_o + \frac{(\Delta_{max} - \Delta_o)^2}{\Delta_{nc} - \Delta_o}\right)(1 - R^2) \qquad (11.31)$$

The evolution of the damage variable can be obtained by relating the damage variable rate to the fracture mechanics crack propagation rate using the following expression:

$$\frac{\partial d}{\partial N} = \frac{\partial d}{\partial a} \times \frac{\partial a}{\partial N} \qquad (11.32)$$

The crack propagation rate is determined from Paris' law, i.e.:

$$\frac{\partial a}{\partial N} = C_1 \times \left(\frac{\Delta G_I}{G_{Ic}}\right)^m \qquad (11.33)$$

The change in damage variable due to a small crack extension can be determined from FEA by performing analyses for crack lengths a_o and $a_i = a_o + \Delta a$:

$$\frac{\partial d}{\partial a} = \frac{\Delta d}{\Delta a} = \frac{d_{i+1} - d_i}{\Delta a} \qquad (11.34)$$

The damage variables d_{i+1}, d_i as well as the range of mode I strain energy release rate, Eqn (11.31), are calculated from the FEA results of the second interface element, denoted by 2 in Figure 11.17. The numerical value of damage evolution rate at crack length a_i is then obtained as:

$$D = \frac{\partial d}{\partial a} \times \frac{\partial a}{\partial N} = \frac{d_{i+1} - d_i}{\Delta a} \times C_1 \times \left(\frac{\Delta G_I}{G_{Ic}}\right)^m \qquad (11.35)$$

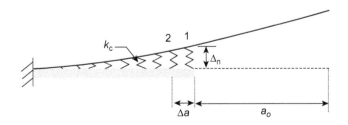

Figure 11.17 CZM elements at a crack.

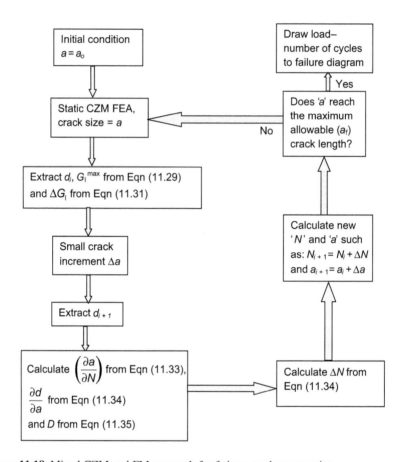

Figure 11.18 Mixed CZM and FM approach for fatigue crack propagation.

The increment in number of cycles is finally calculated as:

$$\Delta N = \frac{1}{D} \times \Delta d \quad (11.36)$$

The steps involves in the mixed CZM and FM approach are summarized in Figure 11.18.

11.6 Conclusions

In this chapter, three techniques for numerical simulation of mode I fatigue crack propagation in adhesively-bonded composite joints have been reviewed: the FM approach, the CZM approach, and a mixed CZM and FM approach. The techniques are based on using FEA combined with either fracture mechanics theory or the cohesive zone model. Material models, which were usually used for adhesive layer and composite substrates, have been briefly presented. FEA approaches for modelling crack propagation, such as re-meshing, node release and XFEM, were also reviewed. The technique based on fracture mechanics requires the calculation of fracture parameters, which can be extracted from FE results using the virtual crack closure technique for a linear elastic material model and the J-integral method for an elastic-plastic material model. A modified Paris' equation can be implemented and integrated within FEA in order to perform crack propagation analysis. The second technique based on the cohesive zone model approach makes use of a cohesive traction−separation law, which is characterized by a bi-linear relationship between the traction vector and displacement separation vector. The mode I crack propagation cohesive fatigue model takes into account the changes in cohesive strength due to damage evolution. The evolution of cohesive stiffness is expressed in terms of number of cycles to failure and cohesive traction. For high cycle fatigue, the CZM cycle-by-cycle analysis is not practical, and a mixed CZM and FM approach is more suitable. This last technique is based on combining a CZM traction−separation law with FM crack growth law in order to produce a damage evolution law, which can be numerically integrated to determine the number of cycles to failure.

References

Abdel Wahab, M. M. (2012). Fatigue in adhesively bonded joints: a review. *Materials Science,* 2012, 1−25.
Barenblatt, G. I. (1961/1962). Mathematical theory of equilibrium cracks in brittle fracture. Journal of Applied Mechanics and Technical Physics, No. 4, pp. 3−56 (1961) (in Russian); Advances in Applied Mechanics, VII, pp. 55−129 (1962) (in English).
Barsoum, R. S. (1976). On the use of isoparametric finite elements in linear elastic fracture mechanics. *International Journal for Numerical Methods in Engineering, 10*(1), 25−37.
Belytschko, T., & Black, T. (1999). Elastic crack growth in finite elements with minimal re-meshing. *International Journal of Fracture Mechanics, 45,* 601−620.
Camacho, G. T., & Ortiz, M. (1996). Computational modeling of impact damage in brittle materials. *International Journal of Solids and Structures, 33,* 2899−2938.
Cook, R. D., Malkus, D. S., & Plesha, M. E. (1989). *Concepts and applications of finite element analysis* (3rd ed.). New York: John Wiley & Sons.
Dugdale, D. S. (1960). Yielding of steel sheets containing slits. *Journal of the Mechanics and Physics of Solids, 8,* 100−104.
Geubelle, P. H., & Baylor, J. (1998). Impact-induced delamination of laminated composites: a 2D simulation. *Composites Part B Engineering, 29*(5), 589−602.
Khoramishad, H., Crocombe, A. D., Katnam, K. B., & Ashcroft, I. A. (2010). Predicting fatigue damage in adhesively bonded joints using a cohesive zone model. *International Journal of Fatigue, 32*(7), 1146−1158.
Maiti, S., & Geubelle, P. H. (2005). A cohesive model for fatigue failure of polymers. *Engineering Fracture Mechanics, 72,* 691−708.
Martin, R. H., & Murri, G. B. (1990). *Characterisation of mode I and mode II delamination growth and thresholds in AS4/PEEK composites.* Philadelphia: American Society for Testing and Materials.

Moroni, F., & Pirondi, A. (2011). A procedure for the simulation of fatigue crack growth in adhesively bonded joints based on the cohesive zone model and different mixed mode propagation criteria. *Engineering Fracture Mechanics*, *78*(8), 1808−1816.
Needleman, A. (1987). A continuum model for void nucleation by inclusion debonding. *Journal of Applied Mechanics*, *54*, 525−531.
Pirondi, A., & Moroni, F. (2010). A progressive damage model for the prediction of fatigue crack growth in bonded joints. *Journal of Adhesion*, *86*(5−6), 501−521.
Pirondi, A., & Nicoletto, G. (2006). Mixed mode I/II fatigue crack growth in adhesive joints. *Engineering Fracture Mechanics*, *73*, 2557−2568.
Ramberg, W., & Osgood, W. R. (1943). *Description of stress-strain curves by three parameters*. Technical Report, Technical Note No. 902. National Advisory Committee on Aeronautics (NACA).
Rice, J. R. (1968a). A path independent integral and the approximate analysis of strain concentration by notches and cracks. *Journal of Applied Mechanics*, *35*, 379−386.
Rice, J. R. (1968b). Mathematical analysis in the mechanics of fracture. In H. Liebowitz (Ed.), *Fracture 2* (pp. 191−311). New York: Academic Press.
Rybicki, E. F., & Kanninen, M. F. (1977). A finite element calculation of stress intensity factors by a modified crack closure integral. *Engineering Fracture Mechanics*, *9*, 931−938.
Sethuraman, R., & Maiti, S. K. (1988). Finite element based computation of strain energy release rate by modified crack closure integral. *Engineering Fracture Mechanics*, *30*, 227−231.
Zienkiewicz, O. C., Taylor, R. L., & Zhu, J. Z. (2005). *The Finite Element Method: Its Basis and Fundamentals*. Elsevier, 733pp.
Tsai, S. W. (1992). *Theory of Composite Design, Think Composites, Dayton, OH*.
Tvergaard, V., & Hutchinson, J. W. (1992). The relation between crack growth resistance and fracture process parameters in elastic−plastic solids. *Journal of the Mechanics and Physics of Solids*, *40*(6), 1377−1397.
Wahab, M. M. A., Ashcroft, I. A., Crocombe, A. D., & Smith, P. A. (2002). Numerical prediction of fatigue crack propagation lifetime in adhesively bonded structures. *International Journal of Fatigue*, *24*(6), 705−709.
Wahab, M. M. A., Ashcroft, I. A., Crocombe, A. D., & Smith, P. A. (2004). Finite element prediction of fatigue crack propagation lifetime in composite bonded joints. *Composites Part A*, *35*(2), 213−222.
Zienkiewicz, O. C., Taylor, R. L., & Nithiarasu, P. (2005). *The Finite Element Method for Fluid Dynamics*. Elsevier, 435pp.

Simulating the effect of fiber bridging and asymmetry on the fracture behavior of adhesively-bonded composite joints

12

M. Shahverdi, A.P. Vassilopoulos, T. Keller
École Polytechnique Fédérale de Lausanne (EPFL), Lausanne, Switzerland

12.1 Introduction

The Mode I fracture behavior of unidirectional (UD) and multidirectional (MD) composite laminates has already been extensively studied (Ashcroft, Hughes, & Shaw, 2001; Brunner, 2000; Choi, Kinloch, & Willams, 1999; Hojo, Kageyama, & Tanaka, 1995; Williams, 1988). The most commonly used specimen for this type of investigation is the double cantilever beam (DCB). Nevertheless, although the experimental procedure for the delamination of unidirectional laminates is standardized (ASTM D5528-01, 2007), there is no standard process for investigation of the fatigue/fracture behaviors of non-unidirectional laminates or adhesively-bonded joints composed of composite adherends and a paste adhesive. In these cases the crack propagates along paths away from the symmetry plane and is usually accompanied by significant fiber bridging (Shahverdi, Vassilopoulos, & Keller, 2011; Zhang, Vassilopoulos, & Keller, 2010). Modeling of the effects of the asymmetry and fiber bridging on fatigue/fracture behavior and quantification of their contribution to fracture energy calculations are essential for an accurate description of the fatigue/fracture behavior of non-standardized laminates and adhesively-bonded joints.

12.1.1 Effect of asymmetry

The effect of the asymmetry is usually investigated experimentally by using the asymmetric double cantilever beam (ADCB) specimen, in which the two arms may differ in thickness and constituent material. There are in fact two types of asymmetric specimens: those that present a geometrical asymmetry (e.g., Ducept, Gamby, & Davies, 1999; Mollón, Bonhomme, Viña, & Argüelles, 2010b) with the same material in different thicknesses above and below the crack, and those that present material asymmetry (e.g., Sundararaman & Davidson, 1997; Xiao, Hui, & Kramer, 1993) with different materials, sometimes also in different thicknesses (e.g., Shahverdi et al., 2011; Zhang et al., 2010) above and below the crack. In all of these cases, pure Mode I fracture no longer exists and a Mode II component is introduced.

Analytical and experimental works concerning calculation of the Mode II percentage in the resulting mixed-mode fracture obtained from an ADCB specimen exist (e.g., Bennati, Colleluori, Corigliano, & Valvo, 2009; Ducept et al., 1999; Hutchinson & Suo, 1992; Mollón, Bonhomme, Viña, & Argüelles, 2010a). The main analytical approaches regarding the mode partitioning in the literature are the global approach, based on the beam theory as proposed by Williams (1988), and the local approach, based on the stress intensity factor calculation around the crack tip, proposed by Hutchinson and Suo (1992). Ducept et al. (1999) carried out experiments on ADCB glass fiber reinforced epoxy composite samples and estimated that a high Mode II fracture component can be introduced owing to the geometrical asymmetry in a DCB specimen. They analyzed the experimental results using both the local and global approaches. The results obtained from these two approaches were compared with numerical results derived by the authors Ducept et al. (1999). The Mode II fracture component obtained from the local approach and the numerical models were of the same order, although the global approach resulted in pure Mode I for all ADCB specimens examined.

In another publication, Zhang et al. (2010) experimentally and numerically investigated the Mode I fracture behavior of pultruded ADCB joints such as those examined in the present chapter and proved that the through-thickness relative displacement between the two arms of the joint, which is related to the Mode I fracture component, was considerably greater than the corresponding in-plane relative displacement, corresponding to the Mode II fracture component. According to the results reported in Zhang et al. (2010) for this type of ADCB specimen, the induced Mode II fracture component is limited to 1% of Mode I.

Fracture behavior modeling and calculation of the fracture components of an ADCB joint can also be carried out by finite element (FE) analyses according to the virtual crack closure technique (VCCT) (De Morais & Pereira, 2006; Krueger, 2004; Rybicki & Kanninen, 1977; Xie & Biggers, 2006). This method is based on the assumption that the amount of energy released by a crack propagation of length Δa is equal to the energy required to close the crack faces back to the same length. This is an accurate method for calculation of the fracture energy at the crack tip, especially when homogeneous materials are being examined. However, when the crack path lies in a bi-material interface, the VCCT results concerning the mode partitioning become sensitive to the crack extension length, Δa (Agrawal & Karlsson, 2006; Beuth, 1996; Dattaguru, Venkatesha, Ramamurthy, & Buchholz, 1994; Raju, Crews, & Aminpour, 1988; Sun & Qian, 1997). To overcome this problem, Atkinson (1977) proposed a method for analyzing isotropic fracture problems regarding a bi-material crack interface. This method involves inserting a thin resin layer between the layers forming the interface and placing the crack within it. Because the crack tip is fully embedded in the resin layer, mode-mixity is not sensitive to the crack extension length.

12.1.2 Effect of fiber bridging

In principle, the strain energy release rate (G) of composite materials is considered equal to that of their matrix; however, an increased G is caused by fiber bridging that delays the propagation of delamination in the case of most fiber architectures and types of loading. This increasing G is usually described by the resistance curve (R-curve) describing the relationship between the crack length and the corresponding G. The VCCT is able to calculate only the G at the crack tip; however, the fiber bridging that occurs behind the crack tip is accompanied by the fracture behavior of nearly all types of fibrous composite materials (Bao & Suo, 1992; Sørensen & Jacobsen, 1998; Sorensen, Botsis, Gmür, & Humbert, 2008; Spearing & Evans, 1992). Fiber bridging results in an increase in the G of the examined composite materials and this increasing G is usually described by the resistance curve (R-curve). Bridging fibers are always observed during the fracture of composite laminates, although their number varies according to the fiber volume fraction, fiber misalignment, fiber architecture, and crack opening. The ideal R-curve follows an initially increasing trend before reaching a plateau. The crack length up to the plateau is equal to the fiber-bridging length. After its development, the bridging length "moves" together with the crack tip during crack propagation, whereas fibers behind the bridging length are broken or pulled out and no longer contribute to the fracture energy of the specimen. The fiber-bridging zone can be considered part of the fracture process zone where the fracture energy is released. According to the literature, the total fracture energy, G_{tot}, of a composite material is composed of a fiber-bridging component, G_{br}, and a tip component, G_{tip}.

Many efforts have been made to model the effects of fiber bridging (e.g., Sørensen & Jacobsen, 1998; Sorensen et al., 2008), and separate the two G components, mainly by FE modeling. A comprehensive review concerning crack-bridging mechanisms proposing the bridging constitutive relationship concepts for different bridging mechanisms was presented by Bao and Suo (1992). The cohesive zone model (CZM) approach is the one most commonly used for determination of the G_{br} (see, e.g., Spearing & Evans, 1992). The behavior of the cohesive element is based on a traction—separation law that defines the stresses at a particular location in a prescribed cohesive zone as a function of the opening displacement of the zone at that location. Cohesive models in FE modeling have been used extensively during recent years (Blackman, Hadavinia, Kinloch, & Williams, 2003; Chandra, Li, Shet, & Ghonem, 2002; Shet & Chandra, 2002; Xu & Needleman, 1994). The applicability of the CZM technique for modeling fiber bridging using a single layer of zero-thickness cohesive elements (COH_2D_4 in ABAQUS) along the delamination plane was demonstrated by Sorensen et al. (2008).

12.1.3 Content of the chapter

The main objective of this chapter is the modeling of fiber bridging in adhesively-bonded pultruded glass fiber reinforced polymer (GFRP) joints used in civil engineering structures. The effect of a combined (geometrical and material) asymmetry of DCB joints on their fracture behavior is investigated for different crack paths. Asymmetric pultruded bonded joints are common in civil engineering applications, e.g., in bridge and building structures. Therefore, understanding their fracture behavior and the mechanisms leading to their failure is compulsory for the design of reliable structures. Their design is usually based on quasi-static loading cases, although it is more probable that they will face fatigue loads during their operational lifetime.

In this chapter, the results of an experimental investigation of DCB adhesively-bonded pultruded GFRP joints published in Shahverdi et al. (2011) are analyzed. The VCCT is used to calculate the fracture components at the crack tip and a CZM is established for simulation and quantification of the fiber-bridging effect. The fracture components and resulting R-curves are compared with those resulting from the experimental investigation. Using the VCCT and CZM allows the investigation of the effects of both asymmetry and fiber bridging on the fracture energy of the examined joints.

12.2 Experimental investigation of asymmetry and fiber-bridging effects

12.2.1 Adherend and adhesive materials

Adhesively-bonded pultruded GFRP DCB joints were examined under quasi-static loading. The laminates, supplied by Fiberline A/S, Denmark, consisted of E-glass fibers embedded in isophthalic polyester resin and had a width of 40 mm and thickness of 6.0 mm. The laminates were composed of two outer combined mat layers and a roving layer in the symmetry plane. One combined mat consisted of two outer chopped strand mats, CSM, and an inner woven 0°/90° fabric, all stitched together. On the outside, a 40-g/m^2 polyester surface veil was added to protect against environmental attack. The fiber architecture of the laminates is shown in Figure 12.1.

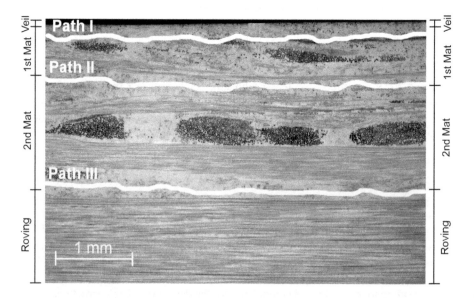

Figure 12.1 Fiber architecture of upper half of the laminate (section parallel to pultrusion direction) and observed crack propagation paths.

Estimation of the nominal thickness of each layer derived by optical microscopy is also shown in Figure 12.1. The fiber content, determined by burn-off according to ASTM D3171 (2001), was 43.3 vol.% based on the fiber density of 2560.0 kg/m^3 specified by the manufacturer and the assumption that no voids were present; the fiber fractions are described in Chapter 6 of this volume.

The weight of the second combined mat layer was almost double that of the first mat layer and the proportion of woven fabrics was much higher. The longitudinal strength and Young's modulus of the GFRP laminate were obtained from tensile experiments, according to ASTM D3433 (2005), as being 307.5 MPa and 25.0 GPa, respectively (Shahverdi et al., 2011).

A two-component epoxy adhesive system was used, Sikadur 330, supplied by Sika AG Switzerland, as the bonding material. The tensile strength of the adhesive was 39.0 MPa and the longitudinal Young's modulus was 4.6 GPa. The epoxy exhibited an almost elastic behavior and brittle failure under quasi-static tensile loading (De Castro & Keller, 2008).

12.2.2 Experimental setup and procedure

The geometry of the DCB specimens is shown in Figure 12.2. The specimen length was 250 mm. All surfaces subjected to bonding were mechanically abraded by approximately 0.3 mm to increase roughness and then chemically degreased using acetone. To study the effect of crack depth, different specimen configurations with different depths of pre-crack were examined (Shahverdi et al., 2011). The pre-crack depth, H, was determined according to the layer thicknesses (see Figure 12.1). Additional

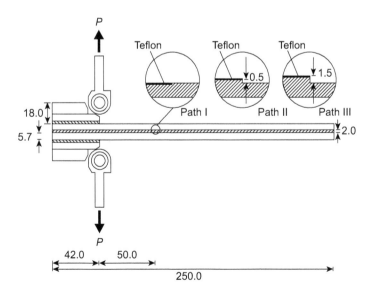

Figure 12.2 Specimen configuration, dimensions in millimeters.

mechanical sanding was required to reach different depths corresponding to different paths. By removing an additional 0.5 mm of material, the crack tip was located between the two mat layers of the laminate (Path II), whereas a sanding of 1.5 mm was necessary to reach Path III between the second mat and the roving layer (Shahverdi et al., 2011). A Teflon film of 0.05-mm thickness was placed between the upper arm and the adhesive layer to introduce the pre-crack. The length of the pre-crack was 50 mm measured from the loading line. An aluminum frame was used to assist the alignment of the two pultruded laminates. The 2-mm thickness of the adhesive was controlled by using spacers embedded in the bonding area. In-house developed piano hinges were bonded, using the same epoxy adhesive, at the end of both specimen arms to allow load application. After preparation of the configuration, the specimens were kept under laboratory conditions for 24 h and then placed in a conditioning chamber at 35 °C and 50 ± 10% relative humidity (RH) for 24 h to ensure full curing of the adhesive. The resulting thickness of the DCB specimens was 13.4 mm and the crack was located 1.0, 1.5, or 2.5 mm above the center axis of the joints because of the presence of the adhesive layer and crack depth.

In total, 16 ADCB experiments were performed on a testing machine of 5-kN capacity, under displacement control at a constant rate of 1 mm/min under laboratory conditions, 23 ± 5 °C and 50 ± 10% RH.

12.2.3 Experimental results

12.2.3.1 Failure modes

In all 16 examined specimens the observed failure mode, according to ASTM D5573 (1999), was a fiber-tear failure or light fiber-tear failure. The crack paths were located between the adhesive and the first mat layer (Path I), between the first and second mat layers (Path II), or between the second mat and the roving layer (Path III), as planned by the selected depths of the pre-crack (see also Figure 12.1 and Shahverdi et al., 2011). Fiber bridging started to develop with increasing crack opening displacement. Fibers from both arms of the specimen bridged the crack, transferring the load from one side to the other. At a certain crack opening displacement, fibers far from the crack

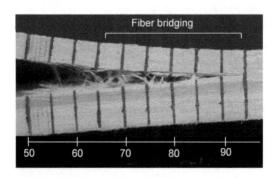

Figure 12.3 Representative side view of fiber bridging (in millimeters).

tip were broken or pulled out (see crack length between 50 and 65 mm in Figure 12.3). The length along which fibers were not broken or pulled out is designated the "fiber-bridging length" and is kept constant, following the crack tip for the rest of the fracture process (see e.g., Figure 12.3 crack length of about 65−95 mm).

The fiber-bridging length varied between 15 and 30 mm in the examined specimens (Shahverdi et al., 2011). The bridging lengths for observed paths in examined specimens are presented in Table 12.1.

12.2.3.2 Fracture data analysis

Representative load-opening displacement responses are shown in Figure 12.4 for three selected specimens with cracks propagating along the three different crack paths.

Specimen DCB-04 is representative of specimens with a crack propagating along Path I and DCB-16 is an example of Path II, whereas DCB-13 is representative of Path III. For all examined cases, the load increased until a maximum value was reached and then gradually decreased. The highest maximum load was achieved by specimens exhibiting Path II cracks, followed by those with Path III cracks, whereas the fracture of specimens with cracks propagating along Path I occurred under the lowest loads.

Table 12.1 Bridging lengths along different paths

Specimen code	Bridging length (mm)		
	Path I	Path II	Path III
DCB-01	15		
DCB-02		20	
DCB-03	25		
DCB-04	30		
DCB-05	20	25	
DCB-06		25	
DCB-07			30
DCB-08		25	
DCB-09		20	
DCB-10		15	
DCB-11		20	
DCB-12		25	
DCB-13			30
DCB-14			30
DCB-15		30	
DCB-16		30	

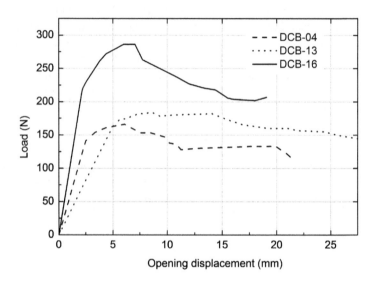

Figure 12.4 Load versus opening displacement of paths I (DCB-04), II (DCB-16), and III (DCB-13).

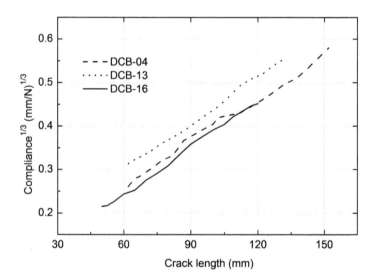

Figure 12.5 Compliance versus crack length of paths I (DCB-04), II (DCB-16), and III (DCB-13).

Corresponding crack length versus compliance plots for the same three specimens are shown in Figure 12.5. Specimens representing Path I and Path II crack propagation showed similar compliance. The slight difference can be attributed to less fiber bridging exhibited by specimens with the crack propagating along Path I. However, in a specimen with a crack propagating along Path III, one arm of the joint was much stiffer than the other, and therefore the joint exhibited overall lower stiffness (higher compliance) compared with the Path I and Path II configurations.

The strain energy release rate of the DCB joints can be calculated based on linear-elastic fracture mechanics (Zhang et al., 2010). According to this theory, for a DCB joint with width B and crack length a, the G is a function of the applied load, P, and the compliance change rate, dC/da:

$$G = \frac{P^2}{2B} \frac{dC}{da} \quad (12.1)$$

Typical methods for the calculation of G are based on this equation, the difference between them basically being the way in which the derivative dC/da is obtained. A thorough analysis of the applicability of several methods (simple beam theory, corrected beam theory, experimental compliance method, and the modified compliance calibration method) for the G calculation of similar pultruded GFRP DCB joints is presented in Zhang et al. (2010). It was concluded that all methods give similar results with the exception of simple beam theory. The total strain energy release rate is the sum of the contributions of the matrix and fiber bridging. In general, matrix fracture at the tip is always accompanied by fiber bridging whereas the contribution of the fiber bridging depends on the crack paths. Because significant fiber bridging was observed, the experimental compliance method (ECM) was selected because the effect of fiber bridging is included in the measured compliance. According to the ECM, the measured compliance is fitted to the measured crack length by the power law equation $C = ka^n$. The G can then be calculated as:

$$G = \frac{P^2}{2B} \frac{dC}{da} = \frac{P^2}{2B} nka^{n-1} = \frac{P^2}{2B} \frac{a}{a} nka^{n-1} = \frac{nP^2}{2B} \frac{C}{a} = \frac{nP^2}{2B} \frac{\delta/P}{a} = \frac{nP\delta}{2Ba} \quad (12.2)$$

where P and δ denote the load and opening displacement and B the specimen width. Correction factors for the loading blocks and moments resulting from large displacements were applied according to ASTM D5528-01 (2007).

Strain energy release rate values for each specimen were calculated and corresponding R-curves were established (see Figure 12.6).

The mean value of the visually determined plateau, taking the typical scatter of this type of material into account, was assumed to represent the G for propagation (Shahverdi et al., 2011). The highest G corresponded to Path II crack propagation whereas those obtained from Path III were higher than those from Path I.

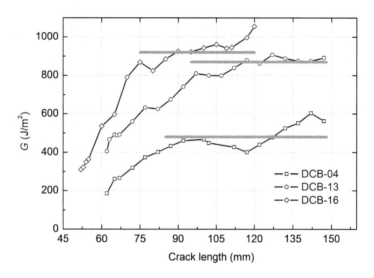

Figure 12.6 G versus crack length of paths I (DCB-04), II (DCB-16), and III (DCB-13).

12.3 Finite element modeling

12.3.1 Overview

Two-dimensional plane-strain models were developed in ANSYS (academic version 13.0) to model the effects of fiber bridging and asymmetry on the fracture behavior of the examined joints. All layers of the laminates (veil, mat, and roving) were modeled according to the thicknesses estimated by optical microscopy. The material properties are given in Table 12.2.

The element PLANE182, a four-node structural solid, was used to model the different layers and a manual mesh with controlled element size was used (see Figure 12.7).

Fiber bridging along the crack faces was modeled in ANSYS by using a single layer of zero-thickness cohesive elements with an exponential cohesive model, INTER202. This element type is a two-dimensional, four-node interface element with two degrees of freedom at each node. The boundary conditions of the experimental configuration were simulated at nodes A and B, as presented schematically in Figure 12.8.

All translational degrees of freedom (in the nodal X and Y directions) of the lower node (node A) were constrained to simulate the fixed piano hinge, whereas only the degree of freedom in the X-direction was constrained at node B. Experimental opening displacement values at node B corresponding to arbitrarily selected crack lengths were used as input for calculation of the corresponding nodal displacements and forces. Linear-elastic analysis for models without cohesive elements (without fiber bridging) and nonlinear analysis for models with cohesive elements were performed, allowing calculation of the specimen deformation, nodal forces, and nodal displacements.

Table 12.2 Properties used for FE modeling

Material data	First combined mat	Second combined mat	Roving	Veil	Adhesive
E_{11} (GPa)	12.8	15.1	38.9	3.2	4.6
E_{22} (GPa)	12.8	15.1	3.2	3.2	4.6
E_{33} (GPa)	3.2	3.2	3.2	3.2	4.6
G_{12} (GPa)	6.2	6.7	2.7	1.2	1.7
G_{23} (GPa)	1.4	1.4	1.4	1.2	1.7
G_{31} (GPa)	1.4	1.4	2.7	1.2	1.7
ν_{12}	0.27	0.27	0.32	0.38	0.37
ν_{23}	0.36	0.36	0.27	0.38	0.37
ν_{31}	0.36	0.36	0.35	0.38	0.37

12.3.2 Cohesive elements

Cohesive elements were inserted along the crack plane to model the fracture behavior of the fiber-bridging zone. In a CZM, cohesive element behavior is based on a traction–separation law that defines the stresses at a particular location as a function of the opening displacement. The traction–separation relationship is such that with increasing opening displacement the traction across the interface reaches a maximum, σ_{max}, then decreases and eventually vanishes, permitting a complete separation at an opening displacement of δ_f, as shown in Figure 12.9.

The area under the $\sigma-\delta$ curve represents the amount of energy dissipated during crack propagation in the cohesive zone, the cohesive energy. The three parameters, cohesive energy, Φ, maximum traction, σ_{max}, and maximum opening displacement, δ_f, are interdependent and therefore the CZM can be described by two of them (Chandra et al., 2002).

The traction–separation cohesive law model can be linear, polynomial, exponential, or user-defined. In this study for modeling fiber bridging an exponential law is used, which, according to Sorensen et al. (2008), can model this effect better than the others. The applied exponential law is Xu and Needleman (1994):

$$\sigma_{br} = e\sigma_{br,max} \frac{\delta}{\bar{\delta}} e^{-\frac{\delta}{\bar{\delta}}} \qquad (12.3)$$

where σ_{br} is the fiber-bridging traction, δ is the opening displacement along the cohesive zone, and $\bar{\delta}$ is the opening displacement at the maximum traction, i.e., $\sigma_{br,max}$ (see Figure 12.9). The length of the fiber-bridging zone and the maximum opening displacement, δ_f, were obtained from the experimental investigation (Shahverdi et al.,

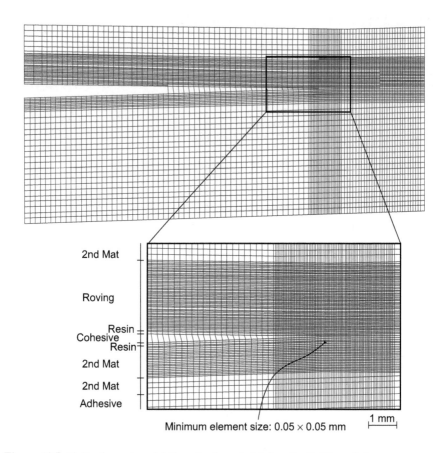

Figure 12.7 Finite element model discretization at vicinity of path III crack tip.

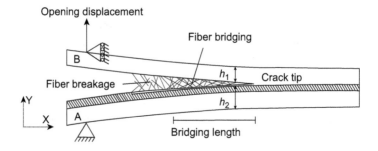

Figure 12.8 Loading and boundary conditions.

Simulating the effect of fiber bridging and asymmetry on fracture behavior 357

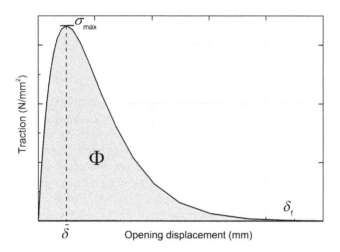

Figure 12.9 Schematic illustration of cohesive traction—separation law.

2011). The values of $\bar{\delta}$ and $\sigma_{br,max}$ required by the CZM were estimated by an iterative procedure aiming to fit the FE output to the corresponding experimental data. The selected $\bar{\delta}$ is the one that allows the FE model to predict an opening displacement equal to δ_f, that has been derived experimentally. Accordingly, selected $\sigma_{br,max}$ values were those that resulted in the same loads computed by FE models as those obtained from the experiments, both corresponding to identical displacements and crack lengths. The same process was followed for indicative specimens of each different crack path configurations. The estimated cohesive element model parameters for different paths are listed in Table 12.3.

The following equation represents the amount of energy dissipated in the crack-bridging zone, G_{br} according to the CZM approach (Sorensen et al., 2008).

$$G_{br} = \int_0^{\delta_f} \sigma_{br} d\delta = \sum_{i=1}^{n} \sigma_{br,i} \delta_i \qquad (12.4)$$

Table 12.3 Traction—separation cohesive model parameters for different paths

Specimen (path)	$\sigma_{br,max}$ (MPa)	$\bar{\delta}$ (mm)	δ_f (mm)
DCB-04 (Path I)	0.40	0.28	2.5
DCB-16 (Path II)	0.75	0.30	3.0
DCB-13 (Path III)	0.45	0.40	5.0

where $\sigma_{br,i}$ is the bridging traction and δ_i is the relative opening displacement of a node i along the fiber-bridging length from the upper and lower arms. In Eqn (12.4) the bridging traction is obtained from the nodal forces in the FE models along the bridging length.

12.3.3 Virtual crack closure technique

The VCCT can be used to calculate the fracture parameters at the crack tip (Atkinson, 1977; Rybicki & Kanninen, 1977; Xie & Biggers, 2006; Zhang et al., 2010). In a two-dimensional finite element plane stress, or plane-strain model, the crack is represented as a one-dimensional discontinuity. Nodes at the top and bottom surfaces of the discontinuity have the same coordinates but are not connected with each other, as shown in Figure 12.10. The element contains two sets of node groups: the top set (nodes 1, 3, and 5) and the bottom set (nodes 2 and 4). Nodes 1 and 2 are linked together with a stiff spring to compute the nodal forces at the crack tip. Nodes 3, 4, and 5 are introduced to extract information concerning opening displacement and crack extension length. The opening displacements are:

$$\delta_X = u_{X,3} - u_{X,4}, \quad \delta_Y = u_{Y,3} - u_{Y,4} \qquad (12.5)$$

where $(u_{X,3}, u_{Y,3})$ and $(u_{X,4}, u_{Y,4})$ are the displacement components for nodes 3 and 4, respectively, in the global coordinate system (X,Y). The crack extension length is the distance between nodes 1 and 5 and is therefore calculated by:

$$\Delta a = \sqrt{(X_5 - X_1)^2 + (Y_5 - Y_1)^2} \qquad (12.6)$$

where (X_1,Y_1) and (X_5,Y_5) are the global coordinates for nodes 1 and 5, respectively. In the present study, the crack extension length is always equal to the element size in the vicinity of the crack tip.

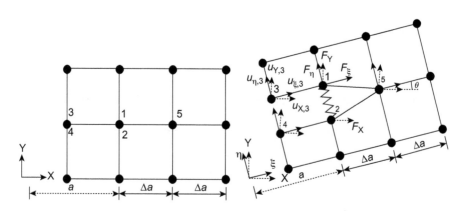

Figure 12.10 Definition of node numbering of VCCT before and after rotation of crack tip.

To partition the fracture modes (Modes and II), the strain energy release rates (G_I and G_{II}) must be computed with respect to the local coordinate system ξ and η attached to the crack tip, as shown in Figure 12.10. The angle, θ, between ξ and X can be determined by:

$$\sin \theta = \frac{u_{Y,5} - u_{Y,1}}{\Delta a} \qquad (12.7)$$

where $u_{Y,5}$ and $u_{Y,1}$ are the displacement components for nodes 5 and 1, respectively, in the global coordinate system (X,Y).

The nodal forces, F_X and F_Y, and the opening displacements, δ_X and δ_Y, are projected onto the local coordinate system ξ and η as:

$$\begin{cases} F_\xi = F_X\cos\theta + F_Y\sin\theta \\ F_\eta = -F_X\sin\theta + F_Y\cos\theta \end{cases} \qquad (12.8)$$

and

$$\begin{cases} \delta_\xi = \delta_X\cos\theta + \delta_Y\sin\theta \\ \delta_\eta = -\delta_X\sin\theta + \delta_Y\cos\theta \end{cases} \qquad (12.9)$$

The projections of the parameters are mandatory to take into account the rotation of the joint owing to the asymmetry configuration. In a two-dimensional model, the strain energy release rates can be approximated as the product of the nodal forces at the crack tip and the nodal opening displacements behind the crack tip:

$$G_I = \frac{F_\eta \delta_\eta}{2B\Delta a}; \quad G_{II} = \frac{F_\xi \delta_\xi}{2B\Delta a} \qquad (12.10)$$

where B is the width of the specimen.

12.3.4 Mesh sensitivity

Bi-material interfaces are present in all specimens with cracks propagating in one of the three different paths as previously discussed. Therefore, the calculated G_I and G_{II} components and the calculated mode-mixity ratios, G_{II}/G_I, depended on the crack extension length and did not represent the actual fracture development. The obtained mode-mixity was sensitive to the Δa, equal to one-element size, and did not converge to any particular value when Δa approached infinitesimal values (see the dashed line in Figure 12.11). As shown, the mode-mixity is insensitive to the crack extension length only for large element sizes. However, selecting this element size does not allow correct modeling of thin layers. The presented mode-mixity values are for a crack along Path III where a bi-material interface crack existed (crack between the second mat layer and the roving layer).

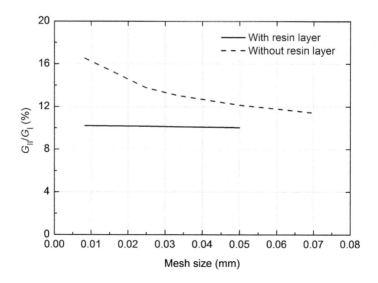

Figure 12.11 Mesh sensitivity analysis for crack along path III; $a = 110$ mm.

The approach proposed by Atkinson (1977) was applied in the present study. In this method, a thin layer, designated the resin interlayer, was inserted that had the average properties of the adjacent layers of the interface. The thickness of the resin interlayer was selected to be 0.1 mm, as a compromise resulting in almost no changes in the stiffness of the model ($<1\%$) and introducing a reasonable number of elements into the FE model.

A mesh sensitivity analysis was performed for the new FE model, including the resin interlayer. For the resin interlayer thickness of 0.10 mm, the mesh size was gradually varied from 0.0500 to 0.0083 mm, representing from two to 12 elements through the resin interlayer. As shown in Figure 12.11, the G_{II} component obtained for a crack along Path III with a resin interlayer is independent of the crack extension length. This configuration diminishes the sensitivity of the calculated mode-mixity to the crack extension length (see Figure 12.11).

12.4 Results and discussion of asymmetry and fiber-bridging effects

12.4.1 Effects of specimen asymmetry

Mode I and II fracture components were calculated for the three different crack paths using the VCCT method. A Mode II component was introduced in all cases (see Figure 12.12), which was negligible for Path I and Path II cracks (about 1% of Mode I) but became more significant (about 10%) for cracks propagating along Path III.

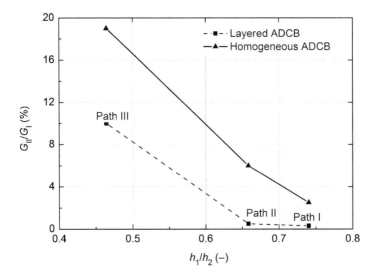

Figure 12.12 Mode II introduced owing to asymmetry along different paths in a layered ADCB joint and homogeneous ADCB.

Results obtained from a homogeneous model (all materials comprising the joint have the same properties) are presented in Figure 12.12. In the homogeneous model, the induced Mode II fracture clearly depended on the ratio between the thicknesses of the upper and lower arms. As the thickness ratio decreased, the induced Mode II fracture increased. This is attributed to the direct relationship between the bending stiffness ratio and the thickness ratio of the two arms:

$$\frac{(E_f I)_1}{(E_f I)_2} = \left(\frac{h_1}{h_2}\right)^3 \tag{12.11}$$

where h_1 and h_2 are the specimen arm thicknesses. However, the bending stiffness of the joint in the layered model results from the presence of fibers along the longitudinal direction. For Path I and II crack propagation, both arms of the specimen contain the same quantity of the second mat layer and roving layer and there is therefore no significant difference between the bending stiffness of the two arms. For Path III crack propagation, however, the upper arm contains far fewer longitudinal fibers than the lower and is therefore much less stiff. As a result of this asymmetry, a Mode II component as high as 10% of the corresponding Mode I component, independent of the crack length, can be introduced in this case.

12.4.2 Effect of fiber bridging

The results from the numerical models without considering the fiber bridging were significantly different from those obtained experimentally. In Figure 12.13 the

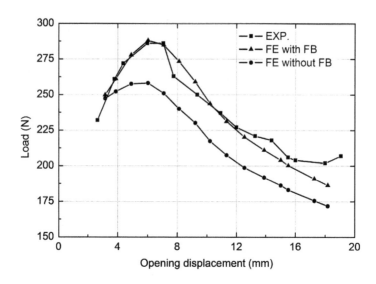

Figure 12.13 Load versus opening displacement from FE with and without fiber bridging (FB) and experiment, DCB-16.

computed load from the FE model for DCB-16 is compared with the experimental values, and it can be seen that the numerical values indicated by dots are lower than those derived experimentally.

This difference is due to the fiber-bridging effect that is not considered in the FE model. Similar conclusions can be drawn from observation of the compliance of the modeled joint (Figure 12.14).

As expected, the FE model without fiber bridging is less stiff than the actual joint. The G values calculated by VCCT for the FE model without considering the fiber-bridging effect are also compared with the experimentally derived G values in Figure 12.15. The R-curve obtained according to the FE results shows lower G values for crack propagation (the plateau of the R-curve) compared with those derived based on the experimental results.

Nevertheless, the numerical result is improved when the fracture energy owing to fiber bridging is also considered. The G derived from the experimental investigation is the sum of the energy released rate at the crack tip, G_{tip}, and the energy released rate due to fiber bridging, G_{br}, components, i.e., $G_{total} = G_{tip} + G_{br}$. In this chapter, the VCCT was used to calculate G_{tip}.

Figure 12.15 presents the G values regarding the crack tip and the fiber bridging versus the crack length for the DCB-16. The summation of these two values is also shown as G_{total}-FE with F.B., which is in good agreement with the experimentally derived values.

The R-curves for the representative specimens for Path I and Path III are shown in Figures 12.16 and 12.17, respectively.

According to the presented R-curves, a specific G value for G_{tip} and G_{br} can be assigned to each specimen. For G_{tip}, a value of around 200 J/m² is assigned to

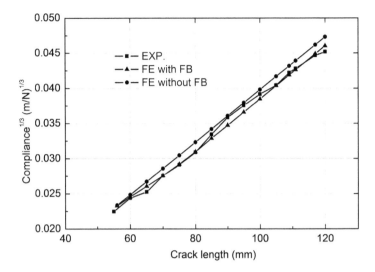

Figure 12.14 Compliance versus crack length from FE with and without fiber bridging (FB) and experiment, DCB-16.

Path I and a value of around 400 J/m² to Paths II and III. In Path I, where the crack propagated between the adherend and the adhesive, lower values for G_{tip} resulted than those obtained in Paths II and III, where the crack propagated between the layers of the adherend. The G_{br} exhibits higher values: around 300 J/m² for Path I, 600 J/m² for Path II, and 500 J/m² for Path III.

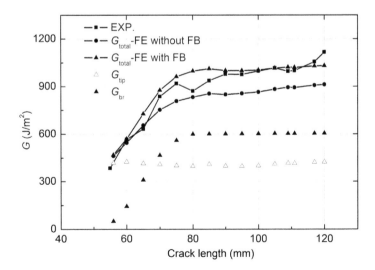

Figure 12.15 Separation of G_{total} into G_{tip} and G_{br}, DCB-16 (Path II).

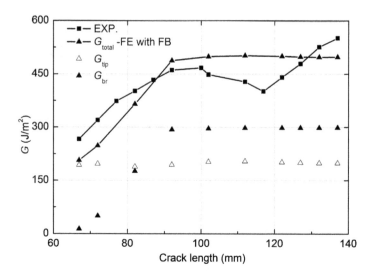

Figure 12.16 Separation of G_{total} into G_{tip} and G_{br}, DCB-04 (Path I).

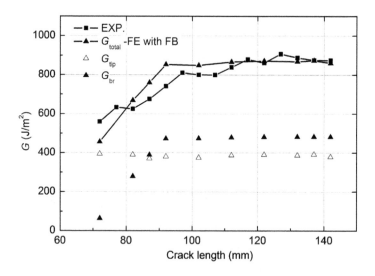

Figure 12.17 Separation of G_{total} into G_{tip} and G_{br}, DCB-13 (Path III).

12.5 Conclusions

The numerical model can be developed to investigate the effects of asymmetry and fiber bridging on the strain energy release rates of pultruded GFRP DCB joints. Zero-thickness cohesive elements can be used to model the fiber-bridging zone.

This chapter shows that an exponential traction—separation description of the CZM is able to model the fiber bridging and its effect on the G. Therefore, separation of the fracture parameters, G_{tip} and G_{br}, can be successfully performed by the cohesive zone FE model.

The results showed that the mode-mixity is a function of the crack extension length when the crack propagates in a bi-material interface. Introduction of a resin interlayer with the average properties of the adjacent layers of the interface solved this problem.

Although the CZM developed requires experimental data for calibration of the model, it can subsequently be used to simulate progressive crack propagation in another joint comprising the same adherends and adhesive and exhibiting the same failure modes. Progressive damage modeling is a failure analysis technique that is widely used to predict the fracture behavior and strength of bonded joints based on the evolution of the damage state. FE analyses usually assist numerical simulations of this type. According to this approach, the static behavior of bonded joints can be predicted. To apply the cohesive model approach, the crack path must be known in advance to place the cohesive elements in the model. A nonlinear solution and relatively fine mesh size are required for the cohesive elements to obtain an accurate simulation. Moreover, if undergoing certain modifications, e.g., adoption of a progressive degradation rule for the CZM parameters, the same procedure can be also used to model crack propagation behavior under fatigue loading.

References

Agrawal, A., & Karlsson, A. M. (2006). Obtaining mode mixity for a bimaterial interface crack using the virtual crack closure technique. *International Journal of Fracture, 141*(1—2), 75—98.
Ashcroft, I. A., Hughes, D. J., & Shaw, S. J. (2001). Mode I fracture of epoxy bonded composite joints: part 1. Quasi-static loading. *International Journal of Adhesion and Adhesives, 21*(2), 87—99.
ASTM D3171. (2001). Standard test methods for constituent content of composite materials. In *Annual book of ASTM standards: Adhesive section 15.03.*
ASTM D5573. (1999). Standard practice for classifying failure modes in fiber-reinforced-plastic (FRP) joints. In *Annual book of ASTM standards: Adhesive section 15.06.*
ASTM D3433-99. (2005). Standard test method for fracture strength in cleavage of adhesives in bonded metal joints. In *Annual book of ASTM standards: Adhesive section 15.06.*
ASTM D5528-01. (2007). Standard test method for mode I interlaminar fracture toughness of unidirectional fiber-reinforced polymer matrix composites. In *Annual book of ASTM standards: Adhesive section 15.03.*
Atkinson, C. (1977). On stress singularities and interfaces in linear elastic fracture mechanics. *International Journal of Fracture, 13*(6), 807—820.
Bao, G., & Suo, Z. (1992). Remarks on crack-bridging concepts. *Applied Mechanics Reviews, 45*(8), 355—366.
Bennati, S., Colleluori, M., Corigliano, D., & Valvo, P. S. (2009). An enhanced beam-theory model of the asymmetric double cantilever beam (ADCB) test for composite laminates. *Composites Science and Technology, 69*(11—12), 1735—1745.

Beuth, J. L. (1996). Separation of crack extension modes in orthotropic delamination models. *International Journal of Fracture*, 77(4), 305−321.
Blackman, B. R. K., Hadavinia, H., Kinloch, A. J., & Williams, J. G. (2003). The use of a cohesive zone model to study the fracture of fibre composites and adhesively-bonded joints. *International Journal of Fracture*, 119(1), 25−46.
Brunner, A. J. (2000). Experimental aspects of mode I and mode II fracture toughness testing of fiber-reinforced polymer−matrix composites. *Computer Methods in Applied Mechanics and Engineering*, 185(2−4), 161−172.
Chandra, N., Li, H., Shet, C., & Ghonem, H. (2002). Some issues in the application of cohesive zone models for metal−ceramic interfaces. *International Journal of Solids and Structures*, 39(10), 2827−2855.
Choi, N. S., Kinloch, A. J., & Willams, J. G. (1999). Delamination fracture of multidirectional carbon-fiber/epoxy composites under mode I, mode II and mixed mode I/II loading. *Journal of Composite Materials*, 31(1), 73−100.
Dattaguru, B., Venkatesha, K. S., Ramamurthy, T. S., & Buchholz, F. G. (1994). Finite element estimates of strain energy release rate components at the tip of an interface crack under mode I loading. *Engineering Fracture Mechanics*, 39(3), 451−463.
De Castro, J., & Keller, T. (2008). Ductile double-lap joints from brittle GFRP laminates and ductile adhesives, part I: experimental investigation. *Composites, Part B: Engineering*, 39(2), 271−281.
De Morais, A. B., & Pereira, A. B. (2006). Mixed mode I + II interlaminar fracture of glass/epoxy multidirectional laminates − part 1: analysis. *Composites Science and Technology*, 66(13), 1889−1895.
Ducept, F., Gamby, D., & Davies, P. (1999). A mixed-mode failure criterion derived from tests on symmetric and asymmetric specimens. *Composites Science and Technology*, 59(4), 609−619.
Hojo, M., Kageyama, K., & Tanaka, K. (1995). Pre-standardization study on mode I interlaminar fracture toughness test for CFRP in Japan. *Composites*, 26(4), 243−255.
Hutchinson, J. W., & Suo, Z. (1992). Mixed mode cracking in layered materials. *Advances in Applied Mechanics*, 29(C), 64−122.
Krueger, R. (2004). The virtual crack closure technique: history, approach and applications. *Applied Mechanics Reviews*, 57(2), 109−143.
Mollón, V., Bonhomme, J., Viña, J., & Argüelles, A. (2010a). Mixed mode fracture toughness: an empirical formulation for G_I/G_{II} determination in asymmetric DCB specimens. *Engineering Structures*, 32(11), 3699−3703.
Mollón, V., Bonhomme, J., Viña, J., & Argüelles, A. (2010b). Theoretical and experimental analysis of carbon epoxy asymmetric DCB specimens to characterize mixed mode fracture toughness. *Polymer Testing*, 29(6), 766−770.
Raju, I. S., Crews, J. H., & Aminpour, M. A. (1988). Convergence of strain energy release rate components for edge delaminated composite materials. *Engineering Fracture Mechanics*, 30(3), 383−396.
Rybicki, E. F., & Kanninen, M. F. (1977). A finite element calculation of stress intensity factors by a modified crack closure integral. *Engineering Fracture Mechanics*, 9(4), 931−938.
Shahverdi, M., Vassilopoulos, A. P., & Keller, T. (2011). A phenomenological analysis of mode I fracture of adhesively-bonded pultruded GFRP joints. *Engineering Fracture Mechanics*, 78(10), 2161−2173.
Shet, C., & Chandra, N. (2002). Analysis of energy balance when using cohesive zone models to simulate fracture processes. *Trans ASME: Journal of Engineering Materials and Technology*, 124(4), 440−450.

Sørensen, B., & Jacobsen, T. (1998). Large-scale bridging in composites: R-curves and bridging laws. *Composites, Part A: Applied Science and Manufacturing, 29*(11), 1443−1451.

Sorensen, L., Botsis, J., Gmür, Th, & Humbert, L. (2008). Bridging tractions in mode I delamination: measurements and simulations. *Composites Science and Technology, 68*(12), 2350−2358.

Spearing, S., & Evans, A. (1992). The role of fiber bridging in the delamination resistance of fiber-reinforced composites. *Acta Metallurgica et Materialia, 40*(9), 2191−2199.

Sun, C. T., & Qian, W. (1997). The use of finite extension strain energy release rates in fracture of interfacial cracks. *International Journal of Solids and Structures, 34*(20), 2595−2609.

Sundararaman, V., & Davidson, B. D. (1997). An unsymmetric double cantilever beam test for interfacial fracture toughness determination. *International Journal of Solids and Structures, 34*(7), 799−817.

Williams, J. G. (1988). On the calculation of energy release rates for cracked laminates. *International Journal of Fracture, 36*(2), 101−119.

Xiao, F., Hui, C. Y., & Kramer, E. J. (1993). Analysis of a mixed mode fracture specimen: the asymmetric double cantilever beam. *Journal of Materials Science, 28*(20), 5620−5629.

Xie, D., & Biggers, S. B. (2006). Progressive crack growth analysis using interface element based on the virtual crack closure technique. *Finite Elements in Analysis and Design, 42*(11), 977−984.

Xu, X. P., & Needleman, A. (1994). Numerical simulations of fast crack growth in brittle solids. *Journal of the Mechanics and Physics of Solids, 42*(9), 1397−1434.

Zhang, Y., Vassilopoulos, A. P., & Keller, T. (2010). Mode I and II fracture behavior of adhesively-bonded pultruded GFRP joints. *Engineering Fracture Mechanics, 77*(1), 128−143.

Simulating the mixed-mode fatigue delamination/debonding in adhesively-bonded composite joints

A. Pirondi[1], G. Giuliese[1], F. Moroni[1], A. Bernasconi[2], A. Jamil[2]
[1] Università di Parma, Parma, Italy; [2] Politecnico di Milano, Milan, Italy

13.1 Introduction to the simulation of fatigue delamination/debonding

Composite materials and structural adhesively-bonded joints were first applied in the aerospace industry, but thanks to continuous improvements in performance and reduced costs, many more industry fields are approaching the use of this type of materials and structural components. The extensive use of composites requires more sophisticated capability to simulate and predict their mechanical behavior. For this purpose, analytical methods are being progressively integrated or replaced by the finite element method. In engineering applications, it is well established that fatigue is the root cause of many structural failures. In the case of composite laminates, fatigue life is related to the initiation and propagation of delamination defects started at free edges, holes, and joining regions. Especially in the case of damage-tolerant or fail-safe design, it is necessary to know how cracks, or defects in general, propagate during the service life of a component. A relationship between the applied stress intensity factor and the fatigue crack growth (FCG) rate of a defect is generally expressed as a power law (Paris & Erdogan, 1961). In the case of polymers, adhesives, and composites, the relationship is traditionally written as a function of the range of SERR (ΔG) as

$$\frac{da}{dN} = B \Delta G^d \tag{13.1}$$

where B and d are the parameters depending on the material and load mixity ratio and a is the defect length. In this simple form, the presence of a FCG threshold and an upper limit to ΔG for fracture are not represented, although when they are needed, expressions accounting for these limits are easily found (see, for example, Curley, Hadavinia, Kinloch, & Taylor, 2000). In the same way, the influence of the stress ratio, R, on the FCG rate can be introduced into Eqn (13.1) by a term derived from extensions of the Paris law expressed in terms of the range of stress intensity factor, ΔK (Forman, Kearnay, & Engle, 1967).

When the SERR can be analytically defined and updated as the crack length increases, the framework is simple. Primarily, a critical crack length (a_f) criterion has to be defined based on stress or strain (ductile adhesives) or fracture toughness (brittle adhesives) (Pirondi & Moroni, 2009). Then, the procedure for the prediction becomes a simple numerical integration between the initial crack length (a_0) and the final crack length (a_f) of the inverse of the crack growth rate:

$$N_f = \int_{a_0}^{a_f} \frac{1}{da/dN} da \tag{13.2}$$

The procedure becomes more complicated when the SERR cannot be computed simply by using an analytical relationship. In real applications the SERR can only be computed numerically by using, for example, finite element (FE) simulations. The prediction of crack growth can be carried out by a stepwise analysis, each step of which corresponds to a user-defined crack growth increment, which may require a large amount of time. Hence, the number of cycles can be obtained by manually integrating the crack growth rate computed from the Paris law.

13.1.1 Virtual crack closure technique (VCCT)

To speed up the process described previously, in some FE software, this procedure is integrated in special features (for example the *Debonding procedure in ABAQUS®), in which the SERR is obtained using the contour integral or the virtual crack closure technique (VCCT).

Based on linear elastic fracture mechanics, VCCT is a widely used technique for the evaluation of SERR and mode-mixity for cracks in homogeneous materials. It is based on the equality between the strain energy released when a crack is extended by a certain amount δa and the work done by crack tip nodal forces to virtually close it with the same amount of δa. In addition, self-similar crack growth is assumed. With the help of this assumption, the same model is used for the extraction of reaction forces and displacements required to close the crack by δa, and thus the two-step crack closure method reduces to a one-step virtual crack closure method (Figure 13.1).

To calculate the SERR rates using two-dimensional (2D) FE models under either plane stress or plane strain conditions, an advancing crack is considered with an initial crack front at point l; point l splits into two points, l_1 and l_2, forming a new crack front at point i, as seen in Figure 13.2. If u and u' are the displacements in the local x-direction and v and v' are the displacements in the y-direction of points l_1 and l_2, respectively, SERR G based on VCCT may evaluated as:

$$G_I = \frac{1}{2t\delta a} F_y(v - v')$$
$$G_{II} = \frac{1}{2t\delta a} F_x(u - u') \tag{13.3}$$

The total energy release rate is

$$G = G_I + G_{II} \quad (13.4)$$

where,

t = element thickness
δa = element length
F_x = force per unit length on node i in x-direction
F_y = force per unit length on node i in y-direction
δu = difference of displacements between nodes l_1 and l_2
δv = difference of displacements between nodes l_1 and l_2

Initially proposed by Rybicki and Kanninen (1977) for four-noded elements and extended to higher-order elements by Raju (1987) and 3D cracked bodies by Shivakumar, Tan, and Newman (1988), in recent years VCCT has been successfully implemented in commercial FE codes, both in 2D and 3D, and has emerged as a promising tool in energy release rate calculations in Mode I, Mode II, and in Mixed-Mode fracture problems.

Among the early works, Sun and Jih (1987) investigated the stress fields near the crack tip at a bi-material using the VCCT and described the oscillatory behavior of the SERR. Whitcomb (1992) was one of the first to introduce the use of the VCCT to determine SERR distributions for circular delamination. Hutchinson and Suo (1991) proposed the use of G_{TOT} for delamination growth under mixed-mode conditions.

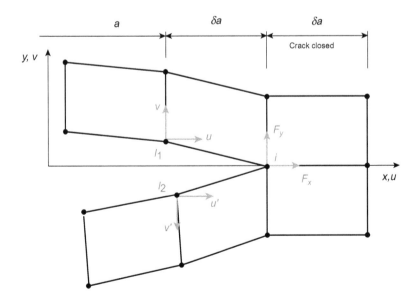

Figure 13.1 Modified crack closure method with a single step (one-step VCCT).

Harbert and Hogan (1992, pp. 107–112) described a methodology for modeling delamination growth using VCCT in composite notched tensile bars, Salpekar, O'Brien, and Shivakumar (1996) showed the independence of the method from orthogonal meshes in modeling delaminations of graphite/epoxy laminates, and Fawaz (1998) applied 3D VCCT with non-orthogonal meshes on elliptical crack front in riveted lap-splice joints of transport aircraft fuselages. Beuth and Narayan (1997, pp. 324–342) demonstrated techniques for the minimization of oscillator behavior of near-tip stresses while implementing VCCT, whereas the importance of the fiber orientation of the plies in the delamination part was demonstrated by Shen, Lee, and Tay (2001). In the same year, O'Brien (2001) presented the state of art for characterizing, analyzing, and predicting delamination growth in composite materials and structures using VCCT, and demonstrated fatigue life prediction in composite rotor hub flexbeams and stiffener pull-off behavior in skin-stiffener reinforced composites. Zou, Reid, Soden, and Li (2001) used transverse shear deformable laminate theory along with VCCT with the aim of avoiding oscillatory singular stresses around the delamination tip with the use of both G_{TOT} and individual components of SERR.

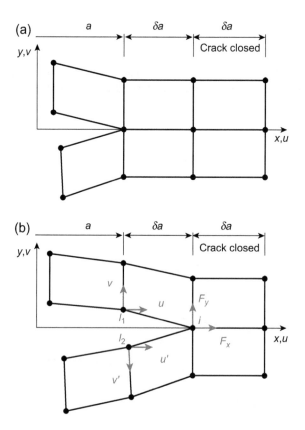

Figure 13.2 Extension of the crack from point l to point i: (a) before and (b) after extension.

Cheuk, Tong, Wang, Baker, and Chalkley (2002) used VCCT in modeling fatigue crack growing along the first ply of the composite in metal-to-composite bonded double-lap joints and proposed an equivalent strain—energy release rate ΔG_{eq}, accounting for the effect of mode ratio on FCG rates.

A comprehensive review of VCCT formulas for different element types was given by Krueger (2004), and Okada, Higashi, Kikuchi, Fukui, and Kumazawa (2005) proposed corrections for skewed and non-symmetric mesh arrangements at the crack front.

Murri and Schaff (2006) developed a VCCT-based 2D FE model of flexbeam geometry that formed a full-size composite helicopter rotor hub using both ANSYS® and ABAQUS®. Xie, Waas, Shahwan, Schroeder, and Boeman (2004) proposed the use of VCCT for the evaluation of SERR using user element subroutines UEL in commercial FEA code ABAQUS® for kinking cracks, and in Xie et al. (2005) for failure analysis of adhesively-bonded structures. Xie and Biggers (2006) introduced an interface element tailored for VCCT ensuring seamless integration using user element subroutines UEL in commercial FEA code ABAQUS®.

Leski (2007) presented the general conditions for applying VCCT in conjunction with commercial programs. Marannano, Mistretta, Cirello, and Pasta (2008) implemented 2D VCCT by user-defined subroutines for the characterization of SERR at adhesive-adherent interface in bonded joints under mixed-Mode I/II, by considering a mixed-mode end-loaded split (MMELS) specimen in FCG analysis. Krueger (2010) presented a benchmark example for cyclic delamination growth prediction for the commercial code of ABAQUS® based on the VCCT in modeling a double cantilever beam (DCB) specimen and described procedures for modeling the delamination onset and growth under cyclic loading along with the effects of different input parameters.

Pietropaoli and Riccio (2010) introduced a *SMART-TIME* and *SMART-CORNER* fail release approach able to cope with the problems of robustness resulting from mesh and load step size dependency of VCCT and demonstrated the effectiveness of the technique by comparing the results with experimental data. The same authors in 2011 (Pietropaoli & Riccio, 2011), presented a front-tracing algorithm and suitable expressions for the evaluation of SERR using VCCT when dealing with non-smoothed delamination fronts.

Liu et al. (2011) made a comparatively study using ABAQUS® on typical failure criteria for predicting crack propagation along with the effects of different mesh sizes and preexisting crack length on the delamination growth and postbuckling properties of composite flat laminates. Chang, Shi, and Cheng (2012) developed a post-processing user-defined subroutine *UEXTERNALDB* by integrating XFEM and VCCT in ABAQUS® to simulate crack propagation and predict the effect of reinforcing particles on the crack propagation behavior of an $Al_2O_3/Al6061$ particle-reinforced metal matrix.

13.1.2 Cohesive zone model

An alternative way to deal with FCG problems is to use a cohesive zone model (CZM). This model is largely used to simulate quasi-static fracture problems, especially in the case of interface cracks such as delamination in composites (Blackman, Hadavinia,

Kinloch, & Williams, 2003; Hutchinson & Evans, 2000; Li, Thouless, Waas, Schroeder, & Zavattieri, 2005 among others). The possibility of simulating the growth of a defect without remeshing requirements and the relatively easy possibility of manipulating the constitutive law of the cohesive elements make the CZM attractive for FCG simulation. In fact, most work in which the CZM is used to simulate FCG deals with interfaces, in particular delamination in composite materials. The first approach analyzed is the one proposed by Maiti and Geubelle (2005), who defined a cohesive model for fatigue simulation in polymers in which the damage of the cohesive element is related both to monotonic quasi-static loading and the number of cycles. In particular, fatigue cycling affects the tensile stiffness, K_{22}, that is postulated to evolve as:

$$K_{22} = \frac{d\sigma_{22}}{d\delta_{22}} = -\gamma(N_f)\sigma_{22} = -\frac{N_f^{-\beta}}{\alpha}\sigma_{22} \qquad (13.5)$$

where N_f represents the number of cycles to damage initiation in the cohesive element, and β and α are the two parameters that can be calibrated by comparing FE modeling and FCG experiments.

Yang, Thouless, and Ward (1999) worked on quasi-brittle materials and developed a CZM based on the boundary element method in which the material is damaged by reducing the stiffness in both loading and unloading paths. This allows simulation of fatigue growth with no imposition of another law of growth within the cohesive model. The evolution law of stiffness in particular is defined in the form of a polynomial expansion and its parameter can be experimentally obtained by measuring the traction—crack displacement jump during cyclic loading.

Concerning interfaces, Roe and Siegmund (2003) introduced cyclic degradation of monotonic cohesive strength based on a damage variable, D, representing the ratio between the effective (damaged) and nominal (undamaged) cross-section of a representative interface element. At the same time, the damage variable D relates the cohesive zone traction vector (T_{CZ}) with the effective cohesive zone traction vector $\tilde{T}_{CZ}$, by the equation

$$\tilde{T}_{CZ} = \frac{T_{CZ}}{1-D} \qquad (13.6)$$

The cyclic damage evolution law is

$$\dot{D} = \frac{|\Delta \dot{\bar{u}}|}{\delta_\Sigma}\left[\frac{\overline{T}}{\sigma_{max}} - \frac{\sigma_f}{\sigma_{max,0}}\right] \qquad (13.7)$$

where $\Delta \dot{\bar{u}}$ is the mixed-mode equivalent displacement jump between crack surfaces, $\overline{T}$ is the equivalent traction, $\sigma_{max} = \sigma_{max,0}(1-D)$ is the maximum stress of the damaged cohesive law, δ_Σ is the accumulated cohesive length, σ_f is the cohesive zone endurance limit, and $\sigma_{max,0}$ is the maximum stress of the cohesive law before damage. In this formulation, the two parameters, δ_Σ and σ_f, have to be calibrated by FCG experiments.

An approach similar to that of Roe and Siegmund (2003) was developed in Muñoz, Galvanetto, and Robinson (2006) in which the robustness of the model in predicting the crack growth rate was demonstrated, with an upper bound for the cohesive element length and number of cycles per increment to preserve accuracy.

A different approach was proposed by Turon, Costa, Camanho, and Dávila (2007). In this model, calibration of a cohesive parameter for cyclic loading is not required. In fact, a damage homogenization criterion is used to relate the experimental FCG rate, represented by Eqn (13.1), with the damage evolution of the cohesive elements. In this way a cycle-by-cycle FE analysis is not necessary for integration of the damage rate, which means significant computational time savings. However, only simple geometries in which the SERR is not dependent on the crack length were treated.

In the work of Khoramishad, Crocombe, Katnam, and Ashcroft (2010), Khoramishad, Crocombe, Katnam, and Ashcroft (2011), the damage (D) evolution with respect to the number of cycles is expressed in terms of strain (or crack opening) by the equation

$$\frac{\Delta D}{\Delta N} = \begin{cases} \alpha(\varepsilon_{max} - \varepsilon_{th})^\beta & \varepsilon_{max} > \varepsilon_{th} \\ 0 & \varepsilon_{max} \leq \varepsilon_{th} \end{cases} \quad (13.8)$$

where ε_{max} is the maximum principal strain in the cohesive element (therefore a combination of the normal and shear components of strain), ε_{th} is the threshold strain (value of strain below which no damage occurs), and α and β are the material constants. The set of parameters, ε_{th}, α, and β, has to be calibrated by comparison with experimental tests. The fatigue degradation does not affect the stiffness of the cohesive element, but rather the maximum stress, as shown in Figure 13.3.

Naghipour, Bartsch, and Voggenreiter (2011) (see also Chapter 15 of this volume) revisited the model of Turon et al. (2007), improving the cohesive zone area definition under mixed-Mode I/II loading and integration scheme of the cohesive law in the user-defined element (UEL) developed in FE analysis software ABAQUS®. This work

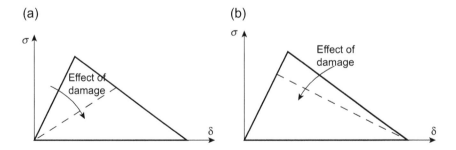

Figure 13.3 Example of different effects of damage: (a) damage affects the stiffness (Turon et al., 2007); (b) damage affects the maximum stress (Khoramishad et al., 2010, 2011).

yielded better agreement between the FCG rate (B and d parameters in Eqn (13.1)) input to the analysis and the FCG rates in output with respect to the work of Turon et al. (2007).

In Roe and Siegmund (2003) and Muñoz et al. (2006), damage evolution is simulated on a cycle-by-cycle basis, whereas the schemes proposed in Maiti and Geubelle (2005), Turon et al. (2007), Naghipour et al. (2011), Khoramishad et al. (2010, 2011) as in the scheme described in the current chapter, work incrementally on cycles only and are therefore much less expensive from a computational point of view. On the other hand, a cycle-by-cycle simulation allows the definition of a more complex damage evolution, and this may help predict experimental features such as crack growth retardation after an overload (Roe & Siegmund, 2003).

The model presented here was developed by some of the authors (Moroni & Pirondi, 2012a) starting from the framework proposed by Turon et al. (2007). The main differences with respect to that work are that damage D is related directly to its effect on stiffness and not to the ratio between the energy dissipated during the damage process and the cohesive energy and then, in turn, to the stiffness; and the process zone size A_{CZ} is defined as the sum of A_e of the cohesive elements for which the difference in opening between the maximum and minimum load of the fatigue cycle, $\Delta\delta = \delta_{max} - \delta_{min}$, is higher than a threshold value $\Delta\delta^{th}$; therefore, it is evaluated by FE analysis during the simulation and is not derived from a theoretical model. Moreover, the SERR is calculated using the contour integral method over the cohesive process zone and the model is implemented as a user-defined field subroutine (USDFLD) in ABAQUS acting on standard cohesive elements, instead of a user element. In Moroni and Pirondi (2012a), it was demonstrated that the FCG rates coming from the simulation with this model are as expected from the values of B and d (Eqn (13.1)) given in input.

13.2 Cohesive zone and virtual crack closure technique (VCCT) model formulation

In this section, the fundamental concepts used to formulate the models used in this chapter are given.

13.2.1 Cohesive zone model

The model developed by some of the authors (Moroni & Pirondi, 2012a) and reported in the following was devoted to fatigue debonding of composite adhesive joints, but it can be readily applied to fatigue delamination. In that model, the concept proposed in Turon et al. (2007) is retained, whereas a different relationship between damage and stiffness has been proposed and the size of the process zone (i.e., the zone where the damage process will take place, A_{CZ}) has been defined and evaluated in a different way.

Although different and complicated shapes of the cohesive law are proposed in the literature, the triangular one (Figure 13.4) is often good enough to describe crack growth behavior, and it was demonstrated that this kind of law is appropriate for untoughened adhesives.

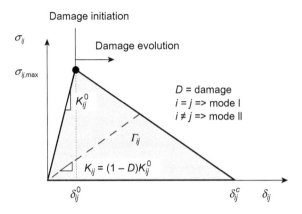

Figure 13.4 Example of a triangular cohesive law.

Considering a representative surface element (represented in the simulation by a cohesive element section, Figure 13.5) with a nominal surface equal to A_e, the accumulated damage can be related to the damaged area owing to micro voids or crack (A_d) according to Lemaitre (1985):

$$D = \frac{A_d}{A_e} \qquad (13.9)$$

In Turon et al. (2007), D is related to the ratio between the energy dissipated during the damage process and the cohesive energy (Γ_1 in Figure 13.4) and then in turn to the stiffness. In this work, instead, D acts directly on stiffness, as in Lemaitre (1985). Referring to a Mode I loading case, when the opening is relatively small, the cohesive element behaves linearly; this happens until a given value of displacement, $\delta_{22,0}$ (or, equivalently, until a certain value of stress $\sigma_{22,0}$). This initial step is characterized by stiffness $K_{22,0}$, which remains constant until $\delta_{22,0}$. Beyond this limit the stiffness is progressively reduced by D, until the final fracture in $\delta_{22,C}$, where the two surfaces are completely separated. Between $\delta_{22,0}$ and $\delta_{22,C}$, stiffness K_{22} can be computed as

$$K_{22} = K_{22,0}(1 - D) \qquad (13.10)$$

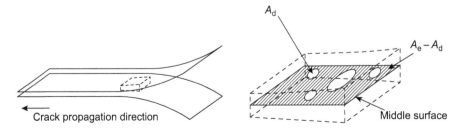

Figure 13.5 Nominal and damaged area in a representative surface element (RSE).

The area Γ_1 underling the cohesive law is the energy to make the defect grow to an area equal to the element cross-section and it is therefore representative of the fracture toughness, G_{IC}.

$$\Gamma_1 = \int_0^{\delta_C} \sigma_{22} d\delta_{22} \tag{13.11}$$

In the monotonic case, the damage variable D can be written as a function of the opening (δ_{22}) and of the damage initiation and critical opening (respectively, $\delta_{22,0}$ and $\delta_{22,C}$):

$$D = \frac{\delta_{22,C}(\delta_{22} - \delta_{22,0})}{\delta_{22}(\delta_{22,C} - \delta_{22,0})} \tag{13.12}$$

When the element is unloaded, the damage cannot be healed; therefore, looking at Figure 13.4, the unloading and subsequent loadings will follow the dashed line until further damage is attained. This simple model is able to describe monotonic damage in the case of Mode I loading.

Considering the entire cohesive layer, the crack extension (A) can be computed as the sum of damaged areas of all the cohesive elements (A_d) (Turon et al., 2007)

$$A = \sum A_d \tag{13.13}$$

When the fatigue damage is considered from the previous equation, crack growth (dA) can be written as a function of the increment of the damage area of all cohesive elements (dA_d); therefore:

$$dA = \sum dA_d \tag{13.14}$$

However, the damage increment would not concern the whole cohesive layer but would be concentrated in a relatively small process zone close to the crack tip. To estimate the size of A_{CZ}, analytical relationships can be found in the literature (Harper & Hallett, 2008), where the size per unit thickness is defined as the distance from the crack tip to the point where $\sigma_{22,0}$ is attained. In this work, different definition and evaluation method are proposed: A_{CZ} corresponds to the sum of the nominal sections of the cohesive elements where the difference in opening between the maximum and minimum load of the fatigue cycle, $\Delta\delta_{22} = \delta_{22,max} - \delta_{22,min}$, is higher than a threshold value $\Delta\delta_{22}^{th}$. The value $\Delta\delta_{22}^{th}$ is supposed to be the highest value of $\Delta\delta_{22}$ in the cohesive layer when ΔG in the simulation equals ΔG_{th} experimentally obtained by FCG tests. In this way FCG may take place even at $\delta_{22,max} \leq \delta_{22,0}$, which is a condition that should be accounted for because $\delta_{22,0}$ results from calibration of the cohesive zone on fracture

Simulating the mixed-mode fatigue delamination/debonding

tests and may not be representative of a threshold for FCG. The process zone size A_{CZ} therefore has to be evaluated by FE analysis while performing the FCG simulation but it does not need to be assumed from a theoretical model.

Eqn (13.14) can be therefore rewritten as (Turon et al., 2007)

$$dA = \sum_{i \in A_{CZ}} dA_d^i \tag{13.15}$$

where only the elements lying in the process zone (named A_{CZ}) are considered.

To represent crack growth due to fatigue (dA/dN), local damage of the cohesive elements (D) has to be related to the number of cycles (N). This is done using the equation

$$\frac{dD}{dN} = \frac{dD}{dA_d} \frac{dA_d}{dN} \tag{13.16}$$

The first part of Eqn (13.16) can be easily obtained, deriving Eqn (13.9); therefore,

$$\frac{dD}{dA_d} = \frac{1}{A_e} \tag{13.17}$$

The process to obtain the second part is more complicated. The derivative of Eqn (13.15) with respect to the number of cycles is

$$\frac{dA}{dN} = \sum_{i \in A_{CZ}} \frac{dA_d^i}{dN} \tag{13.18}$$

At this point, an assumption is introduced: The increment of damage per cycle is supposed to be the same for all elements lying in the process zone. Therefore, value dA_d/dN is assumed to be the average value of the damaged area growth rate dA_d^i/dN for all elements in the process zone.

Hence, the crack growth rate can be rewritten as (Turon et al., 2007):

$$\frac{dA}{dN} = \sum_{i \in A_{CZ}} \frac{dA_d}{dN} = n_{CZ} \frac{dA_d}{dN} \tag{13.19}$$

where n_{CZ} is the number of elements lying on the process area A_{CZ}. n_{CZ} can be written as the ratio between the process zone extension (A_{CZ}) and the nominal cross-section area (A_e) leading to the equation

$$\frac{dA}{dN} = \frac{A_{CZ}}{A_e} \frac{dA_d}{dN} \tag{13.20}$$

The second part of Eqn (13.16) can be therefore written as:

$$\frac{dA_d}{dN} = \frac{dA}{dN}\frac{A_e}{A_{CZ}} \quad (13.21)$$

The crack growth rate can be finally expressed as a function of the applied SERR, in the simplest version using Eqn (13.1)

$$\frac{dD}{dN} = \frac{1}{A_{CZ}}B\Delta G^d \quad (13.22)$$

13.2.1.1 Strain energy release rate computation

In the previous section, a relationship between the applied SERR and the increase in damage in the cohesive zone was defined. To simulate FCG a general method is therefore required to calculate the value of the SERR as a function of crack length. The most common methods for SERR evaluation using the FE method are the contour integral (J) and the VCCT. These two methods are usually available in FE software but in general VCCT is intended to be an alternative to using cohesive elements, and the software used in this work (ABAQUS®) did not output the contour integral for an integration path including a cohesive element.

To compute the J-integral, a path surrounding the crack has to be selected. Considering, for example, the crack in Figure 13.6, the path (Ω) is displayed by a dashed line and is represented by all top and bottom nodes of the cohesive elements.

The J-integral definition (Rice, 1968) is

$$J = \int_\Omega \mathbf{n}[H]\mathbf{q}\, d\Omega \quad (13.23)$$

where $\mathbf{n}$ is a vector normal to the path, $\mathbf{q}$ is a vector lying on the crack propagation direction, and $[H]$ is defined as

$$[H] = W[I] - [\sigma_{ij}]\left[\frac{\partial u_{ij}}{\partial x_{ij}}\right] \quad (13.24)$$

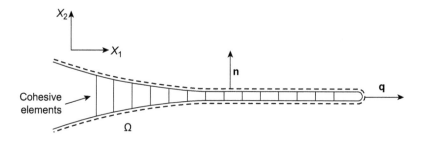

Figure 13.6 Example of J-integral surrounding the cohesive element layer.

Simulating the mixed-mode fatigue delamination/debonding

where W is the strain energy density, $[\sigma_{ij}]$ is the stress matrix, and u_i is the displacements of points lying on the path.

Neglecting geometrical nonlinearity, vector $\mathbf{q}$ can be assumed to be perpendicular to the direction x_2 along the whole path; therefore, the J-integral can be rewritten as:

$$J = \int_\Omega \left(-\sigma_{12} \frac{\partial u_1}{\partial x_1} - \sigma_{22} \frac{\partial u_2}{\partial x_1} \right) d\Gamma \qquad (13.25)$$

Extracting the opening/sliding and the stresses in the cohesive elements at the beginning of the increment, the SERR is then computed. An interesting feature of this approach is that the Mode I and Mode II components of the J-integral can be obtained by integrating separately the second or the first components of the integral in Eqn (13.25), respectively.

This method can be easily implemented for a 2D problem because there is only one possible path. In the case of a 3D problem implementation is more difficult because several paths can be identified along the crack width, and their definition is troublesome, especially when dealing with irregular meshes. A 3D version was implemented by Moroni and Pirondi (2012b) in the case of planar crack geometries and regular cohesive mesh. In this case, Eqn (13.25) is evaluated on several parallel contours to obtain the J-integral along the crack front.

13.2.1.2 Finite element implementation

The theoretical framework described in Section 13.1.2 and the SERR calculation procedure are implemented using suitable Fortran subroutines in the commercial software ABAQUS®. In particular the USDFLD ABAQUS® subroutine is used to modify cohesive element stiffness by means of a field variable that accounts for damage, whereas the URDFIL subroutine is used to obtain the result in terms of stresses, displacements, and energies. Fatigue analysis is carried out as a simple static analysis divided into a certain number of increments. Each increment corresponds to a given number of cycles.

Assuming that the fatigue cycle load varies from a maximum value F_{max} to a minimum value F_{min}, the analysis is carried out applying the maximum load F_{max} to the model. The load ratio is defined as the ratio between the minimum and maximum load applied:

$$R = \frac{F_{max}}{F_{min}} \qquad (13.26)$$

The SERR amplitude is therefore

$$\Delta G = \left(1 - R^2\right) G_{max} \qquad (13.27)$$

This latter is compared with the SERR threshold ΔG_{th}. If $\Delta G > \Delta G_{th}$ the analysis starts (or it continues if the increment is not the first); otherwise the analysis is stopped.

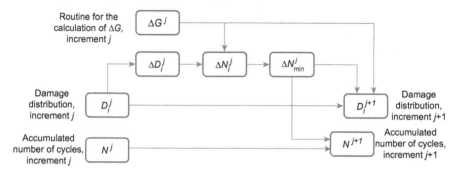

Figure 13.7 Flow diagram of the automatic procedure for crack growth rate prediction.

The flow diagram in Figure 13.7 shows the operations performed within each increment.

At the beginning of increment j the number of cycles (N^j) and the damage variable for each of the i elements (D_i^j) are known. Now for each element the maximum possible damage change within the increment (ΔD_i^j) is computed. If $\Delta D_{\max}$ is the maximum allowable variation in a single increment (it is a user-defined value and is used to ensure smooth crack growth) ΔD_i^j is calculated as follows:

$$\Delta D_i^j = \Delta D_{\max} \quad \text{if } 1 - D_i^j > \Delta D_{\max}$$
$$\Delta D_i^j = 1 - D_i^n \quad \text{if } 1 - D_i^j < \Delta D_{\max}$$
(13.28)

In other words, ΔD_i^j is the minimum between the $\Delta D_{\max}$ and the amount needed for D to reach unity. Therefore, for each element the amount of cycles ΔN_i^j to produce ΔD_i^j is calculated by integrating Eqn (13.22) using the ΔG evaluated at the beginning of the increment, as described in the previous paragraph. After that, the routine searches for the minimum value among the calculated ΔN_i^j within the cohesive zone. This value, $\Delta N_{\min}^j$, is assumed to be the number of cycles of the increment, ΔN^j. Finally, the number of cycle is updated (ΔN^{j+1}), and using Eqn (13.22) this time to calculate the ΔD_i^j corresponding to ΔN^j the new damage distribution (ΔD_i^{j+1}) is determined for all elements belonging to the process zone.

The procedure is fully automated, i.e., the simulation is performed in a unique run without stopping.

13.2.1.3 Mixed-mode loading

With the aim of extending the model to mixed-Mode I/II conditions, a mixed-mode cohesive law has to be defined. This is done according to the scheme shown in Figure 13.8 from knowledge of pure Mode I and pure Mode II cohesive laws (index 22 refers to opening or Mode I direction; index 12 refers to the sliding or Mode II direction).

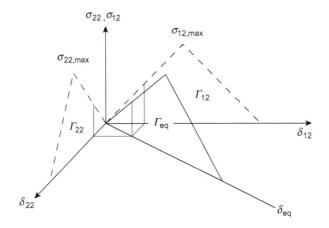

Figure 13.8 Example of cohesive law in the case of mixed-mode conditions.

First, the mixed-mode equivalent opening has to be defined. This is done using the relationship

$$\delta_{eq} = \sqrt{\left(\frac{\delta_{22} + |\delta_{22}|}{2}\right)^2 + (\delta_{12})^2} \qquad (13.29)$$

In the case of pure Mode I, this equation is given as δ_{eq}, the value of δ_{22} in the case of positive δ_{22}, whereas it is 0 in the case of negative δ_{22}. This is done because it is supposed that compression stresses do not lead to damage of the adhesive layer. Of course, δ_{22} assumes only positive values if the crack surface compenetration is properly prevented in the model.

Moreover, the mixed-mode cohesive law is defined in terms of the initial stiffness ($K_{eq,0}$), damage initiation equivalent opening ($\delta_{eq,0}$), and critical equivalent opening ($\delta_{eq,C}$).

The equivalent initial stiffness is obtained by equating the equivalent strain energy (U_{EQ}) to the total strain energy (U_{TOT}), which in turn is equal to the sum of the strain energy in Mode I (U_{22}) and Mode II (U_{12})

$$U_{EQ} = U_{TOT} = U_{22} + U_{12} = \frac{1}{2} \cdot \delta_{eq}^2 \cdot K_{eq}^0$$
$$= \frac{1}{2} \cdot (\delta_{22} + |\delta_{22}|)^2 \cdot K_{22}^0 + \frac{1}{2} \cdot \delta_{12}^2 \cdot K_{12}^0 \qquad (13.30)$$

$K_{22,0}$ and $K_{12,0}$ represent the initial stiffness of the Mode I and Mode II cohesive laws, respectively.

A further relationship is needed to define damage initiation. This is done using the quadratic failure criterion (Ungsuwarungsru & Knauss, 1987)

$$\left(\frac{\sigma_{22}}{\sigma_{22,\max}}\right)^2 + \left(\frac{\sigma_{12}}{\sigma_{12,\max}}\right)^2 = 1. \quad (13.31)$$

The last relationship needed regards the definition of the critical equivalent opening. Because the area underlying the cohesive law represents the critical SERR, using the Kenane and Benzeggagh (KB) theory (Kenane & Benzeggagh, 1996), the area underlying the mixed-mode equivalent cohesive law (Γ_{eq}) can be computed as

$$\Gamma_{eq} = \Gamma_{22} + (\Gamma_{12} - \Gamma_{22}) \, \mathrm{MM}^{m_m} \quad (13.32)$$

where (Γ_{22}) and (Γ_{12}) are the areas underling the Mode I and Mode II cohesive laws, respectively, m_m is a mixed-mode coefficient depending on the adhesive, and MM is the mixed-mode ratio defined as a function of the Mode I and Mode II SERRs as follows:

$$\mathrm{MM} = \frac{G_{II}}{G_I + G_{II}}. \quad (13.33)$$

The KB mixed-mode fatigue crack propagation model (Kenane & Benzeggagh, 1997) is the first considered, because it is the most general law that can be found in the literature. The FCG rate is given by Eqn (13.1) where this time B and d are the functions of the mixed-mode ratio MM:

$$d = d_I + (d_{II} - d_I) \, (\mathrm{MM})^{n_d} \quad (13.34)$$

$$\ln B = \ln B_{II} + (\ln B_I - \ln B_{II})(1 - \mathrm{MM})^{n_B} \quad (13.35)$$

where d_I, B_I, and d_{II}, B_{II} are, respectively, the parameters of the Paris law in Mode I and Mode II and n_d, n_B are the material parameters. Other approaches from the literature were implemented in Moroni and Pirondi (2012a) but they are not considered here for the sake of comparison with ABAQUS® VCCT fatigue delamination, where KB is the only mixed-mode loading FCG model. Moreover, updating of B and d with MM during propagation has been deactivated because it is not a feature available in ABAQUS®.

13.2.2 Virtual crack closure technique (VCCT)

The VCCT is well implemented in ABAQUS® for both 2D and 3D. In a 2D problem, the crack is represented as a 1D discontinuity formed by a line of nodes with the bulk material located on both sides of the discontinuity, as seen in Figure 13.9. The bulk material is modeled in the form of two distinct parts joined together by means of a contact pair along the discontinuity, with either of the coinciding edges as a master surface

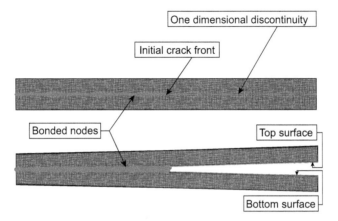

Figure 13.9 Representation of constituents of a VCCT model.

and the other as a slave surface. The nodes on the discontinuity share the same coordinates and have an important role in the definition of pre-cracked region, the crack front, and the crack path.

The nodes on the discontinuity, which are not bonded and are free to move away from each other, represent the pre-crack region, whereas the nodes that are bonded and stick to each other, referred as bonded nodes, define the crack propagation path and the point of transition of bonded and unbounded nodes forms the crack front. It is possible to define a completely bonded interface; however, at least a single node has to be kept unbounded to identify the crack front. Normal surface behavior is specified for the contact pair with pressure-overclosure=HARD and the initial conditions of the contact pair are set to bonded, over the bonded nodes by means of the following lines inserted in the input file before the definition of the step.

```
*INITIAL CONDITIONS, TYPE =CONTACT
<Slave Surface>, <Master Surface>, <Bonded Node Set>
```

A structured mesh with an aspect ratio of 1 is preferred in the meshing of the region forming the crack path and is done in such a way that the nodes on the contact edge of one side of the bulk material have the same coordinates as the nodes on the contact edge of the other side of the bulk material along the crack path. The loading cycle is represented by means of the *AMPLITUDE term, which may be periodic, tabular, and so forth, depending on the loading history, the R-ratio, etc. In the current study, a sinusoidal loading history was implemented and the corresponding parameters were defined.

In the definition of the step, at the end of every increment, the SERRs are calculated using the *DEBOND command, which is used to specify that crack propagation may occur between two surfaces that are initially partially bonded. This is done by inserting the following lines into the Step module:

```
*DEBOND, SLAVE=<Slave Surface>, MASTER=<Master Surface>
```

This is followed by the definition of the criterion, using the *FRACTURE CRITERION command, which governs the fracture of the bonded region by releasing the bonded nodes and letting the crack propagate along the crack path. For the case of static crack propagation, this criterion is set to TYPE = VCCT by inserting the following lines:

```
*FRACTURE CRITERION, TYPE=VCCT
<GIc>, <GIIc>, <GIIIc>, <eta>
```

in which the second line denotes the material parameters.

For the case of fatigue crack propagation, Equation (13.1) is followed by setting this criterion to TYPE = FATIGUE, in the direct cyclic step, and the following lines are inserted into the input file:

```
*FRACTURE CRITERION, TYPE=FATIGUE
<c1>, <c2>, <c3>, <c4>, <r1>, <r2>, <GIc>, <GIIc>,
<GIIIc>, <eta>
```

in which $<c_1>$, $<c_2>$ represent the fatigue crack onset parameters, $<c_3>$, $<c_4>$ represent the Paris parameters of the fatigue crack propagation, $<r_1>$ represents the definitions of the threshold regions of the Paris curve given by ($r_1 = G_{thresh}/G_c$), $<r_2>$ represents the definition of the unstable region of the Paris curve and is given by ($r_2 = G_{pl}/G_c$), and the rest of the parameters signify the material parameters. The advancement of the crack is determined by applying the Paris law, which is based on the total SERR G_{TOT} in a direct cyclic analysis.

13.2.2.1 Direct cyclic analysis

Direct cyclic analysis, as implemented in ABAQUS®, is a quasi-static analysis that uses a combination of Fourier series and time integration of the nonlinear material by iteratively using the modified Newton method, with the elastic stiffness matrix at the beginning of the analysis step serving as the Jacobian, to obtain the stabilized response of an elastic–plastic structure subjected to constant amplitude cyclic loading. It effectively provides the cyclic response of the structure directly by neglecting the pre-stability loading cycles of a transient analysis, which are numerically expensive. The workflow of the procedure is described in Figure 13.10. The method is based on the development of a displacement function $F(t)$, which describes the structural response at all moments of time t, in a loading cycle, within a given time period T. This function is represented in the following way:

$$U(t) = U_0 + \sum\nolimits_{k=1}^{n} \left[U_k^s \sin k\omega t + U_k^c \cos k\omega t \right] \qquad (13.36)$$

where n represents the number of terms in the Fourier series, ω is the angular frequency, and U_0, U_k^s, and U_k^c are the coefficients of displacement corresponding to each degree of freedom. The residual vectors are of the same form as the displacement function and are represented by

$$R(t) = R_0 + \sum\nolimits_{k=1}^{n} \left[R_k^s \sin k\omega t + R_k^c \cos k\omega t \right] \qquad (13.37)$$

Simulating the mixed-mode fatigue delamination/debonding

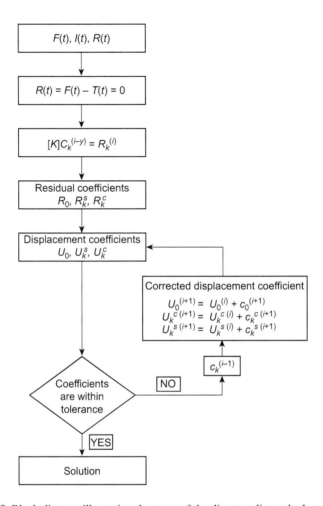

Figure 13.10 Block diagram illustrating the steps of the direct cyclic method.

where R_0, R_k^s, and R_k^c have the same correspondence with the displacement coefficients U_0, U_k^s, and U_k^c, respectively, and this vector $R(t)$ is tracked for each instance of time in the loading cycle using element-to-element calculations. Integration of this function $R(t)$ over the entire cycle yields the following Fourier coefficients:

$$R_0 = \frac{2}{T} \int_0^T R(t) \, dt$$

$$R_k^s = \frac{2}{T} \int_0^T R(t) \sin k\omega t \, dt \qquad (13.38)$$

$$R_k^s = \frac{2}{T} \int_0^T R(t)\cos k\omega t\, dt$$

These coefficients correspond to the displacement coefficients and are compared with the tolerances defined in the step to achieve convergence. If the tolerance is met, convergence is achieved and the solution is obtained for that loading cycle. However, when these residuals are larger than the tolerance parameters, correction parameter c_k is evaluated in which corrections to the displacement coefficients c_0, c_k^s, and c_k^c are made in the following way:

$$U_0^{(i+1)} = U_0^{(i)} + c_0^{(i+1)}$$

$$U_k^{c(i+1)} = U_k^{c(i)} + c_k^{c(i+1)} \quad (13.39)$$

$$U_k^{s(i+1)} = U_k^{s(i)} + c_k^{s(i+1)}$$

The updated displacement coefficients are used in the next iteration to obtain displacements at each instant in time. This process is repeated until convergence is obtained. Each pass through the complete load cycle can therefore be thought of as a single iteration of the solution to the nonlinear problem.

The general syntax of the direct cyclic analysis pertaining to fatigue may be represented as:

*DIRECT CYCLIC, FATIGUE.
I_0, T_s,,, F_i, F_{max}, ΔF, i_{max},
N_{min}, N_{max}, N_{TOT},,

where I_0 represents the initial time increment size and if unspecified a default value equal to 0.1 times the single loading cycle period is assumed; T_s is the time of single loading cycle; and the next two blank values are, respectively, minimum and maximum time increments allowed, which are generally kept unspecified and a default of 10^{-5} times T_s for the first parameter (minimum time increment allowed) and a default of 0.1 times T_s, for the second parameter (maximum time increment allowed) unless the *CETOL* or *DELTMX* parameter is specified. F_i represents the initial number of terms in the Fourier series with a default of 11; F_{max} represents the maximum number of terms in the Fourier series with a default value of 25; ΔF represents the increment in the number of terms in Fourier series with a default of 5, and i_{max} represents the maximum number of iterations with 200 as the default value. The second line is composed of minimum N_{min} and maximum increment N_{max} in the number of cycles over which damage is extrapolated forward. The default values of N_{min} and N_{max} are 100 and 1000, respectively. N_{TOT} represents the total number of cycles allowed in a step, which if skipped is assigned by default a value of $(1 + N_{max}/2)$.

13.2.3 Finite element models

Fatigue delamination models are tested on various joint geometries characterized by varying mixed-mode ratios to verify accuracy, robustness, and performance in terms

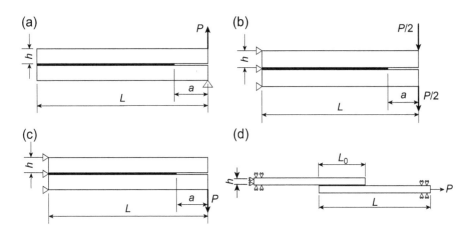

Figure 13.11 Simulated geometries: (a) DCB, (b) ELS, (c) MMELS, and (d) SLJ.

of computational time. In particular, pure Mode I loading is simulated with a DCB geometry, pure Mode II loading with an end-loaded split (ELS) geometry, and mixed-Mode I/II loading with an MMELS geometry, as shown in Figure 13.11. In addition, a single-lap joint (SLJ) was modeled as a representative case of real joint geometry (Figure 13.11). The propagation of the crack in the SLJ was allowed only on one side to simplify comparison of the models' results. The elastic properties are taken from Bernasconi, Jamil, Moroni, and Pirondi (2013) whereas the cohesive law and FCG behavior are taken from Turon et al. (2007). All properties are summarized in Table 13.1 together with the applied load and specimens' dimensions. In all simulations, a load ratio of $R = 0.05$ is assumed. The element type and mesh size are reported in Table 13.2 and represent good balance between convergence on the strain release rate and computational cost. Other parameters to be set, specifically for each FCG model, are as follows:

- a maximum damage increment, $\Delta D_{max} = 0.2$ was used for cohesive zone (CZ) (Pirondi & Moroni, 2010);
- a number of Fourier series terms equal to 49 and time increment 0.001 were set for VCCT, except for SLJ, for which the time increment was set to 0.01 owing to finer mesh.

The choice of a small time increment in the VCCT solution followed from a convergence study. Indeed, a strong influence of time integration points was observed for results obtained by VCCT and different values of SERR were obtained when the time integration points were varied from 10 to 1000. As a result of these variations in SERR, there were significant differences in estimation in the number of cycles owing to the presence of high values of the exponent in the Paris law. Therefore, 1000 time integration points with an initial time increment of 0.001 were used to evaluate accurate results; however, this increased the computational time drastically.

An initial crack length of 0.1 mm (one element) was specified for the SLJ when simulated using VCCT, whereas no initial crack length was needed in the case of CZM.

Table 13.1 **Elastic properties (Bernasconi et al., 2013), cohesive zone parameters, and FCG behavior for pure Mode I and pure Mode II, and mixed-Mode I/II (Turon et al., 2007), together with specimen dimensions and applied load for unit thickness**

Parameter	Mode I	Mode II
Γ (N/mm)	0.266	1.002
σ_{max} (MPa)	30	30
δ_0 (mm)	0.003	0.003
δ_C (mm)	0.0173	0.066
B	0.0616	4.23
d	5.4	4.5

Parameter	Value
m_m	2.6
m_d	1.85
m_B	0.35

Parameter	Value
E_{11} (MPa)	54000
E_{22} (MPa)	8000
ν_{12} (MPa)	0.25
G_{12} (MPa)	2750

	DCB	ELS	MMELS	SLJ
P (N/mm)	10	20	15	200
a_0 (mm)	20	20	20	–
h (mm)	5	5	5	10.56
L (mm)	175	175	175	285.8
L_0 (mm)	–	–	–	110.8

DCB, double cantilever beam; ELS, end-loaded split; MMELS, mixed-mode end-loaded split; SLJ, single-lap joint.

The increment in crack length is fixed in the case of VCCT, i.e., equal to the element size along the delaminating—debonding interface (0.1 mm for the SLJ and 0.5 mm elsewhere), whereas in the case of CZ it results from the increment in damage ΔD; therefore, it is not generally constant because ΔD may vary from increment to increment according to Eqn (13.28). However, the average increment in crack length in the case of CZM ranged from 0.1 to 0.5 mm in the various cases simulated in this work.

Table 13.2 Mesh size and element type of various FE models

	Composite laminate		Cohesive zone	
	Element type	Size	Element type	Size
DCB	Four-node bilinear plane stress quadrilateral, reduced integration	0.5 mm	Four-node 2D cohesive element	0.2 mm
ELS	Four-node bilinear plane stress quadrilateral, reduced integration	0.5 mm	Four-node 2D cohesive element	0.5 mm
MMELS	Four-node bilinear plane stress quadrilateral, reduced integration	0.5 mm	Four-node 2D cohesive element	0.2 mm
SLJ	Four-node bilinear plane stress quadrilateral	0.1 mm (next to cohesive elements)	Four-node 2D cohesive element	0.1 mm

DCB, double cantilever beam; ELS, end-loaded split; MMELS, mixed-mode end-loaded split; SLJ, single-lap joint.

13.3 Comparison of cohesive zone and VCCT on fatigue delamination/debonding

The two methods are compared with respect to: (1) agreement with each other, (2) agreement with numerical integration in Eqn (13.1); and (3) calculation time. Concerning (2), numerical integration was done using ΔG as a function of crack length coming from the FE simulations. Because ΔG is known by FE analysis, the trapezoidal rule (i.e., using the mean ΔG over the increment) was used. In this way, a closer estimate of the number of cycles at failure should be obtained with respect to both the CZ and VCCT, where for numerical reasons the ΔG at the beginning of the increment is used. Because Eqn (13.1) represents the best fit of experimental data (not shown here, but taken from Turon et al. (2007)), the level of agreement between the number of cycles output by the models and the numerical integration of Eqn (13.1) also represents the level of agreement between experimental data and the simulations. Regarding (3), the time the analyst has to wait for the crack to reach the knee of the a-N diagram is close to fracture. In the cases studied here, this means a crack length of 40 mm for all geometries, except SLJ, for which the analyses have stopped at 40 mm of crack

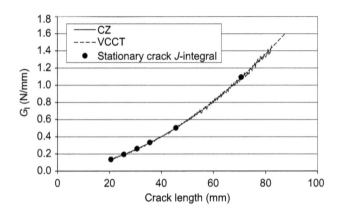

Figure 13.12 Comparison of G_I obtained by CZ, VCCT, and J-integral (stationary crack) in the case of DCB.

length even though they are still far from fracture. Only the outputs strictly necessary for each model were required, to minimize time spent in storing data. The PC used for calculations is an Athlon X2 Dual Core 2 GHz CPU, with 2 GB RAM and a 200-GB hard drive (7200 rpm, 8 MB cache).

13.3.1 Mode I loading (DCB)

Figure 13.12 shows the values of G_I obtained by CZ, VCCT, and J-integral (stationary crack). The three sets show good correspondence with each other, as expected, with only some small oscillation in the SERR calculated using the subroutine in the case of CZ.

The main result in terms of crack length versus number of cycles is shown in Figure 13.13, where a small difference of about 2.5% is evident. Another small difference is the gradient in a-N trend while approaching G_{Ic}, which is much steeper (almost discontinuous) in the case of VCCT. Both CZ and VCCT yielded a higher number of cycles with respect to the numerical integration of Eqn (13.1), with a difference of 2.3% in the case of CZ and 1.8% in the case of VCCT, which is acceptable in engineering terms.

13.3.2 Mode II loading (ELS)

Figure 13.14 shows the values of G_{II} obtained by CZ, VCCT, and J-integral (stationary crack). The three sets show a good correspondence with each other, especially until 50 mm crack length, whereas for longer cracks the CZ G_{II} is lower than the VCCT one and the J-integral lies in between.

The main result in terms of crack length versus number of cycles is shown in Figure 13.15, in which a difference of about 10% is evident. The number of cycles

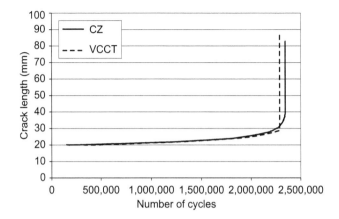

Figure 13.13 Comparison of $a-N$ values in the case of DCB obtained by CZ and VCCT.

at failure was also obtained by integrating Eqn (13.1) using the trapezoidal rule and the ΔG as a function of crack length coming from the FE simulations. Both CZ and VCCT yielded a higher number of cycles with respect to the numerical integration of Eqn (13.1), with a negligible difference both in the case of CZ and VCCT.

13.3.3 Mixed-mode I/II loading (MMELS)

Figure 13.16 shows the values of G_I and G_{II} obtained by CZ and VCCT. The values obtained with the two methods show good correspondence with each other in the case

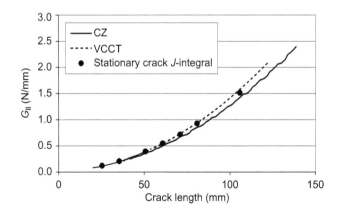

Figure 13.14 Comparison of G_{II} obtained by CZ, VCCT, and J-integral (stationary crack) in the case of ELS.

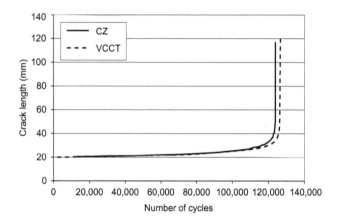

Figure 13.15 Comparison of $a-N$ values in the case of ELS obtained by CZ and VCCT.

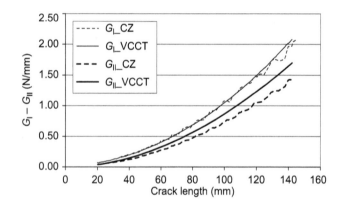

Figure 13.16 Comparison of G_I and G_{II} obtained by CZ and VCCT in the case of MMELS.

of the Mode I component, as for the DCB geometry. Under Mode II, the agreement is good, especially until 50 mm crack length, as in the ELS, whereas for longer cracks the CZ G_{II} is lower than the VCCT one.

The main result in terms of crack length versus number of cycles is shown in Figure 13.17, where a small difference of about 4% is evident. The number of cycles at failure was obtained by integrating Eqn (13.1) using the trapezoidal rule and ΔG as a function of crack length coming from the FE simulations. Both in the case of CZ and VCCT the numerical integration of Eqn (13.1) yielded a lower number of cycles, with a difference of 5.5% in the case of CZ and 8.3% in the case of VCCT, which may be still acceptable in engineering terms.

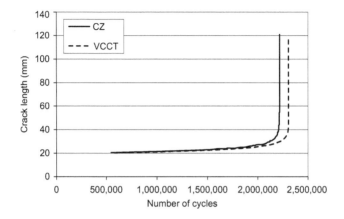

Figure 13.17 Comparison of $a-N$ values in the case of MMELS obtained by CZ and VCCT.

13.3.4 Single-lap joint (SLJ)

Figure 13.18 shows the values for G_I and G_{II} obtained by CZ and VCCT. The values obtained with the two methods show good correspondence with each other in the case of both mode components in the first millimeters of propagation, whereas at longer cracks the CZ values are lower than the VCCT ones. A higher difference is noticed in the case of the Mode II component, similar to Mode II and mixed-Mode I/II loading. However, in those cases the difference in the number of cycles between the two models to failure was affected to a limited extent, whereas in the case of SLJ the discrepancy is much higher (Figure 13.19). This discrepancy, however, occurs because, to date,

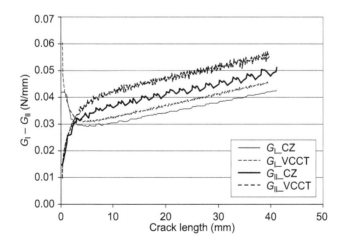

Figure 13.18 Comparison of G_I and G_{II} obtained by CZ and VCCT in the case of SLJ.

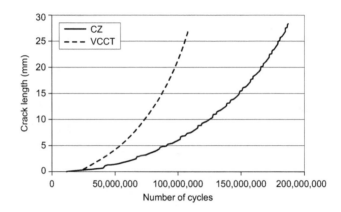

Figure 13.19 Comparison of $a-N$ values in the case of SLJ obtained by CZ and VCCT.

VCCT does not allow modifying the coefficient (B) and exponent (d) of Eqn (13.1) according to the mixed-mode ratio MM (Eqns (13.33)−(13.35), Kenane & Benzeggagh, 1997) as CZ instead does. In the case of SLJ, the MM ratio increases steeply in the first 5 mm of propagation and then becomes almost stationary (ranges between 0.55 and 0.56) (Figure 13.20) and the VCCT simulation was performed in this case using the stationary MM value. Both in the case of CZ and VCCT, the numerical integration of Eqn (13.1) yielded a lower number of cycles, with a difference (after 48 mm of crack propagation) of 2.6% in the case of CZ and 1.1% in the case of VCCT, which is absolutely acceptable in engineering terms.

13.3.5 Calculation time

The calculation times are reported in Table 13.3. The CZ resulted on average two orders of magnitude quicker than VCCT, with calculation times on the order of

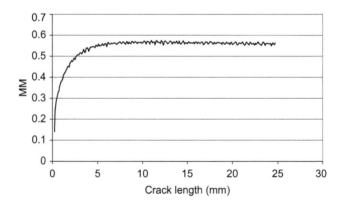

Figure 13.20 MM ratio as a function of crack length in the case of SLJ.

Table 13.3 **Calculation time (min)**

	DCB	ELS	MMELS	SLJ
CZ	9.1	4.6	4.5	21.4
VCCT	676.2	688.6	727.5	2796.8

DCB, double cantilever beam; ELS, end-loaded split; MMELS, mixed-mode end-loaded split; SLJ, single-lap joint; CZ, cohesive zone; VCCT, virtual crack closure technique.

minutes instead of hours. In the case of SLJ, the increase in calculation time is related to the finer mesh, but the time required by VCCT is so important that high-performance computing may be needed if model complexity increases further.

The origin of this large difference in performance between the in-house CZ subroutine and the built-in VCCT, both run using the ABAQUS solver, can be found at least partly in the direct cyclic procedure associated with VCCT in ABAQUS. Indeed, this procedure requires a large number of iterations to satisfy convergence on the ΔG value. On the other hand, relaxing the convergence on ΔG may affect the number of cycles to failure unpredictably.

13.4 Conclusions

Comparison of the performance of the CZM presented in Moroni and Pirondi (2012) and the VCCT embedded in the software ABAQUS on Mode I, Mode II, and mixed-Mode I/II—loaded cracks in composite assemblies yielded the following results:

- The two models agree with each other to within 4%, except in the case of SLJ, where VCCT currently does not allow modifying the coefficient (B) and exponent (d) of Eqn (13.1) according to the mixed-mode ratio MM as CZ does. Therefore, the rapid increase in MM in the first millimeters of propagation generates a large discrepancy between the two models. In this sense, the CZM offers an additional feature with respect to ABAQUS VCCT.
- Whereas the modeling effort is higher (there is a need to introduce a layer of cohesive elements), the CZM results in easier use (there is no need to identify the proper number of Fourier terms and time increment to represent cyclic loading). At the same time, it is more efficient because the computation is lower by about two orders of magnitude, even though the origin of this large difference in performance can be at least partly found in the direct cyclic procedure associated with VCCT in ABAQUS.

References

Bernasconi, A., Jamil, A., Moroni, F., & Pirondi, A. (2013). A study on fatigue crack propagation in thick composite adhesively bonded joints. *International Journal of Fatigue, 50*, 18—25.

Beuth, J. L., & Narayan, S. H. (1997). *Separation of crack extension modes in composite delamination problems* (Vol. 1285). ASTM Special Technical Publication.

Blackman, B. R. K., Hadavinia, H., Kinloch, A. J., & Williams, J. G. (2003). The use of a cohesive zone model to study the fracture of fibre composites and adhesively-bonded joints. *International Journal of Fracture, 119*, 25–46.

Chang, Y., Shi, J., & Cheng, G. J. (2012). An eXtended finite element method (XFEM) study on the effect of reinforcing particles on the crack propagation behavior in a metal–matrix composite. *International Journal of Fatigue, 44*, 151–156.

Cheuk, P. T., Tong, L., Wang, C. H., Baker, A., & Chalkley, P. (2002). Fatigue crack growth in adhesively bonded composite-metal double-lap joints. *Composite Structures, 57*, 109–115.

Curley, A. J., Hadavinia, A., Kinloch, A. J., & Taylor, A. C. (2000). Predicting the service-life of adhesively-bonded joints. *International Journal of Fracture, 103*, 41–69.

Fawaz, S. A. (1998). Application of the virtual crack closure technique to calculate stress intensity factors for through cracks with an elliptical crack front. *Engineering Fracture Mechanics, 59*(3), 327–342.

Forman, F. G., Kearnay, V. E., & Engle, R. M. (1967). Numerical analysis of crack propagation in cyclic loaded structures. *Journal of Basic Engineering, Transactions of ASME, 89*, 885.

Harbert, S. J., & Hogan, H. A. (1992). Modelling delamination growth in composite notched tensile bars using constrained plate finite elements. *American Society of Mechanical Engineers, Petroleum Division (Publication) PD, 45*, 107–112.

Harper, W. P., & Hallett, S. R. (2008). Cohesive zone length in numerical simulations of composite delamination. *Engineering Fracture Mechanics, 75*, 4774–4792.

Hutchinson, J. W., & Evans, A. G. (2000). Mechanics of materials: top-down approaches to fracture. *Acta Materialia, 48*, 125–135.

Hutchinson, J. W., & Suo, Z. (1991). Mixed mode cracking in layered materials. *Advances in Applied Mechanics, 29*, 63–191.

Kenane, M., & Benzeggagh, M. L. (1996). Measurement of mixed mode delamination fracture toughness of unidirectional glass/epoxy composites with mixed mode bending apparatus. *Composites Science and Technology, 56*, 439–449.

Kenane, M., & Benzeggagh, M. L. (1997). Mixed mode delamination fracture toughness of unidirectional glass/epoxy composites under fatigue loading. *Composites Science and Technology, 57*, 597–605.

Khoramishad, H., Crocombe, A. D., Katnam, K. B., & Ashcroft, I. A. (2010). Predicting fatigue damage in adhesively bonded joints using a cohesive zone model. *International Journal of Fatigue, 32*, 1146–1158.

Khoramishad, H., Crocombe, A. D., Katnam, K. B., & Ashcroft, I. A. (2011). Fatigue damage modelling of adhesively bonded joints under variable amplitude loading using a cohesive zone model. *Engineering Fracture Mechanics, 78*, 3212–3225.

Krueger, R. (2004). Virtual crack closure technique: history, approach, and applications. *Applied Mechanics Reviews, 57*, 109–143.

Krueger, R. (2010). *Development of a benchmark example for delamination fatigue growth prediction*. NASA/CR-2010–216723, NIA Report No. 2010–04.

Lemaitre, J. (1985). Continuous damage mechanics model for ductile fracture. *Journal of Engineering Materials and Technology, 107*, 83–89.

Leski, A. (2007). Implementation of the virtual crack closure technique in engineering FE calculations. *Finite Elements in Analysis and Design, 43*, 261–268.

Li, S., Thouless, M. D., Waas, A. M., Schroeder, J. A., & Zavattieri, P. D. (2005). Use of mode-I cohesive-zone models to describe the fracture of an adhesively-bonded polymer-matrix composite. *Composites Science and Technology, 65*, 281–293.

Liu, P. F., Hou, S. J., Chu, J. K., Hub, X. Y., Zhou, C. L., Liu, Y. L., et al. (2011). Finite element analysis of postbuckling and delamination of composite laminates using virtual crack closure technique. *Composite Structures*, *93*, 1549−1560.
Maiti, S., & Geubelle, P. H. (2005). A cohesive model for fatigue failure of polymers. *Engineering Fracture Mechanics*, *72*, 691−708.
Marannano, G. V., Mistretta, L., Cirello, A., & Pasta, S. (2008). Crack growth analysis at adhesive−adherent interface in bonded joints under mixed mode I/II. *Engineering Fracture Mechanics*, *75*, 5122−5133.
Moroni, F., & Pirondi, A. (2012a). A procedure for the simulation of fatigue crack growth in adhesively bonded joints based on a cohesive zone model and various mixed-mode propagation criteria. *Engineering Fracture Mechanics*, *89*, 129−138.
Moroni, F., & Pirondi, A. (2012b). Simulation of 3D fatigue debonding/delamination in composites using cohesive zone. In *ECCM15−15th European conference on composite materials*, Venice, Italy (pp. 24−28). June 2012.
Muñoz, J. J., Galvanetto, U., & Robinson, P. (2006). On the numerical simulation of fatigue driven delamination with interface element. *International Journal of Fatigue*, *28*, 1136−1146.
Murri, G. B., & Schaff, J. R. (2006). Fatigue life methodology for tapered hybrid composite flexbeams. *Composites Science and Technology*, *66*, 499−508.
Naghipour, P., Bartsch, M., & Voggenreiter, H. (2011). Simulation and experimental validation of mixed mode delamination in multidirectional CF/PEEK laminates under fatigue loading. *International Journal of Solids and Structures*, *48*, 1070−1081.
O'Brien, T. K. (2001). *Characterization, analysis and prediction of delamination in composites using fracture mechanics*. ICF10, Honolulu, HI.
Okada, H., Higashi, M., Kikuchi, M., Fukui, Y., & Kumazawa, N. (2005). Three dimensional virtual crack closure-integral method (VCCM) with skewed and non-symmetric mesh arrangement at the crack front. *Engineering Fracture Mechanics*, *72*, 1717−1737. ISSN 0013−7944.
Paris, P., & Erdogan, F. (1961). A critical analysis of crack propagation laws. *Journal of Basic Engineering*, *85*, 528−534.
Pietropaoli, E., & Riccio, A. (2010). On the robustness of finite element procedures based on virtual crack closure technique and fail release approach for delamination growth phenomena definition and assessment of a novel methodology. *Composites Science and Technology*, *70*, 1288−1300.
Pietropaoli, E., & Riccio, A. (2011). Formulation and assessment of an enhanced finite element procedure for the analysis of delamination growth phenomena in composite structures. *Composites Science and Technology*, *71*, 836−846.
Pirondi, A., & Moroni, F. (2009). An investigation of fatigue failure prediction of adhesively bonded metal/metal joints. *International Journal of Adhesion and Adhesives*, *29*, 796−805.
Pirondi, A., & Moroni, F. (2010). A progressive damage model for the prediction of fatigue crack growth in bonded joints. *Journal of Adhesion*, *86*, 1−21.
Raju, I. S. (1987). Calculation of strain energy release rates with higher order and singular finite elements. *Engineering Fracture Mechanics*, *28*, 251−274.
Rice, J. R. (1968). A path independent integral and the approximate analysis of strain concentration by notches and cracks. *Journal of Applied Mechanics*, *35*, 379−386.
Roe, K. L., & Siegmund, T. (2003). An irreversible cohesive zone model for interface fatigue crack growth simulation. *Engineering Fracture Mechanics*, *70*, 209−232.
Rybicki, E. F., & Kanninen, M. F. (1977). A finite element calculation of stress intensity factors by a modified crack closure integral. *Engineering Fracture Mechanics*, *9*, 931−938.

Salpekar, S. A., O'Brien, T. K., & Shivakumar, K. N. (1996). Analysis of local delaminations caused by angle ply matrix crack. *Journal of Composite Materials, 30*(4), 418–440.
Shen, F., Lee, K. H., & Tay, T. E. (2001). Modeling delamination growth in laminated composites. *Composites Science and Technology, 61*, 1239–1251.
Shivakumar, K. N., Tan, P. W., & Newman, J. C., Jr. (1988). A virtual crack-closure technique for calculating stress intensity factors for cracked three dimensional bodies. *International Journal of Fracture, 36*, 43–50.
Sun, C. T., & Jih, C. J. (1987). On strain energy release rates for interfacial cracks in bi-material media. *Engineering Fracture Mechanics, 28*, 13–20.
Turon, A., Costa, J., Camanho, P. P., & Dávila, C. G. (2007). Simulation of delamination in composites under high-cycle fatigue. *Composites Part A, 38*, 2270–2282.
Ungsuwarungsru, T., & Knauss, W. G. (1987). The role of damage-softened material behaviour in the fracture of composites and adhesives. *International Journal of Fracture, 35*, 221–241.
Whitcomb, J. D. (1992). Analysis of a laminate with a postbuckled embedded delamination, including contact effects. *Journal of Composite Materials, 26*, 1523–1535. ISSN 0021–9983.
Xie, D., & Biggers, S. B., Jr. (2006). Progressive crack growth analysis using interface element based on the virtual crack closure technique. *Finite Elements in Analysis and Design, 42*, 977–984.
Xie, D., Chung, J., Waas, A. M., Shahwan, K. W., Schroeder, J. A., Boeman, R. G., et al. (2005). Failure analysis of adhesively bonded structures: from coupon level data to structure level predictions and verification. *International Journal of Fracture, 134*, 231–250.
Xie, D., Waas, A. M., Shahwan, K. W., Schroeder, J. A., & Boeman, R. G. (2004). Computation of strain energy release rate for kinking cracks based on virtual crack closure technique. *CMES: Computer Modeling in Engineering & Sciences, 6*, 515–524.
Yang, Q. D., Thouless, M. D., & Ward, S. M. (1999). Numerical simulations of adhesively-bonded beams failing with extensive plastic deformation. *Journal of Mechanics and Physics of Solids, 47*, 1337–1353.
Zou, Z., Reid, S. R., Soden, P. D., & Li, S. (2001). Mode separation for energy release rate for delamination in composite laminates using sublaminates. *Internation Journal of Solids and Structures, 38*, 2597–2613.

Predicting the fatigue life of adhesively-bonded composite joints under mode I fracture conditions

14

T.A. Hafiz[1], M.M. Abdel Wahab[2]
[1] University of Bristol, Bristol, UK; [2] Ghent University, Zwijnaarde, Belgium

14.1 Introduction

Adhesive-bonding has been gaining much more importance during the last few decades, because it is used for joining a wide range of similar and dissimilar metallic, non-metallic and composite components with different shapes, sizes and thicknesses. The advantages of adhesive-bonding over traditional joining techniques are well known and accepted by the adhesive-bonding community. Adhesives provide greater design flexibility, distribute load over a much wider area, reduce stress concentrations and increase fatigue as well as corrosion resistance. In addition, they also provide weight savings to the whole structure, improving the appearance of the bond (Hafiz, 2011). There is, therefore, no wonder that adhesive-bonding is the primary joining technique for carbon fibre-reinforced polymer (CFRP) used in the aerospace industry. Many other industries make use of adhesives, e.g. civil engineering, transportation, biomechanical, marine, electronics, etc. Fatigue is undoubtedly a very important type of loading for many structural components that contain adhesive-bonding systems. In a fatigue-loading regime, a structure may fail at a small percentage of static strength. Therefore, fatigue analysis and fatigue strength prediction are highly required, especially for the case of failsafe or damage tolerance design. Accurate prediction of fatigue life is a challenge due to the complicated nature of fatigue crack initiation and propagation, geometry of bonded joints and complex material behaviour under loading and unloading regimes (Abdel Wahab, 2012).

A particular issue with the integrity of adhesive joints is the presence of cracks and flaws in the as-manufactured adhesive bondline. The presence of these defects, at least at some scale, appears inevitable and the propagation of such cracks/flaws has the potential to affect the service life of the adhesively-bonded joints and even to cause catastrophic failure of bonded structures in service. Hence, a better understanding of crack behaviour under realistic types of service loading is an important aspect of evaluating the potential performance of adhesively-bonded joints (Hafiz, Abdel Wahab, Crocombe, & Smith, 2010). The knowledge of the crack behaviour is essential for material development and selection and for design and life-prediction studies. Much work has been published to characterize the performance of adhesive joints under mode I loading, but the fatigue process in adhesive joints is poorly characterized. Hence, the estimation of bonded joint fatigue life becomes very difficult. Few efforts have been devoted to characterize fatigue life modelling and prediction of bonded joints under mode I fracture. Therefore, the need for this work becomes unavoidable. The structure of the chapter is as follows. In the next section, the analytical approach is discussed, together with the mode I strain energy release rate and the integration of the crack growth law. Then, the finite element (FE) approach is presented. Finally, a validation of the FE approach using experimental data and concluding remarks are presented.

14.2 Characterization of fatigue in bonded joints

The term 'fatigue' is used when materials loose structural integrity under repeated stresses. Fatigue failure can be described as failure after multiple load applications, which would not have caused failure if applied individually. The fatigue mechanism in polymers is different from metals due to factors such as moisture, temperature and the visco-elastic nature of some polymers at modest temperatures affecting the response to cyclic stresses. There are two main approaches, which have been extensively used in the fatigue characterization of bonded joints: the stress-life approach and fatigue crack growth (FCG) approach.

14.2.1 Stress-life approach

Safe-life design is the main theme of this approach. The philosophy is that the component is flaw-free and is replaced after a fixed service life. In the stress-life approach, a number of samples are tested at different fatigue loads, generally under constant-amplitude fatigue loading. A relationship between stress (S) and the number of cycles (N) is then plotted. A schematic $S-N$ curve is shown in Figure 14.1.

A fatigue threshold, σ_{th}, is defined as the stress amplitude below which there is no fatigue failure and thus an infinite fatigue life of the joint is predicted. When presenting fatigue data for bonded joints in this way, there is an issue of what stress value to use,

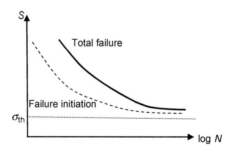

Figure 14.1 Schematic $S-N$ curve.

and therefore the distribution of stress and the mode of failure are not analysed in this approach.

14.2.2 FCG approach

The FCG approach is based on fracture mechanics (FM). In the FCG approach, a very small size initial flaw distribution is assumed, and the knowledge of the conditions and rates at which these cracks will grow along with the detailed inspection programme is utilized in a failsafe design approach.

The FCG method is the correlation between the rate of FCG per cycle (da/dN, where a represents the crack length at a certain number of cycles N) and the change of one fracture parameter over time. The plotted expression of these two factors in a logarithmic scale has a sigmoidal shape that has been previously observed in studies of FCG in metals and polymers (Ashcroft & Shaw, 2002), and for a large range it follows a power law. Figure 14.2 shows a typical propagation curve, which characterizes the properties above a certain fatigue threshold value and below the fracture toughness of the adhesive. From Figure 14.2, three zones can be easily identified: (1) threshold region, within which the crack growth tends to zero; (2) linear region, in which a linear growth is noticed and a Paris-type relation fits well and (3) fast fracture region, in which the crack becomes unstable and is characterized by its rapid and catastrophic growth.

The fatigue crack propagation rate (da/dN) is measured and related to FM parameters such as the stress intensity factor, K, strain energy release rate, G, or the J-integral. In these tests, the sample is usually pre-cracked and only the propagation phase is studied. This approach is well established for characterizing fatigue in metals where log da/dN is plotted against log ΔK (Bannantine, Comer, & Hand-rock, 1990). In the case of bonded joints, strain energy release rate, G, is generally the mostly widely used FM parameter (Dessureault & Spelt, 1997). The key parameters involved in FCG are mode I strain energy release rate, G_I, and da/dN. In the next two sections, we highlight the approaches used to calculate the fatigue life of bonded joints.

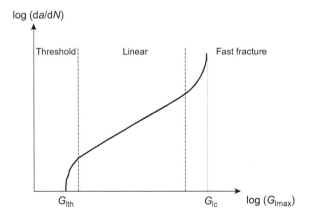

Figure 14.2 Fatigue crack propagation curve.

14.3 Analytical approach to fatigue life prediction of adhesively-bonded joints

14.3.1 Calculation of mode I strain energy release rate

Several methods and approaches have been applied to collect fatigue debond rate data as a function of the applied strain energy release rate for mode I fracture specimens such as double cantilever beams (DCBs). DCB is one of the most popular specimen geometries for characterizing FCG in bonded joints. A DCB specimen is loaded by applying symmetrical opening tensile forces at the end of the beam, as shown in Figure 14.3. This enables the tensile opening mode (mode I) of the fracture energy, G_{IC}, to be measured. However, the use of this relatively very simple FM specimen appears to have given rise to surprising differences in reported values of G_{IC}. Published fracture toughness, G_{IC}, values range from 1400 to 2000 J/m² (Hafiz et al., 2010; Shenoy, Ashcroft, Critchlow, Crocombe, & Abdel Wahab, 2009) to a mean value of around 2800 J/m² from a number of studies (Johnson, Butkus, & Valentin, 1998; Ripling, Crosley, & Johnson, 1988) and a value of 3700 J/m² (Jastrzebski, Sinclair, Raizenne, & Spelt, 2009). The values are based on various configurations and data reduction techniques. Such apparent differences in the fracture behaviour of the material can readily be reproduced (Hashemi, Kinloch, & Williams, 1989). Therefore, it seems reasonable to summarize different methods used to calculate G_I in the next section.

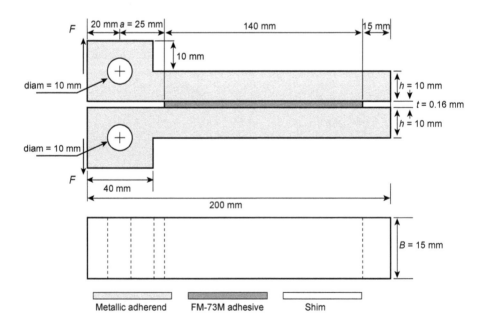

Figure 14.3 Bonded double cantilever beam (DCB) specimen geometry (Hafiz, 2011).

14.3.1.1 Area method

There are different methods for analysing the data contained in a load−displacement trace. In the area method, based on linear elastic fracture mechanics (LEFM), the value of G_{IC} may be defined as:

$$G_{IC} = \frac{\Delta U}{B \Delta a} \quad (14.1)$$

where B is the width of the DCB specimen, ΔU is the area under the load−displacement trace and Δa is the increase in crack length from a_1 to a_2. If the loading and unloading relations are linear; therefore LEFM is applicable and ΔU can be written as (Hashemi et al., 1989):

$$\Delta U = U_2 - U_1 = \frac{1}{2}[F_1 \delta_2 - F_2 \delta_1] \quad (14.2)$$

where F_1 and δ_1 are the load and displacement at crack length a_1, and F_2 and δ_2 are the respective values at a crack length a_2.

14.3.1.2 Compliance method

In this method, the Irwin−Kies method (Irwin & Kies, 1954) is used to evaluate G_{IC}:

$$G_{IC} = \frac{F^2}{2B} \frac{dC}{da} \quad (14.3)$$

where dC/da is the variation of the compliance as a function of the crack length a, and the compliance, C, is given by:

$$C = \frac{\delta}{F} \quad (14.4)$$

where δ is the displacement corresponding to a load F. Compliance is monitored for the peak load. This method is based on Griffith's theory, where the energy needed for a crack to grow is the change in potential energy stored in the specimen due to crack propagation. In other words, the elastic strain energy released, when the crack grows, must be the energy required to create new surfaces. Energy dissipation is considered to be located only in a plane and a linear crack front along the specimen width (B) is assumed (Fernandez, de Moura, da Silva, & Marques, 2011).

14.3.1.3 Load method

If the elastic properties of the adherend material are known, then beam theory can be used to calculate the change in compliance of the DCB as a crack propagates. An expression for the compliance in a DCB, neglecting the contribution of the adhesive, is given by (Hashemi et al., 1989):

$$C = \frac{\delta}{F} = \frac{2a^3}{3EI} \tag{14.5}$$

where I is the second moment of area, which is given by:

$$I = \frac{Bh^3}{12} \tag{14.6}$$

Therefore, the compliance, C, becomes:

$$C = \frac{8a^3}{BEh^3} \tag{14.7}$$

where E and h are the flexural modulus and adherend thickness of DCB specimen, respectively. Substitution of Eqns (14.5) and (14.6) into Eqn (14.3) yields the following expression for strain energy release rate as a function of crack length:

$$G_{IC} = \frac{F^2 a^2}{BEI} \tag{14.8}$$

Timoshenko and Goodier (1970) considered directly the deflection equation for a cantilever beam (i.e. the adherend) loaded at the end to express the load line compliance, C, and strain energy release rate, G_{IC}, as:

$$C = \frac{8}{BE}\left(\frac{a^3}{h^3} + \frac{a}{h}\right) \tag{14.9}$$

$$G_{IC} = \frac{F^2}{2B}\frac{dC}{da} = \frac{F^2 a^2}{BEI}\left(1 + \frac{h^2}{3a^2}\right) \tag{14.10}$$

14.3.1.4 Displacement method

Substitution of F from Eqn (14.5) in Eqn (14.8), gives:

$$G_{IC} = \frac{3F\delta}{2Ba} \tag{14.11}$$

14.3.1.5 Beam on elastic foundation method

Pirondi and Nicoletto (2004), proposed a beam theory based model that accounts for the influence of the adhesive on the joint behaviour. However, this model did not account for the elastic behaviour of the adhesive bondline, which was demonstrated to influence the fracture energy of the joint. To account for the elasticity of the adhesive, they modelled the DCB specimen as a beam on an elastic foundation, where the foundation modulus, k, depends on the elastic constant of the adhesive, E_a, and on the bondline thickness, t. Therefore, G_I is calculated considering the relation between C and a developed by Krenk, Jonsson, and Hansen (1996), where the joint was modelled as a beam on an elastic foundation.

$$G_{IC} = \frac{F^2}{2B}\frac{dC}{da} = \frac{F^2 a^2}{BEI}\left(1 + \frac{1}{\lambda_\sigma a}\right)^2 \qquad (14.12)$$

The load line compliance is:

$$C = \frac{\delta}{F} = \frac{2\lambda_\sigma t}{BE'_a}\left[1 + 2(\lambda_\sigma a) + 2(\lambda_\sigma a)^2 + \frac{2}{3}(\lambda_\sigma a)^3\right] \qquad (14.13)$$

where

$$\lambda_\sigma^4 = \frac{K}{4EI} = \frac{6}{h^3 t}\frac{E'_a}{E} \qquad (14.14)$$

In Eqn (14.14), $1/\lambda_\sigma$ serves as a length scale of the crack problem and varies from about 0.35, at the beginning of the test to about 0.135 at the end of it; therefore it cannot be neglected (Hafiz, 2011). The constant $E'_a = \dfrac{E_a}{(1 - v^2)}$ is the plane strain modulus of the adhesive.

Hafiz et al. (2010) introduced an equation, based on Williams' work (Williams, 1988), that also takes into account the adhesive bondline thickness:

$$G_{IC} = \frac{F^2 a^2}{BEI}\left(1 + \frac{(1+v)}{5}\left(\frac{h}{a}\right)^2\right)\left(1 + \frac{1}{\lambda_\sigma a}\right)^2 \qquad (14.15)$$

14.3.1.6 Compliance-based beam method

This method is based on the beam theory, specimen compliance and crack equivalent concept. It overcomes the inaccuracies committed during crack length monitoring caused by crack bowing and tilting. This means that an equivalent crack length (a_e) can be calculated as a function of the specimen compliance, thus avoiding the experimental crack length monitoring during propagation. de Moura, Morais, and Dourado (2008) obtained the specimen compliance (C) taking into account the effects of shear:

$$C = \frac{8a^3}{E_1 B h^3} + \frac{12a}{5BhG_{13}} \qquad (14.16)$$

where E_1 and G_{13} are the longitudinal and shear modulus of the adherends, respectively. However, it is expected that the compliance of the adhesive and its thickness may influence the global compliance of the specimen and should be accounted for. Consequently, an equivalent flexural modulus E_f can be estimated from Eqn (14.16) considering the initial compliance C_0 and the corrected initial crack length $(a_0 + |\Delta|)$ instead of C and a, respectively:

$$E_f = \frac{8(a_0 + |\Delta|)^3}{Bh^3}\left(C_0 - \frac{12(a_0 + |\Delta|)}{5BhG_{13}}\right)^{-1} \quad (14.17)$$

where Δ accounts for the root rotation effect at the crack tip. de Moura et al. (2008) obtained Δ as:

$$\Delta = h\sqrt{\frac{E_f}{11G_{13}}\left[3 - 2\left(\frac{\Gamma}{1+\Gamma}\right)^2\right]} \quad (14.18)$$

where:

$$\Gamma = 1.18\frac{\sqrt{E_f E_3}}{G_{13}} \quad (14.19)$$

An iterative procedure should be used in order to obtain a converged value for E_f. This procedure also minimizes the influence of eventual errors committed on the initial crack length measurements. The variability of the remaining elastic properties, E_3 and G_{13}, leads to (Fernandez et al., 2011):

$$G_{IC} = \frac{6F^2}{B^2h}\left(\frac{2a_e^2}{h^2 E_f} + \frac{1}{5G_{13}}\right) \quad (14.20)$$

14.3.1.7 Polynomial method (compliance calibration method)

When the load, displacement and crack length have been determined experimentally, the Irwin—Kies relation (Eqn (14.3)) can be directly used to calculate strain energy release rate as a function of cycles or crack length. A simple method of achieving this is by fitting a third-order polynomial curve to the full set of experimental compliance values plotted against crack length:

$$C = X_1 + X_2 a + X_3 a^2 + X_4 a^3 \quad (14.21)$$

where X_1, X_2, X_3 and X_4 are the constants of the fitting procedure. The fracture energy can now be obtained by means of Eqn (14.3) and differentiation of Eqn (14.21).

14.3.2 Determination of FCG rate

A number of different approaches have also been used to calculate the crack growth rate, da/dN, from the experimentally determined crack length as a function of cycles.

14.3.2.1 Polynomial method

This method involves incrementally fitting a second-order polynomial to sets of ($2n + 1$) data points throughout the whole data set. The integer n can be selected to ensure an appropriate number of data points for curve fitting. The form of the polynomial equation for the incremental data fitting is:

$$a = X_0 + X_1 \left(\frac{N - Y_1}{Y_2}\right) + X_2 \left(\frac{N - Y_1}{Y_2}\right)^2 \quad (14.22)$$

where X_0, X_1 and X_2 are the regression parameters determined by applying the least square method to the data set. The terms Y_1 and Y_2 are used to scale the input data in order to avoid numerical difficulties in this process, and are given by:

$$Y_1 = \frac{1}{2(N_{i-n} + N_{i+n})} \quad (14.23)$$

$$Y_2 = \frac{1}{2(N_{i+n} - N_{i-n})} \quad (14.24)$$

where $i = (n + 1)$. The crack growth rate, da/dN, is then obtained by differentiating Eqn (14.22) with respect to number of cycles, N:

$$\frac{da}{dN} = \frac{X_1}{Y_2} + 2X_2 \left(\frac{N - Y_1}{Y_2^2}\right) \quad (14.25)$$

14.3.2.2 Secant method

This method consists of evaluating the variation of crack as a function of the number of cycles considering a discrete number of measurements (n) during the fatigue test. The crack growth rate between two consecutive measurements (i and $i + 1$) is given by:

$$\frac{da}{dN} = \frac{a_{i+1} - a_i}{N_{i+1} - N_i} \quad (14.26)$$

where i represents the ith measurement performed during the test ($0 \leq i \leq n$). This gives an average value of the rate in an increment. This is a simple method that accurately represents the data. However, the method is sensitive to scatter in the experimental data, which tends to be magnified in the calculation of da/dN.

More recently, Hafiz, Abdel Wahab, Crocombe, and Smith (2013) selected a three-parameter exponential function for curve fitting the FCG data as:

$$a = e^{(X+Y/(N+Z))} \qquad (14.27)$$

The three coefficients, namely X, Y and Z, are determined using curve-fitting software.

14.3.3 Integration of FCG law

As mentioned earlier, the propagation curve is generally composed of three different regions: damage nucleation, stable propagation and abrupt final failure. The second phase corresponding to stable propagation leads to a linear trend on the Paris' law representation (log—log scale) and must be well characterized to define the fatigue behaviour of the structure (Figure 14.2). A threshold region of the FCG is associated with a fatigue threshold below which measurable crack growth does not occur, i.e. the material has infinite fatigue life. Therefore, when designing with materials where fatigue crack propagation is to be avoided, the fatigue threshold, G_{th}, becomes of paramount importance. Considerable scatter is often observed in experimental FCG plots and the crack propagation rate in adhesives and composites is usually more sensitive to changes in load than in metals. Therefore fatigue threshold-based designs are more desirable with these materials (Ashcroft & Shaw, 2002). The fatigue threshold depends on the loading configurations and environmental effects and is also used in the fatigue characterization of metals (Taylor, 1989). The Paris' law, which fits well to the data in linear region of the FCG curve in most cases, is given by:

$$\frac{da}{dN} = C(G_I)^m \qquad (14.28)$$

where C and m are the material constants. The value of m indicates the load sensitivity of the crack propagation rate and lies generally between 3 and 4 for metals but is higher for adhesives. Abdel Wahab, Ashcroft, Crocombe, and Smith (2004) obtained 5 and 13.4 for Cytec 4535A paste adhesive and AS4/8552 carbon fibre composite, respectively. When fatigue life is dominated by crack propagation in a linear region, then rearranging Eqn (14.28) provides the fatigue lifetime, i.e.:

$$N_f = \int_{a_i}^{a_f} \frac{da}{C(G_I)^m} \qquad (14.29)$$

where a_i and a_f are the initial and final crack lengths, respectively. The fast fracture region of the FCG curve is related to unstable crack growth as G_I approaches the critical strain energy release rate, G_C, in quasi-static loading. In most of the cases, the fast fracture region does not affect the total propagation life of the systems significantly and can be ignored in the prediction of cycles to failure (Ashcroft & Shaw, 2002). However the full sigmoidal curve is described empirically (Martin & Murri, 1990):

$$\frac{da}{dN} = C(G_I)^m \left[\frac{1 - (G_{th}/G_I)^{m_1}}{1 - (G_I/G_C)^{m_2}} \right] \quad (14.30)$$

where m_1 and m_2 are the additional material constants, G_{th} is the fatigue threshold and G_C is the fracture toughness. The constants m, m_1 and m_2 can be obtained by fitting Eqn (14.30) to experimental data. The difference between the above expression and the classical Paris' law is that Eqn (14.30) describes the full da/dN versus G_I curve, i.e. including the threshold and accelerating crack growth regions. Fatigue sensitivity can be obtained from the ratio of G_{th}/G_C. The number of cycles to failure (N_f) can be obtained by integrating Eqn (14.30) from an initial crack length (a_i) to a final crack length (a_f):

$$N_f = \int_{a_i}^{a_f} \frac{1}{CG_I^n} \left[\frac{1 - (G_I/G_C)^{m_2}}{1 - (G_{th}/G_I)^{m_1}} \right] da \quad (14.31)$$

14.3.4 Effect of fatigue control mode

The fatigue control mode affects the crack growth in constant amplitude (CA) fatigue tests. There are two main controlling modes, namely load control and displacement control.

In load control (fixed loading profile), G_I, increases with increase in crack length, while in displacement control (fixed-grip position profile), G_I decreases as the crack length increases as shown in Figure 14.4. In displacement control, if crack growth rate is related to G_I, rapid crack propagation is initially predicted, and then it slows down as the crack grows and stops before failure if the fatigue threshold is reached. This behaviour is very important, as the complete fatigue crack propagation curve can be generated from a single specimen. However, in load control mode, the phenomenon is the opposite. The crack growth is very slow in the beginning but accelerates rapidly until complete failure. Results in the literature show that the relationship

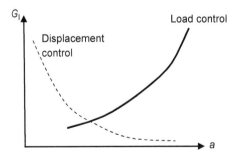

Figure 14.4 G_I versus crack length for different fatigue testing controlling modes. Adapted from Ashcroft and Shaw (2002).

between FCG rate and strain energy release rate remains unaffected, whether a sample is tested under displacement control or load control (Mall, Ramamurthy, & Rezaizdeh, 1987).

14.4 Finite element analysis approach to fatigue life prediction of adhesively-bonded joints

Finite element analysis (FEA) is a very important technique for modelling and hence predicting the crack propagation rate in complex structures such as bonded joints. Numerical integration methodologies have been developed based on Paris' law.

14.4.1 Modelling DCB

A two-dimensional (2D) linear elastic FEA of the bonded DCB geometry was carried out to validate the analytical expressions for G_I introduced in the analytical approach section. The study used second-order eight-noded quadrilateral elements, as these elements generate more accurate results for quadrilateral-triangular automatic meshes. Plane strain conditions were assumed to represent the mid-plane of the bonded joint. The mesh was refined until there was no further change in the results. The overall mesh and mesh around the crack tip element are shown in Figure 14.5 and Figure 14.6, respectively. Load was applied at the left end of the joint, with the joint constrained in the vertical direction at its right end, as shown in Figure 14.7. The material responses were taken as linear elastic and isotropic. The analysis confirmed that the stresses in the adherends were below yield and there was no evidence of adherend plasticity during the experiments.

14.4.2 Mode I strain energy release rate

The strain energy release rate can be calculated using the virtual crack closure technique (VCCT), a well-established FM approach for analysing progressive crack growth in linear elastic problems. This approach (Sethuraman & Maiti, 1988) was implemented using crack tip singularity elements (Barsoum, 1976). The boundary conditions and crack length along with other specifications were kept exactly the same as those used in the experiments. The crack was introduced on the interface (between lower adherend and adhesive), as Abdel Wahab (2000) showed that in most of the

Figure 14.5 Overall mesh of finite element analysis (FEA) model of the bonded double cantilever beam (DCB) geometry.

Figure 14.6 Finite element analysis (FEA) model of the bonded double cantilever beam (DCB) specimen showing mesh around the crack tip element.

Figure 14.7 Finite element analysis (FEA) model of the bonded double cantilever beam (DCB) specimen.

cases interfacial failure is observed for a very thin adhesive layer (<0.5 mm) and in the present case the adhesive layer thickness is 0.16 mm. In this way, the analysis determines the strain energy release rate for crack growth in the adhesive, close to the adherend–adhesive interface. Figure 14.8 shows the reference points in and around the crack tip and the corresponding terminology for the local forces and displacements. The following expression (Sethuraman & Maiti, 1988) is used to calculate the mode I strain energy release rate:

$$G_\mathrm{I} = \frac{\left(U_{yk} - U'_{yk}\right)}{\Delta a} \left[F_{yj} + (1.5\pi - 4)F_{yi}\right] \quad (14.32)$$

where Δa is equal to the crack tip element length, (F_{xi}, F_{yi}), (F_{xj}, F_{yj}) are the crack closure forces at nodes i and j, respectively, and (U_{xk}, U_{yk}), (U'_{xk}, U'_{yk}) are the nodal displacements behind the crack, as shown in Figure 14.8.

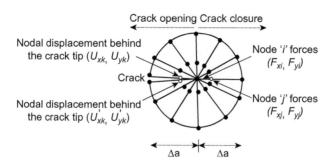

Figure 14.8 Nodes and elements around a crack tip. After Sethuraman and Maiti (1988).

14.4.3 Numerical integration of crack growth law

The main goal of this technique is to develop a methodology to predict an $L-N$ (load versus number of cycles) fatigue curve using FEA for an adhesive joint of a general configuration using the crack growth data generated by a FM test (e.g. DCB specimens). The approach is based on automatic embedded numerical integration of the Paris' law response. The first step is to determine the number of crack increments required in the numerical integration. This can be achieved by determining the number of crack increments for which the number of cycles to failure converges. The second step is to analyse the crack propagation for different loads in order to produce the $L-N$ curves and then compare the numerical predictions to the experimental results. The number of cycles to failure, N_f, is calculated by integrating a FCG law between initial and final crack lengths. This crack growth law is formulated in terms of G_I, which can be determined, at any crack length, from FEA. The complete process is implemented within the FE code enabling automated calculation of the fatigue life for a given set of boundary conditions (Abdel Wahab et al., 2004). The use of a numerical procedure to integrate Eqn (14.28) eliminates the need for empirical curve fitting of G_I against a and also entails less effort once a suitable routine has been programmed within the FE code. Furthermore, this method is applicable to any structural configuration, and the value of a_f can be determined more accurately from the numerical model. The crack length was defined as a parameter that varied between the initial crack length a_i and final crack length a_f. The final crack length can be defined in the FE model as the crack length at which G_I reached the (static) fracture toughness, with a limited value. The FE mesh for the joint is constructed in a parametric manner using the parametric design language available in ANSYS (Abdel Wahab, Ashcroft, Crocombe, & Smith, 2002). Abdel Wahab et al. (2004) applied Paris' law determined from the DCB samples to predict the load-life response of single-lap and double-lap joints and obtained good agreement between the predicted and experimental load-life ($L-N_f$) plots. They stated that this technique will tend to underestimate fatigue life because of neglecting of any initiation phase as well as crack growth at very small crack lengths.

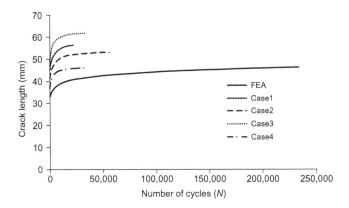

Figure 14.9 Crack length versus number of cycles.

14.5 Validation of the finite element approach

The numerical simulations for fatigue crack propagation explained in the previous sections are validated using the experimental results carried out by Hafiz (2011) using DCB test specimens. The DCB specimens consist of two mild steel substrates bonded together with FM-73M One-Side Tacky (OST) toughened epoxy film adhesive supplied by Cytec®. The substrates had the dimensions 200 × 15 × 10 mm and had been grit-blasted prior to bonding. A bondline thickness having a nominal value of 0.16 mm was produced using metal shims. In total, four DCB specimens made of steel substrates were tested under fatigue displacement control, and the crack length as a function of number of cycles was monitored. Due to the large experimental scatter between the results of these four specimens, average values for Paris' constants, Eqn (14.28), are calculated as $C = 10^{-21}$ and $m = 5.33$ (Hafiz et al., 2013) and are used in the FE simulations. The fatigue threshold is determined from the fatigue tests as 300 J/m² (Hafiz et al., 2013) and the fracture toughness is determined from quasistatic tests as 1348 J/m² (Hafiz et al., 2010). These two values are required as input parameters for the FE numerical integration of Paris' law. The crack length versus number of cycles is plotted in Figure 14.9 for the four DCB specimens and compared to FEA. It can be seen from Figure 14.9 that FEA results are within the experimental scatter of the four specimens.

14.6 Conclusions

In this chapter, techniques for simulating and predicting fatigue lifetime in adhesively-bonded joints under mode I fracture were reviewed. The chapter concentrated on the FCG approach, in which the crack growth rate was integrated from an initial crack length to a final crack length in order to estimate the fatigue lifetime. As the integration of crack growth rate required the calculation of a fracture parameter, such as G_I, analytical and numerical approaches for calculating G_I were presented. In the analytical approach, different techniques were reviewed: the area method, compliance method, beam theory and beam on elastic foundation method. FCG laws, in which the FCG rate is expressed in terms of G_I, which are often used in adhesive joints, were presented. In the numerical approach, FEA was used to model a DCB test specimen under mode I fatigue loading. G_I was extracted from FEA results using a VCCT and was expressed in terms of nodal displacements and forces. The crack growth law was integrated within the FE code using a numerical integration scheme so that the number of cycles due to crack propagation could be computed. Comparison with experimental results showed that FEA results were within the experimental scatter.

References

Abdel Wahab, M. M. (2000). On the use of fracture mechanics in designing a single lap adhesive joint. *Journal of Adhesion Science and Technology, 14*, 851−865.
Abdel Wahab, M. M. (2012). Fatigue in adhesively bonded joints: a review. *Materials Science, 2012*, 1−25.
Abdel Wahab, M. M., Ashcroft, I. A., Crocombe, A. D., & Smith, P. A. (2002). Numerical prediction of fatigue crack propagation lifetime in adhesively bonded structures. *International Journal of Fatigue, 24*, 705−709.
Abdel Wahab, M. M., Ashcroft, I. A., Crocombe, A. D., & Smith, P. A. (2004). Finite element prediction of fatigue crack propagation lifetime in composite bonded joints. *Composites Part A: Applied Science and Manufacturing, 35*, 213−222.
Ashcroft, I. A., & Shaw, S. J. (2002). Mode I fracture of epoxy bonded composite joints 2. Fatigue loading. *International Journal of Adhesion and Adhesives, 22*, 151−167.
Bannantine, J. A., Comer, J. J., & Hand-rock, J. L. (1990). *Fundamentals of metal fatigue analysis*. Englewood Cliffs, NJ: Prentice Hall.
Barsoum, R. S. (1976). On the use of isoparametric finite elements in linear fracture mechanics. *International Journal for Numerical Methods in Engineering, 10*, 551−564.
Dessureault, M., & Spelt, J. K. (1997). Observation of fatigue crack initiation and propagation in an epoxy adhesive. *International Journal of Adhesion and Adhesives, 17*, 183−195.
Fernandez, M. V., de Moura, M. F. S. F., da Silva, L. F. M., & Marques, A. T. (2011). Composite bonded joints under mode I fatigue loading. *International Journal of Adhesion and Adhesives, 31*, 280−285.
Hafiz, T. A. (2011). *Mixed-mode fracture in adhesively bonded joints under quasi-static and fatigue loading (Ph.D. thesis)*. UK: University of Surrey.
Hafiz, T. A., Abdel Wahab, M. M., Crocombe, A. D., & Smith, P. A. (2010). Mixed-mode fracture of adhesively bonded metallic joints under quasi-static loading. *Engineering Fracture Mechanics, 77*, 3434−3445.
Hafiz, T. A., Abdel Wahab, M. M., Crocombe, A. D., & Smith, P. A. (2013). Mixed-mode fatigue crack growth in FM73 bonded joints. *International Journal of Adhesion and Adhesives, 40*, 188−196.
Hashemi, S., Kinloch, A. J., & Williams, J. G. (1989). Corrections needed in double-cantilever beam tests for assessing the interlaminar failure of fibre-composites. *Journal of Materials Science Letters, 8*, 125−129.

Irwin, G. R., & Kies, J. A. (1954). Critical energy rate analysis of fracture strength. *The Welding Journal, 33*, 193−198.
Jastrzebski, M. U., Sinclair, A. N., Raizenne, D. D., & Spelt, J. K. (2009). Development of adhesive bonds with reduced fracture strength as NDE benchmarks. *International Journal of Adhesion and Adhesives, 29*, 372−379.
Johnson, W. S., Butkus, L. M., & Valentin, R. V. (1998). *Application of fracture mechanics to the durability of bonded composite joints*. Report No. DOT/FAA/AR-97/56.
Krenk, S., Jonsson, J., & Hansen, L. P. (1996). Fatigue analysis and testing of adhesive joints. *Engineering Fracture Mechanics, 53*, 859−872.
Mall, S., Ramamurthy, G., & Rezaizdeh, M. A. (1987). Stress−ratio effect on cyclic de-bonding in adhesively bonded composite joints. *Composite Structures, 8*, 31−45.
Martin, R. H., & Murri, G. B. (1990). Characterisation of mode I and mode II delamination growth and thresholds in AS4/PEEK composites. In S. P. Garbo (Ed.), *Composite materials: testing and design (Vol. 9). ASTM STP 1057* (pp. 251−270). Philadelphia, PA: American Society for Testing and Materials.
de Moura, M. F. S. F., Morais, J., & Dourado, N. (2008). A new data reduction scheme for mode I wood fracture characterization using the double cantilever beam test. *Engineering Fracture Mechanics, 75*, 3852−3865.
Pirondi, A., & Nicoletto, G. (2004). Fatigue crack growth in bonded DCB specimens. *Engineering Fracture Mechanics, 71*, 859−871.
Ripling, E. J., Crosley, P. B., & Johnson, W. S. (1998). A comparison of pure mode I and mixed mode I-III cracking of an adhesive containing an open knit cloth carrier. In W. S. Johnson (Ed.), *Adhesively bonded joints: testing, analysis and design, ASTM STP 981* (pp. 163−182). Philadelphia, USA: American Society for Testing and Materials.
Sethuraman, R., & Maiti, S. K. (1988). Finite element based computation of strain energy release rate by modified crack closure integral. *Engineering Fracture Mechanics, 30*, 227−331.
Shenoy, V., Ashcroft, I. A., Critchlow, G. W., Crocombe, A. D., & Abdel Wahab, M. M. (2009). An evaluation of strength wearout models for the lifetime prediction of adhesive joints subjected to variable amplitude fatigue. *International Journal of Adhesion and Adhesives, 29*, 639−649.
Taylor, D. (1989). *Fatigue thresholds*. London: Butterworths.
Timoshenko, S. P., & Goodier, J. N. (1970). *Theory of elasticity* (3rd ed.). New York, NY: McGraw Hill.
Williams, J. G. (1988). On the calculation of energy release rates for cracked laminates. *International Journal of Fracture, 36*, 101−119.

Predicting the fatigue life of adhesively-bonded composite joints under mixed-mode fracture conditions

P. Naghipour
NASA Glenn Research Center, Cleveland, OH, USA

15.1 Introduction

15.1.1 Why modeling of mixed-mode fatigue delamination of composites is important

The use of composite materials and their structural components is attractive in aircraft industry since they enable reduced airframe weight and therefore better fuel economy and lower operating costs. In aircraft structures, carbon fiber—reinforced plastic (CFRP) composites may be used in control surfaces, wings, and numerous parts of the fuselage. Together with the growing use of CFRP in aircraft industry, reliable prediction and analysis of their failure mechanisms under various loading conditions must be studied extensively. Cyclic loading is one of the most significant loading conditions to which aircraft structures are subjected throughout their life. The growth of the cyclic damage under subsequent loading may lead to final catastrophic failure of the structural component. Thus, predicting the extent of fatigue damage growth through development of an accurate fatigue damage model is essential to the continued employment of CFRP structures into even more demanding aerospace applications. Delamination is particularly important for the structural integrity of composite structures because it is difficult to detect during inspections. Furthermore, delamination drastically reduces the bending stiffness of a composite structure and, when compressive loads are present, promotes local buckling that can compromise the global stability of the structure. Cyclic delamination failures in CFRP laminates generally arise under mixed-mode stresses, which combine mode I (normal) and mode II (shear) stresses. Because of the complex interactions of normal and shear modes, analysis of cyclic mixed-mode delamination in CFRP laminates is a challenging task. Costly cyclic experiments, lack of testing standards, and data reduction procedures for the evaluation of multidirectional laminates are the main motives for the development of a numeric analysis tool that provides a reliable estimation of the remaining load-bearing capacity of the CFRP structure subjected to successive delamination cycles, and the development of this numeric analysis tool is the main objective of this chapter.

15.1.2 Novelties of the mixed-mode fatigue delamination numerical tool

Few studies of single- or mixed-mode fatigue delamination have been reported in literature, but no attempts to numerically model the fatigue delamination in multi-directional CFRPs with varying fiber orientations at the delamination interface have been undertaken. Accordingly, addressing the deficiencies in the available literature, this chapter focuses on the development of a precise predictive numerical tool able to successfully estimate the successive loss of load-bearing capacity and predict damage modes occurring in multidirectional CFRPs subjected to cyclic mixed-mode delamination. A comprehensive combination of numerical and experimental analysis tools, mainly emphasizing the effect of multidirectional interfaces on mixed-mode fatigue delamination of CFRP laminates, also is addressed thoroughly in this chapter. For modeling the mixed-mode delamination growth of cracks under cyclic loading, the proposed numerical approach incorporates the interface formulation suggested by Turon, Costa, Camanho, and Dávila (2007), further elaborated by two remarkable improvements: redefining the interfacial constitutive law and increasing fidelity of the model. The functionality of the numerical model then is validated by predicting the interacting damage states and by reproducing load reduction in successive cycles of the conducted cyclic mixed-mode bending (MMB) experiments.

15.2 Diverse approaches to modeling fatigue life of composite materials

There are several approaches to describe the mixed-mode fatigue and fracture (delamination) phenomena in composite materials. Among the most representative approaches for describing the experimental fatigue behavior are fatigue life models, which predict the number of cycles (N) corresponding to fatigue failure under fixed loading conditions (S) using $S-N$ curves (Andersons, 1994; Reifsnider, 1991; Suresh, 1991; Talreja, 1999). Fatigue life models predict the number of cycles until fatigue-induced failure under fixed loading conditions using $S-N$ curves or coupled with a fatigue failure criterion, which is generally a function of the ultimate strengths. The second approach can be classified as fatigue-related fracture models, which basically study the rate of crack growth under cyclic loading (e.g., mixed-mode cyclic delamination). Fracture mechanics models relate the variation of the energy to form two new crack surfaces with the crack growth (Dowling & Begley, 1976; Ewalds, 1984; McDowell, 1997; Paris & Erdogan, 1963; Paris, Gomez, & Anderson, 1961; Rice, 1980). Crack propagation rate under fatigue is denoted by dA/dN, where A is the crack area, or da/dN, where a is the characteristic crack length. The correlation of the fatigue-related crack growth rate with the amplitude of the energy release rate, ΔG (or stress intensity factor, ΔK), is commonly represented in a log–log diagram known as a Paris plot (Paris & Erdogan, 1963; Paris et al., 1961). The Paris law, describing crack growth versus energy release rate, is widely used and accepted among other empirical or semiempirical crack growth laws. According to this law, the crack growth rate is related to the energy release rate range by a power law that can be expressed as

$$\frac{\partial A}{\partial N} = C\left(\frac{\Delta G}{G_c}\right)^m \quad (15.1)$$

The parameters C and m (Paris plot parameters) must be determined experimentally. The energy release rate range, ΔG, depends on the loading conditions, and G_c is the critical energy release rate of the material. The third general approach, used throughout this chapter, is damage mechanics models, in which the deterioration of material's mechanical response is characterized by a dimensionless field variable (damage variable, d). Assuming that the fracture process in CFRP takes place in an infinite thin plane, the plane of delamination cracking, the evolution of damage until separation of the material under cyclic loading can be described by cohesive zone models, which can be extended from cohesive laws for quasi-static loading into forms suitable for cyclic loading. The cohesive zone technique, first suggested by Dugdale (1960) and Barenblatt (1962), is based on using an interface element technique to predict crack initiation and propagation (Alfano & Crisfield, 2001; Allix & Blanchard, 2006; Harper & Hallett, 2008; Naghipour, Schulze, Hausman, & Bartsch, 2012; Turon, Dávila, Camanho, & Costa, 2007). Subjected to cyclic loading, the constitutive law of the interface element must be reformulated to account for subcritical damage accumulation and stiffness degradation within subsequent unloading—reloading steps (Maiti & Geubelle, 2005; Munoz, Galvanetto, & Robinson, 2006; Nguyen, Repetto, Ortiz, & Radovitzky, 2001; Peerlings, Brekelmans, de Borst, & Geers, 2000; Roe & Siegmund, 2003; Serebrinsky & Ortiz, 2005; Turon, Costa, et al., 2007; Yang, Mall, & Ravi-Chandar, 2001). Although there is a wealth of experimental results on fatigue-related crack propagation in quasi-brittle materials such as fiber-reinforced composites, few attempts to numerically model fatigue-induced crack propagation using a cohesive technique have been made. A short overview of the studies available in the literature is given in the following section.

15.3 Various cohesive zone models for cyclic delamination

To capture the effect of cyclic crack growth, it has been identified that a distinction needs to be made between the loading and unloading paths allowing for hysteresis. This physical phenomenon is represented mathematically by incorporating a cyclic damage variable, which evolves with the number of cycles. Various numerical approaches to study the cyclic failure phenomena using a cohesive zone technique are briefly described here.

15.3.1 Yang, Mall, and Ravi-Chandar (2001)

Yang et al. (2001) modeled fatigue-induced crack growth in quasi-brittle materials using a cohesive zone model incorporating irreversible damage, which is assumed to accumulate not only along the damage locus but also during any unloading–reloading path. This idea makes it possible to predict the subcritical crack growth caused by cyclic loading. Therefore, the fatigue damage behavior of a material can be studied under any arbitrary loading condition provided that the properties of the cohesive zone are specified correctly. Yang et al. also proposed a cohesive law for a general polynomial form, representing different stiffness, K, expressions for unloading and reloading paths. The predicted reduction of stiffness due to each cycle is given in the following Eqn (15.2)

$$\frac{\mathrm{d}\ln k}{\mathrm{d}N} = \frac{\left(\sum_{l=1}^{L} \alpha_l (w_c/w_d)^l - \sum_{m=1}^{M} \beta_m (w_c/w_d)^m\right)}{1 + \sum_{m=1}^{M} \beta_m (w_c/w_d)^m} \tag{15.2}$$

where β, m, α, and l are the user-defined parameters representing a polynomial of degree L or M. N, w_c, and w_d represent the number of cycles, the displacement jump, and the damage parameter, respectively. Detailed information on the above-mentioned cohesive fatigue damage model is given by Yang et al. (2001).

15.3.2 Roe and Siegmund (2003)

According to Roe and Siegmund (2003), the delamination damage process of a structure is viewed as a result of progressive material deterioration in the cohesive zone and the interaction thereof with the surrounding continuum. When subjected to monotonic loading, the cohesive law developed to describe material separation is given by a potential (energy) function, which is motivated by interatomic potentials. The derivatives of the defined potential with respect to separations provide cohesive traction in normal T_n and shear T_s modes under monotonic loading (e is the Euler number).

$$T_n = \sigma_{\max,0} e \exp\left(\frac{-\Delta u_n}{\delta_0}\right) \left\{ \frac{\Delta u_n}{\delta_0} \exp\left(\frac{-\Delta u_t^2}{\delta_0^2}\right) \right.$$
$$\left. + (1-q)\frac{\Delta u_n}{\delta_0}\left[1 - \exp\left(\frac{-\Delta u_t^2}{\delta_0^2}\right)\right]\right\} \tag{15.3}$$

$$T_t = 2\sigma_{\max,0} eq\left(\frac{\Delta u_t}{\delta_0}\right)\left\{\left(1 + \frac{\Delta u_n}{\delta_0}\right)\exp\left(-\frac{\Delta u_n}{\delta_0}\right)\exp\left(-\frac{\Delta u_t^2}{\delta_0^2}\right)\right\}$$

One of the material parameters in these constitutive relations is the initial cohesive strength under monotonic loading, $\sigma_{\max,0}$, that is, the maximum normal traction reached under pure normal loading. The second material parameter, the cohesive length, δ_0, is the displacement where initial separation occurs corresponding to the cohesive normal strength. Δ_{ut} and Δ_{un} stand for shear and normal displacement jumps (separations), and q is the ratio of shear to normal cohesive surface energies.

Under cyclic loading, the constitutive relation for a cohesive zone model accounting for damage accumulation in every cycle is given by replacing the tractions with effective tractions. In other words, during each unloading and reloading cycle, the mentioned tractions are degraded by the factor $(1 - D_c)$, where D_c $(0 < D_c < 1)$ stands for the damage parameter. The evolution equation for damage of the cohesive zone under cyclic loading, D_c, is given by Roe and Siegmund (2003) as:

$$\dot{D}_c = \frac{|\Delta \bar{u}|}{\delta_\Sigma} \left[\frac{T_{CZ}^{eff}}{\sigma_{max}} - C_f \right] H(\Delta \bar{u} - \delta_0) \qquad (15.4)$$

In unloading—reloading conditions the suggested Eqn (15.4) for the irreversible degradation of the cohesive zone above incorporates the effects of accumulation of damage during subcritical cyclic loading. H is the Heaviside function, C_f is a material constant, Δu is the resultant separation, and T_{eff} is the effective cohesive traction. δ_Σ determines the amount of accumulated effective separation necessary to fail the cohesive zone (the displacement where final separation occurs) and is a multiple of δ_0. Further information about this cohesive damage model is given by Roe and Siegmund (2003).

15.3.3 Maiti and Guebelle (2005)

Maiti and Geubelle (2005) developed a model that relies on the combination of a bilinear cohesive failure law used for fracture simulations under monotonic loading and a damage evolution law relating the cyclic degradation of the cohesive stiffness with the rate of crack opening displacement and the number of cycles since the onset of failure. The fatigue component of the cohesive model involves two parameters that can be readily calibrated based on the classical log—log Paris crack growth curve between the crack advance per cycle and the amplitude of the stress intensity factor applied. The cohesive model, leading to similar unloading and reloading paths in the traction—separation curve, prevents crack growth under subcritical cyclic loading because of the progressive degradation of the cohesive properties in the failure zone. This limitation suggests the need for an evolution law to describe the changes incurred by the cohesive strength under fatigue. A phenomenological model of such processes involves the progressive degradation of the cohesive zone's strength during reloading events. Hence, under cyclic loading, the evolution law of the instantaneous cohesive stiffness K_c, that is, the ratio of the cohesive traction, T_n, to the displacement jump, Δu, during reloading is expressed as shown in Eqn (15.5). The cohesive strength decays exponentially, and the rate of decay is controlled by the parameter λ.

Maiti and Geubelle define a power law relation as a function of number of cycles, N, for γ, with mathematical constants α and β describing the degradation of the cohesive failure properties.

$$k_c = \frac{dT_n}{d\Delta_n} = -\gamma(N)T_n = \underbrace{\frac{1}{\alpha}N^{-\beta}}_{\gamma} T_n \quad (15.5)$$

The proposed evolution law for the cohesive model can also be expressed in terms of the rate of change of K_c and then discretized in time steps, as follows:

$$k_c^{i+1} = k_c^i + \left(-\frac{1}{\alpha}N^{-\beta}k_c^i\left(\Delta_n^{i+1} - \Delta_n^i\right)\right) \quad (15.6)$$

The superscripts i and $i+1$ stand for loading steps i and $i+1$, respectively. As shown in the above-mentioned relations, during the reloading phase, the cohesive stiffness at each material point along the cohesive zone gradually decreases in proportion to the increment in displacement of the crack opening. This proportionality factor γ evolves with the number of cycles N and thus gives a measure of the total damage accumulated during the degradation process.

15.3.4 Serebrinsky and Ortiz (2005)

Yet another similar approach presented by Serebrinsky and Ortiz (2005) and Nguyen et al. (2001) states that under monotonic loading the cohesive tractions decrease linearly with the opening displacement and eventually reduce to zero upon attaining a critical loading displacement. The formation of the new surface entails the expenditure of a well-defined energy per unit area, known variously as the strain energy release rate. For fatigue applications, specifying the monotonic loading envelope is not enough and, therefore, the counterpiece is a cohesive law with unloading—reloading hysteresis. In materials that show no plastic deformation in a process zone in front of the crack tip, degradation mechanisms (behind the crack tip) in the crack wake might prevail. For example, upon unloading and subsequent reloading, interlocking asperities in a material may rub against each other, and this frictional interaction dissipates energy. This repeated rubbing of asperities may result in wear or smoothening of the contact surfaces, resulting in a steady weakening of the cohesive response. Therefore, it can be assumed that the interfacial stiffness during unloading—reloading degrades with the number of cycles. Similar to the previously defined damage laws, for a monotonically increasing opening displacement, the traction across the cohesive surface is governed by a monotonic envelope. For fatigue applications, as mentioned earlier, specification of the monotonic cohesive envelope is not enough, and the material stiffness degradation in each unloading—reloading cycle must be considered. The cohesive interface is cycled at amplitudes smaller than the cohesive envelope. Therefore, a simple phenomenological model, which embodies this assumption, is obtained by assuming different incremental stiffness values depending on whether the cohesive surface opens (is reloaded) or closes (is unloaded). Accordingly, for a cyclically applied opening, the reloading stiffness is assumed to evolve in accordance with the following kinetic relation in each unloading–reloading cycle:

$$K_{N+1}^{+} = \left(\frac{\delta}{\delta_0}\left(1 - e^{-\delta_0/\delta}\right)^2 + e^{-2\delta_0/\delta}\right)^{N+1} K_0^{+} \quad (15.7)$$

K and K_0 are the incremental and initial stiffness, respectively, and δ and δ_0 stand for the opening and initial opening displacement, respectively.

15.3.5 Munoz, Galvanetto, and Robinson (2006)

The next approach to cyclic damage simulation is given by Munoz et al. (2006). It is stated that an alternative approach for the simulation of fatigue driven delamination growth has to incorporate fatigue degradation into the interface element technique used to model crack propagation. The fatigue damage component of the interface model given by Munoz et al. (2006) is adapted from Peerling's law (Peerlings et al., 2000) and rewritten for the mixed-mode delamination. The evolution law for the cyclic damage in coupled mode is written as

$$\Delta D_{i,f} = \frac{\Delta N C e^{\lambda D_i}}{1 + \beta} \left(\frac{1 + \gamma}{1 + \gamma_c}\right)^{\beta+1} \quad (15.8)$$

$$\gamma_c = \left(\frac{\delta_{cI}}{\delta_{0I}}\right)^{\alpha} + \left(\frac{\delta_{cII}}{\delta_{0II}}\right)^{\alpha} - 1$$

The relative displacements, δ_{0i} and δ_{ci}, are the elastic limit and the failure limit of the relative displacement, respectively, and for $\delta > \delta_{ci}$ the two surfaces of the interface element are considered completely disconnected. The subscript i refers to modes of failure: mode I and mode II. The definition of γ also provides the single-mode delamination by assuming that one of the two relative displacements is zero. C, λ, α, and β are parameters of the model that have to be determined by comparison with experimental data. The above-mentioned fatigue damage law allows the damage to increase with the number of cycles even if the initial damage of the interface is zero. Thus, using this law, a crack can grow and propagate even in an initially undamaged interface. By using some numerical examples, the computational robustness of the formulation described by Munoz et al. (2006) is tested and detailed information about this constitutive model is available in the corresponding reference.

15.4 Cohesive zone model for cyclic delamination incorporating the Paris fatigue law

Turon, Costa, et al. (2007) also proposed a damage model for simulating delamination propagation under high-cycle fatigue loading. Similar to the works mentioned works in the previous section, the basis for the formulation is a cohesive law that links fracture and damage mechanics to establish the evolution of the damage variable in terms of the crack growth rate dA/dN. Turon et al. implemented the present model as a user-written finite element in ABAQUS (2006) by adding the fatigue damage model to the constitutive behavior of a cohesive element previously developed by them (Turon, Dávila, et al., 2007). One of the major improvements added by Naghipour (2011) is the enhancement in the element fidelity. To reduce the computational time, a quadratic cohesive element was formulated instead of the linear ones used by Turon, Dávila, et al. (2007), Turon, Costa, et al. (2007). Next, to account for cyclic loading, a methodology similar to the one presented by Turon, Costa, et al. (2007) is followed but with a different definition of the cohesive zone area. A cyclic damage parameter evolving based on number of cycles has to be defined to account for interfacial cyclic degradation. The evolution of the damage variable, d, is related to the crack growth rate dA/dN as follows:

$$d_{\text{cyclic}} = \frac{\partial d}{\partial N} = \frac{\partial d}{\partial A_d} \frac{\partial A_d}{\partial N} \tag{15.9}$$

The first term, $\partial d/\partial A_d$, is obtained using the formulation, which relates the damage parameter d to the damaged area (dissipated energy region), A_d, in the cohesive section. In the context of the mechanics of finite element damage, the ratio of the energy dissipated during the fracture process, A_d, with respect to the total area of the cohesive response, A_{tot} (Figure 15.1(a)), is given by A_d/A_{tot} (Eqn (15.10)). Meanwhile, the damage variable d is defined as a function of maximum mixed-mode relative displacement in the previous history, $\delta_m^{\max}$, initial displacement at delamination onset, δ_m^0, and final separation displacement, δ_m^f.

$$\frac{A_d}{A_{\text{tot}}} = 1 - (1 - d)\frac{\delta_m^{\max}}{\delta_m^0} \tag{15.10}$$

$$d = \frac{\delta_m^f \left(\delta_m^{\max} - \delta_m^0\right)}{\delta_m^{\max} \left(\delta_m^f - \delta_m^0\right)} \tag{15.11}$$

Corresponding normal and shear tractions, τ_n and τ_{shear}, are related elastically to displacements up to the onset of delamination (Figure 15.1(b)). Once the delamination has started, it evolves based on an energy-based propagation criterion introduced by Benzeggagh and Kenane (B-K) (1996) until a final separation point, δ_m^f, is reached. This corresponds physically to the totally damaged state of the interface element.

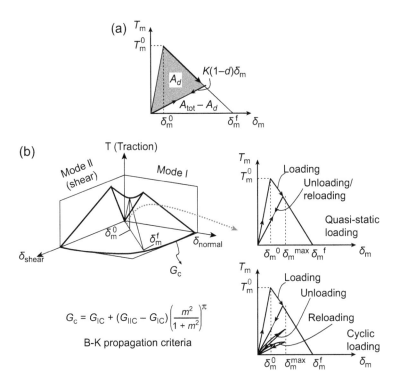

Figure 15.1 (a) Traction displacement laws describing the numerical constitutive equations of the cohesive zone models after Turon, Dávila, et al. (2007). (b) Cohesive law for mixed-mode delamination with linear softening.

G_{IC} and G_{IIC} and η appearing in the B-K criterion (Figure 15.1(b)) stand for mode I (normal mode) and mode II (shear mode) critical strain energy release rates and the parameter defining the failure locus under mixed-mode loading, respectively. These parameters have to be determined by mixed-mode experiments (Benzeggagh & Kenane, 1996).

By solving Eqn (15.10) for δ_{max} and substituting it in the damaged area ratio (Eqn (15.11)), $\partial d/\partial A_d$ can then be derived using the chain rule:

$$\frac{\partial d}{\partial A_d} = \frac{1}{A_e} \frac{\left[\delta_m^f(1-d) + d\delta_m^0\right]^2}{\delta_m^0 \delta_m^f} \qquad (15.12)$$

where A_e represents the area of a cohesive element. The second term, $\partial A_d/\partial N$, represents the mean value of the damaged area growth rate per element per cycle in the cohesive zone. The width of the delamination front (numerical damaged area) is assumed to be constant. Therefore the growth of the damage area correlates directly with the growth of the crack length. According to Turon, Costa, et al. (2007) the crack growth rate ($\partial A/\partial N$) can be assumed to be equal to the sum of the damaged area growth rates of all damaged elements ahead of the crack tip:

$$\frac{\partial A}{\partial N} = \sum_{e \in A_{cz}} \frac{\partial A_d^e}{\partial N} = \frac{A_{CZ}}{A^e} \frac{\partial A_d}{\partial N} \qquad (15.13)$$

where

$\dfrac{A_{CZ}}{A^e}$ = number of cohesive elements in the cohesive zone.

The area of the cohesive element can be varied by changing the element length or, in other words, by changing the mesh size in the cohesive zone. Previous publication by Naghipour, Schneider, Bartsch, Hausmann, and Voggenreiter (2009) shows the sensitivity of the model to cohesive element length. According to this work, comparisons with experimentally obtained results indicate that as long as the size of the interface element is <1 mm, a better solution convergence can be achieved in the case of mixed-mode loading. Therefore, here in this work, the cohesive element length is taken as 0.6 mm in all the simulations.

To define the rate of crack growth under fatigue loading, $\partial A/\partial N$, the Paris law given in Eqn (15.14a) is embedded in Eqn (15.9). Accordingly, rewriting Eqn (15.9) gives:

$$\frac{\partial A}{\partial N} = C\left(\frac{\Delta G}{G_c}\right)^m \qquad (15.14a)$$

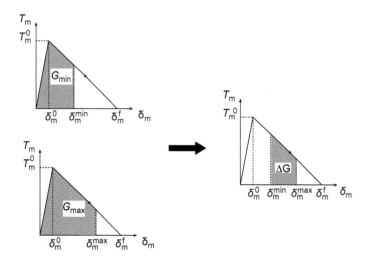

Figure 15.2 Variation of the energy release rate after Turon, Dávila, et al. (2007).

$$\frac{\partial d}{\partial N} = \left\{ \begin{array}{ll} \dfrac{1}{A_{cz}} \dfrac{\left(\delta_m^f(1-d) + d\delta_m^0\right)^2}{\delta_m^f \delta_m^0} C\left(\dfrac{\Delta G}{G_c}\right)^m & G_{th} < G_{max} < G_c \\ 0 & 0 \end{array} \right\} \quad (15.14b)$$

C, m, and G_{th} are the Paris plot parameters that are obtained by plotting $\partial A/\partial N$ versus cyclic variation of the energy release rate ΔG on a log–log scale. G_c is the total mixed-mode fracture toughness under a specific mode ratio. In contrast to the work of Turon, Costa, et al. (2007), required Paris parameters were calculated directly using the experimental outputs of this work. As mentioned by Blanco, Gamstedt, Asp, and Costa (2004), there are several expressions where the cyclic crack propagation rates are expressed by relative contributions of mode I and mode II parameters. However, it has not yet been proven that either of these expressions or the one used by Turon et al. provides the best fit for all types of composite materials. Therefore, the parameters are extracted directly from the related MMB experiments in this work. The maximum energy release rate G_{max} and cyclic variation in the energy release rate ΔG can be computed by the constitutive law of the cohesive zone model using the symmetry of triangles in Figure 15.2:

$$G_{max} = \frac{\tau_m^0 \left(\delta_m^f\right)}{2} - \left(\frac{\delta_m^f - \delta_m^{max}}{\delta_m^f - \delta_m^0}\right)^2 \frac{\tau_m^0 \left(\delta_m^f - \delta_m^0\right)}{2}$$

$$G_{max} = \frac{\tau_m^0}{2} \left(\delta_m^f - \frac{\left(\delta_m^f - \delta_m^{max}\right)^2}{\delta_m^f - \delta_m^0} \right)$$

(15.15)

In Eqn (15.15) G_{max} is the area of the trapezoid expanding from origin to δ_m^{max} (Figure 15.2). Assuming the load ratio as $R^2 = G_{min}/G_{max}$, ΔG can be written as:

$$\Delta G = \frac{\tau_m^0}{2}\left(\delta_m^f - \frac{\left(\delta_m^f - \delta_m^{max}\right)^2}{\delta_m^f - \delta_m^0}\right)(1 - R^2) \quad (15.16)$$

Finally, the final form of evolution of the damage parameter with subsequent cycles, Eqn (15.9), can be rewritten as:

$$\frac{\partial d}{\partial N} = \left\{ \begin{array}{ll} \dfrac{1}{A_{cz}} \dfrac{\left(\delta_m^f(1-d) + d\delta_m^0\right)^2}{\delta_m^f \delta_m^0} C\left(\dfrac{\dfrac{\tau_m^0}{2}\left(\delta_m^0 + \dfrac{\left(\delta_m^f - \delta_m^{max}\right)^2}{\delta_m^f - \delta_m^0}\right)(1-R^2)}{G_c}\right)^m & G_{th} < G_{max} < G_c \\ 0 & 0 \end{array} \right\}$$

(15.17)

In Eqns (15.14)–(15.17), A_{cz} stands for the cohesive zone area, defined as the area from the crack tip to the point where the maximum cohesive traction is attained. Turon, Costa, et al. (2007) defined the A_{cz} based on the closed form equation described by Rice (1980). However, this equation was developed for pure mode I loading, and the model tends to be less accurate when subjected to mixed-mode loading with higher mode II domination. To improve the functionality and accuracy of the model, in this work the estimation of the cohesive zone area is further improved for a mixed-mode load case.

15.5 Cohesive zone model for cyclic delamination incorporating the Paris fatigue law and a mixed-mode cohesive area

Cohesive zone length and, consequently, the cohesive zone area are defined as structural and material properties. Different models to estimate the length of the cohesive zone have been proposed. The first estimation, suggested by Dugdale (1960), is based on the size of the yield zone ahead of a mode I crack by idealizing the plastic region as a narrow strip extending ahead of the crack tip. Analogous to Dugdale (1960), Barenblatt (1962) provided a similar estimation for ideally brittle materials. Rice (1980) estimated the length of the cohesive zone as a function of the crack growth velocity. The expressions resulting from these models can be found in the literature (Barenblatt, 1962; Dugdale, 1960; Rice, 1980). Under plane stress conditions, for an isotropic material these models have a general form:

$$A_{cz} = bE\frac{G_c}{\tau_0^2} \quad (15.18)$$

Modified versions of Eqn (15.18) have been developed (Harper & Hallett, 2008) and for mode I and mode II components of mixed-mode loading can be written as:

$$A_{cz,I} = bE'\frac{G_{Ic}}{\left(\tau_n^0\right)^2} \quad (15.19a)$$

$$A_{cz,II} = bE'\frac{G_{IIc}}{\left(\tau_s^0\right)^2} \quad (15.19b)$$

E' is an equivalent elastic modulus for an orthotropic material, whose value depends on longitudinal and transverse modulus. For the transversely isotropic laminate here, the value of E' is assumed to be equal to the elastic modulus in thickness direction (E_{33}). b is the specimen width, G_{IC} and G_{IIC} are the critical energy release rate for mode I and mode II components of the mixed-mode loading, and τ_n^0, τ_s^0 are the maximum interfacial strength of the cohesive element in normal and shear directions respectively.

Based on the detailed parametric studies conducted by Harper and Hallett (2008), the most reasonable mixed-mode cohesive area is predicted as the minimum possible area of the fully developed cohesive zone multiplied by a scaling factor M, using the formula:

$$A_{cz,mixed} = M[\min (\text{Eqn (15.19a) and (15.19b)})] \qquad (15.20)$$

The applied scaling factor of 0.5 was chosen by Harper and Hallett (2008) and 0.65 in this work because they provide the best correlation between numerical and experimental results when compared to each other. $A_{cz,mixed}$, obtained through Eqn (15.20), is taken as the effective cohesive area in all the subsequent calculations in this chapter. The improvement achieved via reformulation of the cohesive zone area is shown schematically in the following section ("Numerical simulations of cyclic MMB experiments"). It is worth mentioning that in all the different approaches considered for the evaluation of cyclic damage growth in the literature (Maiti & Geubelle, 2005; Munoz et al., 2006; Nguyen et al., 2001; Peerlings et al., 2000; Roe & Siegmund, 2003; Serebrinsky & Ortiz, 2005; Turon, Costa, et al., 2007; Yang et al., 2001), as in this work, either the degradation of the interface stiffness K, or the evolution of the damage parameter d, per cycle is defined explicitly (see Table 15.1 for a summary).

15.6 Modeling cyclic mixed-mode delamination using the developed cohesive zone technique

15.6.1 One element tests

Several single-element tests were performed to verify the response of the fatigue damage model. The finite element model shown in Figure 15.3 is composed of two quadratic plane stress elements connected by a 16-node cohesive element representing the interface. The material properties corresponding to the PEEK/AS4 carbon fiber—reinforced laminate, and the Paris law coefficients used in the simulation were taken from experimental results available in the literature. The load was applied in two steps. The first loading step was a quasi-static step and the second step was the cyclic loading with a predefined amplitude. The evolution of the interface traction in the constitutive equation for a displacement-controlled cyclic loading is shown in Figure 15.4. Moreover, Figure 15.5 clearly indicates that without implementing the fatigue damage law, no traction degradation is observed within successive cycles, which supports the importance of considering and implementing the cyclic damage model. The evolution of the interface traction with the number of cycles is shown in Figure 15.5, and both results designate that fatigue damage causes a reduction of the stiffness and the interfacial traction.

Table 15.1 **Different approaches to evaluating the cyclic damage parameter found in the literature**

$\dfrac{d\ln k}{dN} = \dfrac{\left(\sum_{l=1}^{L}\alpha_l(w_c/w_d)^l - \sum_{m=1}^{M}\beta_m(w_c/w_d)^m\right)}{1+\sum_{m=1}^{M}\beta_m(w_c/w_d)^m}$ β, m, α, and L: fitting parameters to represent a polynomial of degree L; N: number of cycles; w: displacement jump; K: initial cohesive stiffness	Yang et al. (2001)
$k_c^{i+1} = k_c^i + \left(-\dfrac{1}{\alpha}N^{-\beta}k_c^i\left(\Delta_n^{i+1} - \Delta_n^i\right)\right)$ β, α: fitting parameters to be determined from experiments; k_c: (cohesive stiffness) degrades in each unloading–reloading step	Maiti and Geubelle (2005)
$K_{N+1}^+ = \left(\dfrac{\delta}{\delta_0}(1-e^{-\delta_0/\delta})^2 + e^{-2\delta_0/\delta}\right)^{N+1} K_0^+$ K: cohesive stiffness decays exponentially during each cyclic step; δ: equivalent displacement jump	Serebrinsky and Ortiz (2005)
$\dot{D}_c = \dfrac{|\Delta \bar{u}|}{\delta_\Sigma}\left[\dfrac{T_{CZ}^{eff}}{\sigma_{max}} - C_f\right]H(\Delta\bar{u} - \delta_0)$ T_{CZ}^{eff}: effective cohesive zone traction, Δu: equivalent displacement jump; C_f: material constant to be determined from experiments; H: heaviside function, δ_Σ: cohesive zone length	Roe and Siegmund (2003)
$\Delta D_{i,f} = \dfrac{\Delta NCe^{\lambda D_i}}{1+\beta}\left(\dfrac{1+\gamma}{1+\gamma_c}\right)^{\beta+1}$ (derived from Peerling's law) γ: equivalent displacement jump; C, λ, and β: fitting parameters to be determined from experiments	Munoz et al. (2006)
$\dfrac{\partial d}{\partial N} = \dfrac{\partial d}{\partial A}\underbrace{\dfrac{\partial A}{\partial N}}_{\text{Paris law}}$ $\dfrac{\partial d}{\partial N} = \dfrac{1}{A_{cz}}\dfrac{\left(\delta^f(1-d)+d\delta^0\right)^2}{\delta^f\delta^0}\dfrac{\partial A}{\partial N}$ $\dfrac{\partial A}{\partial N} = C\left(\dfrac{\Delta G}{G_c}\right)^m = C\left(\dfrac{\dfrac{\tau_0}{2}\left(\delta^0 + \dfrac{(\delta^f-\delta^m)^2}{\delta^f-\delta^0}\right)}{G_c}\right)^m$ C, m: fitting parameters to be determined from experiments; A_{cz}: area of the cohesive zone	Turon, Costa, et al. (2007)

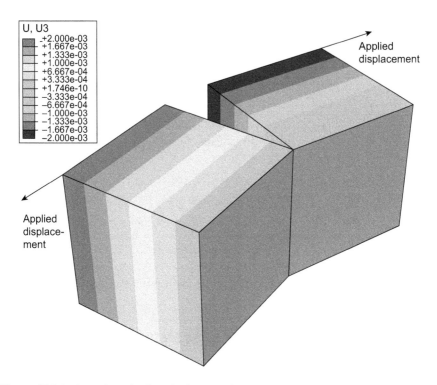

Figure 15.3 Deformed mesh of a cohesive one-element test.

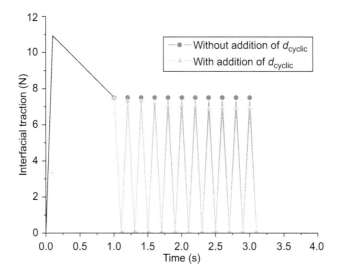

Figure 15.4 Evolution of the interfacial traction in the constitutive equation for a displacement-controlled cyclic loading test.

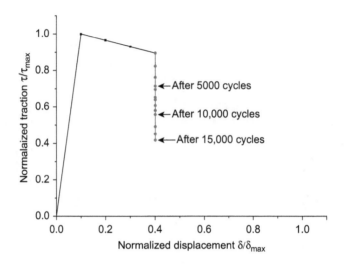

Figure 15.5 Evolution of the interface traction with the number of cycles for a one-element test.

15.6.2 Numerical simulations of cyclic MMB experiments

Numerical simulations of 50% MMB tests under cyclic loading, with the added fatigue damage law, were performed to demonstrate that the constitutive damage model can be used in a structural analysis and successfully reproduces the response of the test specimens. The numerical model is a combination of 24 individual plies with specified orientations (Table 15.2), together with interface elements, placed in the midplane of the laminate to capture the delamination behavior. Each ply is assumed as an orthotropic continuum under plane stress modeled using a reinforced ply model (Hashin, 1981) with quadratic, reduced integration shell elements. Interface elements are zero thickness, 16-node, mixed-mode cohesive elements, implemented as a User Element in ABAQUS (2006), as described earlier in Section 15.4. Loading boundary conditions (displacements) are applied directly to middle and end supports (Figure 15.6).

The material properties required for the numerical cyclic MMB simulation are given in Tables 15.3 and 15.4. For each lamina, X, Y, and S stand for ultimate in-plane strength in fiber, transverse, and shear directions, with E_{11}, E_{22}, and G_{12}

Table 15.2 **Multidirectional carbon fiber-reinforced plastic specimen configurations**

Layups to be considered	Layup name
$(+22.5/-22.5)_{12}$	Layup 22.5
Quasi-isotropic ($[0/\pm 45/90]_6$)	Layup QI

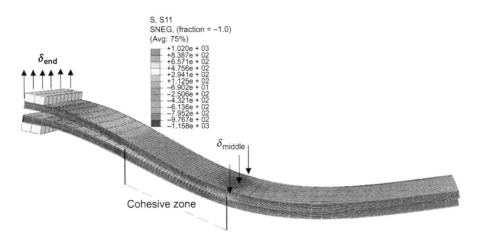

Figure 15.6 Schematic view of the numerical model and applied boundary conditions.

Table 15.3 Mechanical properties of lamina

Property		Property	
E_{11} (MPa)	138,000	X_t (MPa)	2070
E_{22} (MPa)	10,500	X_c (MPa)	1360
ν_{12}	0.3	Y_t (MPa)	86
G_{12} (MPa)	6300	Y_c (MPa)	196
G_{23} (MPa)	3500	S (MPa)	147

t: tension; c: compression.
Source: Kohgruber (1997).

Table 15.4 Mechanical properties of interface (layups 22.5 and QI)

Layup	τ_n^0 (MPa)	$\tau_s^0 = \tau_t^0$ (MPa)	K (N/mm³)	G_{Ic} (mJ/mm²)	G_{IIc} (mJ/mm²)	η
22.5	75	80	10^7	1.74	2.89	2.3
QI	75	80	10^7	1.36	2.21	2.25

Paris plot parameters for layup 22.5: $C = 0.0015$ mm/cycle, $m = 5.5$; $G_{th} = 0.08$ mJ/mm²; QI: $C = 0.000959$ mm/cycle; $m = 5.3$; $G_{th} = 0.06$ mJ/mm².

representing the corresponding moduli, respectively. The laminar properties mentioned are determined from standard tension and compression coupon tests in fiber, matrix, and shear directions (Kohlgruber, 1997). For the interface element, G_{Ic} and G_{IIc} are obtained using the corrected beam theory data reduction scheme described by

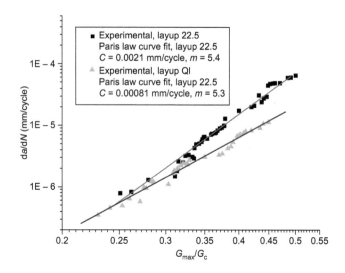

Figure 15.7 Paris plot for a mode mix of 50% showing two exemplary experimental results and the linear regression.

Naghipour, Bartsch, Chernova, Hausmann, and Voggenreiter (2009). Normal and shear interfacial strengths are estimated to be 70–80% of resin strength, and the mathematical penalty value for the initial interfacial stiffness K is approximated to be 10^7 N/mm^3. The fatigue-related Paris plot parameters are extracted from cyclic MMB experimental data (Figure 15.7). Since the loading lever is not simulated, specified displacement increments are applied directly to middle and end supports, as shown in Figure 15.6. The loading is defined in two steps: the first analysis loading step is quasi-static and it ends at the maximum applied displacement. It is assumed that no fatigue damage accumulates during this step. Next, a second loading and unloading step, in which the maximum displacement is held constant during the cycle, is applied; the step time increment is assumed to be 0.1 so that 10 successive cycles can be simulated in a time step.

The results obtained from these simulations and the experimental data are shown in Figure 15.8. It can be observed that the constitutive model successfully predicts the reduction of the applied load during successive cycles. For both layups, the degradation starts with a moderate rate in the beginning of the second (cyclic) step and slows down again within the final cycles. By adding the cyclic damage law, that is, the cyclic damage parameter (d_{cyclic}), reduction of the applied load through successive cycles can be approximated with <10% error, which implies a reliable predictive capability of the numerical model under cyclic mixed-mode loading. Redefining the cohesive area, according to Eqn (15.20), is one of the key factors improving the model's predictive capability under mixed modes, whereas some inaccuracy was reported under mixed-mode conditions by Turon, Costa, et al. (2007). The crack propagation rate is faster in layup 22.5 compared to layup QI,

Predicting fatigue life under mixed-mode fracture conditions 437

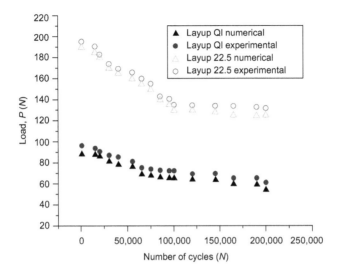

Figure 15.8 Comparison of the load reduction (P) within successive cycles observed in experiments and determined numerically with the cyclic damage law.

which can also be observed in the Paris plot (Figure 15.7). Numerically, this might result in higher growth rate of d_{cyclic} and higher degradation rate of the applied force (P) in layup 22.5, as supported by Figure 15.8.

As observed in Figure 15.8, the stacking sequence of the composite layups has a significant effect on cyclic fracture, which arises from the various stress fields in the fracture zone. If a higher resistance to cyclic failure is the major requirement to be fulfilled by the laminate, the stacking sequence must be optimized accordingly. For the two plies examined in this study (see Table 15.2 for the stacking sequence), when comparing the damage initiation profiles in the adjacent ply to the delamination plane, the matrix tension damage initiation criterion was fully satisfied, and the value of the damage initiation parameter reached its maximum, whereas in layup 22.5 it remained close to zero. This numerical result indicates the initiation of matrix ply damage in layup QI, which is in accordance with the experimental observation of a lower cyclic fracture load and the slower degradation rate of the corresponding load. Micromechanical observations (Naghipour, 2011) also prove the same point. The appearance of broken fiber or fiber pull outs (generally in form of fiber bundles) are important fracture surface characteristics observed in cyclic mixed-mode loadings for layup 22.5. However, the fracture surface of layup QI is dominated by matrix fracture areas tilted slightly in relation to the overall fracture surface. Few broken fibers are present in the fracture surface compared to layup 22.5. Ridge and valley markings observed in the micrograph are recognized as characteristic of a combination of peel and shear failures during matrix fracture. Although a few broken fibers of the adjacent ply appear at some locations of the valleys, their number remains small compared with layup 22.5, where bundles of broken fibers appear at the fracture's surface. The effect of stacking sequence on mixed-mode cyclic failure pattern of various laminates is reaffirmed when crack tip failure stresses in the ply adjacent to the delamination plane are compared for layup 22.5 and layup QI. In layup 22.5 the maximum value of the longitudinal stress is about 30% higher than in layup QI, but the maximum values of the in-plane shear stress and the transverse stress are much lower. This in turn leads to a different plane of maximal principal stress in layup 22.5 and layup QI, corresponding to the fracture plane, along which the crack tends to propagate. The stress components acting in layup QI result in a theoretical fracture angle (angle between the fracture plane and the delamination plane of the precrack) of ≈ 15, which may be the driving force behind transverse matrix cracks rather than longitudinal growth (parallel to fibers) of microcracks. In contrast, the fracture angle in layup 22.5 remains very close to zero and leads to a totally different crack propagation path. Therefore, if a higher resistance to cyclic failure is a main design criterion, the stacking sequence should be chosen wisely and accordingly.

The estimation of A_{cz} in this work is further improved for a mixed-mode load case using the formulation described by Harper and Hallett (2008), as discussed earlier. According to Harper and Hallett, different approximations of A_{cz} can be achieved by varying the scaling factor M in Eqn (15.20). In Figure 15.9 numerically obtained load degradation curves for various A_{cz} approximations are compared with the experimental result. It is observed that under a mixed-mode loading case the A_{cz} obtained with the scaling factor of $M = 0.65$ provides a superior correlation with the experimental result compared to the A_{cz} suggested by Turon, Costa, et al. (2007).

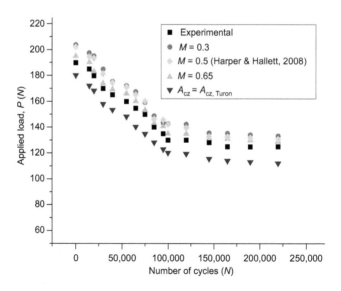

Figure 15.9 Effect of A_{cz} estimation on the numerical prediction of load reduction (layup 22.5, 50% mode mix).

Superior accuracy of using higher-order elements and integration schemes, especially in the case of material or geometric nonlinearity, is mentioned by various authors in literature (Alfano & Crisfield, 2001; Bathe, 2001; Szabó & Babuška, 1991). Since the cohesive law is highly nonlinear, a quadratic element with the 3×3 Newton–Cotes integration provides much more accurate and reliable results than lower scheme integrations and therefore has been adopted in this numerical model.

15.7 Conclusions and future trends

The research on cyclic mixed-mode delamination failure in multidirectional composites still has notable deficiencies to be addressed. Although there have been some numerical efforts to replace the costly fatigue experiments with corresponding numerical simulations, there are still some shortcomings, such as model-specific parameters, to be studied more in detail. Modeling fatigue delamination is still a challenging task for the research community because various model-specific parameters and several failure phenomena interact with each other, leading to final failure. This chapter presents a successfully validated numerical tool for simulating the cyclic delamination behavior of CFRP materials, with various multidirectional stacking architectures under mixed-mode loading for any arbitrary mode mix. However, model-specific variables, such as the cohesive zone area and Paris parameters, have to be calibrated from individual models or from characterization experiments. Similar to other works available in the literature, this part has to be improved to achieve a fully numerical fatigue delamination analysis tool independent of costly characterization experiments.

References

ABAQUS 6.6 user's manuals. (2006). Pawtucket, USA: Hibbitt, Karlsson and Sorensen.
Alfano, G., & Crisfield, M. A. (2001). Finite element interface models for delamination analysis of laminated composites: mechanical and computational issues. *International Journal for Numerical Methods in Engineering, 50,* 1701–1736.
Allix, O., & Blanchard, L. (2006). Mesomodeling of delamination: towards industrial applications. *Composites Science and Technology, 66,* 731–744.
Andersons, J. (1994). Methods of fatigue prediction for composite laminates: a review. *Mechanics of Composite Materials, 6,* 545–554.
Bathe, K. J. (2001). *Finite element procedures.* Berlin: Springer.
Barenblatt, G. I. (1962). The mathematical theory of equilibrium cracks in brittle fracture. *Advances in Applied Mechanics, 7,* 55–129.
Benzeggagh, M. L., & Kenane, M. (1996). Measurement of mixed-mode delamination fracture toughness of unidirectional glass/epoxy composites with mixed-mode bending apparatus. *Composites Science and Technology, 56,* 439–449.
Blanco, N., Gamstedt, E. K., Asp, L. E., & Costa, J. (2004). *International Journal of Solids and Structures, 4,* 4219–4235.
Dowling, N., & Begley, J. (1976). Fatigue crack growth during gross plasticity and the J-integral. *ASTM STP, 590,* 82–103.

Dugdale, D. S. (1960). Yielding of steel sheets containing slits. *Nonlinear Dynamics, 8*, 100–104.
Ewalds, H. L. (1984). *Fracture mechanics*. London: Edward Arnold.
Harper, P. W., & Hallett, S. R. (2008). Cohesive zone length in numerical simulations of composite delamination. *Engineering Fracture Mechanics, 75*, 4774–4792.
Hashin, Z. (1981). Failure criteria for unidirectional fiber composites. *Journal of Applied Mechanics, 20*, 329–334.
Kohlgruber, D. (1997). *Internal report: Mechanical properties of PEEK/AS4*. Source: CYTEC/DLR. Stuttgart: Institute of Structures and Design, German Aerospace centre -DLR.
Maiti, S., & Geubelle, P. (2005). A cohesive model for fatigue failure of polymers. *Engineering Fracture Mechanics, 72*, 691–708.
McDowell, D. (1997). An engineering model for propagation of small cracks in fatigue. *Engineering Fracture Mechanics, 56*, 357–377.
Munoz, J. J., Galvanetto, U., & Robinson, P. (2006). On the numerical simulation of fatigue driven delamination with interface elements. *International Journal of Fatigue, 28*, 1136–1146.
Naghipour, P. (2011). *Numerical simulation and experimental investigation on quasi-static and cyclic mixed mode delamination of multidirectional CFRP laminates* (Ph.D. thesis). University of Stuttgart, Germany.
Naghipour, P., Bartsch, M., Chernova, L., Hausmann, J., & Voggenreiter, H. (2009). Effect of fiber angle orientation and stacking sequence on mixed mode fracture toughness of carbon fiber reinforced plastics; numerical and experimental investigations. *Materials Science and Engineering: A, 527*, 509–517.
Naghipour, P., Schneider, J., Bartsch, M., Hausmann, J., & Voggenreiter, H. (2009). Fracture simulation of CFRP laminates in mixed mode bending. *Engineering Fracture Mechanics, 76*, 2821–2833.
Naghipour, P., Schulze, K., Hausman, J., & Bartsch, M. (2012). Numerical and experimental investigation on lap shear fracture of Al/CFRP laminates. *Composites Science and Technology, 72*, 1718–1724.
Nguyen, O., Repetto, E. A., Ortiz, M., & Radovitzky, R. A. (2001). A cohesive model of fatigue crack growth. *International Journal of Fracture, 110*, 351–369.
Paris, P., & Erdogan, F. (1963). A critical analysis of propagation laws. *Journal of Basic Engineering, 85*, 528–534.
Paris, P., Gomez, M., & Anderson, W. (1961). A rational analytical theory of fatigue. *Trends in Engineering, 13*, 9–14.
Peerlings, R. H. J., Brekelmans, W. A. M., de Borst, R., & Geers, M. G. D. (2000). Gradient-enhanced damage modelling of high-cycle fatigue. *International Journal for Numerical Methods in Engineering, 49*, 1547–1569.
Reifsnider, K. L. (1991). *Fatigue of composite materials*. London, UK: Elsevier.
Rice, J. R. (1980). The mechanics of earthquake rupture. In A. M. Dziewonski, & E. Boschhi (Eds.), *Physics of the earth's interior, proceedings of the international school of physics* (pp. 555–649). Amsterdam: Italian Physical Society/North-Holland.
Roe, K., & Siegmund, T. (2003). An irreversible cohesive zone model for interface fatigue crack growth simulation. *Engineering Fracture Mechanics, 70*, 209–232.
Serebrinsky, S., & Ortiz, M. (2005). A hysteretic cohesive-law model of fatigue-crack nucleation. *Scripta Materialia, 53*, 1193–1196.
Suresh, S. (1991). *Fatigue of materials*. Cambridge, UK: Cambridge University Press.
Szabó, B., & Babuška, I. (1991). *Introduction to finite element analysis, formulation, verification and validation*. New York: John Wiley and Sons Inc.

Talreja, R. (1999). Damage mechanics and fatigue life assessment of composite materials. *International Journal of Damage Mechanics, 8*(4), 339−354.
Turon, A., Costa, J., Camanho, P. P., & Dávila, C. G. (2007). Simulation of delamination in composites under high-cycle fatigue. *Composites Part A, 38*(11), 2270−2282.
Turon, A., Dávila, C. G., Camanho, P. P., & Costa, J. (2007). An engineering solution for mesh size effects in the simulation of delamination using cohesive zone models. *Engineering Fracture Mechanics, 74*, 1665−1682.
Yang, B., Mall, S., & Ravi-Chandar, K. (2001). A cohesive zone model for fatigue crack growth in quasi-brittle materials. *International Journal of Solids and Structures, 38*, 3927−3944.

Predicting the fatigue life of adhesively-bonded structural composite joints

A.P. Vassilopoulos
École Polytechnique Fédérale de Lausanne (EPFL), Lausanne, Switzerland

16.1 Introduction

16.1.1 Fatigue life modeling

Significant research efforts have been devoted to understanding the fatigue behavior of fiber-reinforced polymer (FRP) composite materials and composite structural elements, such as adhesively-bonded or bolted composite joints. One of the most explicit and straightforward ways to represent experimental fatigue data is the $S-N$ diagram. It is preferred to other approaches for the modeling of the fatigue life of FRP composite materials and structures, e.g., those based on stiffness degradation, or crack propagation measurements during lifetime, since it requires input data (applied load and corresponding cycles to failure) that can be collected using very simple recording devices (Vassilopoulos & Keller, 2011).

Usually, fatigue data for preliminary design purposes are gathered in the region of fatigue cycles ranging between 10^3 and 10^7. However, depending on the application, high- or low-cycle fatigue regimes can be of interest. Additional data are needed in such cases to avoid the risk of poor modeling due to extrapolation into unknown spaces. Although for the high-cycle fatigue (HCF) regime long-term and time-consuming fatigue data must be acquired, the situation seems easier for low-cycle fatigue (LCF) where static strength data can apparently be used in combination with the fatigue data. However, when the static strength data are considered in the analysis, other problems arise. As reported in Nijssen, Krause, and Philippidis (2004) and Vassilopoulos and Nijssen (2010) for typical composite material systems, static strength data should not be a part of the $S-N$ curve, especially when they have been acquired under strain rates much lower than those used in fatigue loading. The use of static data in the regression leads to incorrect slopes of the $S-N$ curves as presented by Nijssen et al. (2004). On the other hand, although excluding static strength data improves the description of the fatigue behavior, it introduces errors in the lifetime predictions when the low-cycle regime is important, as for example for loading spectra with a low number of high-load cycles. Therefore, the use of static strength data for the derivation of fatigue curves (such as fatigue data for 1 or $^1/_4$ cycle) is arguable.

The choice of a particular (an appropriate) fatigue theory for composite materials and structures is based on the material's behavior under the given loading pattern and the experience of the user. For uniaxial loading cases, the established $S-N$ curves for metals were initially adopted for composites as well. Finding the appropriate $S-N$ curve type for the examined material is not simple, since there is no rule governing this selection process. The best $S-N$ curve type is the one that can best fit the available experimental data (Vassilopoulos & Keller, 2011). A common $S-N$ behavior for FRP laminates has been described by the wear-out model in Sendeckyj (1981, pp. 245—260) as a power curve that flattens at high applied cyclic stress levels. In another work by Salkind (1972, pp. 143—169) this change of slope at the LCF regime is attributed to the sensitivity of FRP composites to high strain ranges that occurred under high stress levels. A comparative study by Mandell (1990) on short fiber—reinforced composites showed that the $S-N$ data for chopped glass strand polyester laminates follow a different trend. The behavior can be appropriately simulated by a linear curve in the semilogarithmic scale, although at the HCF regime the presence of run-outs causes a decrease in the slope of the $S-N$ curve. Detailed analysis of fatigue data with computational tools, like artificial neural networks and genetic programming (see, e.g., Vassilopoulos, Georgopoulos, & Dionysopoulos, 2007; Vassilopoulos, Georgopoulos, & Keller, 2008), showed that a multislope curve fits better the fatigue behavior of typical composite laminates.

A review of articles on the fatigue behavior of composite materials shows that the mechanism of fatigue failure can alter with changes in the cyclic stress level (Aymerich & Found, 2000; Bakis, Simonds, Vick, & Stinchcomb, 1990, pp. 349—370; Mandell, McGarry, Huang, & Li, 1983; Miyano, Nakada, & Muki, 1997; Philippidis & Vassilopoulos, 2001) explaining the variation in the $S-N$ curve slope. Different fatigue behavior was identified at high and low stress levels for an injection-molded polysulfone matrix composite reinforced by short glass and carbon fibers (Mandell et al., 1983). The experimental results showed a significant change in the $S-N$ curves at around 10^3 to 10^5 cycles. Different fatigue responses under high- and low-stress levels has also been reported in Aymerich and Found (2000) and Bakis et al. (1990, pp. 349—370) for carbon/PEEK laminates where the dominant failure changes from fiber to matrix damage when the cyclic load level is decreased. Investigation of the fatigue behavior of glass/polyester $[0/(\pm45)_2/0]_T$ composite laminates (Philippidis & Vassilopoulos, 2001) showed a significant difference in the stiffness degradation at failure (it was found to be higher for lower stress levels), although no difference was identified in the fracture surfaces. Miyano et al. (1997) reported different failure modes under low- and high-stress levels in conically-shaped FRP joint systems, proved by observation of the fracture surfaces and a lower slope of the $S-N$ curve at high stress levels.

With this variety of behaviors exhibited by composite materials, the selection of the fatigue model that is established by fitting a mathematical equation to the experimental data becomes of paramount importance for any fatigue analysis. The fatigue model reflects the behavior of the experimental data in theoretical equations which are subsequently used during design calculations. A number of different types of fatigue model (or types of $S-N$ curves) have been presented in the literature, with the most "famous" being the empirical exponential (lin—log) and power (log—log) relationships. Based on these, it is assumed that the logarithm of the loading cycles is linearly dependent on the cyclic stress parameter, or its logarithm. Fatigue models determined in this way do not take different stress ratios or frequencies into account, i.e., different model parameters should be determined for different loading conditions. Also, they do not take into account any of the failure mechanisms that develop during the failure process. Other more sophisticated fatigue formulations that also take the influence of stress ratio and/or frequency into account have also been reported (Adam, Fernando, Dickson, Reiter, & Harris, 1989; Epaarachchi & Clausen, 2003). A unified fatigue function that permits the representation of fatigue data under different loading conditions (different R-ratios) in a single two-parameter fatigue curve was proposed by Adam et al. (1989). In another work by Epaarachchi and Clausen (2003), an empirical model that takes into account the influence of the stress ratio and loading frequency was presented and validated against experimental data for different glass fiber—reinforced composites. Although these models seem promising, their empirical nature is a disadvantage as their predictive ability is strongly affected by the selection of a number of parameters that must be estimated or even, in some cases, assumed.

Experimental evidence showed that the commonly used models are not appropriate for fitting material behavior from the LCF to the HCF region (see for example Harik, Klinger, & Bogetti, 2002; Sarkani, Michaelov, Kihl, & Beach, 1999). Sarkani et al. (1999) reported a deviation between the power $S-N$ curve and the experimental data points obtained for bonded and bolted FRP joints. Similar experimental evidence is provided by Harik et al. (2002) for glass—fiber—reinforced polymer (GFRP) matrix laminates. Bilinear models, in the logarithmic (Sarkani et al., 1999) and semilogarithmic (Harik et al., 2002) scales, were introduced in both works to separately fit the material fatigue behavior in different (LCF and HCF) regions. The disadvantage of these approaches is the need for fatigue data in all regions for the fitting of model parameters and therefore their inability to extrapolate any result, since they simply resemble fitting procedures. In addition to this, the resulting $S-N$ curve equations are not continuous since different model parameters are estimated for LCF and HCF, and the selection of the data subsets corresponding to each group has to be performed based on the experience of the user, and they are therefore not practical for integration into design methodologies.

Other models, able to derive multislope $S-N$ curves, are available. However, they require more data in all the examined life regions, since they are only fitting equations that simulate material behavior by adjusting a number of fitting parameters. Mu, Wan, and Zhao (2011) proposed a multislope model comprising three parameters for modeling the fatigue behavior of composite materials. However, the model is based purely on the fitting of a logistic function to the experimental data and therefore its results cannot be extrapolated outside the range of existing experimental data.

Methods based on damage mechanics, therefore having a physical background, also exist. A typical example is the wear-out model adopted by Sendeckyj (1981, pp. 245—260). The wear-out model was initially introduced by Halpin, Jerina, and Johnson (1973, pp. 5—64) for composite materials based on metal crack growth concepts. However, owing to objections to the dominant crack assumption for composites, the model has been reviewed and modified by a number of authors considering residual strength as the damage metric. The form of the wear-out model adopted by Sendeckyj is based on the "Strength Life Equal Rank Assumption" or SLERA, introduced by Hahn and Kim (1975), stating that a specimen of a certain rank in the static strength probability distribution has the same rank in the fatigue life distribution. In other words, application of the wear-out model is valid as long as no competing failure modes are observed during fatigue life, or even between the fatigue and static loading.

Methods for the $S-N$ curve modeling of composite materials, also appropriate for the derivation of $S-N$ curves that take into account the probabilistic nature of the fatigue properties of composite materials, have been established to permit the derivation of $S-N$ curves with a given statistical significance based on limited data sets (e.g., Sendeckyj, 1981, pp. 245—260; Whitney, 1981). These statistical methods presented in detail in Vassilopoulos and Keller (2011) are also based on a deterministic $S-N$ equation for representation of the fatigue data; however, a more complicated process, compared to the simple regression analysis, is followed for the estimation of model parameters, which in one of the presented models (wear-out) leads to a multislope $S-N$ curve.

16.1.2 Influence of loading parameters on the fatigue life

The effect of several critical parameters on the fatigue life of a material under a certain loading condition can be examined when experimental work is performed under this condition. Although the result is useful for the analysis of the examined loading scenario, the experimental effort should be repeated for any other applied loading spectrum. This practice is costly and cannot be followed in practice where numerous different loading patterns are applied on a structural element. Therefore, experimental databases are derived for basic loading conditions (e.g., constant amplitude fatigue loading) and appropriate modeling is performed for extrapolation of the experimental evidence to predict the life under other, more complicated loading conditions.

The stress ratio, the ratio of the minimum to maximum applied cyclic stress ($R = \sigma_{min}/\sigma_{max}$), is used to specify the loading type; $0 < R < 1$ expresses tension—tension (T—T) fatigue, $1 < R < +\infty$ represents compression—compression (C—C) fatigue, while $-\infty < R < 0$ denotes mixed tension-compression (T—C) fatigue loading that can be tension- or compression-dominated. It is well documented that for a given maximum stress in a tension—tension case, the fatigue life of the composite increases with increasing magnitude of R. In compression—compression loading, increasing the magnitude of R reduces the fatigue life of the examined composite (Abd Allah, Abdin, Selmy, & Khashaba, 1997; Ellyin & El-Kadi, 1994; Mallick & Zhou, 2004; Mandell & Meier, 1982, pp. 55—77; Mandell & Samborsky, 2010).

16.1.3 Constant life diagrams

The influence of the R-ratio on the fatigue behavior of composite materials has been the subject of numerous investigations in the past (e.g., Kawai & Koizumi, 2007; Petermann & Schulte, 2002). This effect is assessed by using constant life diagrams (CLDs). CLDs reflect the combined effect of mean stress and material anisotropy on fatigue life, and can be used for estimation of the fatigue life of the material under loading patterns for which no experimental data exist. The main parameters that define a CLD are the cyclic mean stress, stress amplitude and number of fatigue cycles.

In previous papers (Beheshty & Harris, 1998; Gathercole, Reiter, Adam, & Harris, 1994; Kawai & Koizumi, 2007; Mandell, Samborsky, Wang, & Wahl, 2003; Sutherland & Mandell, 2005; Vassilopoulos, Manshadi, & Keller, 2010a) it has been proved that, although the classic linear Goodman diagram is the most commonly used, particularly for metals, it is not suitable for composite materials, mainly because of the variation in their tensile and compressive strengths that they exhibit. As a result, a typical CLD for composite materials is shifted to the right-hand side and the highest point is located away from the line corresponding to zero mean stress, $\sigma_m = 0$, or else the line representing the reversed $S-N$ curve with a ratio between the minimum and maximum applied stress, $R = -1$. Moreover, the damage mechanisms under tension are different from those under compression. In tension, the composite material properties are generally governed by the fibers, while in compression the properties are mainly determined by the matrix and matrix—fiber interaction. Therefore, straight lines connecting the ultimate tensile stress (UTS) and the ultimate compressive stress (UCS) with points on the $R = -1$ line for different numbers of cycles are not capable of describing the actual fatigue behavior of composite materials.

As mentioned earlier, several CLD models have been presented in the literature (Boestra, 2007; Harris, 2003; Kawai, 2007; Vassilopoulos, Manshadi, & Keller, 2010b) in order to cope with the aforementioned characteristics of composite materials. A comprehensive evaluation of the fatigue life predicting ability of the most commonly used and most recent CLD models is given in Vassilopoulos et al. (2010a). In addition to the presented analytical methods, novel computational techniques have also been employed during the last decade for modeling the fatigue behavior of composite materials and deriving CLD based on limited amounts of experimental data (e.g., Silverio Freire, Dória Neto, & De Aquino, 2009; Vassilopoulos et al., 2007, 2008). These methods offer a means of representing the fatigue behavior of the examined composite materials that is not biased by any damage mechanisms and not restricted by any mathematical model description. They are data-driven techniques, and their modeling quality depends on the quality of the available experimental data.

Since the introduction of the CLD concept by Gerber and Goodman back in the nineteenth century (Gerber, 1874; Goodman, 1899), all presented methods have two common features—they represent the fatigue data on the $\sigma_m-\sigma_a$ plane and their formulation is based on the fitting of available fatigue data for specified R-ratios or the interpolation between them. The same concept has been followed for the derivation of CLDs for composite materials. Already in 1972, Salkind presented fatigue data of boron reinforced aluminum on the $\sigma_m-\sigma_a$ plane in order to compare their fatigue resistance against unreinforced aluminum alloys. CLD-like diagrams were also plotted by Hahn (1979) in order to prove that the fatigue strength of a B/Ep laminate is compressive dominated at low cycles and becomes tension dominated as the high cycle region is approached. Another example in the same reference presents that the fatigue failure of a Gr/Ep laminate in the low-cycle region is controlled by the maximum tensile stress, while in the high cycle region is controlled by the alternating stress.

In 2007, Boerstra explicitly proposed an alternative method of constructing CLDs for composite materials without the need to define specific R-ratios. The Boerstra model is based on the sparse fatigue data available for several different loading cases. The drawback of this method is the need for a complicated solution of a seven-parameter optimization problem. Furthermore, as proved in Vassilopoulos et al. (2010a), although this independence of the R-ratio offered simplicity in testing, the resulting CLD was not more accurate than other methods such as the piecewise linear and/or Harris' CLD (Beheshty & Harris, 1998; Gathercole et al., 1994). An innovative formulation was later on introduced in Vassilopoulos et al. (2010b) for predicting an asymmetric, piecewise nonlinear CLD. It was based on the relationship between the stress ratio and the stress amplitude and expressed by simple phenomenological equations, showing remarkable accuracy.

16.1.4 Content of the chapter

This chapter aims to provide an overview of the commonly used $S-N$ curves and the available CLD formulations for the simulation of the fatigue behavior of composite materials and adhesively-bonded composite joints. Several $S-N$ models, including the "conventional" models that are based on linear regression analysis of the existing fatigue data, are presented and their modeling accuracy is compared. The most commonly used and most recently introduced CLD for composite materials and adhesively-bonded composite joints are also presented in this chapter. Their predictive ability is evaluated by using the experimental database (for adhesively-bonded pultruded GFRP joints) presented in Chapter 8 of this volume.

16.2 $S-N$ formulations for composites and adhesively-bonded composite joints

Extended literature review presented in Sarfaraz, Vassilopoulos, and Keller (2013) revealed that there is no universal theoretical model able to accurately describe the constant amplitude fatigue behavior of adhesively-bonded fiber-reinforced joints. The power law formulation, initially established by Basquin in 1910, Eqn (16.1), usually fits the data well when a high number of cycles is involved (Sims & Brogdon, 1977, pp. 185–205), although its performance is often poor in the LCF region.

$$\sigma = bN_f^{-a} \tag{16.1}$$

Equations (16.2) and (16.3) have been established for metals in order to improve the Basquin formulation for the entire fatigue life range by introducing the endurance limit (σ_e) and a constant parameter (B) which control the $S-N$ curve in the HCF and LCF regions, respectively.

$$\sigma = bN_f^{-a} + \sigma_e \quad (16.2)$$

$$\sigma - \sigma_e = b(N_f + B)^{-a} \quad (16.3)$$

Several phenomenological models that take into account parameters affecting the fatigue behavior such as the stress ratio, frequency, temperature, and fiber direction, as well as models that consider the probabilistic nature of the fatigue properties of composite materials were introduced. Jarosch and Stepan (1970) incorporated an additional term into the power law formulation that considers the stress ratio effect on fatigue life, Eqn (16.4), to characterize the fatigue life of glass fiber—reinforced epoxy composite used in rotor blades.

$$\sigma = \frac{a + b \big/ N_f^x - c}{(1 - R/1 + R)^y} \quad (16.4)$$

This model was used by Sims and Brogdon (1977, pp. 185—205) to characterize the fatigue behavior of S-glass/epoxy and graphite/epoxy laminates. The effect of the stress ratio on fatigue life was also addressed by Bach (1996), who suggested a model for GFRP laminates, Eqn (16.5).

$$\sigma = \text{UTS}(1 - D \log N_f), \quad D = f(\text{material}, R) \quad (16.5)$$

Parameter D is a function of material and R is a function of the stress ratio and is different for each examined material. Appel and Olthoff (Bach, 1996) established an estimation of parameter D after analyzing data from literature, Eqn (16.6).

$$\sigma = \text{UTS}(1 - D \log N_f), \quad D = 0.015(1 - R) + 0.08 \quad (16.6)$$

However, the evaluation of this model showed that predictions are reliable for tension—tension fatigue, but become optimistic for higher load levels under reversed loading.

Attempts to introduce damage mechanics into the process of $S-N$ derivation were also reported. Strength degradation due to fatigue loading was considered a valid measure for the development of $S-N$ curve formulations (Chou & Croman, 1979, pp. 431—454; D'Amore, Caprino, Stupak, Zhou, & Nicolais, 1996; Epaarachchi & Clausen, 2003; Qiao & Yang, 2006; Sendeckyj, 1981, pp. 245—260). Based on this method, fatigue failure is assumed when the residual strength decreases to the same level as the maximum applied cyclic stress. Sendeckyj (1990) established a model on this basis in the form of Eqn (16.7).

$$\sigma_{eq} = \sigma_a \big(1 + (N_f - 1)f\big)^S \quad (16.7)$$

The model consists of two parameters, S (the slope of the $S-N$ curve) and f (the asymptotic slope of the $S-N$ curve at low stress levels), which can be constants or functions of stress ratio. Different alternatives for S and f provide a variety of models such as the classic power law model for S and $f = 1$, the wear-out model for constant S and f or a model in which both S and f are the functions of the stress ratio. Subsequent models were also developed based on strength degradation. The effect of the stress ratio was included in the model developed by D'Amore et al. (1996), Eqn (16.8).

$$N_f = \left(1 + \frac{1}{\alpha(1-R)}\left(\frac{\beta}{\sigma_{max}}|\ln[1-P_N(N)]|^{\frac{1}{\alpha_f}} - 1\right)\right)^{1/\delta} \qquad (16.8)$$

The effect of frequency was incorporated in Epaarachchi and Clausen (2003) and Qiao and Yang (2006) in the form of Eqns (16.9) and (16.10), respectively.

$$(\sigma_u/\sigma_{max} - 1)(\sigma_u/\sigma_{max})^{0.6-\psi|\sin\theta|}\left(fr^\delta \big/ (1-\psi)^{1.6-\psi|\sin\theta|}\right) = \alpha\left(N_f^\delta - 1\right) \qquad (16.9)$$

$$N_f^\delta - 1 = \frac{1}{\alpha}(\Delta\sigma/\sigma_u)^{-r}\left(\rho A_{cr} fr^2/E\right)^{-m}(1 - \sigma_{max}/\sigma_u) \qquad (16.10)$$

Fatigue models have also been developed based on the stiffness degradation of the material. Hwang and Han (1986) for example established an $S-N$ model based on the fatigue modulus concept:

$$N_f = (B(1 - \sigma_a/\sigma_u))^c \qquad (16.11)$$

Philippidis and Vassilopoulos (2000) introduced a stiffness-based $S-N$ curve, designated Sc-N curve, Eqn (16.12), for the modeling of GFRP laminates. The concept of stiffness controlled curves has been later on successfully used for the fatigue life modeling and development of design allowables for adhesively-bonded FRP joints (Zhang, Vassilopoulos, & Keller, 2008; Zhang, Vassilopoulos, & Keller, 2010).

$$\sigma_a = E_0((1 - E_N/E_1)/KN)^{1/c} \qquad (16.12)$$

Bilinear and sigmoid $S-N$ formulations were also introduced (e.g., 14, 15, 28, 36, 37), for the modeling of experimental data sets showing different behaviors in LCF and HCF regimes. Harik et al. (2002) and Harik and Bogetti (2003) proposed a bilinear model for unidirectional GFRP laminates to describe the LCF and HCF regions:

$$\sigma = \frac{\sigma_{max}}{\sigma_u} = \begin{cases} a_{LCF} + b_{LCF} \log N_f, & \text{for}: N_{cyclic} \leq N_f \leq N_{LCF} \\ a_{HCF} + b_{HCF} \log N_f, & \text{for}: N_{LCF} \leq N_f \leq N_f \end{cases} \qquad (16.13)$$

According to Eqn (16.13) the fatigue data must be divided into two groups and the fitting is performed separately for each group. Sarkani et al. (1999) reported a deviation between the $S-N$ data and the power law model for FRP bonded and bolted joints. The authors found a good match between the experimental data and a two-segment power law equation, however, still based on the subjective classification of the fatigue data into the LCF and HCF regimes. Sigmoid models like the one developed by Xiong and Shenoi (2004), Eqn (16.14), and Mu et al. (2011), Eqn (16.15), provide algebraic fitting equations that cover the entire fatigue life range, but require greater numbers of fatigue data to adequately estimate the increased number of model parameters.

$$F_\sigma(\sigma) = 1 - \exp\left(-\left(\left(10^{c(\log N_f)^m}(\sigma - S_0) + S_0\right)\big/\beta\right)^{\alpha_f}\right) \quad (16.14)$$

$$\sigma = (1-c)\big/\left((1-a) + ae^{-b(\log N_f)}\right) + c \quad (16.15)$$

16.2.1 The hybrid S–N formulation

Recently, the hybrid $S-N$ formulation has been introduced (Sarfaraz, Vassilopoulos, & Keller, 2012b). It resembles a semiempirical formulation based on the commonly used exponential and power law models presented in Eqns (16.16) and (16.17).

$$\sigma = A + B\log(N_f) \quad \text{(Exponential fatigue model)} \quad (16.16)$$

$$\sigma = bN_f^{-a} \quad \text{(Power fatigue model)} \quad (16.17)$$

where σ is the stress parameter (it can be maximum cyclic stress, amplitude or stress range), N_f denotes the number of cycles to failure, while A, B, b, and a are the fatigue model parameters that can be derived by linear regression analysis, after fitting the equations to the experimental fatigue data.

The introduced formulation resembles the exponential model for LCF, while it is mutated to the power model, through a transition region, for longer lifetimes. This combination of the exponential and power models derives a formulation that adequately simulates the fatigue behavior of the examined composite material across the entire lifetime, from LCF to HCF. The hybrid $S-N$ formulation is:

$$\sigma = f_1(N_f)\left[A + B\log(N_f)\right] + f_2(N_f)\left[bN_f^{-a}\right] \quad (16.18)$$

where, in addition to the aforementioned parameters, two weighting functions, $f_1(N_f)$ and $f_2(N_f)$, are included to control the transition from the exponential to the power curve model:

$$f_1(N_f) = \left[\frac{1}{1 + (N_f/N_{\text{trans}})^2}\right] \quad (16.19)$$

$$f_2(N_f) = \left[\frac{-1}{1 + (N_f/N_{\text{trans}})^2} + 1\right] \quad (16.20)$$

A schematic plot of the fluctuations of these two empirical functions versus the logarithm of the number of cycles is shown in Figure 16.1. These sigmoid functions define the time and the rate of transition between the two basic fatigue models in order to avoid any singularity between the LCF and HCF regions. The transition number of cycles, N_{trans}, controls the moment of transition and is defined as the shortest lifetime among the examined sample of constant amplitude fatigue data. A sensitivity analysis was performed in order to assess the effect of the N_{trans} selection (between the shortest and the longest life at the highest stress level) on the model accuracy. The derived hybrid curves with different values for N_{trans}, showed that the accuracy of the model is not sensitive to the assumed value of N_{trans}. Although differences were observed, depending on the scatter of the experimental data, they were limited to the transition region. These differences did not affect the LCF and HCF regions and consequently the overall accuracy of the model. Therefore, for consistency, the shortest lifetime was used in all examined cases, and it is recommended to use this value for the model derivation.

Substitution of the weighting functions, Eqns (16.19) and (16.20), into Eqn (16.18) and the rearranging of the latter results in the hybrid model formulation:

$$\sigma = bN_f^{-a} + \left[\frac{1}{1 + \left(N_f/N_{\text{trans}}\right)^2}\right]\left[A + B\log(N_f) - bN_f^{-a}\right] \qquad (16.21)$$

with parameters A, B, b, and a being the same as those resulting from fitting the exponential and power models separately and N_{trans} equal to the shortest experimentally derived fatigue life, as defined earlier.

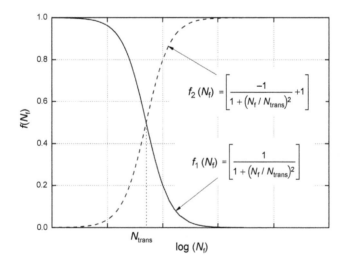

Figure 16.1 Applied weighting functions in the hybrid model.

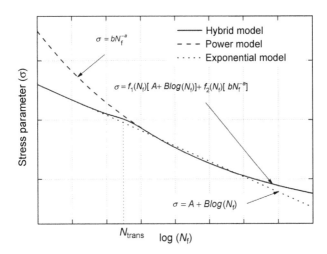

Figure 16.2 Scheme of hybrid model in comparison with exponential and power models.

The form of the hybrid model is schematically presented in Figure 16.2 and compared to the form of the exponential and power models. As described earlier, for low numbers of cycles, the hybrid model coincides with the exponential fatigue model and gradually mutates to the power fatigue model via a transition region determined by the weighting functions $f_1(N_f)$ and $f_2(N_f)$.

16.2.2 Disadvantages of commonly used S–N models

The disadvantages of the common $S-N$ formulations are summarized and demonstrated in the following by modeling the fatigue behavior of a wide variety of composite material systems. An alternative $S-N$ formulation is established based on the commonly used exponential and power $S-N$ fatigue models in order to resolve their shortcomings and appropriately simulate the fatigue life of several composite materials and composite structural elements from the LCF to the HCF region. The modeling ability of the introduced formulation is evaluated by comparison of the derived $S-N$ curves with the existing fatigue data and resulting curves from popular existing fatigue models.

The exponential and power fatigue models are commonly used for the interpretation of composite material fatigue data due to the nature of these models that are straightforward, not based on any assumptions, and easily applied, even on limited databases, in order to derive a reliable estimate of fatigue life. The estimation of the model parameters is based on linear regression analysis that can be performed by simple hand calculations.

Comparison between the $S-N$ curves derived from the two fatigue models shows that the performance of the exponential model is superior to that of the power model in the LCF region, while the power model is more accurate in the HCF region. In the LCF region, the power model curve almost always overestimates the fatigue strength of the material. On the other hand, the exponential fatigue model underestimates the lifetime for large numbers of cycles and occasionally estimates finite life as being at zero stress level. This behavior is shown in Figure 16.3 for the typical constant amplitude fatigue data of composite materials—in this case E-glass/polyester laminates (Post, 2008).

The aforementioned low accuracy of the power model in the LCF region can be improved by including the static strength data in derivation of the power $S-N$ curve, considering them as fatigue data. However, the simultaneous treatment of static and fatigue data may result in an incorrect slope that affects the entire range of the lifetime, as presented in Figure 16.4.

The wear-out model (Sendeckyj, 1981, pp. 245—260) is capable of deriving multi-slope reliability-based $S-N$ curves that fit the available static strength and fatigue data quite accurately. However, it has the drawback of the quite complicated optimization process required in order to estimate the model parameters. A detailed analysis reveals several other disadvantages including the sensitivity of the model parameter estimation to the exhibited scatter of the fatigue data and the inability of the model to extrapolate the life modeling to the LCF region when no available data (static strength or LCF fatigue data) exist.

This drawback is demonstrated by modeling a set of constant amplitude fatigue data for $[\pm 45]_{2S}$ T300/5208 graphite/epoxy composite laminates with 57 fatigue data points distributed at five stress levels (Lee, Yang, & Sheu, 1993). The wear-out model was applied using all the available data and also after censoring the two data points (at the 140-MPa stress level, open symbols in Figures 16.5 and 16.6) that do not seem to follow the observed trend. These two points were assumed as being outliers in the given data set. The derived $S-N$ curves based on the wear-out

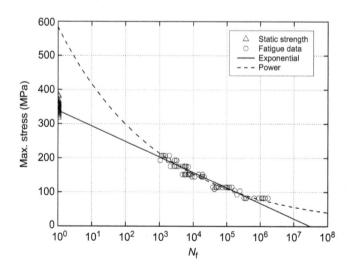

Figure 16.3 Comparison of exponential and power curves for $[0/+45/90/-45/0]_S$ E-glass/vinylester laminate fatigue data (Post, 2008) (excluding static data).

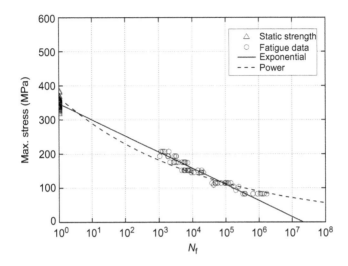

Figure 16.4 Comparison of exponential and power curves for [0/+45/90/−45/0]$_S$ E-glass/vinylester laminate fatigue data (Post, 2008) (including static data).

model, presented in Figure 16.5, are very different in both the LCF and the HCF regions. The slope of the $S-N$ curve across the intermediate cycle fatigue region (c. 10^3-10^6) was also considerably altered by censoring only these two of the 57 available data points. On the other hand, no significant changes were made to the $S-N$ curve derived by using the power fatigue model (see Figure 16.6) when the

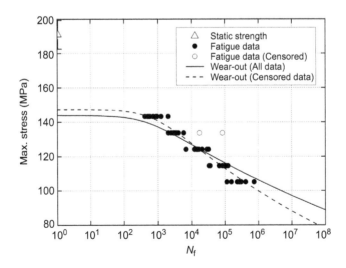

Figure 16.5 Comparison of $S-N$ curves derived by wear-out model for [±45]$_{2S}$ graphite/epoxy laminates (Lee et al., 1993) (censored and uncensored fatigue data).

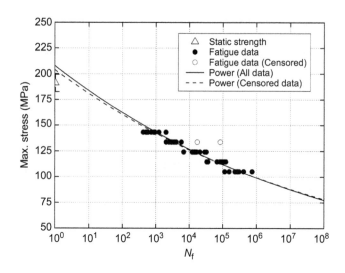

Figure 16.6 Comparison of S–N curves derived by power model for [±45]$_{2S}$ graphite/epoxy laminates (Lee et al., 1993) (censored and uncensored fatigue data).

entire or the censored data set was used. The average ultimate tensile strength, 191 MPa, with a standard deviation of ±8.4 MPa is shown in the figure at $N_f = 1$, but was not used for estimation of the model parameters.

The new hybrid model was developed taking into account this remarkable stability of the linear regression models (exponential and power). The aim was to retain their advantages, but at the same time improve their weaknesses to appropriately model the fatigue life of composite materials in the LCF and the HCF regions.

16.3 Comparison of existing fatigue models

A comparison performed on a wide range of composite materials and adhesively-bonded composite joints in Sarfaraz et al. (2012b) is used here in order to demonstrate the modeling ability of some of the commonly used S–N formulations.

The examined material databases in Sarfaraz et al. (2012b) include continuous and short, glass and carbon, fiber laminates combined with various polymer resins, and composite structures. The exponential and power curve models that were combined in order to derive the hybrid formulation were used for comparisons. In addition, the wear-out fatigue model (Sendeckyj, 1981, pp. 245–260), able to derive multislope S–N curves, was used.

The comparison of the results was based on graphical comparison of the derived S–N curves and a quantification of the fitting quality using the sum of squared errors (SSE). The SSE was calculated for each curve by defining the error as the difference between the logarithms of the estimated and experimental cyclic stress values.

The SSE values were normalized by the maximum error value for each material as presented in Table 16.1 (Sarfaraz et al., 2012b).

The SSE is calculated by the following equation:

$$\text{SSE} = \sum_{i=1}^{n} \left(\log(\sigma_{\text{exp}}) - \log(\sigma_{\text{est}})\right)^2 \qquad (16.22)$$

where σ_{exp} and σ_{est} are the experimental and estimated cyclic stresses corresponding to the same number of cycles, and n is the number of experimental data.

The static data were not included in the processes for the derivation of the $S-N$ curves. However, since fatigue strength data in the LCF region do not exist in most of the available databases, static data were used for calculation of the SSE, considered as being fatigue strength data corresponding to specimens that failed after one cycle ($N_f = 1$). The software recently developed by the authors, *CCfatigue* (Vassilopoulos, Sarfaraz, Manshadi, & Keller, 2010), was used for estimation of the fatigue model parameters and derivation of the $S-N$ curves.

16.3.1 Glass fiber-reinforced laminates

Numerous fatigue databases for glass fiber-reinforced laminates with different matrix resins and stacking sequences exist in the literature. Typical cases for which adequate

Table 16.1 **Calculated normalized sum of squared errors for derived $S-N$ curves**

	Normalized sum of squared errors (SSE)			
	Exponential	Power	Wear-out	Hybrid
Figure 16.7	0.080	1.000	0.090	0.037
Figure 16.8	0.598	0.250	1.000	0.311
Figure 16.9	1.000	0.201	0.034	0.016
Figure 16.10	0.120	0.232	1.000	0.116
Figure 16.11	0.317	1.000	0.247	0.109
Figure 16.12	0.333	0.152	1.000	0.130
Figure 16.13	0.392	0.922	1.000	0.369
Figure 16.14	0.618	0.702	1.000	0.305
Figure 16.15	1.000	0.444	0.987	0.717
Figure 16.16	0.348	1.000	0.562	0.329
Average	0.481	0.590	0.692	0.244
Standard deviation	±0.323	±0.370	±0.418	±0.210

constant amplitude fatigue data exist were selected and are examined here. The first examined database comprises data from constant amplitude fatigue loading applied to woven roving E-glass/vinylester laminates with a stacking sequence of [0/+45/90/−45/0]$_S$ (according to the fabric warp direction in each layer) (Post, 2008). The comparison of the used $S-N$ curves is shown in Figure 16.7 for the fatigue data obtained under tension−tension fatigue, at stress ratio $R = \sigma_{min}/\sigma_{max} = 0.1$. It is evident that the hybrid $S-N$ curve is the most appropriate, since it adequately simulates the fatigue behavior, covering all regions, from very LCF to HCF. This figure also demonstrates the problems presented by the other methods. The exponential model results in a linear $S-N$ curve on the semilogarithmic plane, which is very conservative at the HCF regime. In this case, it predicts failure under zero load after around 3×10^7 cycles, which is physically impossible. On the other hand, the power model does not exhibit this behavior in the HCF region, but, overestimates the fatigue strength at the LCF regime. For example, at $N_f = 10$ the fatigue strength estimated by the power model is 417.5 MPa and for one cycle approaches 600.0 MPa, while the tensile strength of the examined material equals 346.8 ± 15.8 MPa. The $S-N$ curve derived based on the wear-out model also underestimates the life in the LCF region. Nevertheless, this result is to be expected when this model is applied without the use of the static strength data. Comparison of the calculated error indices (SSE) given in Table 16.1 also shows that the hybrid formulation (SSE = 0.037) performs better than the other models.

Three data sets were selected from the DOE/MSU composite material fatigue database (Mandell & Samborsky, 2010). The first comprises results from the constant amplitude fatigue loading of multidirectional E-glass/polyester laminates with stacking sequence of [90/0/±45/0]$_S$, encoded in DOE/MSU as DD16 material. The subset

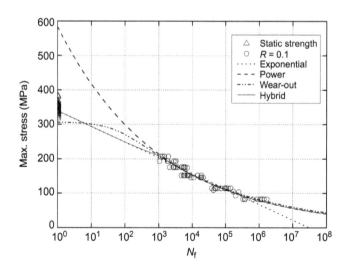

Figure 16.7 Comparison of hybrid model with other $S-N$ curves for [0/+45/90/−45/0]$_S$ E-glass/vinylester laminate fatigue data (Post, 2008).

containing results from experiments performed under $R = 0.1$ was used for the demonstration of the selected fatigue model application. The comparison of the modeling accuracy of the examined fatigue models, shown in Figure 16.8, reveals a significant deviation between the experimental results and the wear-out models at the LCF regimes and the exponential model in the HCF. The hybrid and power curves follow the trend of the experimental data at HCF and are in good agreement with static strength data, although the power model overestimates and the hybrid underestimates the life in the LCF region.

The second material system from the DOE/MSU database is for $[0/\pm45]_4$ E-glass/polyester laminates tested at $R = 0.1$ designated as "Material N." Although a limited number of fatigue data (16 data) is available, they cover a very large range of fatigue lifetime, between 20 and 8 million cycles. The derived $S-N$ curves, shown in Figure 16.9, prove the inability of the exponential fatigue model to properly simulate the real trend of the fatigue data. In the range between 10^3 and 10^5 cycles, the exponential model overestimates the fatigue life, while for cycles of more than 5×10^5 it significantly underestimates the fatigue life, also indicating failure under zero loads after 10^7 cycles. On the other hand, the power fatigue model is appropriate, but only for cycle numbers larger than 10^2. Below this limit, in the LCF region, the power model considerably overestimates the fatigue strength as also shown in Figure 16.9. The $S-N$ curve derived according to the wear-out method and the newly introduced hybrid model almost coincides in this case, showing differences only at the LCF regime where the hybrid model is more accurate, as it converges to the static strength data for very low-cycle numbers. The calculated statistical indices shown in Table 16.1 confirm that the hybrid model provides the most accurate $S-N$ curve.

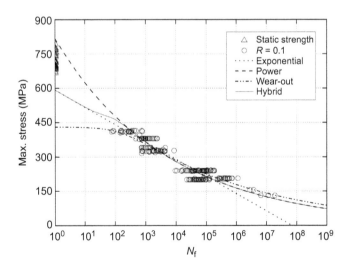

Figure 16.8 Comparison of hybrid model with other $S-N$ curves for $[90/0/\pm45/0]_S$ E-glass/polyester laminate fatigue data (Mandell & Samborsky, 2010).

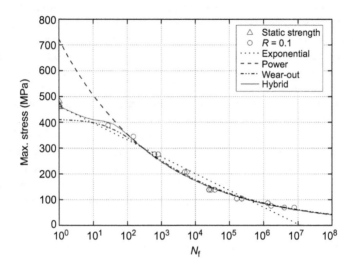

Figure 16.9 Comparison of hybrid model with other $S-N$ curves for $[0/\pm 45]_4$ E-glass/polyester laminate fatigue data (Mandell & Samborsky, 2010).

The third data set from the DOE/MSU database relates to a material designated 208 Fiber Strand and composed of 45 S-glass fibers and iso-polyester resin. This data set includes data points covering a very wide cycle range (10^3-10^{10} cycles). The same comments as for Figure 16.9 concerning the modeling of the HCF apply to Figure 16.10. All models, except the exponential one, simulate well the exhibited

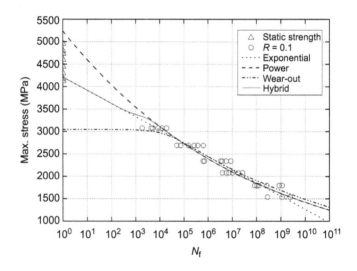

Figure 16.10 Comparison of hybrid model with other $S-N$ curves for fiber strand glass/iso-polyester fatigue data (Mandell & Samborsky, 2010).

fatigue behavior in this region. In the LCF region, the wear-out model underestimates and the power model overestimates the fatigue behavior. On the other hand, the hybrid and exponential models predict static strength values well within the range of the experimental scatter.

The fatigue models were also used for the simulation of the fatigue behavior of $[(\pm 45)_8/0_7]_S$ E-glass/polyester laminates under tension−tension ($R = 0.1$) loading (see Figure 16.11), retrieved from the FACT database (Nijssen, 2006). The data set includes fatigue data points corresponding to a life of only a few cycles to more than 100 million cycles. Again, the hybrid model is fairly accurate across the entire range of cycles, while the power and wear-out models fail to accurately simulate fatigue life in the LCF region and the exponential model provides only a rough averaging of the experimental data in this case. The lowest error index (SSE $= 0.109$) was obtained for the hybrid formulation.

The $S-N$ curves derived for a $[\pm 45]_{2S}$ glass/epoxy material system loaded under $R = 0.1$ (Sendeckyj, 1990) are compared in Figure 16.12. The hybrid and power fatigue models perform well in this case—the former resulting in a lower SSE (0.130) than the latter (0.152). However, the wear-out model overestimates the fatigue strength in the LCF region in contrast to the results shown in Figures 16.7−16.11 while the exponential model performs well only until c. 10^5 cycles and then becomes conservative.

16.3.2 Carbon fiber-reinforced laminates

The constant amplitude fatigue data for graphite/epoxy $[90/+45/-45/0]_S$ laminates (Yang, Yang, & Jones, 1989), was also selected for the evaluation of the hybrid model.

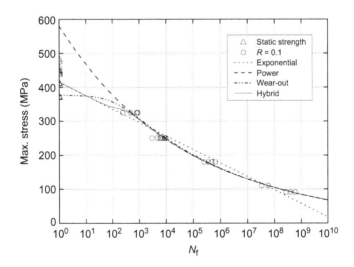

Figure 16.11 Comparison of hybrid model with other $S-N$ curves for $[(\pm 45)_8/0_7]_S$ E-glass/polyester laminate fatigue data (Nijssen, 2006).

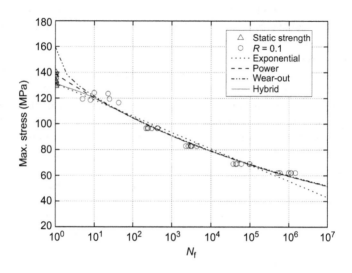

Figure 16.12 Comparison of hybrid model with other $S-N$ curves for $[\pm 45]_{2S}$ glass/epoxy laminate fatigue data (Sendeckyj, 1990).

The fatigue experiments were performed under tension—tension fatigue, $R = 0.1$. As shown in Figure 16.13 the comment regarding Figure 16.10 also applies in this case and therefore similar SSE indices (see Table 16.1) were calculated for the exponential and hybrid models due to the similarity of the curves at HCF where fatigue data are available. As already pointed out, the wear-out model exhibits high sensitivity to

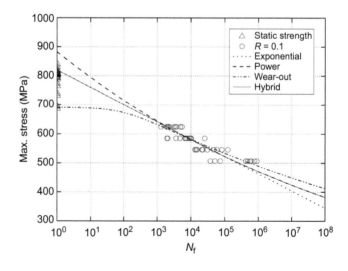

Figure 16.13 Comparison of hybrid model with other $S-N$ curves for $[90/+45/-45/0]_S$ graphite/epoxy laminate fatigue data (Yang et al., 1989).

the scatter of fatigue data and fails to accurately simulate the behavior in the LCF region when static strength and LCF data are not used for estimation of the model parameters. In the examined case the wear-out model significantly underestimates the fatigue life in the LCF region.

16.3.3 Hybrid glass–carbon fiber-reinforced laminates

Hybrid glass–carbon fiber-reinforced laminates are used in various engineering domains, such as wind turbine rotor blades, in order to improve the modulus and strength and also decrease density and fatigue sensitivity compared to glass–reinforced polymers (Mandell et al., 2003). A set of fatigue data from hybrid glass–carbon/epoxy laminates was used for the assessment of the new model (Mandell & Samborsky, 2010). The laminates were composed of 0° carbon and ±45 glass prepreg laminae with stacking sequence of $[\pm 45/0_4]_S$ tested in the transverse direction under stress ratio $R = -1$ (coded P2BT in DOE database). The differences between the derived $S-N$ curves at HCF are not significant (see Figure 16.14) except for that derived by the exponential model, which is conservative. However, it can be observed that only the hybrid and exponential models converge with the static strength data, while the power and wear-out models predict higher and lower static strength, respectively.

16.3.4 Short fiber-reinforced laminates

Short fiber-reinforced composites are used for the fabrication of numerous parts used in the automotive and aerospace industries, for example (Mandell, 1990).

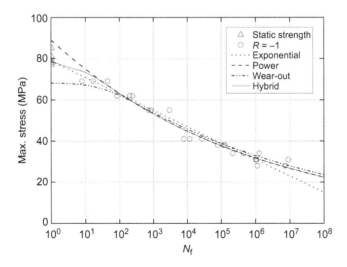

Figure 16.14 Comparison of hybrid model with other $S-N$ curves for hybrid $[\pm 45/0_{4C}]_S$ glass-carbon/epoxy laminate fatigue data (Mandell & Samborsky, 2010).

The applicability of the hybrid formulation was examined by modeling the fatigue behavior of injection-molded fiber glass—reinforced PEEK thermoplastic polymer composite laminates (Mandell, 1990). The resulting $S-N$ curves, shown in Figure 16.15, prove that the hybrid, wear-out, and power models appropriately simulate the behavior in the HCF region. Only the power fatigue model converges with the static strength data, with the wear-out model being more conservative than the rest. The minimum value for the SSE, 0.444, was estimated for the power model. The maximum SSE is calculated for the exponential model, which again estimates very low fatigue strength in the HCF region.

16.3.5 Pultruded fiber-reinforced adhesively-bonded joints

The selected $S-N$ formulations have also been applied for the modeling of the fatigue behavior of the pultruded FRP adhesively-bonded joints examined in Chapters 8 and 9 of this volume. An extended database has been recently created containing tension—tension, compression—compression and tension-compression constant amplitude fatigue data and static strength values obtained under high strain rates, similar to those for the fatigue loading (Sarfaraz, Vassilopoulos, & Keller, 2011). The data set selected for the demonstration of fatigue model applicability relates to the tension—tension fatigue results ($R = 0.1$). The resulting $S-N$ curves according to the different fatigue models are shown in Figure 16.16. All the above-mentioned disadvantages of the examined models are clearly apparent in Figure 16.16. The power model overestimates the life in the LCF region, the exponential underestimates the life at the HCF region, while the wear-out model fails to

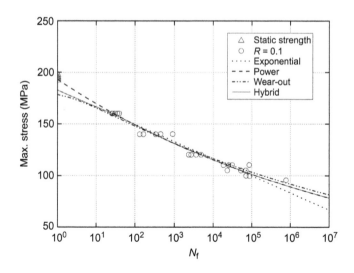

Figure 16.15 Comparison of hybrid model with other $S-N$ curves for short fiber 30% glass/PEEK fatigue data (Mandell, 1990).

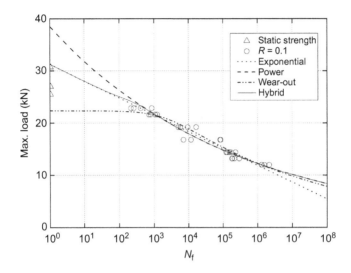

Figure 16.16 Comparison of hybrid model with other $S-N$ curves for double-lap joints (Sarfaraz et al., 2011).

converge with the static strength data and is unable to extrapolate any prediction to the LCF region. The hybrid fatigue model produces the most accurate $S-N$ formulation.

16.4 Discussion on the $S-N$ formulations

The applicability of the selected $S-N$ formulations to a wide range of typical composite materials and structural elements for different applications, ranging from the wind turbine rotor blade industry to construction, or automotive and aerospace structures, has been validated in the previous section. The performance of the fatigue models was graphically and quantitatively (by using the SSE index) compared. A thorough comparison based on several qualitative and quantitative criteria will be made in this section. The appropriateness of each examined fatigue model can be evaluated based on accuracy of modeling, ability to extrapolate and interpolate, number of model parameters, accuracy of parameter estimation, implemented assumptions, sensitivity to the available experimental data, etc.

As shown in Table 16.1, the calculated SSEs for the hybrid formulation are the lowest except for two cases (Figures 16.8 and 16.15) where the power model produced the most accurate $S-N$ curve, as it accurately predicted the static strength of the examined material. On average the hybrid model is the most accurate showing substantially lower SSE indices, 50% lower than the next best one, the exponential model. The similar SSE values obtained for the hybrid formulation and linear regression models were in some cases anticipated since the hybrid formulation was developed based on these models. Accordingly, the results of the hybrid and exponential fatigue models are comparable when the difference between the linear regression models at HCF is small (see, e.g., Figures 16.10 and 16.13). The difference between the hybrid and power results is negligible when the difference between the linear regression models in the LCF region is low (Figure 16.12).

Each of the linear regression models is unable to extrapolate both to the LCF and HCF regions. The exponential model was proved more accurate in the LCF region and the power in the HCF region. The wear-out model was proved incapable of extrapolating any lifetime predictions in the LCF region in three out of the 10 examined materials when no static strength data was used. On the other hand, the hybrid model, by combining the exponential and power models, is able to accurately simulate the fatigue behavior in both the LCF and HCF regions and successfully extrapolate results in order to estimate the static strength of the examined materials.

With regard to the number of model parameters, the exponential and power models are the simplest, since they both require the estimation of two parameters by simple linear regression analysis. Five parameters describing the hybrid model are shown in Eqn (16.21). However, the complexity of their estimation is practically the same as for the two aforementioned fatigue models. As mentioned in Section 16.2.1, one of the model parameters, N_{trans}, is directly defined by the experimental data, while the remaining four correspond to the parameters of the linear and power models and can be estimated by following the same process as for the individual linear regression models. The wear-out model is the most demanding, since it requires the estimation of four independent parameters via a multiparameter optimization process: the Weibull shape and scale parameters and the fatigue model parameters.

The accuracy of the estimated parameters of the hybrid model is the same as for the power and exponential fatigue models. For all three methods, a straightforward and simple linear regression analysis is performed, resulting in a very precise estimation of the model parameters. On the other hand, the estimation of the model parameters for the wear-out fatigue model is quite complicated. The parameter estimation is based on the multiparameter optimization process whose convergence cannot be guaranteed when the static strength data are not considered in the analysis. As presented in Figures 16.11 and 16.13, the wear-out $S-N$ curves converge with the static data while in Figures 16.7−16.10 and Figures 16.14−16.16 the curves intersect the ordinate at lower and in Figure 16.12 at higher values compared to static strengths.

The wear-out model is based on several assumptions, including the stress life equal rank approach, which compromises the validity of the model when competing failure modes are observed during fatigue life, or between fatigue and static loading. However, the other examined methods resemble mathematical formulations, without any physical meaning, that are fitted to the experimental data without the need to satisfy any assumptions.

Finally, as presented in Section 16.2.2, the wear-out model is sensitive to the selection of the fatigue data and the presence of data points that do not follow the same trend as the fatigue data. This makes the process sensitive to subjective modeling according to the experience of the user.

16.5 Constant life diagram (CLD) formulations for composites and adhesively-bonded composite joints

CLDs reflect the combined effect of mean stress and material anisotropy on the fatigue life of the examined composite material. Furthermore, they offer a predictive tool for the estimation of the fatigue life of the material under loading patterns for which no experimental data exist. The main parameters that define a CLD are the mean cyclic stress, σ_m, the cyclic stress amplitude, σ_a, and the R-ratio defined as the ratio between the minimum and maximum cyclic stress, $R = \sigma_{min}/\sigma_{max}$. A typical CLD annotation is presented in Figure 16.17.

As shown, the positive $(\sigma_m - \sigma_a)$-half-plane is divided into three sectors, the central one comprising combined tensile and compressive loading. The Tension–Tension (T–T) sector is bounded by the radial lines, representing the $S-N$ curves at $R = 1$ and $R = 0$, the former corresponding to static fatigue and the latter to tensile cycling with $\sigma_{min} = 0$. $S-N$ curves belonging to this sector have positive R-values less than unity. Similar comments regarding the other sectors can be derived from the annotations shown in Figure 16.17. Every radial line with $0 < R < 1$, i.e., in the T–T sector, has a corresponding symmetric line with respect to the σ_a-axis, which lies in the compression–compression (C–C) sector and whose R-value is the inverse of the tensile one, e.g., $R = 0.1$ and $R = 10$.

Radial lines emanating from the origin are expressed by:

$$\sigma_a = \left(\frac{1-R}{1+R}\right)\sigma_m, \tag{16.23}$$

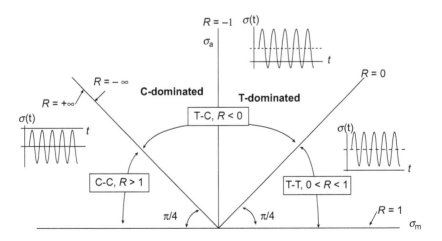

Figure 16.17 Constant life diagram annotation for $(\sigma_m - \sigma_a)$ plane.

and represent a single $S-N$ curve. Points along these lines are points of the $S-N$ curve for that particular stress ratio. CLDs are formed by joining points of consecutive radial lines, all corresponding to a certain value of cycles.

Although from a theoretical point of view the aforementioned representation of the CLD is rational, it presents a deficiency when seen from the engineering point of view. This deficiency is related to the region close to the horizontal axis, which represents loading under very low stress amplitude and high mean values with a culmination for zero stress amplitude ($R = 1$). The classic CLD formulations require that the constant life lines converge to the ultimate tensile stress (UTS) and the ultimate compressive stress (UCS), regardless of the number of loading cycles. However, this is an arbitrary simplification originating from the lack of information about the fatigue behavior of the material when no amplitude is applied. In fact, this type of loading cannot be considered fatigue loading; but rather creep of the material (constant static load over a short or long period). Although modifications to account for the time-dependent material strength have been introduced, their integration into CLD formulations requires the adoption of additional assumptions (see, e.g., Awerbuch & Hahn, 1981, pp. 243−273; Sutherland & Mandell, 2005). A new CLD formulation for adhesively-bonded joints has been presented in a recent publication (Sarfaraz, Vassilopoulos, & Keller, 2012a). This phenomenological CLD formulation has been developed for modeling the R-ratio effect taking into account the creep damage with little experimental data required.

16.5.1 Linear constant life diagram

The linear CLD model (Amijima, Tanimoto, & Matsuoka, 1982; Brondsted, Andersen, & Lilholt, 1997; Dover, 1979) is based on a single $S-N$ curve that should be experimentally derived. All other $S-N$ curves can be determined from the given one by simple calculations. This simplified formulation assumes that the failure mechanism is identical in tension and in compression when the load amplitude is the same. In the ($\sigma_m-\sigma_a$)-plane, the above-mentioned assumption implies that any constant life line forms an isosceles triangle, subtending $\pi/4$ angles with the axes (Passipoularidis & Philippidis, 2009). Any constant life line can be calculated by:

$$\frac{\sigma_a}{\sigma_0} + \frac{\sigma_m}{\sigma_0} = N^{-1/k} \tag{16.24}$$

where k and σ_0 are the parameters of the power law equation which describes the $S-N$ curve at the selected R-value.

16.5.2 Piecewise linear constant life diagram

The piecewise linear CLD (Philippidis & Vassilopoulos, 2004) is derived by linear interpolation between known values in the ($\sigma_m-\sigma_a$)-plane. This CLD model requires a limited number of experimentally determined $S-N$ curves along with the ultimate tensile and compressive stresses of the materials. $S-N$ curves representing the entire range of possible loading are commonly used for the construction of piecewise linear CLDs, normally at $R = 0.1$ for T−T loading, $R = -1$ for T−C loading and $R = 10$ for C−C loading patterns. Constant life lines connect data points of the same number of cycles on various $S-N$ curves. Unknown $S-N$ curves are calculated by linear interpolation between known values of fatigue and static strength data.

Analytical expressions were developed for the description of each region of the piecewise linear CLD in Philippidis and Vassilopoulos (2004).

1. If R' is in the T−T sector of the CLD, and between $R = 1$ and the first known R-ratio on the $(\sigma_m - \sigma_a)$-plane when moving counterclockwise, R_{1TT}, then

$$\sigma'_a = \frac{\text{UTS}}{\frac{\text{UTS}}{\sigma_{a,1TT}} + r' - r_{1TT}}, \tag{16.25}$$

where σ'_a and $\sigma_{a,1TT}$ are the stress amplitudes corresponding to R' and R_{1TT}, respectively, and $r_i = (1 + R_i)/(1 - R_i)$, $r' = (1 + R')/(1 - R')$.

2. If R' is located between any of two known R-ratios, R_i and R_{i+1},

$$\sigma'_a = \frac{\sigma_{a,i}(r_i - r_{i+1})}{(r_i - r')\frac{\sigma_{a,i}}{\sigma_{a,i+1}} + (r' - r_{i+1})} \tag{16.26}$$

3. If R' lies in the C−C region of the CLD, and between $R = 1$ and the first known R-ratio in the compression region, R_{1CC},

$$\sigma'_a = \frac{\text{UCS}}{\frac{\text{UCS}}{\sigma_{a,1CC}} - r' + r_{1CC}}, \tag{16.27}$$

where σ'_a and $\sigma_{a,1CC}$ are the stress amplitudes corresponding to R' and R_{1CC}, respectively.

16.5.3 Harris constant life diagram

Harris and his coworkers (Beheshty & Harris, 1998; Gathercole et al., 1994; Harris, 2003) developed a semiempirical equation based on fatigue test data obtained from a range of carbon- and glass-fiber composites:

$$a = f(1 - m)^u (c + m)^v \tag{16.28}$$

where a is the normalized stress amplitude component, σ_a/UTS, m is the normalized mean stress component, σ_m/UTS, and c is the normalized compression strength, UCS/UTS. In this equation, f, u and v are three adjustable parameters that are functions of fatigue life. Early studies (Gathercole et al., 1994; Harris, 2003) showed that parameter f mainly controls the height of the curve, and is a function of the ratio of the compressive to the tensile strength, while the exponents u and v determine the shapes of the two "wings" of the bell-shaped curve. Initially, the model was established with two simplified forms of Eqn (16.28) where $u = v = 1$ and $u = v$ for a family of carbon/Kevlar unidirectional hybrid composites (Gathercole et al., 1994; Harris, 2003). However, this model was not accurate for different material systems. Therefore, the general form of Harris' model was implemented in the sequel. In the general form, parameters f, u, and v were considered as functions of fatigue life. Depending on the examined material, and the quality of the fatigue data, these parameters were found to depend linearly on the logarithm of fatigue life, log (N), for a wide range of FRP materials (Beheshty & Harris, 1998):

$$f = A_1 \log N + B_1$$
$$u = A_2 \log N + B_2 \quad (16.29)$$
$$v = A_3 \log N + B_3$$

where the parameters A_i and B_i are determined by fitting Eqn (16.29) to the available experimental data for different loading cycles. Beheshty and Harris (1998) showed that the selection of this empirical form for the parameters u and v, can be employed for a wide range of materials, especially carbon—fiber—reinforced polymer (CFRP) matrix laminates. However, parameter f is extremely sensitive to the examined material and its values vary considerably between GFRP and CFRP laminates. Since, the modeling accuracy of the Harris' CLD is significantly dependent on the quality of the fitting of these parameters, Harris and his coworkers established different formulations for the estimation of parameter f based on experimental evidence obtained from a number of different composite material systems. The most recent proposal for the estimation of parameter f is the following equation:

$$f = Ac^{-p} \quad (16.30)$$

where A and p are the functions of log N as well. However, experimental evidence proved that values of $A = 0.71$ and $p = 1.05$ can be used in order to produce acceptable results for a wide range of CFRP and GFRP laminates (Harris, 2003).

16.5.4 Kawai's constant life diagram

Kawai's group (Kawai, 2007; Kawai & Koizumi, 2007) developed a formula that describes an asymmetric CLD, designated the anisomorphic constant fatigue life (CFL) diagram in Kawai and Koizumi (2007). The basic characteristic of this formulation is that it can be constructed by using only one experimentally derived $S-N$, which is called the critical $S-N$ curve. The R-ratio of this $S-N$ curve is defined as the ratio of the ultimate compressive over the ultimate tensile stress of the examined material. The formulation is based on three main assumptions: (1) The stress amplitude, σ_a, for a given constant value of fatigue life N is greatest at the critical stress ratio, (2) the shape of the CFL curves changes progressively from a straight line to a parabola with increasing fatigue life, and (3) the diagram is bounded by the static failure envelope, i.e., two straight lines connecting the ultimate tensile and ultimate compressive stresses with the maximum σ_a on the critical $S-N$ curve.

The CFL formulation depends on the position of the mean stress on the $(\sigma_m - \sigma_a)$-plane, whether it is in the tensile or the compressive region. The mathematical formulation reads:

$$\frac{\sigma_a^\chi - \sigma_a}{\sigma_a^\chi} = \begin{cases} \left(\dfrac{\sigma_m - \sigma_m^\chi}{UTS - \sigma_m^\chi}\right)^{(2-\psi_\chi)}, & UTS \geq \sigma_m \geq \sigma_m^\chi \\ \left(\dfrac{\sigma_m - \sigma_m^\chi}{UCS - \sigma_m^\chi}\right)^{(2-\psi_\chi)}, & UCS \leq \sigma_m \leq \sigma_m^\chi \end{cases} \quad (16.31)$$

where σ_m^χ and σ_a^χ represent the mean and cyclic stress amplitude for a given constant value of life N under fatigue loading at the critical stress ratio. ψ_χ denotes the fatigue strength ratio and is defined as:

$$\psi_\chi = \frac{\sigma_{max}^\chi}{\sigma_B} \quad (16.32)$$

where σ_{max}^χ is the maximum fatigue stress for a given constant value of life N under fatigue loading at the critical stress ratio. $\sigma_B(>0)$ is the reference strength (the absolute maximum between UTS and UCS) of the material that defines the peak of the static failure envelope. Therefore this normalization guarantees that ψ_χ always varies in the range (0, 1) and the exponents $(2 - \psi_\chi)$ in Eqn (16.31) are always greater than unity. Subsequently, linear $(2 - \psi_\chi = 1)$ or parabolic $(2 - \psi_\chi > 1)$ curves can be obtained from Eqn (16.31).

The critical fatigue strength ratio represents the normalized cyclic stress, and its relation to the number of loading cycles defines the normalized critical $S-N$ curve:

$$\psi_\chi = f^{-1}(2N_f) \quad (16.33)$$

After determining the critical $S-N$ curve by fitting to the available fatigue data, the CFL diagram can be constructed on the basis of the static strengths, UTS and UCS, and the reference $S-N$ relationship.

Kawai introduced the anisomorphic CFL diagram for the description of the fatigue behavior of CFRP materials. Later on, Kawai and Matsuda (2012) and Kawai, Matsuda, and Yoshimura (2012) extended the anisomorphic CLD formulation into a more general methodology that can deal with the mean stress sensitivity in fatigue of composites at different temperatures. The temperature dependence of the anisomorphic CFL diagram for a given composite is characterized by the temperature dependence of the static strengths in tension and compression and of the reference $S-N$ relationship for a critical stress ratio (Kawai et al., 2012).

16.5.5 Boerstra's constant life diagram

Boerstra (2007) proposed an alternative formulation for CLD that can be applied on random fatigue data, which do not necessarily belong to an $S-N$ curve. In this way, the R-ratio is not considered a parameter in the analysis and the model can be applied to describe the behavior of the examined material under loads with continuously changing mean and amplitude values. Boerstra's model constitutes a modification of the Gerber line. The exponent was replaced by a variable also including the difference in tension and compression. The general formulae of the model are:

$$\text{For } \sigma_m > 0: \quad \sigma_{ap} = \sigma_{AP}\left(1 - (\sigma_m/\text{UTS})^{\alpha T}\right) \tag{16.34}$$

$$\text{For } \sigma_m < 0: \quad \sigma_{ap} = \sigma_{AP}\left(1 - (\sigma_m/\text{UCS})^{\alpha C}\right) \tag{16.35}$$

where σ_{ap} is the stress amplitude component for a reference number of cycles, N_p, σ_{AP} is an "apex" stress amplitude for N_p and $\sigma_m = 0$, and αT and αC are the two shape parameters of the CLD curves for the tensile and compressive sides, respectively.

The above-mentioned equations represent the CLD lines in the $(\sigma_m-\sigma_a)$-plane. According to Boerstra (2007), existing fatigue data for different kinds of composite materials show steeper $S-N$ curves under tension and less steep under compression. An exponential relationship with the mean stress can be a good description for the slope $(1/m)$ of $S-N$ lines as follows:

$$m = m_0 e^{(-\sigma_m/D)} \tag{16.36}$$

where m_0 is a measure for the slope of the $S-N$ curve on the log–log scale for $\sigma_m = 0$ and D is the skewness parameter for the dependency of m.

Equations (16.34)–(16.36) suggest that five parameters, m_0, D, N_p, αT, and αC, must be defined in order to construct the CLD model. However, the estimation of the parameters requires a multiobjective optimization process. The aim of this optimization is to estimate the parameters allowing the calculation of the shortest distance between each measuring point and the $S-N$ line for its particular mean stress. The procedure is as follows:

1. The static strengths UTS and UCS are determined and some fatigue test data on coupons with various values of stress amplitude, σ_a, and mean stress, σ_m, should also be available.
2. The desired value of N_p is chosen and an initial set of values for parameters m_0, D, N_p, aT, and aC is assumed (Boerstra, 2007).
3. The slope of the $S-N$ line, m, is calculated for each measured σ_m using Eqn (16.34).
4. The σ_a corresponding to each σ_m is projected to the $(\sigma_m-\sigma_a)$-plane for the selected number of cycles N_p by $\sigma_{ap} = \sigma_a(N/N_p)^{(1/m)}$.
5. σ_{AP}, is calculated for each pair of σ_{ap} and σ_m using Eqns (16.34) and (16.35).
6. A modified stress amplitude, $\sigma_{ap,mod}$, is calculated by feeding back the average value of σ_{AP} and the measured mean stress value, σ_m, into Eqns (16.34) and (16.35).
7. The difference between the logarithms of the measured stress amplitude and the modified stress amplitude is then computed as: $\Delta\sigma_a = \ln(\sigma_{ap}) - \ln(\sigma_{ap,mod})$.

8. The theoretical number of cycles, N_e, corresponding to the $\sigma_{ap,mod}$ stress amplitude and the measured mean stress, σ_m, can be calculated by solving the equation:

$$N_e = N_p \left(\frac{\sigma_{a,mod}}{\sigma_a}\right)^m \quad (16.37)$$

9. The difference between the measured number of cycles, N, and the theoretical number of cycles, N_e, is defined by $\Delta n = \ln(N) - \ln(N_e)$.
10. The shortest distance between each independent point and the $S-N$ lines in the $\sigma_m-\sigma_a-N$ space is expressed by: $\Delta t = \text{sign}(\Delta\sigma_a)\sqrt{(1/(1/\Delta\sigma_a^2 + 1/\Delta n^2))}$. The sum of all Δt's is designated the total standard deviation, SDt. Minimization of the SDt results in the estimation of the optimal m_o, D, N_p, aT, and aC parameters.

16.5.6 Kassapoglou's constant life diagram

A very simple model was recently proposed by Kassapoglou (2007). Although the model was proposed for the derivation of $S-N$ curves under different R-ratios, it can potentially be used for the construction of piecewise nonlinear CLD. The basic assumption of the model is that the probability of failure of the material during a cycle is constant and independent of the current state or number of cycles up to this point. The assumption that the statistical distribution that describes failure under static loading can be used to describe failure under fatigue loading patterns oversimplifies the reality and masks the effect of the different damage mechanisms that develop under fatigue loading and static loading. However, this model requires no fatigue testing, no empirically determined parameters and no detailed modeling of damage mechanisms.

The model comprises the following equations for calculation of maximum cyclic stress as a function of number of cycles:

$$\sigma_{max} = \frac{\beta_T}{(N)^{\frac{1}{a_T}}} \quad \text{for} \quad 0 \leq R < 1 \quad (16.38)$$

$$\sigma_{max} = \frac{\beta_C}{(N)^{\frac{1}{a_C}}} \quad \text{for} \quad R > 1 \quad (16.39)$$

While for $R < 0$ the following equation should be solved numerically:

$$N = \frac{1}{\left(\frac{\sigma_{max}}{\beta_T}\right)^{a_T} + \left(\frac{\sigma_{min}}{\beta_C}\right)^{a_C}} \quad (16.40)$$

Parameters a_i and b_i ($i = T$ or C) denote the scale and shape of a two-parameter Weibull distribution that can describe the static data in tension and compression, respectively.

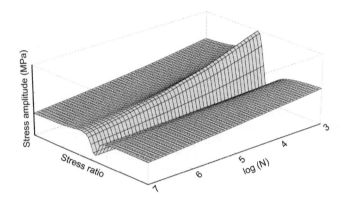

Figure 16.18 Representation of relationship between fatigue parameters $\sigma_a - R - \log(N)$.

16.5.7 The piecewise nonlinear constant life diagram

The piecewise nonlinear model has been introduced by Vassilopoulos et al. (2010b). The model has been established on the basis of the relationship between the stress ratio and the stress amplitude, and simple phenomenological equations were derived from this relationship, without the need for the adoption of any assumptions.

All previous CLD formulations are based on the fitting of linear or nonlinear equations to existing experimental fatigue data on the $\sigma_m - \sigma_a$ plane. However, there is no rational explanation for the selection of the two aforementioned stress parameters. Any other combination of $\sigma_a - \sigma_m - R$ can just as well be used for the derivation of a CLD. A plot of stress amplitude against stress ratio for different numbers of loading cycles is presented in Figure 16.18.

The surface of Figure 16.18 also represents the fatigue failure locus of the examined material. Any loading combination above the surface causes failure. A projection of this surface on the $R - \sigma_a$ plane can be considered as a CLD, see Figure 16.19.

In Figure 16.19, the x-axis represents the R-ratio and ranges from $-\infty$ to $+\infty$ without any singularity. The y-axis represents the stress amplitude and has positive values. $S-N$ curves for any stress ratio, R, are represented by vertical lines emanating from the corresponding value of R on the x-axis. The above-mentioned diagram can be divided into four distinct regions, each corresponding to different loading conditions. Part I for tension-compression (T—C) loading under $-\infty \leq R \leq -1$, Part II for T—C loading under $-1 \leq R \leq 0$, Part III corresponding to T—T loading under $0 \leq R \leq 1$ and Part IV for C—C loading under $1 \leq R \leq +\infty$. The behavior of each of the aforementioned parts can be described by simple phenomenological equations and model parameters can be estimated by using appropriate boundary conditions for each part of the diagram and known values of σ_a, σ_m and R, as described in the following:

Parts I and IV: $-\infty \leq R \leq -1$ and $1 \leq R \leq +\infty$

$$\sigma_a = (1 - R)\left(\frac{A_{\text{I, or IV}}}{R} + \frac{B_{\text{I, or IV}}}{R^2}\right) \tag{16.41}$$

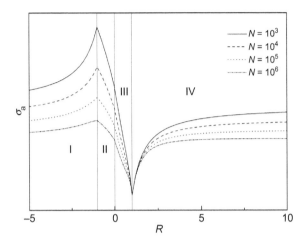

Figure 16.19 Constant life diagram schematic representation on $(R-\sigma_a)$ plane.

where A_I, B_I, A_{IV}, and B_{IV} are the parameters that can be easily determined by fitting to the available experimental data.

The process is based on the selection of the boundary conditions for each part of the CLD. For Parts I and IV, described by Eqn (16.42), the boundary conditions are the following:

$$\text{for } R = -1, \quad \sigma_a = \sigma_a^{-1}, \text{ and } \sigma_m = 0,$$
$$\text{for } R = \pm\infty, \quad \sigma_a = \sigma_a^{\pm\infty}, \text{ and} \qquad (16.42)$$
$$\text{for } R = 1, \quad \sigma_a = 0, \text{ and } \sigma_m = \text{UCS}$$

where stress parameter superscripts denote the corresponding stress ratio, e.g., σ_α^{-1} is the stress amplitude for $R = -1$.

By applying these three boundary conditions, Eqn (16.41) becomes:

$$\text{for } R = -1 \rightarrow \sigma_a^{-1} = 2 \times (-A_I + B_I), \text{ or}$$

$$\text{for } R = 1 \rightarrow \text{UCS} = 2 \times (A_{IV} + B_{IV}), \text{ or}$$

$$\text{for } R = \pm\infty \rightarrow \sigma_a^{\pm\infty} = \lim_{R \to \pm\infty}(1-R)\left(\frac{A_{I, \text{ or IV}}}{R} + \frac{B_{I, \text{ or IV}}}{R^2}\right) = -A_{I, \text{ or IV}}$$

(16.43)

and the four parameters A_I, B_I, A_{IV}, and B_{IV} can be defined as:

$$A_{I, \text{ or IV}} = -\sigma_a^{\pm\infty}$$

$$B_I = \frac{\sigma_a^{-1}}{2} - \sigma_a^{\pm\infty} \qquad (16.44)$$

$$B_{IV} = \frac{\text{UCS}}{2} + \sigma_a^{\pm\infty}$$

When the S–N curve under $R = 10$ is available instead of that under $R = \pm\infty$, the boundary conditions should be adjusted accordingly.

Parts II and III: $-1 \leq R \leq 0$ and $0 \leq R \leq 1$, the following equation is used:

$$\sigma_a = \frac{1-R}{A_{\text{II, or III}} R^n + B_{\text{II, or III}}} \tag{16.45}$$

with parameter n being equal to 1 for Part II and equal to 3 for Part III, based on the experimental evidence. The boundary conditions are the following:

$$\begin{aligned} &\text{for } R = 1, \ \sigma_a = 0, \ \text{and } \sigma_m = \text{UTS} \\ &\text{for } R = -1, \ \sigma_a = \sigma_a^{-1}, \ \text{and } \sigma_m = 0 \end{aligned} \tag{16.46}$$

when no reference S–N curve is used between $R = -1$ and $R = 1$. Implementing the above-mentioned boundary conditions results in:

$$\begin{aligned} A_{\text{II, or III}} &= \frac{1}{\text{UTS}} - \frac{1}{\sigma_a^{-1}} \\ B_{\text{II, or III}} &= \frac{1}{\text{UTS}} + \frac{1}{\sigma_a^{-1}} \end{aligned} \tag{16.47}$$

However, if the S–N curve under $R = 0$ is considered as well, the boundary conditions (Eqn (16.46)) are supplemented by:

$$\sigma_a = \sigma_a^0, \quad \text{for } R = 0 \tag{16.48}$$

By applying the boundary conditions for Parts II and III in Eqn (16.45), parameters A_{II}, B_{II}, A_{III}, and B_{III} obtain the following values:

$$\begin{aligned} A_{\text{II}} &= \frac{1}{\sigma_a^0} - \frac{2}{\sigma_a^{-1}} \\ A_{\text{III}} &= \frac{2}{\text{UTS}} - \frac{1}{\sigma_a^0} \\ B_{\text{II, or III}} &= \frac{1}{\sigma_a^0} \end{aligned} \tag{16.49}$$

Similar to Parts I and IV, when the S–N curve under $R = 0.1$ is available instead of that under $R = 0$, the boundary conditions are modified accordingly. More S–N curves may be used to improve the accuracy of the model. However, as shown in the next paragraphs, the use of only two or three S–N curves, under $R = -1$, $R = \pm\infty$ (alternatively $R = 10$), and $R = 0$ (alternatively $R = 0.1$) suffices to produce an accurate model.

16.5.8 The polynomial constant life diagram

The asymmetry of the CLD and the observed fatigue-creep interactions close to the $R = 1$ domains prevent the possibility of accurate modeling using simple linear diagram formulations such as the classic Goodman diagram. Based on the characteristics of the CLD for the examined bonded joints (see Chapters 8 and 9 of this volume), a semiempirical CLD formulation was proposed (Sarfaraz et al., 2012a):

$$\sigma_a - \sigma_{(R=-1)} = \sigma_m \left(\alpha \sigma_m^3 + \beta \sigma_m^2 + \gamma \sigma_m + \delta \right) \qquad (16.50)$$

where $\sigma_{(R=-1)}$ is the fatigue strength of joints at $R = -1$ and α, β, γ, and δ are the model parameters. This formulation incorporates both the mean and the amplitude components of the cyclic loading. Using the classic power law relationship, Eqn (16.1), to simulate the fatigue behavior, Eqn (16.50) is also a direct function of the number of cycles as shown in:

$$bN^{-a} - \sigma_{(R=-1)} = \sigma_m \left(\alpha \sigma_m^3 + \beta \sigma_m^2 + \gamma \sigma_m + \delta \right) \qquad (16.51)$$

As deduced from Eqn (16.51), at the fatigue strength is equal to that under reversed loading ($R = -1$). Equation (16.50) is actually a fourth-order polynomial equation describing constant life lines. Nevertheless, four boundary conditions corresponding to the physical meaning of the CLD must be satisfied in order to determine the model parameters as described in the following.

For the common CLD the ultimate compressive and tensile strengths of the material, which are independent of the number of cycles to failure, usually constitute the upper and lower limits of the mean load (σ_m) and iso-life curves. This assumption leads to an inadequate modeling of fatigue behavior at high mean load levels as discussed in Vassilopoulos et al., (2010b). To improve the accuracy of the modeling at R-ratios close to one at zero load amplitude ($\sigma_a = 0$), the two boundaries of the CLD are defined by the creep rupture strength under compression (σ_{cc}) and tension (σ_{ct}) instead of the ultimate compressive and tensile strengths ($\sigma_a = 0$ at $\sigma_m = \sigma_{ct}$ and $\sigma_m = \sigma_{cc}$) as shown in Figure 16.20.

The fourth-order polynomial given in Eqn (16.50) provides the flexibility to simulate the inflection of the iso-life curves described earlier. The parameters of Eqn (16.50) can be determined in such a way that it includes two minima, one maximum and consequently two inflection points bounded by the upper and lower limits of the mean load. To ensure this, since the maximum value of the iso-life curve (polynomial) is between σ_{cc} and σ_{ct}, the minima of the polynomial have to satisfy σ_{cc} and σ_{ct}. Therefore, the first derivative of Eqn (16.50) with respect to σ_m at σ_{ct} and σ_{cc} must be zero, which leads to two more boundary conditions (Crocombe & Richardson, 1999).

These boundary conditions allow the derivation of iso-life curves that simulate the real material behavior as demonstrated by experimental data by shifting them towards the stronger domain. Furthermore, using the creep rupture strength data instead of static strength data improves the existing CLD. The four unknown parameters of Eqn (16.50) can be estimated by applying the aforementioned boundary conditions

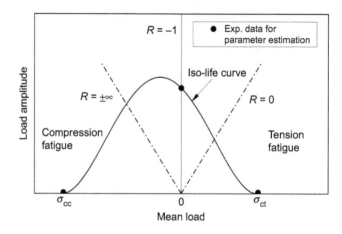

Figure 16.20 Polynomial iso-life curve and imposed boundary conditions.

as a function of the experimental fatigue data under $R = -1$ and the creep rupture strengths as shown in the following equations:

$$\alpha = \frac{\sigma_{(R=-1)}}{(\sigma_{ct}\sigma_{cc})^2} \tag{16.52}$$

$$\beta = \frac{-2(\sigma_{ct} + \sigma_{cc})\sigma_{(R=-1)}}{(\sigma_{ct}\sigma_{cc})^2} \tag{16.53}$$

$$\gamma = \frac{\left[(\sigma_{ct} + \sigma_{cc})^2 + 2\sigma_{ct}\sigma_{cc}\right]\sigma_{(R=-1)}}{(\sigma_{ct}\sigma_{cc})^2} \tag{16.54}$$

$$\delta = \frac{-2(\sigma_{ct} + \sigma_{cc})\sigma_{(R=-1)}}{\sigma_{ct}\sigma_{cc}} \tag{16.55}$$

Obviously the resulting parameters are functions of fatigue life and the CLD formulation given by Eqn (16.50) is valid only in the physically acceptable mean load range between the compressive and tensile creep strengths of the joints ($\sigma_{cc} \leq \sigma_m \leq \sigma_{ct}$). The accuracy of the proposed model is evaluated by comparisons with the available experimental data and the results of commonly used CLD formulations for composite materials in the following section.

16.6 Comparison of existing constant life diagram (CLD) formulations

Most of the aforementioned CLD models were applied to model R-ratio effect on the constant amplitude fatigue behavior of the symmetric adhesively-bonded double-lap joints examined in Chapter 8 of this volume. The experimental results presented in

Chapter 8 of this volume contain representative experimental data in the range between 1 and 10^8 cycles. The fatigue life of the examined joints under the 9 applied R-ratios is plotted against the cyclic load amplitude in Figures 16.21 and 16.22 for the experiments under positive and negative mean loads, respectively.

The effect of the load ratio on the fatigue life of the examined joints was also visualized by using the "experimental" CLD as shown in Figure 16.23. It is obvious

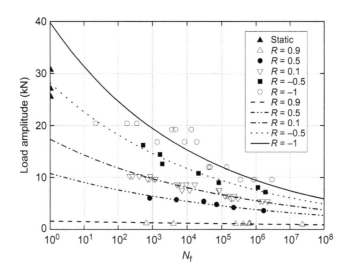

Figure 16.21 $S-N$ data for tension and tension-dominant fatigue loading.

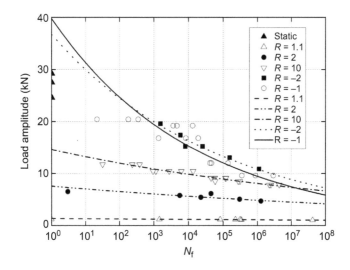

Figure 16.22 $S-N$ data for compression and compression-dominant fatigue loading.

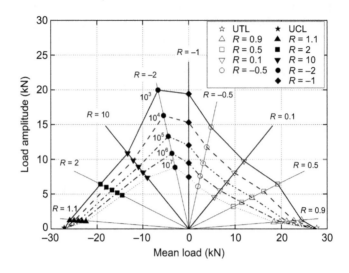

Figure 16.23 Variation of alternating load versus mean load at different fatigue lives.

that the CLD is not symmetric with respect to the zero mean cyclic load axis and shifted somewhat towards the compression-dominated domain with the apex corresponding to the S–N curve under $R = -2$. This behavior can be attributed to the difference in fatigue strength under tension and compression loading as discussed earlier. An inflection in the curvature of the iso-life curves is observed, as they change from concave to convex when the loading condition shifts from T–T or C–C to combined tension-compression fatigue loading. Moreover it is observed that the ultimate tensile and compressive load values (UTL = 27.7 ± 2.17 kN and UCL = −27.1 ± 1.92 kN) are not appropriate for description of the fatigue behavior under zero load amplitude since, as shown in Figure 16.23, a fatigue-creep interaction occurs under R-ratios close to one due to the presence of very low amplitude and high mean values that characterize the cyclic loading in this region.

The selected CLD formulations for the modeling of the constant amplitude fatigue behavior of the examined adhesively-bonded composite joints are the following: the linear (LR), the piecewise linear (PWL), the piecewise nonlinear (PNL), the Harris (HR), the Kawai (KW), the Boerstra (BR), and the Polynomial (POLY).

The linear CLD can be easily derived using only one set of experimental data usually under reversed loading and assumes a similar failure mode under tension and compression loading. A modified form of the linear Goodman diagram, PWL CLD, is used extensively in the wind turbine rotor blade industry and, although it requires quite large experimental databases, it provides high prediction accuracy. An alternative, the PNL model, recently introduced by Vassilopoulos et al. (2010b), can result in more accurate constant life behavior simulations based on fewer experimental data compared to the PWL model.

The Harris CLD requires experimental data comparable with those needed for the application of the PWL model, but its implementation requires more computational effort, and its precision is based on the accurate estimation of the model parameters (Vassilopoulos et al., 2010a). It results in continuous bell-shaped constant life lines calculated by a unified equation that describes the fatigue behavior for both tension and compression loading. The Kawai anisomorphic model can be constructed using only one experimentally derived $S-N$ curve, known as the critical $S-N$ curve. The critical R-ratio is defined as the ratio of the compressive strength to the tensile static strength of the examined materials. The multislope formulation proposed by Boerstra can be applied to random fatigue data, which do not necessarily belong to a certain $S-N$ curve. However, the implementation of this model relies on the solution of a multiparametric optimization problem for estimation of the five model parameters.

The *CCfatigue* software (Vassilopoulos et al., 2010) was used for derivation of all CLDs. The predictive accuracy of the applied CLD models was evaluated by comparing predicted $S-N$ curves with corresponding experimental ones that were not used for estimation of the model parameters. The fatigue data under $R = 0.9$ (T−T) and $R = 1.1$ (C−C) were used in this work to simulate the combined fatigue-creep behavior of the joints under high mean-low amplitude loading conditions. CLD produced by the described models are shown in Figures 16.24−16.30. Data used for the estimation of model parameters are shown by open triangles, while data used for the model validation are shown by closed circles in all figures. A quantification of the predictive ability of each of the applied models was performed. The coefficients of multiple determination (R^2) between the predicted and the experimentally derived $S-N$ curves are shown in Table 16.2.

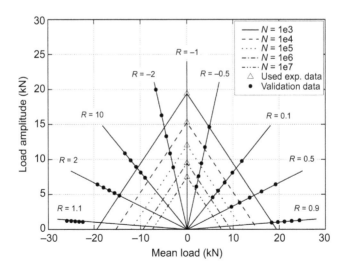

Figure 16.24 Linear constant life diagram.

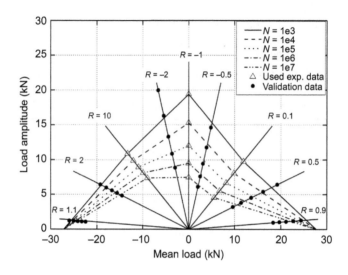

Figure 16.25 Piecewise linear constant life diagram.

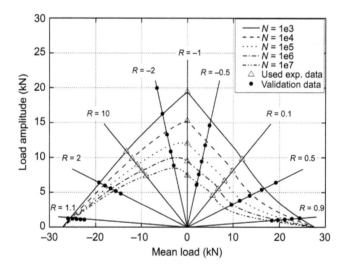

Figure 16.26 Piecewise nonlinear constant life diagram.

The linear model shown in Figure 16.24, using only fatigue data at $R = -1$, underestimates the fatigue life in all regions except for $R = -0.5$. Comparison of the R^2 values in Table 16.2 confirms a good accuracy of the model for this loading condition ($R^2 = 0.977$). The PWL model, employing three sets of fatigue data ($R = 0.1, -1,$ and 10), provides more accurate estimations of fatigue life than the linear model (see

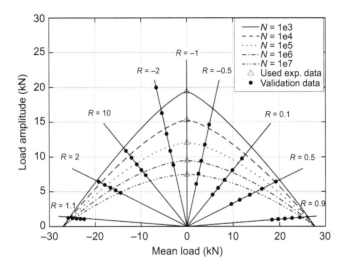

Figure 16.27 Kawai's constant life diagram.

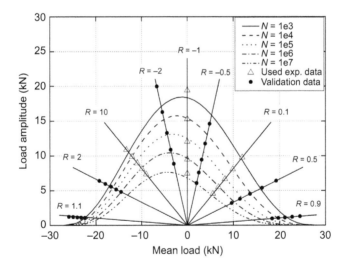

Figure 16.28 Harris constant life diagram.

Figure 16.25). However, the accuracy of the PWL model under high mean loads is poor. The modified form of this model, the PNL formulation, can appropriately describe the concave upward behavior in the region between $R = 1$ and $R = 0.1$ (see Figure 16.26) using the same amount of fatigue data. In this case, the shift of the apex of the iso-life curves towards the negative mean load quadrant at a high number of cycles improves modeling accuracy. On the other hand, the convex shape of the curves between

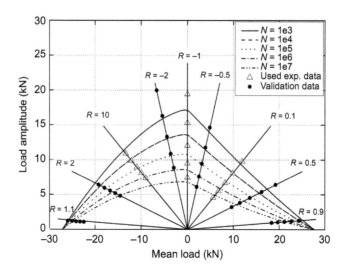

Figure 16.29 Boerstra's constant life diagram.

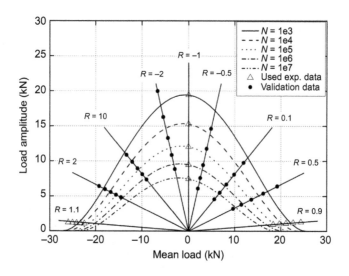

Figure 16.30 Polynomial constant life diagram.

$R = 10$ and $R = 1$ reduces its accuracy compared to the PWL model, i.e., $R^2 = 0.803$ and $R^2 = 0.675$ at $R = 2$ for the PWL and PNL models, respectively.

For Kawai's anisomorphic model the critical R-ratio is calculated as being -0.98. Therefore, the S–N curve under reversed loading ($R = -1$) was considered as the reference one for the derivation of the CLD (see Figure 16.27). The obtained CLD, in contrast to the trend of the experimental data, provides convex constant life curves

Table 16.2 Comparison of predictive ability of applied CLD formulations

R-ratio	R^2						
	POLY	LR	PWL	PNL	KW	HR	BR
0.5	0.930	0.537	0.921	0.839	0.493	0.800	0.940
0.9	0.511	0.224	0.363	0.733	0.104	0.426	0.444
−0.5	0.914	0.977	0.976	0.973	0.870	0.948	0.977
2	0.854	0.485	0.803	0.675	0.687	0.750	0.746
−2	0.905	0.552	0.708	0.815	0.872	0.838	0.723
1.1	0.539	0.328	0.547	0.308	0.669	0.348	0.656
Average	0.775	0.517	0.720	0.724	0.616	0.685	0.748
Standard deviation	0.196	0.259	0.233	0.227	0.288	0.241	0.196

CLD, constant life diagram; POLY, polynomial; LR, linear; PWL, piecewise linear; PNL, piecewise nonlinear; KW, Kawai; HR, Harris; BR, Boerstra.

in both the tension- and the compression-dominated domains. Consequently in most regions, except at around $R = 10$, the model overestimates the fatigue life and especially at high mean loads produces very poor predictions.

More accurate predictions can be obtained by the bell-shaped equation according to the Harris model (see Figure 16.28) than those produced by the PWL model. However, application of the model, based on a nonlinear fitting of a function to the fatigue data, results in constant life lines that do not satisfy any boundary conditions except at $\sigma_a = 0$ where $\sigma_m =$ UCS and UTS. Therefore, some deviations can be seen even between the derived constant life lines and the experimental data used for estimation of the model parameters.

The use of five parameters for the derivation of the BR CLD makes it very flexible and able to accurately model the fatigue behavior of a large number of different material systems. The average accuracy of Boerstra's CLD ($R^2 = 0.748$) is one of the highest among the examined models. Nonetheless, the fatigue life in regions susceptible to creep is overestimated. Also, as with the Harris' CLD, since the model parameters are calculated based on an optimization process, the CLD only satisfies the boundary conditions at zero load amplitude. Consequently the iso-life lines do not necessarily comply with the experimental data employed for modeling, as shown in Figure 16.29.

The polynomial CLD presented in Figure 16.30 shows the capability of the introduced CLD model to follow the trend of the experimental data. It is apparent that incorporating the creep rupture strength data (actually the fatigue data under $R = 0.9$ (T−T) and $R = 1.1$ (C−C)) into the formulation, instead of using the static strength values, improves the accuracy of the model predictions in high mean load regions. Although under the applied boundary conditions the constant life lines must satisfy the fatigue failure condition under reversed loading (when $\sigma_m = 0$, $\sigma_a = \sigma_{(R = -1)}$), this does not impose a maximum value of the CLD on the radial line representing the $S-N$ curve under $R = -1$. The model allows the diagram to move towards the tension- or compression-dominated domain according to the experimental evidence. The highest average R^2 value with the lowest standard deviation obtained for the new CLD formulation, 0.775 ± 0.196 (see Table 16.2), proved the good accuracy and consistency of the model.

16.7 Conclusions

An overview of the commonly used $S-N$ curves and the available CLD formulations for the simulation of the fatigue behavior of composite materials and composite structural elements has been presented in this chapter. It has been proved that the constant amplitude fatigue behavior of composite laminates and composite structural elements under different loading conditions can be modeled by several different formulations that have been proposed in the past. The fatigue behavior under a constant R-ratio can be simulated by the so-called $S-N$ curves, while the effect of the mean stress (also shown in the literature as the R-ratio effect) on the fatigue life can be investigated by the use of the CLD. Although, yet there is no commonly accepted universal formulation for both $S-N$ curves and CLDs able to accurately model the behavior of wide ranges of examined materials, there are certain models that are commonly used because of their simplicity and their modeling accuracy. Some of the characteristics of the commonly used models are exploited in this chapter, while in addition, recently introduced $S-N$ curve and CLD models are presented. The accurate modeling achieved by applying the selected $S-N$ curve models and CLD formulations to different classes of composite materials and structures under different loading conditions proved that the applicability of the models is independent of the material system, loading conditions, failure mode, and range of experimental data.

Comparisons of the modeling ability of several $S-N$ curve models revealed that the power law model in general overestimates the fatigue life in the LCF region while it fits the experimental data in the HCF region fairly well. On the other hand, the exponential model underestimates the fatigue life at HCF whereas it successfully estimates the lifetime in the LCF region. The sensitivity of reliability-based models to the scatter of experimental data was recognized. The strong effect of censoring a few data points on the overall behavior of this type of $S-N$ curve was demonstrated by the analysis of a well-established fatigue database for carbon fiber—reinforced laminates.

Comparison of the newly introduced hybrid formulation with the commonly used fatigue models showed a superiority of the new $S-N$ formulation in terms of fitting accuracy. This formulation is no more complex than the linear regression models, since their parameters are used for the establishment of the hybrid $S-N$ curve. The continuous hybrid $S-N$ curve equation can be used in fatigue life prediction methodologies to provide a modeling tool for the entire fatigue lifetime. In contrast, two-segment power law or exponential models, or even fitted mathematical equations cannot be easily implemented in design codes since they require subjective selection of fatigue model parameters and cannot be used for extrapolation outside the range of the available fatigue data. Continuous $S-N$ curve equations, such as the one described by the hybrid model can be used in fatigue life prediction methodologies to provide a modeling tool for the entire fatigue lifetime.

Based on the above-mentioned conclusions, all the examined $S-N$ curve types are appropriate for modeling the fatigue life of composite materials. The empirical $S-N$ formulations can be used for any preliminary stage of a design process, since the model parameters can be estimated by a few experimental data and even by hand calculations. On the other hand, the statistical methods, such as the wear-out model described here, are able to derive $S-N$ curves for any desired reliability level and are therefore useful for design processes where high reliability levels are desirable. However, as was proved here, they should be used with caution since they are very much sensitive to outline data in the available data set.

CLDs are used in order to model the high dependency of the fatigue strength on the mean load and avoid extensive experimental programs. A large number of CLD formulations have been proposed; the most recent and the most commonly used ones for composite materials and composite structural elements have been presented in this chapter. It has been shown that, in general, the modeling and predicting ability of the existing simulations is sufficient, considering the necessary experimental data requested for their application. Nevertheless, there is no model able to consider the creep—fatigue interaction that is more pronounced for R-ratios close to 1 in both the tension and compression domains. To this end, a new phenomenological model, recently introduced by Sarfaraz et al. (2012a) for the derivation of CLDs for adhesively-bonded composite joints taking into account the creep—fatigue interaction has been also discussed in this chapter.

Acknowledgments

This work was supported by the Swiss National Science Foundation (Grant No 200020-121756), Fiberline Composites A/S, Denmark (supplier of the pultruded laminates), and Sika AG, Zurich (adhesive supplier).

References

Abd Allah, M. H., Abdin, E. M., Selmy, A. I., & Khashaba, U. A. (1997). Effect of mean stress on fatigue behavior of GFRP pultruded rod composites. *Composites Part A, 28*(1), 87—91.
Adam, T., Fernando, G., Dickson, R. F., Reiter, H., & Harris, B. (1989). Fatigue life prediction for hybrid composites. *International Journal of Fatigue, 11*(4), 233—237.
Amijima, S., Tanimoto, T., & Matsuoka, T. (1982). A study on the fatigue life estimation of FRP under random loading. In *Progress in science and engineering of composites* (pp. 701—708). Germany:Berlin.
Awerbuch, J., & Hahn, H. T. (1981). *Off-axis fatigue of graphite/epoxy composite*. Philadelphia: ASTM STP 723, American Society for Testing and Materials.

Aymerich, F., & Found, M. S. (2000). Response of notched carbon/peek and carbon/epoxy laminates subjected to tension fatigue loading. *Fatigue and Fracture of Engineering Materials and Structures, 23*(8), 675—683.

Bach, P. W. (1996). ECN investigations of Gl-UP materials. In *Fatigue of materials and components for wind turbine rotor blades*. European Commission, Directorate-General XII, Science, Research and Development, ISBN 92-827-4361-6.

Bakis, C. E., Simonds, R. A., Vick, L. W., & Stinchcomb, W. W. (1990). *Matrix toughness, long-term behavior, and damage tolerance of notched graphite fiber—reinforced composite materials*. Philadelphia, USA: ASTM STP 1059, American Society for Testing and Materials.

Beheshty, M. H., & Harris, B. (1998). A constant life model of fatigue behavior for carbon fiber composites: the effect of impact damage. *Composites Science and Technology, 58*, 9—18.

Boerstra, G. K. (2007). The multislope model: a new description for the fatigue strength of glass reinforced plastic. *International Journal of Fatigue, 29*(8), 1571—1576.

Brondsted, P., Andersen, S. I., & Lilholt, H. (1997). Fatigue damage accumulation and lifetime prediction of GFRP materials under block loading and stochastic loading. In *Proceedings of the 18th RISOE international symposium on material science*, Roskilde, Denmark (pp. 269—278).

Chou, P. C., & Croman, R. (1979). *Degradation and sudden-death models of fatigue of graphite/epoxy composites*. Philadelphia: ASTM STP 674, American Society for Testing and Materials.

Crocombe, A. D., & Richardson, G. (1999). Assessing stress and mean load effects on the fatigue response of adhesively bonded joints. *International Journal of Adhesion and Adhesives, 19*(1), 19—27.

D'Amore, A., Caprino, G., Stupak, P., Zhou, J., & Nicolais, L. (1996). Effect of stress ratio on the flexural fatigue behaviour of continuous strand mat reinforced plastics. *Science and Engineering of Composite Materials, 5*(1), 1—8.

Dover, W. D. (1979). Variable amplitude fatigue of welded structures. In *Fracture mechanics: current status future prospects* (pp. 125—147). Cambridge: Pergamon Press.

Ellyin, F., & El-Kadi, H. (1994). Effect of stress ratio on the fatigue of unidirectional glass fibre/epoxy composite laminae. *Composites, 25*(10), 917—924.

Epaarachchi, J. A., & Clausen, P. D. (2003). An empirical model for fatigue behavior prediction of glass fibre-reinforced plastic composites for various stress ratios and test frequencies. *Composites Part A: Applied Science and Manufacturing, 34*(4), 313—326.

Gathercole, N., Reiter, H., Adam, T., & Harris, B. (1994). Life prediction for fatigue of T800/5245 carbon fiber composites: I. Constant amplitude loading. *International Journal of Fatigue, 16*, 523—532.

Gerber, W. Z. (1874). Bestimmung der zulässigen spannungen in eisen-constructionen (Calculation of the allowable stresses in iron structures). *Z Bayer Archit Ing-Ver, 6*(6), 101—110.

Goodman, J. (1899). *Mechanics applied to engineering*. Harlow, UK: Longman Green.

Hahn, H. T. (1979). Fatigue behavior and life prediction of composite laminates. In S. W. Tsai (Ed.), *Composite materials: Testing and design (Fifth conference), ASTM STP674*. Philadelphia, PA: ASTM (pp. 383—417)297—311.

Hahn, H. T., & Kim, R. Y. (1975). Proof testing of composite materials. *Journal of Composite Materials, 9*(3), 297—311.

Halpin, J. C., Jerina, K. L., & Johnson, T. A. (1973). *Characterization of composites for the purpose of reliability evaluation*. Philadelphia, USA: ASTM STP 521, American Society for Testing and Materials.

Harik, V. M., & Bogetti, T. A. (2003). Low cycle fatigue of composite laminates: a damage-mode-sensitive model. *Journal of Composite Materials, 37*(7), 597—610.
Harik, V. M., Klinger, J. R., & Bogetti, T. A. (2002). Low-cycle fatigue of unidirectional composites: bi-linear S-N curves. *International Journal of Fatigue, 24*(2—4), 455—462.
Harris, B. (2003). A parametric constant-life model for prediction of the fatigue lives of fibre-reinforced plastics. In B. Harris (Ed.), *Fatigue in composites* (pp. 546—568). Cambridge, England: Woodhead Publishing Limited.
Hwang, W., & Han, K. S. (1986). Fatigue of composites - fatigue modulus concept and life prediction. *Journal of Composite Materials, 20*(2), 154—165.
Jarosch, E., & Stepan, A. (1970). Fatigue properties and test procedures of glass reinforced plastic rotor blades. *Journal of the American Helicopter Society, 15*(1), 33—41.
Kassapoglou, C. (2007). Fatigue life prediction of composite structures under constant amplitude loading. *Journal of Composite Materials, 41*(22), 2737—2754.
Kawai, M. (2007). A method for identifying asymmetric dissimilar constant fatigue life diagrams for CFRP laminates. *Key Engineering Materials, 334—335*, 61—64.
Kawai, M., & Koizumi, M. (2007). Non-linear constant fatigue life diagrams for carbon/epoxy laminates at room temperature. *Composites Part A: Applied Science and Manufacturing, 38*(11), 2342—2353.
Kawai, M., & Matsuda, Y. (2012). Anisomorphic constant fatigue life diagrams for a woven fabric carbon/epoxy laminate at different temperatures. *Composites, Part A: Applied Science and Manufacturing, 43*(4), 647—657.
Kawai, M., Matsuda, Y., & Yoshimura, R. A. (2012). General method for predicting temperature-dependent anisomorphic constant fatigue life diagram for a woven fabric carbon/epoxy laminate. *Composites, Part A: Applied Science and Manufacturing, 43*(6), 915—925.
Lee, L. J., Yang, J. N., & Sheu, D. Y. (1993). Prediction of fatigue life for matrix-dominated composite laminates. *Composites Science and Technology, 46*(1), 21—28.
Mallick, P. K., & Zhou, Y. (2004). Effect of mean stress on the stress-controlled fatigue of a short E-glass fiber reinforced polyamide-6.6. *International Journal of Fatigue, 26*(9), 941—946.
Mandell, J. F. (1990). Fatigue behavior of short fiber composite materials. In K. L. Reifsnider (Ed.), *Fatigue of composite materials*. Amsterdam, the Netherlands: Elsevier science publishers.
Mandell, J. F., McGarry, F. J., Huang, D. D., & Li, C. G. (1983). Some effects of matrix and interface properties on the fatigue of short fiber—reinforced thermoplastics. *Polymer Composites, 4*(1), 32—39.
Mandell, J. F., & Meier, U. (1982). *Effect of stress ratio, frequency and loading time on the tensile fatigue of glass—reinforced epoxy*. Philadelphia, USA: ASTM STP 813, American Society for Testing and Materials.
Mandell, J. F., & Samborsky, D. D. (March 31, 2010). *DOE/MSU composite material fatigue database* (Vol. 19). Sandia National Laboratories, SAND97-3002. http://windpower.sandia.gov/.
Mandell, J. F., Samborsky, D. D., Wang, L., & Wahl, N. K. (2003). New fatigue data for wind turbine blade materials. *Journal of Solar Energy Engineering Transactions ASME, 125*(4), 506—514.
Miyano, Y., Nakada, M., & Muki, R. (1997). Prediction of fatigue life of a conical shaped joint system for reinforced plastics under arbitrary frequency, load ratio and temperature. *Mechanics of Time-Dependent Materials, 1*(2), 143—159.
Mu, P., Wan, X., & Zhao, M. A. (2011). New S-N curve model of fiber reinforced plastic composite. *Key Engineering Materials, 462—463*, 484—488.

Nijssen, R. P. L. (2006). *OptiDAT—fatigue of wind turbine materials database.* http://www.kc-wmc.nl/optimat_blades/index.htm.
Nijssen, R. P. L., Krause, O., & Philippidis, T. P. (2004). *Benchmark of lifetime prediction methodologies, optimat blades technical report.* OB_TG1_R012 rev.001. http://www.wmc.eu/public_docs/10218_001.pdf.
Passipoularidis, V. A, & Philippidis, T. P. (2009). A study of factors affecting life prediction of composites under spectrum loading. *International Journal of Fatigue, 31*(3), 408—417.
Petermann, J., & Schulte, K. (2002). The effects of creep and fatigue stress ratio on the long-term behavior of angle-ply CFRP. *Composite Structures, 57*(1—4), 205—210.
Philippidis, T. P., & Vassilopoulos, A. P. (2000). Fatigue design allowables for GFRP laminates based on stiffness degradation measurements. *Composites Science and Technology, 60*(15), 2819—2828.
Philippidis, T. P., & Vassilopoulos, A. P. (2001). Stiffness reduction of composite laminates under combined cyclic stresses. *Advanced Composites Letters, 10*(3), 113—124.
Philippidis, T. P., & Vassilopoulos, A. P. (2004). Life prediction methodology for GFRP laminates under spectrum loading. *Composites Part A: Applied Science and Manufacturing, 35,* 657—666.
Post, N. L. (2008). *Reliability based design methodology incorporating residual strength prediction of structural fiber reinforced polymer composites under stochastic variable amplitude fatigue loading* (Ph. D. thesis). Virginia Polytechnic Institute and State University.
Qiao, P., & Yang, M. (2006). Fatigue life prediction of pultruded E-glass/polyurethane composites. *Journal of Composite Materials, 40*(9), 815—837.
Salkind, M. J. (1972). *Fatigue of composites.* Philadelphia, USA: ASTM STP 497, American Society for Testing and Materials.
Sarfaraz, R., Vassilopoulos, A. P., & Keller, T. (2011). Experimental investigation of the fatigue behavior of adhesively-bonded pultruded GFRP joints under different load ratios. *International Journal of Fatigue, 33*(11), 1451—1460.
Sarfaraz, R., Vassilopoulos, A. P., & Keller, T. (2012a). Experimental investigation and modeling of mean load effect on fatigue behavior of adhesively-bonded pultruded GFRP joints. *International Journal of Fatigue, 44,* 245—252.
Sarfaraz, R., Vassilopoulos, A. P., & Keller, T. (2012b). A hybrid S-N formulation for fatigue life modeling of composite materials and structures. *Composites Part A: Applied Science and Manufacturing, 43*(3), 445—453.
Sarfaraz, R., Vassilopoulos, A. P., & Keller, T. (2013). Modelling the constant amplitude fatigue behavior of adhesively bonded pultruded GFRP joints. *Journal of Adhesion Science and Technology, 27*(8), 855—878.
Sarkani, S., Michaelov, G., Kihl, D. P., & Beach, J. E. (1999). Stochastic fatigue damage accumulation of FRP laminates and joints. *Journal of Structural Engineering, 125*(12), 1423—1431.
Sendeckyj, G. P. (1981). *Fitting models to composite materials fatigue data.* Philadelphia, USA: ASTM STP 734, American Society for Testing and Materials.
Sendeckyj, G. P. (1990). Life prediction for resin matrix composite materials. In K. L. Reifsnider (Ed.), *Fatigue of composite materials.* Amsterdam, the Netherlands: Elsevier science publishers.
Silverio Freire, R. C., Dória Neto, A. D., & De Aquino, E. M. F. (2009). Comparative study between ANN models and conventional equation in the analysis of fatigue failure of GFRP. *International Journal of Fatigue, 31*(5), 831—839.

Sims, D. F., & Brogdon, V. H. (1977). *Fatigue behavior of composites under different loading modes*. Philadelphia: ASTM STP 636, American society for testing and materials.
Sutherland, H. J., & Mandell, J. F. (2005). Optimized constant life diagram for the analysis of fiberglass composites used in wind turbine blades. *Journal of Solar Energy Engineering Transactions ASME, 127*(4), 563–569.
Vassilopoulos, A. P., Georgopoulos, E. F., & Dionysopoulos, V. (2007). Artificial neural networks in spectrum fatigue life prediction of composite materials. *International Journal of Fatigue, 29*(1), 20–29.
Vassilopoulos, A. P., Georgopoulos, E. F., & Keller, T. (2008). Comparison of genetic programming with conventional methods for fatigue life modeling of FRP composite materials. *International Journal of Fatigue, 30*(9), 1634–1645.
Vassilopoulos, A. P., & Keller, T. (2011). Fatigue of fiber reinforced composites. In *Engineering materials and processes series*. London: Springer.
Vassilopoulos, A. P., Manshadi, B. D., & Keller, T. (2010a). Influence of the constant life diagram formulation on the fatigue life prediction of composite materials. *International Journal of Fatigue, 32*(4), 659–669.
Vassilopoulos, A. P., Manshadi, B. D., & Keller, T. (2010b). Piecewise non-linear constant life diagram formulation for FRP composite materials. *International Journal of Fatigue, 32*(10), 1731–1738.
Vassilopoulos, A. P., & Nijssen, R. P. L. (2010). Fatigue life prediction of composite materials under realistic loading conditions (variable amplitude loading). In *Fatigue life prediction of composites and composite structures*. Oxford, UK: Woodhead Publishing Limited.
Vassilopoulos, A. P., Sarfaraz, R., Manshadi, B. D., & Keller, T. (2010). A computational tool for the life prediction of GFRP laminates under irregular complex stress states: influence of the fatigue failure criterion. *Computational Materials Science, 49*(3), 483–491.
Whitney, J. M. (1981). Fatigue characterization of composite materials. In K. N. Lauraitis (Ed.), *Fatigue of fibrous composite materials* (pp. 133–151). Philadelphia, USA: ASTM STP 723, American Society for Testing and Materials.
Xiong, J. J., & Shenoi, R. A. (2004). Two new practical models for estimating reliability-based fatigue strength of composites. *Journal of Composite Materials, 38*(14), 1187–1209.
Yang, J. N., Yang, S. H., & Jones, D. L. (1989). A stiffness-based statistical model for predicting the fatigue life of graphite/epoxy laminates. *Journal of Composites Technology & Research, 11*(4), 129–134.
Zhang, Y., Vassilopoulos, A. P., & Keller, T. (2008). Stiffness degradation and fatigue life prediction of adhesively-bonded joints for fiber–reinforced polymer composites. *International Journal of Fatigue, 30*(10–11), 1813–1820.
Zhang, Y., Vassilopoulos, A. P., & Keller, T. (2010). Fracture of adhesively-bonded pultruded GFRP joints under constant amplitude fatigue loading. *International Journal of Fatigue, 32*(7), 979–987.

Developing an integrated structural health monitoring and damage prognosis (SHM-DP) framework for predicting the fatigue life of adhesively-bonded composite joints

17

M. Gobbato[1], J.B. Kosmatka[2], J.P. Conte[2]
[1] Risk Management Solutions Inc., Newark, CA, USA; [2] University of California, La Jolla, CA, USA

17.1 Introduction

Integrated structural health monitoring and damage prognosis (SHM-DP) methodologies (Ling & Mahadevan, 2012), coupled with local sensor-based nondestructive evaluation (NDE) techniques (Farrar, Worden, Lieven, & Park, 2010; Giurgiutiu, 2008; Shull, 2002), are becoming fundamental engineering tools for assessing the current structural integrity and predicting the remaining service life of mechanical and structural systems fabricated with lightweight composite materials (Farrar & Worden, 2010; Inman, Farrar, Lopez, & Steffen, 2005). Most importantly, calibrated and validated SHM-DP methodologies can be used to provide cost-efficient reliability-based (or risk-based) inspection and maintenance (RBIM) plans as described in Deodatis, Fujimoto, Ito, Spencer, and Itagaki (1992), Ito, Deodatis, Fujimoto, Asada, and Shinozuka (1992), Deodatis, Asada, and Ito (1996), and Kim, Frangopol, and Soliman (2013). As discussed in recent research (Gobbato, 2011; Gobbato, Conte, Kosmatka, & Farrar, 2012; Rabiei & Modarres, 2013), NDE-based SHM-DP methodologies attempt to achieve the objectives outlined earlier through (1) periodic NDE inspections, (2) a rigorous probabilistic treatment of the NDE inspection results and all pertinent sources of uncertainty, (3) calibrated and validated mechanics-based models of stochastic damage growth, and (4) well-established component and system reliability analysis methods (Ditlevsen & Madsen, 1996). To obtain accurate and meaningful damage prognosis results, the NDE-based SHM-DP system must be capable of detecting all damage locations and correctly identifying all damage mechanisms progressing in time. In this respect, fatigue-driven damage propagation is one of the most unpredictable failure mechanisms for a large variety of lightweight composite systems which are subjected to cyclic and/or random operational loads during their service life. Furthermore, the adhesive joints, bonding together the various composite subassemblies of such systems, are widely recognized as one of the most fatigue-sensitive components (Zhang, 2010). Therefore, fundamental tasks are: (1) monitoring these critical components, (2) periodically assessing their structural integrity, (3) recursively predicting their remaining fatigue life (RFL) as well as the RFL of the overall system, and (4) determining a cost-efficient RBIM program. To this end, the current chapter presents in detail a comprehensive NDE-based SHM-DP methodology capable of recursively predicting the time-varying reliability and RFL of adhesively-bonded composite joints and composite structural systems. According to this methodology, data collected during sensor-based NDE inspections are processed and then used to assess probabilistically the current state of damage of the monitored structural system or sub-assembly (i.e., damage locations, damage mechanisms, and characteristic/equivalent debonding sizes/extents), see Vanniamparambil et al. (2012). Given the processed NDE measurement data, a Bayesian inference scheme is used to recursively update the joint conditional probability distribution function (PDF) of the debonding extents at the inspected damage locations as well as the joint conditional PDF of the fatigue-driven debonding evolution model parameters. Hazard models for future operational loads, together with calibrated and validated mechanics-based debonding evolution models, are then used to stochastically simulate the multisite fatigue-driven debonding propagation. Finally, both local and global performance limit states are considered to compute the time-varying system reliability and the corresponding RFL for the monitored system. These RFL estimates provide the rational basis for the decision-making process to either justify an extended service life of the monitored structural system/component (Gobbato, Conte, & Kosmatka, 2013a) or deploy an intelligent repair or retirement plan by detecting a fault earlier than anticipated and therefore avoiding a costly catastrophic failure.

The material presented in this chapter is organized into 10 sections. After a brief introduction to the topic of NDE-based SHM-DP for adhesively-bonded composite structural systems, a general overview of the proposed reliability-based SHM-DP framework and its three main analysis modules is presented. Afterward, a more in-depth description of each of these modules and their corresponding analysis steps is given in Sections 17.3–17.7 with the aim of guiding the reader through the sequential operations that must be performed each time a new set of NDE inspection results becomes available. These sections start by discussing the first module of the methodology, namely *Bayesian inference*, with special emphasis on the probabilistic treatment of the NDE inspection results and the proposed recursive Bayesian updating scheme. Subsequently, the second module of the proposed framework and its three predictive modeling steps is unveiled and analyzed. These three steps, namely, *load hazard analysis*, *debonding evolution analysis*, and *system performance analysis*, form the core of the proposed methodology and involve multiple applications of the total probability theorem (TPT) in a nested fashion (Gobbato, 2011). Finally, the theoretical formulation of the proposed damage prognosis methodology is completed by presenting in detail its third module, *damage prognosis*. The remaining parts of the chapter illustrate the effectiveness of the proposed SHM-DP methodology in predicting failure and updating the RBIM program, discuss future research directions, summarize the important contributions and findings of this chapter, and provide additional sources of information.

17.2 Proposed reliability-based structural health monitoring and damage prognosis (SHM-DP) framework for fatigue damage prognosis

The reliability-based SHM-DP methodology presented in this chapter aims to recursively predict the time-varying reliability and the RFL of adhesively-bonded composite joints and/or composite structural systems periodically monitored through sensor-based NDE inspections. As illustrated in Figure 17.1, the proposed methodology is composed of three main analysis modules: *Bayesian inference*, *predictive modeling*, and *damage prognosis*. The first module uses processed NDE inspection results and Bayesian updating techniques to update prior knowledge on the current state of damage of the monitored system. This updated information is referred to here as posterior knowledge on the current state of damage; for the specific case considered in this study, it consists of the joint conditional PDF of the actual debonding lengths at the inspected damage locations and the joint conditional PDF of the modeling parameters used to model and simulate the fatigue-driven debonding propagation process along the adhesive joints. The results obtained through the recursive

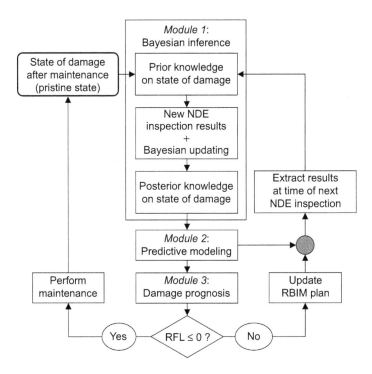

Figure 17.1 Overview of proposed NDE-based SHM-DP framework for recursive RFL predictions and risk-based inspection and maintenance plan updating. NDE, nondestructive evaluation; RFL, remaining fatigue life; RBIM, reliability-based inspection and maintenance.

application of the proposed Bayesian inference analysis are then used in the second module of the framework (predictive modeling) to predict the evolution in time of the multisite debonding propagation process and the overall performance level of the system as damage progresses. Finally, the last analysis module (damage prognosis) uses well-established component and system reliability analysis methods to compute the time-varying system reliability and the corresponding RFL. If a predefined minimum level for the system reliability is down-crossed (i.e., the system's RFL is completely exhausted) either some maintenance must be performed or an intelligent retirement of the whole system must be undertaken. On the other hand, if the predefined minimum level of system reliability is not yet reached or down-crossed, the RBIM plan can be updated accordingly. In other words, based on the predicted time-varying performance of the system, the next maintenance activities can be moved at an earlier time, maintained, or postponed. Similarly, the time of the next NDE inspection can also be revised (Deodatis et al., 1992) and at that point (i.e., when a new set of NDE results becomes available), a portion of the predictive modeling and analysis results is extracted and used as new prior information on the state of damage of the system (Figure 17.1). Therefore, the repetition of this process provides a systematic way to generate recursive predictions for the time-varying system reliability and the corresponding RFL each time a new NDE inspection is performed and new NDE inspection results become available. The inspection times are hereafter denoted as $t_0, t_1, \ldots, t_p, t_{p+1}, \ldots$ with t_0 and t_p being the first and current inspection times, respectively. In more detail, the flowchart shown in Figure 17.2 focuses on a single swipe, at current time t_p, across the three main analysis modules outlined earlier and it illustrates conceptually the process of uncertainty quantification and propagation necessary to predict the time-varying system reliability (for $t \geq t_p$) and RFL of the

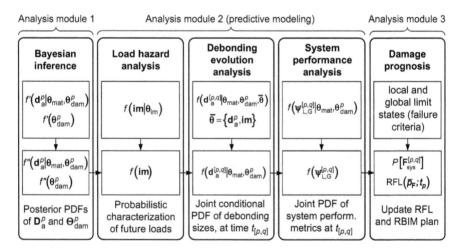

Figure 17.2 Flowchart emphasizing the five analysis steps of the proposed NDE-based SHM-DP framework to be performed at time t_p. All PDFs are conditional on the event $\mathbf{D}_m^{[0,p]} = \mathbf{d}_m^{[0,p]}$. PDF, probability distribution function.

monitored structural system, once the set of NDE inspection results at current time t_p becomes available. In addition, as illustrated in Figure 17.2, the central analysis module is conveniently subdivided into three analysis steps: *load hazard analysis, debonding evolution analysis*, and *system performance analysis*. Consequently, the proposed SHM-DP framework involves a total of five analysis steps.

The set of NDE inspection results at time t_p is represented by the measured (through NDE sensor data processing) debonding length/extent vector, $\mathbf{d}_m^p$, at the inspected locations at time t_p—i.e., a particular realization of the random measured debonding length vector $\mathbf{D}_m^p$. This new information is used in the first step of the methodology (*Bayesian inference*) to compute the posterior joint PDF of the actual debonding length vector at time t_p, $\mathbf{D}_a^p$, conditional on the material and damage model parameter vectors ($\mathbf{\Theta}_{mat}$ and $\mathbf{\Theta}_{dam}^p$), as well as on all the previous $p+1$ NDE inspection results obtained up to time t_p and denoted as $\mathbf{d}_m^{[0,p]} = \{\mathbf{d}_m^0, \mathbf{d}_m^1, \ldots, \mathbf{d}_m^p\}$ — i.e., a particular realization of the vector $\mathbf{D}_m^{[0,p]} = \{\mathbf{D}_m^0, \mathbf{D}_m^1, \ldots, \mathbf{D}_m^p\}$. This posterior joint conditional PDF of $\mathbf{D}_a^p$ is hereafter denoted as $f''_{\mathbf{D}_a^p|\mathbf{\Theta}_{mat},\mathbf{\Theta}_{dam}^p,\mathbf{D}_m^{[0,p]}}\left(\mathbf{d}_a^p\big|\mathbf{\theta}_{mat},\mathbf{\theta}_{dam}^p,\mathbf{d}_m^{[0,p]}\right)$, whereas its prior counterpart (i.e., the joint conditional PDF of $\mathbf{D}_a^p$ before Bayesian updating at time t_p) is expressed as $f'_{\mathbf{D}_a^p|\mathbf{\Theta}_{mat},\mathbf{\Theta}_{dam}^p,\mathbf{D}_m^{[0,p-1]}}\left(\mathbf{d}_a^p\big|\mathbf{\theta}_{mat},\mathbf{\theta}_{dam}^p,\mathbf{d}_m^{[0,p-1]}\right)$. Concurrently, the measured debonding lengths at time t_p, are also used to update the prior joint conditional PDF of the damage evolution model parameter vector, $f'_{\mathbf{\Theta}_{dam}^p|\mathbf{D}_m^{[0,p-1]}}\left(\mathbf{\theta}_{dam}^p\big|\mathbf{d}_m^{[0,p-1]}\right)$, into the corresponding posterior joint conditional PDF defined as $f''_{\mathbf{\Theta}_{dam}^p|\mathbf{D}_m^{[0,p]}}\left(\mathbf{\theta}_{dam}^p\big|\mathbf{d}_m^{[0,p]}\right)$. The random vector $\mathbf{\Theta}_{dam}^p$ (of length n_{dam}^p) quantifies the uncertainty of the parameters of the model used to simulate the fatigue-driven debonding propagation in the pre-identified damageable sub-components—i.e., the adhesively-bonded composite joints in this particular study. On the other hand, the material model parameter vector $\mathbf{\Theta}_{mat}$ (of length n_{mat}) is characterized by its joint PDF $f_{\mathbf{\Theta}_{mat}}(\mathbf{\theta}_{mat})$, which quantifies the uncertainty in the material properties used to model the parts of the structure that are assumed to be non-damageable (Gobbato, 2011). In the proposed NDE-based SHM-DP framework, the two random vectors $\mathbf{\Theta}_{mat}$ and $\mathbf{\Theta}_{dam}^p$ are reasonably assumed to be statistically independent (s.i.) and time-invariant (Gobbato, Conte et al., 2012). In addition, as can be inferred from the notation introduced thus far, the joint PDF of the material model parameter vector, $f_{\mathbf{\Theta}_{mat}}(\mathbf{\theta}_{mat})$, is not recursively updated by the proposed Bayesian inference scheme (i.e., $f_{\mathbf{\Theta}_{mat}}(\mathbf{\theta}_{mat})$ is assumed time-invariant). Multiple debonding locations can be handled by the recursive Bayesian updating procedure presented here, with the actual debonding length/extent at the i-th monitored damage location at time t_p denoted as $D_a^{(i,p)}$. Consequently, the actual debonding length vector can be written as $\mathbf{D}_a^p = \left\{D_a^{(i,p)}, i = 1,\ldots,n_{loc}\right\}$, where n_{loc} denotes the number of inspected debonding locations. In a similar fashion, the vector $\mathbf{D}_m^p$ can be written as

$\mathbf{D}_m^p = \left\{ \mathbf{D}_m^{(i,p)}, i = 1, \ldots, n_{\text{loc}} \right\}$, where the subvector $\mathbf{D}_m^{(i,p)} = \left\{ D_{m,k}^{(i,p)}, k = 1, \ldots, n_m^{(i,p)} \right\}$ collects all the $n_m^{(i,p)}$ NDE measurement results obtained at the i-th inspected damage location at time t_p; if a particular debonding location is not inspected at time t_p, then $n_m^{(i,p)} = 0$. It is also worth mentioning that multiple damage mechanisms (i.e., not only debonding) can potentially be considered and integrated within the proposed framework. This modeling extension is beyond the scope of this study but an exhaustive and detailed treatment of this more general case can be found in (Gobbato, 2011; Gobbato, Conte et al., 2012).

The second analysis step of the proposed SHM-DP framework, *load hazard analysis*, defines the joint PDF of a vector of load intensity measures, **IM**, characterizing in probabilistic terms future service loads and extreme load events imposed on the monitored structural system. As an example, Gobbato, Conte et al. (2012) showed how the vector **IM** can be used to describe the intensity of steady-level flight loads, maneuver-induced loads (e.g., changes in velocity, direction, and altitude), and turbulence-induced loads encountered by an aircraft during flight. Similarly, Gobbato et al. (2013a) demonstrated that the proposed load hazard analysis step can systematically and efficiently account for the uncertainty in the amplitude of the sinusoidal load applied during fatigue tests, propagate this source of uncertainty throughout all subsequent steps, all the way to the RFL predictions. The joint PDF of **IM** is denoted here as $f_{\mathbf{IM}}(\mathbf{im})$ and, in the most general case, is computed by unconditioning the joint conditional PDF $f_{\mathbf{IM}|\boldsymbol{\Theta}_{\text{im}}}(\mathbf{im}|\boldsymbol{\theta}_{\text{im}})$ with respect to (w.r.t.) the distribution parameter vector $\boldsymbol{\Theta}_{\text{im}}$—i.e., the random vector collecting the uncertain distribution parameters used to define the joint PDF of **IM** (see Section 17.4 and Gobbato, Conte et al., 2012).

The third step of the proposed methodology, *debonding evolution analysis*, estimates the joint conditional PDF of the actual debonding size vector $\mathbf{D}_a^{[p,q]}$ at future time $t_{p,q} = t_p + q \cdot \Delta\tau$ with $q \in \{0, 1, 2, \ldots, \bar{q}\}$ (Gobbato, Conte et al., 2012). This joint PDF, conditional on $\boldsymbol{\Theta}_{\text{mat}}$, $\boldsymbol{\Theta}_{\text{dam}}^p$, and all previous NDE inspection results up to current time t_p (i.e., $\mathbf{d}_m^{[0,q]}$), is denoted as $f_{\mathbf{D}_a^{[p,q]} | \boldsymbol{\Theta}_{\text{mat}}, \boldsymbol{\Theta}_{\text{dam}}^p, \mathbf{D}_m^{[0,p]}} \left(\mathbf{d}_a^{[p,q]} \middle| \boldsymbol{\theta}_{\text{mat}}, \boldsymbol{\theta}_{\text{dam}}^p, \mathbf{d}_m^{[0,p]} \right)$ and is computed by unconditioning the joint conditional PDF $f_{\mathbf{D}_a^{[p,q]} | \boldsymbol{\Theta}_{\text{mat}}, \boldsymbol{\Theta}_{\text{dam}}^p, \mathbf{D}_a^p, \mathbf{IM}, \mathbf{D}_m^{[0,p]}} \left(\mathbf{d}_a^{[p,q]} \middle| \boldsymbol{\theta}_{\text{mat}}, \boldsymbol{\theta}_{\text{dam}}^p, \mathbf{d}_a^p, \mathbf{im}, \mathbf{d}_m^{[0,p]} \right)$ w.r.t. $\mathbf{D}_a^p$ and **IM**. This process is carried out through either extensive Monte Carlo (MC) simulations or semi-analytical methods (Gobbato et al., 2013a; Gobbato, Conte, & Kosmatka, 2013b) and it yields a probabilistic characterization of the future state of damage (or performance level) of the structure at the local reliability component level. The quantity $\Delta\tau$, introduced earlier, is a suitable fixed time interval (smaller than the inter-inspection time) related to the time scale of the debonding propagation process of interest, whereas $\bar{q}$ represents the number of damage prognosis evaluations within the prognosis window $[t_p, t_{p,\bar{q}}]$.

The fourth step of the proposed framework, *system performance analysis*, estimates the joint PDF of $\mathbf{D}_a^{[p,q]}$ and a vector of n_G global performance metrics—collected in

$\mathbf{D}_{\mathrm{g}}^{[p,q]} = \left\{ \mathbf{D}_{\mathrm{g},r}^{[p,q]}, \ r = 1, \ldots, n_{\mathrm{g}} \right\}$—at future time $t_{p,q} = t_p + q \cdot \Delta \tau$ with $q \in \{0, 1, 2, \ldots, \overline{q}\}$. This joint conditional PDF is denoted as $f_{\mathbf{\Psi}_{\mathrm{L,G}}^{[p,q]} \mid \mathbf{D}_{\mathrm{m}}^{[0,p]}} \left(\mathbf{\psi}_{\mathrm{L,G}}^{[p,q]} \mid \mathbf{d}_{\mathrm{m}}^{[0,p]} \right)$, with $\mathbf{\Psi}_{\mathrm{L,G}}^{[p,q]} = \left\{ \mathbf{D}_{\mathrm{a}}^{[p,q]}, \mathbf{D}_{\mathrm{g}}^{[p,q]} \right\}$. As an example, $\mathbf{D}_{\mathrm{g}}^{[p,q]}$ can collect a given set of natural frequencies, mode shapes, and modal strains of the monitored structure (Doebling, Farrar, & Prime, 1998; Zou, Tong, & Steven, 2000), the damage-dependent flutter velocity of an aircraft wing (Styuart, Mor, Livne, & Lin, 2007) or wind turbine blade, and so forth. This analysis step thus provides a joint probabilistic characterization of the effects of damage at the local level (through $\mathbf{D}_{\mathrm{a}}^{[p,q]}$) and at the global level (through $\mathbf{D}_{\mathrm{g}}^{[p,q]}$) for the monitored structural system at time $t_{p,q}$.

Once the joint conditional PDF $f_{\mathbf{\Psi}_{\mathrm{L,G}}^{[p,q]} \mid \mathbf{D}_{\mathrm{m}}^{[0,p]}} \left(\mathbf{\psi}_{\mathrm{L,G}}^{[p,q]} \mid \mathbf{d}_{\mathrm{m}}^{[0,p]} \right)$ is determined, local and Global limit states (GLSs) (or failure criteria) are considered and time-varying probabilities of failure at both local and global reliability component levels are computed. A local reliability component is defined by a unique local limit-state function involving one or more random variables collected in $\mathbf{D}_{\mathrm{a}}^{[p,q]}$. On the other hand, a global reliability component is characterized by a unique global limit-state function involving one or more random variables contained in $\mathbf{D}_{\mathrm{g}}^{[p,q]}$. The results obtained at the reliability component levels are then used as basis to assess the time-varying probability of failure and reliability index at the overall system level at time $t_{p,q}$. These two quantities are conditional on all the NDE inspection results up to current time t_p—i.e., $\mathbf{D}_{\mathrm{m}}^{[0,p]} = \mathbf{d}_{\mathrm{m}}^{[0,p]}$—and are denoted as $P\left[\mathrm{F}_{\mathrm{sys}}^{[p,q]} \mid \mathbf{D}_{\mathrm{m}}^{[0,p]} = \mathbf{d}_{\mathrm{m}}^{[0,p]} \right]$ and $\beta_{\mathrm{sys}}^{[p,q]} \left(\mathbf{D}_{\mathrm{m}}^{[0,p]} = \mathbf{d}_{\mathrm{m}}^{[0,p]} \right) = -\Phi^{-1} \left(P\left[\mathrm{F}_{\mathrm{sys}}^{[p,q]} \mid \mathbf{D}_{\mathrm{m}}^{[0,p]} = \mathbf{d}_{\mathrm{m}}^{[0,p]} \right] \right)$, respectively, where $\Phi^{-1}(\cdot)$ denotes the inverse Standard Normal cumulative distribution function (CDF). In addition, the proposed NDE-based SHM-DP methodology can also recursively predict, after each NDE inspection, the conditional RFL of the monitored structural system. This quantity, estimated at current time t_p, is defined as $\mathrm{RFL}\left(\overline{p}_{\mathrm{F}}; t_p, \mathbf{D}_{\mathrm{m}}^{[0,p]} = \mathbf{d}_{\mathrm{m}}^{[0,p]} \right) = t_{\mathrm{F}}\left(\overline{p}_{\mathrm{F}}; t_p, \mathbf{d}_{\mathrm{m}}^{[0,p]} \right) - t_p$, where $t_{\mathrm{F}}\left(\overline{p}_{\mathrm{F}}; t_p, \mathbf{d}_{\mathrm{m}}^{[0,p]} \right)$ represents the predicted time at failure, i.e., the predicted time at which the conditional *probability of system failure*, $P\left[\mathrm{F}_{\mathrm{sys}}^{[p,q]} \mid \mathbf{D}_{\mathrm{m}}^{[0,p]} = \mathbf{d}_{\mathrm{m}}^{[0,p]} \right]$, reaches the critical predefined threshold $\overline{p}_{\mathrm{F}}$. The estimates of $P\left[\mathrm{F}_{\mathrm{sys}}^{[p,q]} \mid \mathbf{D}_{\mathrm{m}}^{[0,p]} = \mathbf{d}_{\mathrm{m}}^{[0,p]} \right]$ and $\beta_{\mathrm{sys}}^{[p,q]} \left(\mathbf{D}_{\mathrm{m}}^{[0,p]} = \mathbf{d}_{\mathrm{m}}^{[0,p]} \right)$ at future times $t_{p,q}$ (with $q = 0, 1, 2, \ldots, \overline{q}$) and of $\mathrm{RFL}\left(\overline{p}_{\mathrm{F}}; t_p, \mathbf{D}_{\mathrm{m}}^{[0,p]} = \mathbf{d}_{\mathrm{m}}^{[0,p]} \right)$ at current time t_p can then be used as rational decision making tools to schedule and/or update the maintenance/repair plan for the monitored structure on the basis of a predefined maximum acceptable threshold (i.e., $\overline{p}_{\mathrm{F}}$) for $P\left[\mathrm{F}_{\mathrm{sys}}^{[p,q]} \mid \mathbf{D}_{\mathrm{m}}^{[0,p]} = \mathbf{d}_{\mathrm{m}}^{[0,p]} \right]$.

It is worth mentioning that if advanced sampling algorithms such as Markov Chain Monte Carlo (MCMC) methods (Gamerman, 1997) and transitional MCMC methods (Ching & Chen, 2007) are used within the proposed framework, the conditional realizations of the random vector $\mathbf{d}_{\mathrm{a}}^p$ are renewed at each NDE inspection according to its updated probability distribution $f''_{\mathbf{D}^p_a|\Theta_{mat},\Theta^p_{dam},\mathbf{D}^{[0,p]}_m}\left(\mathbf{d}^p_a|\theta_{mat},\theta^p_{dam},\mathbf{d}^{[0,p]}_m\right)$. However, the realizations of the random parameter vectors Θ_{mat} and Θ^p_{dam} must be kept constant from the first damage inspection until the time at which the failure probability of the monitored system exceeds an acceptable threshold (Gobbato, 2011; Gobbato, Conte et al., 2012). The random parameter vectors Θ_{mat} and Θ^p_{dam} are said to be non-ergodic (Der Kiureghian, 2005) and their intrinsic nature adds an additional level of complexity when using MCMC or transitional MCMC sampling algorithms.

As a final remark in this overview section, it is worth emphasizing that even though NDE detection and measurement uncertainties are taken into account in the proposed methodology, all probabilistic characterizations (at time t_p) of the quantities mentioned earlier are conditional on the particular realization/scenario of NDE inspection results (i.e., $\mathbf{D}^{[0,p]}_m = \mathbf{d}^{[0,p]}_m$). The application of the proposed framework with a different realization (i.e., a different set) of NDE inspection results would lead to different Bayesian updating and damage prognosis results. Therefore, to provide an unconditional estimate of all probabilistic quantities of interest (e.g., Bayesian updating results, RFL, time-varying probability of failure), it is necessary to perform multiple analyses across an ensemble of NDE inspection results along the life of the monitored structure. These multiple realizations of the NDE inspection results are denoted as $\left\{\mathbf{D}^{[0,p]}_m = \mathbf{d}^{[0,p]}_m\right\}_e$, with $e = 1,\ldots,n_{ens}$ and n_{ens} representing the number of scenarios considered (i.e., the ensemble size). This approach is referred to as "ensemble scenario analysis" and it provides a robust statistical approach for assessing and evaluating the overall performance of the proposed NDE-based SHM-DP methodology (Gobbato et al. (2013b)).

17.3 Recursive Bayesian characterization of the current state of damage

In the NDE-based SHM-DP framework presented in this chapter, it is assumed that a given structural system, or its most critical subassemblies and/or sub-components—such as the adhesively-bonded joints of a composite structure—are monitored through periodic or nearly continuous NDE inspections. A variety of NDE techniques can be employed to achieve this task, such as acoustic emission (Kosmatka & Velazquez, 2008; Velazquez & Kosmatka, 2010a; Velazquez & Kosmatka, 2010b; Velazquez & Kosmatka, 2011; Velazquez & Kosmatka, 2012) and ultrasonic guided waves (Lanza di Scalea et al., 2007). The only requirement to fully embed the collected NDE results in the proposed framework consists of being able to establish a functional relationship, in probabilistic terms, between the actual debonding length vector ($\mathbf{D}_a$) and its measured counterpart ($\mathbf{D}_m$). In this regard, three assumptions are made: (1) an NDE inspection can detect and locate an ongoing debonding propagation process and also quantify the extent of debonding at the time of inspection; (2) systematic and random measurement errors depend on both location and extent of debonding; and (3) detection and measurement of the debonding extent at an inspected damage location depend solely on the actual (unknown) debonding length at that particular location at the time of inspection. It is also worth mentioning that in the case of sensor-based NDE monitoring (Lanza di Scalea et al., 2007) the location-dependent measurement accuracy may arise from the nonuniform sensor distribution and/or the dysfunctional behavior of some network nodes.

Here, the NDE detection capability is quantified by the so-called probability of detection (POD), which, for a given (i, p) combination (i.e., a given debonding location and inspection time), is defined as the probability of detecting a debonding of any length/extent (i.e., $D_m^{(i,p)} > 0$) given that the actual debonding length is equal to $D_a^{(i,p)} = d_a^{(i,p)}$. This definition is expressed in mathematical terms in Eqn (17.1):

$$\text{POD}_{(i,p)}\left(d_a^{(i,p)}\right) = P\left[D_m^{(i,p)} > 0 \big| D_a^{(i,p)} = d_a^{(i,p)}\right], \quad \left(d_a^{(i,p)} > 0\right) \tag{17.1}$$

In addition, the complement of the conditional event $\left\{D_m^{(i,p)} > 0 \big| D_a^{(i,p)} = d_a^{(i,p)}\right\}$ is represented by the event $\{D_m^{(i,p)} = 0 \big| D_a^{(i,p)} = d_a^{(i,p)}\}$ (i.e., debonding not detected given that the actual debonding size is equal to $d_a^{(i,p)}$). The probability of this latter conditional event is referred to as probability of nondetection (PND) and is defined as

$$\begin{aligned}\text{PND}_{(i,p)}\left(d_a^{(i,p)}\right) &= P\left[D_m^{(i,p)} = 0 \big| D_a^{(i,p)} = d_a^{(i,p)}\right]\\ &= 1 - \text{POD}_{(i,p)}\left(d_a^{(i,p)}\right), \quad \left(d_a^{(i,p)} > 0\right)\end{aligned} \tag{17.2}$$

Finally, the probability that a particular NDE result represents a false alarm—i.e., $D_m^{(i,p)} > 0$ when in reality $d_a^{(i,p)} = 0$—is commonly referred to as false-call probability (FCP) in the literature, and is defined as

$$\text{FCP}_{(i,p)} = P\left[D_m^{(i,p)} > 0 \big| D_a^{(i,p)} = 0\right] = \text{POD}_{(i,p)}\left(d_a^{(i,p)} = 0\right) \tag{17.3}$$

These two pieces of information provided in Eqns (17.1) and (17.3) can be viewed as a function of $d_a^{(i,p)}$ and combined together in the POD curve. Several parametric models for defining a POD curve can be found in the literature (Berens, 1989; Heasler, Taylor, & Doctor, 1990; Staat, 1993). Some illustrative examples of possible POD curves are shown in Figure 17.3(a): *POD curve* 1 represents an NDE technique incapable of detecting very small debonding lengths, *POD curve* 2 characterizes an NDE technique that might misdetect even very large debonding sizes, and *POD curve* 3 describes an NDE technique with a nonzero FCP.

Once debonding is detected and its extent measured, for a given (i, p) combination, it is important to question the accuracy of that particular NDE measurement $(D_m^{(i,p)} = d_m^{(i,p)})$ conditional on the actual debonding length $(D_a^{(i,p)} = d_a^{(i,p)})$. In this perspective, to account explicitly for the NDE measurement accuracy, a calibrated sizing model must be used. Several sizing models, each with its intrinsic strengths and weaknesses, have been used in the literature (e.g., Gobbato, Kosmatka,

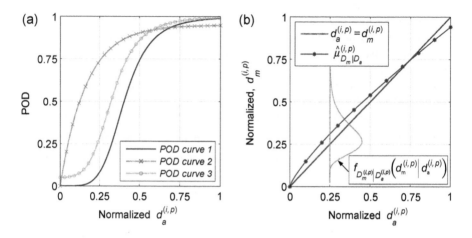

Figure 17.3 (a) Examples of POD curves; (b) Lognormal debonding size measurement model. POD, probability of detection.

& Conte, 2012; Zhang & Mahadevan, 2001; Zheng & Ellingwood, 1998). Here, for the sake of simplicity and for illustration purposes, the following lognormal model is considered (Simola & Pulkkinen, 1998):

$$\ln\left[D_m^{(i,p)}\left(D_a^{(i,p)} = d_a^{(i,p)}\right)\right] = \beta_0^{(i)} + \beta_1^{(i)} \cdot \ln\left(d_a^{(i,p)}\right) + \varepsilon_i \quad (17.4)$$

The terms $\beta_0^{(i)}$ and $\beta_1^{(i)}$ are the location-dependent model coefficients accounting for the systematic measurement errors intrinsic to the NDE technique considered, whereas $\varepsilon_i \sim N(0, \sigma_{\varepsilon_i})$ represents the random measurement error modeled as a Gaussian random variable with zero-mean and standard deviation $\sigma_{\varepsilon i}$ (independent of $D_a^{(i,p)} = d_a^{(i,p)}$). The quantities $\beta_0^{(i)}$, $\beta_1^{(i)}$, and $\sigma_{\varepsilon i}$ are unknown and have to be estimated through linear regression analysis as described in Zhang and Mahadevan (2001) and Gobbato (2011). These estimates of the model coefficients and standard deviation of the random measurement error are hereafter denoted as $\hat{\beta}_0^{(i)}$, $\hat{\beta}_1^{(i)}$, and $\hat{\sigma}_{\varepsilon_i}$, respectively. With this type of information, it is now possible to provide the best estimate for the conditional lognormal PDF of the measured debonding length as

$$f_{D_m|D_a}^{(i,p)}(d_m|d_a) = f_{D_m^{(i,p)}|D_a^{(i,p)}}\left(d_m^{(i,p)}\middle|d_a^{(i,p)}\right)$$

$$= \frac{1}{\sqrt{2\pi}\hat{\sigma}_{\varepsilon_i} d_m^{(i,p)}} \exp\left\{-\frac{1}{2}\left[\frac{\ln\left(d_m^{(i,p)}\right) - \hat{\lambda}_{D_m|D_a}^{(i,p)}}{\hat{\sigma}_{\varepsilon_i}}\right]^2\right\} \quad (17.5)$$

where the term $\hat{\lambda}_{D_m|D_a}^{(i,p)} = \hat{\beta}_0^{(i)} + \hat{\beta}_1^{(i)} \ln(d_a^{(i,p)})$ represents the best estimate of the conditional mean of the natural logarithm of the measured debonding length. In addition, the best estimates of the conditional expected value, $\hat{\mu}_{D_m|D_a}^{(i,p)}$, and standard deviation, $\hat{\sigma}_{D_m|D_a}^{(i,p)}$, of $D_m^{(i,p)}$ can be expressed as shown in Eqns (17.6) and (17.7), respectively:

$$\hat{\mu}_{D_m|D_a}^{(i,p)} = E\left[D_m^{(i,p)}\middle|D_a^{(i,p)}\right] = \exp\left(\hat{\lambda}_{D_m|D_a}^{(i,p)} + \frac{1}{2}\hat{\sigma}_{\varepsilon_i}^2\right) \tag{17.6}$$

$$\hat{\sigma}_{D_m|D_a}^{(i,p)} = E\left[D_m^{(i,p)}\middle|D_a^{(i,p)}\right] = \hat{\mu}_{D_m|D_a}^{(i,p)} \sqrt{\exp\left(\hat{\sigma}_{\varepsilon_i}^2\right) - 1} \tag{17.7}$$

All of these pieces of information are used to build the likelihood function, $L(\mathbf{d}_a^p|\mathbf{d}_m^p)$, needed to recursively update the prior joint conditional PDF, $f'_{\mathbf{D}_a^p|\Theta_{\text{mat}},\Theta_{\text{dam}}^p,\mathbf{D}_m^{[0,p-1]}}\left(\mathbf{d}_a^p\middle|\Theta_{\text{mat}},\Theta_{\text{dam}}^p,\mathbf{d}_m^{[0,p-1]}\right)$, into the posterior joint conditional PDF, $f''_{\mathbf{D}_a^p|\Theta_{\text{mat}},\Theta_{\text{dam}}^p,\mathbf{D}_m^{[0,p]}}\left(\mathbf{d}_a^p\middle|\Theta_{\text{mat}},\Theta_{\text{dam}}^p,\mathbf{d}_m^{[0,p]}\right)$ as new NDE measurement results (i.e., $\mathbf{d}_m^p$) become available at time t_p. This recursive Bayesian updating scheme for $\mathbf{D}_a^p$ is based on previous research—Zheng and Ellingwood (1998), Lin, Du, and Rusk (2000), Zhang and Mahadevan (2001), Kulkarni and Achenbach (2008), and Gobbato (2011)—and can be formally expressed as

$$\begin{aligned}f''_{\mathbf{D}_a^p|\Theta_{\text{mat}},\Theta_{\text{dam}}^p,\mathbf{D}_m^{[0,p]}}\left(\mathbf{d}_a^p\middle|\Theta_{\text{mat}},\Theta_{\text{dam}}^p,\mathbf{d}_m^{[0,p]}\right) \propto \\ L\left(\mathbf{d}_a^p\middle|\mathbf{d}_m^p\right) \times f'_{\mathbf{D}_a^p|\Theta_{\text{mat}},\Theta_{\text{dam}}^p,\mathbf{D}_m^{[0,p-1]}}\left(\mathbf{d}_a^p\middle|\Theta_{\text{mat}},\Theta_{\text{dam}}^p,\mathbf{d}_m^{[0,p-1]}\right)\end{aligned} \tag{17.8}$$

where $\propto$ denotes the proportionality between the right and left hand sides of the equation. Furthermore, by assuming that the conditional NDE measurement results obtained at a given damage location and at a given time are (s.i.) and the conditional NDE measurement results obtained at different damage locations at a given time are (s.i.), the likelihood function in Eqn (17.8) can be rewritten as (Gobbato, Conte, Kosmatka, & Farrar, 2011)

$$L\left(\mathbf{d}_a^p\middle|\mathbf{d}_m^p\right) = \prod_{\substack{i=1 \\ n_m^{(i,p)}>0}}^{n_{\text{loc}}} \prod_{k=1}^{n_m^{(i,p)}} L\left(d_a^{(i,p)}\middle|d_{m,k}^{(i,p)}\right) \tag{17.9}$$

where $L\left(d_a^{(i,p)}\middle|d_{m,k}^{(i,p)}\right)$ represents the likelihood function of $d_a^{(i,p)}$ associated with the k-th NDE measurement result for a given (i, p) combination—i.e., $d_{m,k}^{(i,p)}$. The explicit mathematical form of the likelihood function $L(d_a^{(i,p)}|d_{m,k}^{(i,p)})$ depends on the NDE measurement result, $d_{m,k}^{(i,p)}$, as

$$L\left(d_a^{(i,p)}\middle|d_{m,k}^{(i,p)}\right) = \begin{cases} f_{D_m^{(i,p)}|D_a}(d_{m,k}|d_a) \cdot \text{POD}_{(i,p)}\left(d_a^{(i,p)}\right) & \text{if } d_{m,k}^{(i,p)} > 0 \\ \text{PND}_{(i,p)}\left(d_a^{(i,p)}\right) & \text{if } d_{m,k}^{(i,p)} = 0 \end{cases}$$

(17.10)

It is also worth mentioning that the initial (i.e., before the first NDE inspection at time t_0) PDF model for the debonding length vector, herein denoted as $f'_{\mathbf{D}_a^0}(\mathbf{d}_a^0)$, has to be chosen on the basis of engineering judgment (Gobbato, 2011; Lin et al., 2000). Moreover, the components of the random vector $\mathbf{D}_a^0$, at time t_0, can reasonably be considered mutually s.i. as well as s.i. of Θ_{mat} and Θ_{dam}^0.

As anticipated earlier, the NDE inspection results at time t_p are also used to concurrently provide a recursive estimate of the posterior joint conditional PDF of the damage evolution model parameter vector Θ_{dam}^p, denoted here as $f''_{\Theta_{\text{dam}}^p|\mathbf{D}_m^{[0,p]}}\left(\boldsymbol{\theta}_{\text{dam}}^p\middle|\mathbf{d}_m^{[0,p]}\right)$. The recursive Bayesian updating scheme leading to this result can be formally expressed as

$$f''_{\Theta_{\text{dam}}^p|\mathbf{D}_m^{[0,p]}}\left(\boldsymbol{\theta}_{\text{dam}}^p\middle|\mathbf{d}_m^{[0,p]}\right) \propto f_{\mathbf{D}_m^p|\Theta_{\text{dam}}^p,\mathbf{D}_m^{[0,p-1]}}\left(\mathbf{d}_m^p\middle|\boldsymbol{\theta}_{\text{dam}}^p,\mathbf{d}_m^{[0,p-1]}\right)$$
$$f'_{\Theta_{\text{dam}}^p|\mathbf{D}_m^{[0,p-1]}}\left(\boldsymbol{\theta}_{\text{dam}}^p\middle|\mathbf{d}_m^{[0,p-1]}\right)$$

(17.11)

and, after some mathematical manipulations described in detail in Gobbato (2011), can be conveniently recast as:

$$f''_{\Theta_{\text{dam}}^p|\mathbf{D}_m^{[0,p]}}\left(\boldsymbol{\theta}_{\text{dam}}^p\middle|\mathbf{d}_m^{[0,p]}\right) \propto \left[\int_{\mathbf{D}_a^p} L(\mathbf{d}_a^p|\mathbf{d}_m^p) f'_{\mathbf{D}_a^p|\Theta_{\text{dam}}^p,\mathbf{D}_m^{[0,p-1]}}\left(\mathbf{d}_a^p\middle|\boldsymbol{\theta}_{\text{dam}}^p,\mathbf{d}_m^{[0,p-1]}\right) d(\mathbf{d}_a^p)\right]$$
$$\times f'_{\Theta_{\text{dam}}^p|\mathbf{D}_m^{[0,p-1]}}\left(\boldsymbol{\theta}_{\text{dam}}^p\middle|\mathbf{d}_m^{[0,p-1]}\right)$$

(17.12)

At this stage, by making use of the TPT and by exploiting the results obtained in Eqns (17.8) and (17.12), it is possible to compute the posterior joint conditional PDF $f''_{\mathbf{D}_a^p|\mathbf{D}_m^{[0,p]}}\left(\mathbf{d}_a^p\middle|\mathbf{d}_m^{[0,p]}\right)$ as

$$f''_{\mathbf{D}_a^p | \mathbf{D}_m^{[0,p]}}\left(\mathbf{d}_a^p \Big| \mathbf{d}_m^{[0,p]}\right) = \int_{\Theta_{mat}} \int_{\Theta_{dam}^p} f''_{\mathbf{D}_a^p | \Theta_{mat}, \Theta_{dam}^p, \mathbf{D}_m^{[0,p]}}\left(\mathbf{d}_a^p \Big| \theta_{mat}, \theta_{dam}^p, \mathbf{d}_m^{[0,p]}\right)$$

$$\times f_{\Theta_{mat}}(\theta_{mat}) f''_{\Theta_{dam}^p | \mathbf{D}_m^{[0,p]}}\left(\theta_{dam}^p \Big| \mathbf{d}_m^{[0,p]}\right) d\theta_{mat} d\theta_{dam}^p \tag{17.13}$$

To conclude this section, at time t_0 (i.e., before the first NDE inspection), the prior joint PDF of the damage evolution model parameter vector (Θ_{dam}^0) is denoted as $f'_{\Theta_{dam}^0}(\theta_{dam}^0)$ and, as discussed in more detail in Section 17.5, can be derived from the statistical analysis of experimental fatigue test data (Kotulski, 1998; Ostergaard & Hillberry, 1983; Virkler, Hillberry, & Goel, 1979). Furthermore, when dealing with high-dimensional problems (i.e., a large number of damage/debonding locations and a consequent large number of damage/debonding evolution model parameters), solving analytically Eqns (17.8), (17.12), and (17.13) becomes a challenging task. To overcome this computational barrier, advanced sampling algorithms such as MCMC methods (Gamerman, 1997) and transitional MCMC methods (Ching & Chen, 2007) must be used. Practical applications of these methods for analyzing fatigue damage prognosis problems can be found in Guan, Jha, and Liu (2011, 2012).

17.4 Probabilistic load hazard analysis

Once the posterior conditional PDFs of $\mathbf{D}_a^p$ and Θ_{dam}^p have been computed according to the updating scheme shown in Eqns (17.8) and (17.12), they can be used to analyze probabilistically the debonding propagation processes at the inspected damage locations and assess the overall structural integrity and performance level at future times, i.e., at $t_{p,q} = t_p + q \cdot \Delta\tau$ with $q \in \{0, 1, 2, \ldots, \bar{q}\}$. These tasks are carried out in the predictive modeling module (Figure 17.1) whose first analysis step involves the probabilistic characterization of future operational and/or extreme load events that the monitored structure or structural component will be subjected to during its remaining service life. All probabilistic information necessary to fully embed and execute this analysis step within the proposed SHM-DP framework are contained in the joint PDF of the load intensity measure vector denoted as $f_{\mathbf{IM}}(\mathbf{im})$. This PDF, in the most general case, can be theoretically evaluated by unconditioning the conditional joint PDF $f_{\mathbf{IM}|\Theta_{im}}(\mathbf{im}|\theta_{im})$ w.r.t. the distribution parameter vector Θ_{im}. For example, if a structure is subjected to a sinusoidal load of given frequency, with deterministic mean value and random amplitude (A) assumed to be uniformly distributed between a minimum (A_{min}) and a maximum (A_{max}) amplitude values, the load intensity measure vector would be defined as $\mathbf{IM} = A$ and $f_{\mathbf{IM}|\Theta_{im}}(\mathbf{im}|\theta_{im}) = f_{\mathbf{IM}|A_{min},A_{max}}(\mathbf{im}|a_{min},a_{max}) = (a_{max} - a_{min})^{-1}$. The minimum and maximum amplitude values would be collected in $\Theta_{im} = \{A_{min}, A_{max}\}$ and, in the most general case, would be viewed as random variables (for more details, see Gobbato et al., 2013a).

Gobbato, Conte, et al. (2012), in a study focused on the fatigue damage prognosis of skin-to-spar adhesively-bonded joints of a composite unmanned aerial vehicle (UAV) wing, showed how the vector **IM** can be used to characterize the intensity of steady-level flight loads, maneuver-induced loads (e.g., during takeoff, cruise, landing), and turbulence-induced loads. UAVs as well as other military jet fighters and next-generation commercial aircrafts are examples of how extensively composite materials can be used in aerospace structures. Various damage mechanisms such as fiber breakage, matrix cracking, debonding, and interply delamination can initiate and invisibly propagate up to catastrophic levels in the most damage-sensitive structural components of these vehicles. In particular, the wing skin-to-spar adhesive joints are recognized as the most fatigue-sensitive structural elements of a lightweight composite UAV because the progressive debonding, evolving from the wing-root along these joints, can compromise the global aeroelastic performance of the vehicle (Bauchau & Loewy, 1997; Styuart, Demasi, Livne, & Lin, 2008; Styuart, Demasi, Livne, & Mor, 2011; Styuart et al., 2007; Wang, Inman, & Farrar, 2005). In Gobbato, Conte, et al. (2012), turbulence-induced aerodynamic loads are modeled by directly characterizing the stochasticity of the atmospheric turbulence velocity fields encountered by an aircraft during flight. Atmospheric turbulence is modeled as a zero-mean, isotropic, stationary (in time), and homogeneous (in space) stochastic Gaussian random velocity field (Hoblit, 1988; Van Staveren, 2003) and its intensity is measured by a scalar random variable taken as the root mean square (RMS) value of the wind velocity fluctuations. This RMS value is denoted as Σ_T and its conditional PDF is defined as $f_{\Sigma_T|\Theta_T}(\sigma_T|\theta_T)$, where Θ_T represents the turbulence distribution parameter vector used to characterize the PDF of Σ_T (Gobbato et al., 2011). Furthermore, the random sequence of the turbulence patches encountered by an aircraft during flight is modeled and simulated using homogeneous Poisson rectangular pulse processes (Wen, 1990) with mean rate of occurrence $\lambda_T = 1/\mu_{\Delta S_T}$, where $\mu_{\Delta S_T}$ represents the mean value of the turbulence patch spatial extent (ΔS_T), assumed to be exponentially distributed. Consistent with this particular modeling approach, Gobbato, Conte, et al. (2012) defined the random load intensity measure vector as $\mathbf{IM} = \{\Sigma_T, \Delta S_T\}$ and the load parameter vector as $\Theta_{im} = \{\Theta_T, \mu_{\Delta S_T}\}$.

In another study (Gobbato, 2011), the random dynamic load imposed on a given structural system was conveniently broken down into two components: the mean load intensity, characterized probabilistically by the intensity measure vector $\mathbf{IM}_m$, and the superimposed random stochastic load (or load fluctuations) about the mean-load intensity, probabilistically described by the intensity measure vector $\mathbf{IM}_a$. Therefore, in this particular case study the load intensity measure vector is written as $\mathbf{IM} = \{\mathbf{IM}_m, \mathbf{IM}_a\}$ and the load parameter vector, Θ_{im}, is defined as $\Theta_{im} = \{\Theta_m, \Theta_a\}$. Similar to the case study discussed earlier, the random sequence of the rectangular pulses representing the mean-load intensity is modeled and simulated using a homogeneous Poisson rectangular pulse process with mean rate of occurrence $\lambda_m = 1/\mu_{\Delta T_m}$, with $\mu_{\Delta T_m}$ denoting the average duration of the mean-load pulse. Each occurrence of a Poisson event raises a rectangular load pulse of random intensity P_m, distributed according to its conditional PDF $f_{P_m|\Theta_{Pm}}(p_m|\theta_{Pm})$ and lasting until the next Poisson arrival. Based on these considerations, the mean-load intensity measure vector $\mathbf{IM}_m$ is defined as $\mathbf{IM}_m = \{P_m, \Delta T_m\}$ whereas the corresponding mean-load parameter vector can be written as $\Theta_m = \{\Theta_{P_m}, \mu_{\Delta T_m}\}$. Finally, within each mean-load pulse, the randomness of the superimposed load fluctuation intensity is described by the conditional PDF $f_{\mathbf{IM}_a|P_m,\Theta_a}(\mathbf{im}_a|p_m,\theta_a)$. A practical subset application of this modeling approach, with P_m treated as a deterministic constant, can be found in Gobbato et al. (2013a, 2013b) in which the authors explicitly accounted for the uncertainty in the amplitude of a sinusoidal load and propagated this source of uncertainty throughout all subsequent prognosis analysis steps.

17.5 Probabilistic mechanics-based debonding evolution analysis

The second analysis step of the predictive modeling module is referred to here as debonding evolution analysis. This step aims to compute the joint PDF of the actual debonding size vector $\mathbf{D}_a^{[p,q]}$, conditional on Θ_{mat}, Θ_{dam}^p, and all previous NDE inspection results up to current time t_p—i.e., $\mathbf{d}_m^{[0,q]}$. This joint conditional PDF is denoted as $f_{\mathbf{D}_a^{[p,q]}|\Theta_{mat},\Theta_{dam}^p,\mathbf{D}_m^{[0,p]}}\left(\mathbf{d}_a^{[p,q]}\big|\theta_{mat},\theta_{dam}^p,\mathbf{d}_m^{[0,p]}\right)$ and is obtained by unconditioning the joint conditional PDF $f_{\mathbf{D}_a^{[p,q]}|\Theta_{mat},\Theta_{dam}^p,\mathbf{D}_a^p,\mathbf{IM},\mathbf{D}_m^{[0,p]}}\left(\mathbf{d}_a^{[p,q]}\big|\theta_{mat},\theta_{dam}^p,\mathbf{d}_a^p,\mathbf{im},\mathbf{d}_m^{[0,p]}\right)$ w.r.t. $\mathbf{D}_a^p$ and $\mathbf{IM}$ as (Gobbato, 2011):

$$f_{\mathbf{D}_a^{[p,q]}|\Theta_{mat},\Theta_{dam}^p,\mathbf{D}_m^{[0,p]}}\left(\mathbf{d}_a^{[p,q]}\big|\theta_{mat},\theta_{dam}^p,\mathbf{d}_m^{[0,p]}\right)$$

$$= \int_{A_a^p}\left[\int_{\mathbf{IM}} f_{\mathbf{D}_a^{[p,q]}|\Theta_{mat},\Theta_{dam}^p,\mathbf{D}_a^p,\mathbf{IM},\mathbf{D}_m^{[0,p]}}\left(\mathbf{d}_a^{[p,q]}\big|\theta_{mat},\theta_{dam}^p,\mathbf{d}_a^p,\mathbf{im},\mathbf{d}_m^{[0,p]}\right)\right.$$

$$\left. \times f_{\mathbf{IM}}(\mathbf{im})d(\mathbf{im})\right] f''_{\mathbf{D}_a^p|\Theta_{mat},\Theta_{dam}^p,\mathbf{D}_m^{[0,p]}}\left(\mathbf{d}_a^p\big|\theta_{mat},\theta_{dam}^p,\mathbf{d}_m^{[0,p]}\right)d(\mathbf{d}_a^p) \quad (17.14)$$

Equation (17.14) is used to obtain the joint conditional PDF of the debonding size vector at time t_{p+1}, (i.e., at the time of the next NDE inspection). This quantity is denoted as $f_{\mathbf{D}_a^{[p,p+1]}|\Theta_{mat},\Theta_{dam}^p,\mathbf{D}_m^{[0,p]}}(\mathbf{d}_a^{[p,p+1]}\big|\theta_{mat},\theta_{dam}^p,\mathbf{d}_m^{[0,p]})$ and is used as new prior information to compute the posterior joint conditional PDF of $\mathbf{D}_a^{p+1}$ at time t_{p+1}, i.e.,

$$f_{\mathbf{D}_a^{[p,p+1]}|\Theta_{mat},\Theta_{dam}^p,\mathbf{D}_m^{[0,p]}}\left(\mathbf{d}_a^{[p,p+1]}\big|\theta_{mat},\theta_{dam}^p,\mathbf{d}_m^{[0,p]}\right)$$

$$= f'_{\mathbf{D}_a^{p+1}|\Theta_{mat},\Theta_{dam}^p,\mathbf{D}_m^{[0,p]}}\left(\mathbf{d}_a^{p+1}\big|\theta_{mat},\theta_{dam}^p,\mathbf{d}_m^{[0,p]}\right) \quad (17.15)$$

In addition, the uncertainty of $\mathbf{D}_a^{[p,q]}$ for given/fixed values of $\Theta_{mat} = \theta_{mat}$, $\Theta_{dam}^p = \theta_{dam}^p$, $\mathbf{D}_a^p = \mathbf{d}_a^p$, $\mathbf{IM} = \mathbf{im}$, and $\mathbf{D}_m^{[0,p]} = \mathbf{d}_m^{[0,p]}$—i.e., the uncertainty represented by the joint conditional PDF $f_{\mathbf{D}_a^{[p,q]}|\Theta_{mat},\Theta_{dam}^p,\mathbf{D}_a^p,\mathbf{IM},\mathbf{D}_m^{[0,p]}}\left(\mathbf{d}_a^{[p,q]}\middle|\theta_{mat},\theta_{dam}^p,\mathbf{d}_a^p,\mathbf{im},\mathbf{d}_m^{[0,p]}\right)$ inside the integral in Eqn (17.14)—is driven by the record-to-record variability of the structural response across the ensemble of all possible load patterns (or loading time histories) for a given $\mathbf{IM} = \mathbf{im}$; see Gobbato (2011) and Gobbato, Conte, et al. (2012) for a detailed discussion on this matter. If there is no record-to-record variability for given/fixed values of $\mathbf{IM} = \mathbf{im}$, the joint conditional PDF $f_{\mathbf{D}_a^{[p,q]}|\Theta_{mat},\Theta_{dam}^p,\mathbf{D}_a^p,\mathbf{IM},\mathbf{D}_m^{[0,p]}}\left(\mathbf{d}_a^{[p,q]}\middle|\theta_{mat},\theta_{dam}^p,\mathbf{d}_a^p,\mathbf{im},\mathbf{d}_m^{[0,p]}\right)$ can be simplified as

$$f_{\mathbf{D}_a^{[p,q]}|\Theta_{mat},\Theta_{dam}^p,\mathbf{D}_a^p,\mathbf{IM},\mathbf{D}_m^{[0,p]}}\left(\mathbf{d}_a^{[p,q]}\middle|\theta_{mat},\theta_{dam}^p,\mathbf{d}_a^p,\mathbf{im},\mathbf{d}_m^{[0,p]}\right) = \delta\left(\mathbf{D}_a^{[p,q]} - \hat{\mathbf{d}}_a^{[p,q]}\right)$$
(17.16)

where $\delta(\cdot)$ represents the Dirac delta and $\hat{\mathbf{d}}_a^{[p,q]} = \left(\mathbf{D}_a^{[p,q]}\middle|\theta_{mat},\theta_{dam}^p,\mathbf{a}_a^p,\mathbf{im},\mathbf{d}_m^{[0,p]}\right)$ denotes the deterministic debonding length vector at time $t_{p,q}$ computed for a given/fixed set of the input parameters θ_{mat}, θ_{dam}^p, $\mathbf{d}_a^p$, $\mathbf{im}$, and $\mathbf{d}_m^{[0,p]}$. Practical applications and validation studies in which record-to-record variability was not considered can be found in Gobbato, Kosmatka et al. (2012), Gobbato et al. (2013a).

The quantity $\Delta\tau$, introduced previously, is a suitable fixed time interval related to the time scale of the damage propagation process of interest. However, $\Delta\tau$ needs to be sufficiently short and lead to a satisfactory grid of response evaluations to provide an accurate and reliable prediction for the propagation path of the structural response of interest (i.e., the debonding propagation path in this specific study) as well as for the trend of the reliability index of the structure in $[t_p, t_{p,\bar{q}}]$. Furthermore, the value of $\bar{q}$ needs to guarantee a sufficiently wide prediction window to render the prognosis results meaningful for the decision-making process. In some cases (e.g., structures subjected to harmonic loads or other cyclic loading conditions), it is convenient to express all time related parameters (i.e., t_p, $t_{p,\bar{q}}$, and $\Delta\tau$) in terms of number of load cycles experienced by the structure. Current time, t_p, is replaced by the current number of load cycles, N_p; the time interval between two subsequent damage propagation/prognosis evaluations, $\Delta\tau$, is substituted with the number of load cycles, ΔN; finally, future time, $t_{p,\bar{q}}$, is replaced by the number of load cycles, $N_{p,q} = N_p + q \cdot \Delta N$ with $q \in \{0, 1, 2, ..., \bar{q}\}$ (Gobbato et al., 2013a; Gobbato, Kosmatka et al., 2012).

Experimental research (Quaresimin & Ricotta, 2006) showed that the fatigue life of adhesively-bonded composite joints can be divided into two distinct phases: an initial nucleation phase followed by the debonding growth up to a critical length. The portion of the fatigue life spent in each one of these two stages depends on many factors such as joint geometry, stress and strain distributions, stress ratio, adhesive joint thickness, and environmental conditions. Furthermore, the fatigue-induced disbonds can nucleate and then propagate either within the adhesive layer or along the adhesive/adherend interface; the first case leads to a purely cohesive failure whereas the second scenario is commonly referred to as adhesive failure and can be strongly influenced by the surface preparation of the composite adherends. A pair of typical cohesive failures, observed in pseudo-static end-notched flexure (ENF) tests performed by the authors, is shown in Figure 17.4. The ENF coupons tested were built by bonding together two unidirectional composite adherends with a two-part structural paste adhesive commonly used in the aerospace field. The tests aimed to calibrate, in probabilistic terms, some of the damage evolution model parameters of a particular class of mechanics-based debonding propagation models, known in the literature as cohesive zone models (CZMs) and widely used to simulate the pseudo-static and dynamic (i.e., impact-induced and fatigue-induced) delamination and debonding propagation processes in laminated composite structures (Alfano & Crisfield, 2001; Nguyen, Repetto, Ortiz, & Radovitzky, 2001). The CZM parameters (e.g., mode I and mode II critical fracture energies, peak cohesive stresses, and fatigue degradation parameters) are viewed as random variables and collected in vector Θ^p_{dam}. Similarly, if empirical damage growth models or models based on linear elastic fracture mechanics principles and experimental observations are used to simulate the fatigue-driven debonding propagation processes, Θ^p_{dam} collects the parameters of those particular empirical models (Blanco, Gamstedt, Asp, & Costa, 2004; Degrieck & Paepegem, 2001; Paris & Erdogan, 1963).

Conducting experimental fatigue tests at the coupon level and, when feasible, at the component and/or subassembly levels is a crucial task to properly derive a complete probabilistic characterization of the damage evolution model parameter vector at time t_0 (i.e., Θ^0_{dam}), here in the form of the joint PDF $f'_{\Theta^0_{dam}}(\theta^0_{dam})$. Most important,

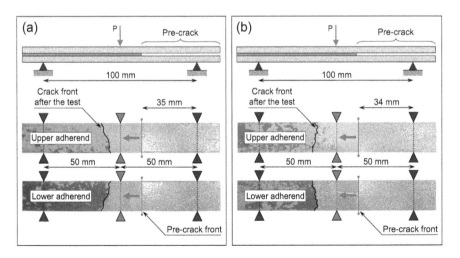

Figure 17.4 Examples of static end-notched flexure tests on adhesively-bonded composite beams showing pure cohesive failure along the adhesive interface and emphasizing the irregularities of the debonding/crack front after the tests.

correct characterization of the full correlation structure of Θ_{dam}^0 represents one of the most important factors to achieve useful and meaningful damage prognosis results (Gobbato et al., 2013a). As an illustration, consider the 68 experimental crack propagation trajectories obtained by Virkler et al. (1979) and shown in Figure 17.5. Each of these trajectories is curve-fit using the well-known Paris–Erdogan law (Paris & Erdogan, 1963), here expressed as $\ln(\dot{d}_a) = \ln C_0 + m_0 \cdot \ln(\Delta K)$, with $\dot{d}_a$ denoting the rate of crack propagation, ΔK representing the range of the stress intensity factor at the crack tip, and $\ln C_0$ and m_0 being the two damage evolution model parameters collected in Θ_{dam}^0 as $\Theta_{dam}^0 = \{\ln C_0, m_0\}$. The prior joint PDF of $\ln C_0$ and m_0—i.e., $f'_{\Theta_{dam}^0}(\theta_{dam}^0) = f'_{\ln C_0, m_0}(\ln C_0, m_0)$—is then derived from the statistical analysis of the 68 pairs of damage evolution model coefficients obtained through the curve-fitting process. The marginal prior PDFs of $\ln C_0$ and m_0 are well represented by normal distributions and the prior joint PDF $f'_{\ln C_0, m_0}(\ln C_0, m_0)$ is well described by a bivariate normal distribution with an estimated correlation coefficient between $\ln C_0$ and m_0 equal to $\rho_{(\ln C_0, m_0)} = -0.9976$ (Gobbato, Kosmatka et al., 2012). The contour plot

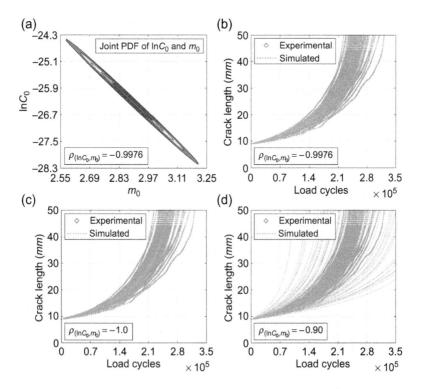

Figure 17.5 (a) Contour map of the joint PDF of the Paris law parameters $\ln C$ and m derived from the statistical analysis of an experimental dataset available in the literature (Virkler et al., 1979). Effect of the degree of statistical correlation between $\ln C$ and m on the dispersion of the simulated crack propagation trajectories: (b) $\rho_{\ln C, m} = -0.9976$, (c) $\rho_{\ln C, m} = -1.0$, and (d) $\rho_{\ln C, m} = -0.90$.

of this joint PDF of $\ln C_0$ and m_0 is shown in Figure 17.5(a), whereas the remaining three subplots in Figure 17.5 illustrate the sensitivity of the crack propagation process to the degree of statistical correlation between the two Paris–Erdogan law parameters, $\ln C_0$ and m_0, by overlapping 68 simulated crack propagation trajectories to the original experimental dataset using three different values for the correlation coefficient $\rho_{(\ln C_0, m_0)}$—i.e., $\rho_{(\ln C_0, m_0)} = -0.9976$ in Figure 17.5(b), $\rho_{(\ln C_0, m_0)} = -1.0$ in Figure 17.5(c), and $\rho_{(\ln C_0, m_0)} = -0.90$ Figure 17.5(d). These 68 simulated trajectories are obtained from the numerical integration of the Paris–Erdogan law by setting the initial crack length to 9.0 mm and by drawing 68 random samples of $\Theta^0_{dam} = \{\ln C_0, m_0\}$ according to its joint PDF $f'_{\Theta^0_{dam}}(\theta^0_{dam})$. The 68 simulated crack propagation trajectories shown in Figure 17.5(b) and obtained by setting $\rho_{(\ln C_0, m_0)} = -0.9976$ reproduce well the scatter-dispersion of the original dataset. On the other hand, the simulated trajectories reported in Figure 17.5(c) and obtained by assuming that $\rho_{(\ln C_0, m_0)} = -1.0$ are far from being able to match the scatter of the original experimental dataset, which demonstrates that considering $\ln C_0$ and m_0 to be perfectly correlated random variables represents a poor and dangerous approximation that leads to misleading damage propagation predictions and erroneous RFL estimations. Finally, as demonstrated in Figure 17.5(d), when the degree of statistical correlation between $\ln C_0$ and m_0 is reduced to $\rho_{(\ln C_0, m_0)} = -0.90$, the dispersion of the simulated crack propagation trajectories is much larger than that of the experimental dataset. This erroneous assumption could lead to numerical convergence issues during the probabilistic fatigue-driven damage propagation analysis aimed at providing an evaluation of Eqn (17.14).

17.6 Probabilistic characterization of global system performance

The fourth step of the proposed framework, *system performance analysis*, aims to provide a joint probabilistic characterization of both local (through $\mathbf{D}_a^{[p,q]}$) and global (through $\mathbf{D}_g^{[p,q]}$) performance levels of the monitored structural system at a future time $t_{p,q} = t_p + q \cdot \Delta\tau$ with $q \in \{0, 1, 2, ..., \bar{q}\}$. This joint probabilistic information is quantified by the conditional joint PDF $f_{\mathbf{\Psi}_{L,G}^{[p,q]} | \mathbf{D}_m^{[0,p]}}\left(\mathbf{\psi}_{L,G}^{[p,q]} \middle| \mathbf{d}_m^{[0,p]}\right)$ with the random vector $\mathbf{\Psi}_{L,G}^{[p,q]}$ defined as $\mathbf{\Psi}_{L,G}^{[p,q]} = \{\mathbf{D}_a^{[p,q]}, \mathbf{D}_g^{[p,q]}\}$. As an illustration, the random vector $\mathbf{D}_g^{[p,q]}$, defined as $\mathbf{D}_g^{[p,q]} = \{\mathbf{D}_{g,r}^{[p,q]}, r = 1, ..., n_G\}$, can collect a given set of natural frequencies, mode shapes, and modal strains of the monitored structure (Doebling et al., 1998; Zou et al., 2000), the damage-dependent flutter velocity of an aircraft wing (Styuart et al., 2007) or wind turbine blade, and so forth. As an illustration, Gobbato (2011) and Gobbato, Conte, et al. (2012) discussed the practical application of the proposed system performance analysis step to characterize the global aeroelastic performance of a composite wing as debonding progresses along the skin-to-spar adhesive joints. The prediction of the flutter onset speed represents a fundamental task in the aerospace field because the loss of dynamic stability may result in unbounded vibrations of the wing structure and lead to the sudden failure of the aircraft's primary structural components. However, in the presence of structural and/or aerodynamic nonlinearities, growth in the amplitude of the structural response quantities of interest (e.g., wing-tip vertical displacement or pitching angle) tends to stabilize/converge to the so-called limit cycle oscillations (LCO) (Lee, Prince, & Wong, 1999; Librescu, Chiocchia, & Marzocca, 2003). These types of nonlinear phenomena can lead to excessive fatigue of the airframe as well as unacceptable workloads for pilots, thereby increasing the risk of incorrectly performing critical tasks. Based on these considerations, Gobbato (2011) and Gobbato, Conte et al. (2012) defined the n_G-dimensional global performance limit-state vector as $\mathbf{D}_g^{[p,q]} = \left\{ V_f^{[p,q]}, V_{\text{lco},1}^{[p,q]}, V_{\text{lco},2}^{[p,q]}, \ldots \right\}$ where $V_f^{[p,q]}$ denotes the damage-dependent flutter speed and $V_{\text{lco},r}^{[p,q]}$, with $r = 1,\ldots,(n_G - 1)$, symbolizes the velocity at which the r-th LCO reaches a predefined amplitude threshold beyond which the structural integrity of the aircraft is compromised.

The n_G random components of $\mathbf{D}_g^{[p,q]}$ are clearly mutually statistically dependent as well as statistically dependent on $\mathbf{D}_a^{[p,q]}$. Therefore, to formally derive the joint conditional PDF $f_{\boldsymbol{\Psi}_{L,G}^{[p,q]} | \mathbf{D}_m^{[0,p]}}\left(\boldsymbol{\psi}_{L,G}^{[p,q]} \middle| \mathbf{d}_m^{[0,p]} \right)$, a four-step procedure must be followed. The first substep computes the joint conditional PDF $f_{\mathbf{D}_g^{[p,q]} | \boldsymbol{\Theta}_{\text{mat}}, \boldsymbol{\Theta}_{\text{dam}}^p, \mathbf{D}_a^{[p,q]}, \mathbf{D}_m^{[0,p]}}\left(\mathbf{d}_g^{[p,q]} \middle| \boldsymbol{\theta}_{\text{mat}}, \boldsymbol{\theta}_{\text{dam}}^p, \mathbf{d}_a^{[p,q]}, \mathbf{d}_m^{[0,p]} \right)$. The second substep derives the joint conditional PDF $f_{\mathbf{D}_g^{[p,q]} | \mathbf{D}_a^{[p,q]}, \mathbf{D}_m^{[0,p]}}\left(\mathbf{d}_g^{[p,q]} \middle| \mathbf{d}_a^{[p,q]}, \mathbf{d}_m^{[0,p]} \right)$ by unconditioning the joint conditional PDF $f_{\mathbf{D}_g^{[p,q]} | \boldsymbol{\Theta}_{\text{mat}}, \boldsymbol{\Theta}_{\text{dam}}^p, \mathbf{D}_a^{[p,q]}, \mathbf{D}_m^{[0,p]}}\left(\mathbf{d}_g^{[p,q]} \middle| \boldsymbol{\theta}_{\text{mat}}, \boldsymbol{\theta}_{\text{dam}}^p, \mathbf{d}_a^{[p,q]}, \mathbf{d}_m^{[0,p]} \right)$ w.r.t. $\boldsymbol{\Theta}_{\text{mat}}$ and $\boldsymbol{\Theta}_{\text{dam}}^p$ as

$$f_{\mathbf{D}_g^{[p,q]} | \mathbf{D}_a^{[p,q]}, \mathbf{D}_m^{[0,p]}}\left(\mathbf{d}_g^{[p,q]} \middle| \mathbf{d}_a^{[p,q]}, \mathbf{d}_m^{[0,p]} \right)$$
$$= \int_{\boldsymbol{\Theta}_{\text{mat}}} \int_{\boldsymbol{\Theta}_{\text{dam}}^p} f_{\mathbf{D}_g^{[p,q]} | \boldsymbol{\Theta}_{\text{mat}}, \boldsymbol{\Theta}_{\text{dam}}^p, \mathbf{D}_a^{[p,q]}, \mathbf{D}_m^{[0,p]}}\left(\mathbf{d}_g^{[p,q]} \middle| \boldsymbol{\theta}_{\text{mat}}, \boldsymbol{\theta}_{\text{dam}}^p, \mathbf{d}_a^{[p,q]}, \mathbf{d}_m^{[0,p]} \right) \quad (17.17)$$
$$\times f_{\boldsymbol{\Theta}_{\text{mat}}}(\boldsymbol{\theta}_{\text{mat}}) f''_{\boldsymbol{\Theta}_{\text{dam}}^p | \mathbf{D}_m^{[0,p]}}\left(\boldsymbol{\theta}_{\text{dam}}^p \middle| \mathbf{d}_m^{[0,p]} \right) d\boldsymbol{\theta}_{\text{mat}} \, d\boldsymbol{\theta}_{\text{dam}}^p$$

Similarly, the third substep determines the joint conditional PDF of the predicted debonding size vector at time $t_{p,q}$ — i.e., $f_{\mathbf{D}_a^{[p,q]} | \mathbf{D}_m^{[0,p]}}\left(\mathbf{d}_a^{[p,q]} \middle| \mathbf{d}_m^{[0,p]} \right)$ as

$$f_{\mathbf{D}_a^{[p,q]} | \mathbf{D}_m^{[0,p]}}\left(\mathbf{d}_a^{[p,q]} \middle| \mathbf{d}_m^{[0,p]} \right) = \int_{\boldsymbol{\Theta}_{\text{mat}}} \int_{\boldsymbol{\Theta}_{\text{dam}}^p} f_{\mathbf{D}_a^{[p,q]} | \boldsymbol{\Theta}_{\text{mat}}, \boldsymbol{\Theta}_{\text{dam}}^p, \mathbf{D}_m^{[0,p]}}\left(\mathbf{d}_a^{[p,q]} \middle| \boldsymbol{\theta}_{\text{mat}}, \boldsymbol{\theta}_{\text{dam}}^p, \mathbf{d}_m^{[0,p]} \right)$$
$$\times f_{\boldsymbol{\Theta}_{\text{mat}}}(\boldsymbol{\theta}_{\text{mat}}) f''_{\boldsymbol{\Theta}_{\text{dam}}^p | \mathbf{D}_m^{[0,p]}}\left(\boldsymbol{\theta}_{\text{dam}}^p \middle| \mathbf{d}_m^{[0,p]} \right) d\boldsymbol{\theta}_{\text{mat}} \, d\boldsymbol{\theta}_{\text{dam}}^p$$
$$(17.18)$$

Finally, the fourth substep formally derives the joint conditional PDF $f_{\mathbf{\Psi}_{L,G}^{[p,q]} | \mathbf{D}_m^{[0,p]}} \left(\mathbf{\psi}_{L,G}^{[p,q]} \middle| \mathbf{d}_m^{[0,p]} \right)$ as

$$f_{\mathbf{\Psi}_{L,G}^{[p,q]} | \mathbf{D}_m^{[0,p]}} \left(\mathbf{\psi}_{L,G}^{[p,q]} \middle| \mathbf{d}_m^{[0,p]} \right) = f_{\mathbf{D}_g^{[p,q]} | \mathbf{D}_a^{[p,q]}, \mathbf{D}_m^{[0,p]}} \left(\mathbf{d}_g^{[p,q]} \middle| \mathbf{d}_a^{[p,q]}, \mathbf{d}_m^{[0,p]} \right)$$
$$\times f_{\mathbf{D}_a^{[p,q]} | \mathbf{D}_m^{[0,p]}} \left(\mathbf{d}_a^{[p,q]} \middle| \mathbf{d}_m^{[0,p]} \right) \quad (17.19)$$

As a final remark to this section, when $q = 0$ and thus $t_{p,q} = t_p$, the joint conditional PDF $f_{\mathbf{D}_a^{[p,q]} | \mathbf{\Theta}_{mat}, \mathbf{\Theta}_{dam}^p, \mathbf{D}_m^{[0,p]}} \left(\mathbf{d}_a^{[p,q]} \middle| \mathbf{\theta}_{mat}, \mathbf{\theta}_{dam}^p, \mathbf{d}_m^{[0,p]} \right)$ simplifies to $f''_{\mathbf{D}_a^p | \mathbf{\Theta}_{mat}, \mathbf{\Theta}_{dam}^p, \mathbf{D}_m^{[0,p]}} \left(\mathbf{D}_a^p \middle| \mathbf{\theta}_{mat}, \mathbf{\theta}_{dam}^p, \mathbf{d}_m^{[0,p]} \right)$ and Eqn (17.18) reduces to Eqn (17.13) derived earlier in Section 17.3. Furthermore, when dealing with high-dimensional problems (i.e., with a significantly large number of damage/debonding locations and global performance metrics), estimating the full joint conditional PDF $f_{\mathbf{\Psi}_{L,G}^{[p,q]} | \mathbf{D}_m^{[0,p]}} \left(\mathbf{\psi}_{L,G}^{[p,q]} \middle| \mathbf{d}_m^{[0,p]} \right)$ becomes a formidable task. To overcome this computational burden, in certain cases, determining the marginal probabilistic characterizations of $\mathbf{D}_a^{[p,q]}$ and $\mathbf{D}_g^{[p,q]}$, can be sufficient. As an illustration, if only unimodal bounds (see Section 17.7 and Ditlevsen & Madsen, 1996) of the conditional probability of system failure $P\left[F_{sys}^{[p,q]} \middle| \mathbf{D}_m^{[0,p]} = \mathbf{d}_m^{[0,p]} \right]$ and the conditional RFL $\left(\overline{p}_F; t_p, \mathbf{D}_m^{[0,p]} = \mathbf{d}_m^{[0,p]} \right)$ have to be determined, the marginal PDFs of the n_{loc} individual components of the debonding size vector $\mathbf{D}_a^{[p,q]}$ the marginal PDFs of the n_G global performance metrics, collected in $\mathbf{D}_g^{[p,q]}$, are sufficient (Gobbato, Conte, et al., 2012).

17.7 Damage prognosis analysis

Once the probabilistic system performance analysis step is completed and the joint conditional PDF $f_{\mathbf{\Psi}_{L,G}^{[p,q]} | \mathbf{D}_m^{[0,p]}} \left(\mathbf{\psi}_{L,G}^{[p,q]} \middle| \mathbf{d}_m^{[0,p]} \right)$ is determined, component and system reliability analyses are carried out to estimate the time-varying probability of failure and reliability index of the monitored system. The system time-varying probability of failure and reliability index are denoted here as $P\left[F_{sys}^{[p,q]} \middle| \mathbf{D}_m^{[0,p]} = \mathbf{d}_m^{[0,p]} \right]$ and $\beta_{sys}^{[p,q]} \left(\mathbf{D}_m^{[0,p]} = \mathbf{d}_m^{[0,p]} \right) = -\Phi^{-1}\left(P\left[F_{sys}^{[p,q]} \middle| \mathbf{D}_m^{[0,p]} = \mathbf{d}_m^{[0,p]} \right] \right)$, respectively, and are updated every time a new set of NDE inspection results at time t_p becomes available and is assimilated by the proposed SHM-DP framework. Both conditional quantities, $P\left[F_{sys}^{[p,q]} \middle| \mathbf{D}_m^{[0,p]} = \mathbf{d}_m^{[0,p]} \right]$ and $\beta_{sys}^{[p,q]} \left(\mathbf{D}_m^{[0,p]} = \mathbf{d}_m^{[0,p]} \right)$, are evaluated at $\overline{q} + 1$ evenly spaced points in time defined as $t_{p,q} = t_p + q \cdot \Delta\tau$ (with $q = 0, 1, 2, ..., \overline{q}$). These results can then be interpolated to derive (for $t \geq t_p$) the continuous functions defined in Eqns (17.20) and (17.21)

$$G_{\text{sys}}\left(t; t_p, \mathbf{d}_{\text{m}}^{[0,p]}\right) : [t_p, t_{p,\bar{q}}] \to [0, 1],$$
$$G_{\text{sys}}\left(t; t_p, \mathbf{d}_{\text{m}}^{[0,p]}\right) = P\left[\mathbf{F}_{\text{sys}}^{[p,t]} \middle| \mathbf{D}_{\text{m}}^{[0,p]} = \mathbf{d}_{\text{m}}^{[0,p]}\right]$$
(17.20)

$$R_{\text{sys}}\left(t; t_p, \mathbf{d}_{\text{m}}^{[0,p]}\right) : [t_p, t_{p,\bar{q}}] \to [-\infty, +\infty],$$
$$R_{\text{sys}}\left(t; t_p, \mathbf{d}_{\text{m}}^{[0,p]}\right) = \beta_{\text{sys}}^{[p,t]}\left(\mathbf{D}_{\text{m}}^{[0,p]} = \mathbf{d}_{\text{m}}^{[0,p]}\right)$$
(17.21)

In these equations, $G_{\text{sys}}\left(t; t_p, \mathbf{d}_{\text{m}}^{[0,p]}\right)$ denotes the monotonically increasing function (of time) defining the time-varying probability of system failure predicted at current time t_p, whereas $R_{\text{sys}}\left(t; t_p, \mathbf{d}_{\text{m}}^{[0,p]}\right)$ represents the monotonically decreasing function (of time) defining the time-varying system reliability index predicted at current time t_p and defined as $\beta_{\text{sys}}^{[p,t]}\left(\mathbf{D}_{\text{m}}^{[0,p]} = \mathbf{d}_{\text{m}}^{[0,p]}\right) = -\Phi^{-1}\left(P\left[\mathbf{F}_{\text{sys}}^{[p,t]} \middle| \mathbf{D}_{\text{m}}^{[0,p]} = \mathbf{d}_{\text{m}}^{[0,p]}\right]\right)$. Concurrently, this last analysis step also estimates the conditional RFL of the system being monitored. This conditional quantity is defined as $\text{RFL}(\bar{p}_{\text{F}}; t_p, \mathbf{D}_{\text{m}}^{[0,p]} = \mathbf{d}_{\text{m}}^{[0,p]}) = t_{\text{F}}\left(\bar{p}_{\text{F}}; t_p, \mathbf{d}_{\text{m}}^{[0,p]}\right) - t_p$, where $t_{\text{F}}\left(\bar{p}_{\text{F}}; t_p, \mathbf{d}_{\text{m}}^{[0,p]}\right)$ represents the predicted time at failure, i.e., the predicted time at which the conditional *probability of system failure*, $G_{\text{sys}}\left(t; t_p, \mathbf{d}_{\text{m}}^{[0,p]}\right) = P\left[\mathbf{F}_{\text{sys}}^{[p,t]} \middle| \mathbf{D}_{\text{m}}^{[0,p]} = \mathbf{d}_{\text{m}}^{[0,p]}\right]$, reaches the predefined critical threshold $\bar{p}_{\text{F}}$.

To derive these results, the structural system of interest is abstracted to a collection of local and global reliability components linked together as a series system, i.e., a system that is considered failed when at least one of its reliability components (either local or global) has failed. A local reliability component is defined by a single local limit state (LLS) function involving one or more random variables in $\mathbf{D}_a^{[p,q]} = \left\{\mathbf{D}_a^{(i,[p,q])}, \ i = 1, \ldots, n_{\text{loc}}\right\}$. On the other hand, a global reliability component is characterized by a single GLS function involving one or more random variables in $\mathbf{D}_g^{[p,q]} = \left\{\mathbf{D}_{g,r}^{[p,q]}, \ r = 1, \ldots, n_g\right\}$. Each reliability component is considered failed when the corresponding limit-state is reached or exceeded, and therefore component and system failure events do not necessarily represent a physical failure of the real structural system, but rather a violation of a mathematical constraint defining a limit-state. The estimates of $P\left[\mathbf{F}_{\text{sys}}^{[p,q]} \middle| \mathbf{D}_{\text{m}}^{[0,p]} = \mathbf{d}_{\text{m}}^{[0,p]}\right]$, $\beta_{\text{sys}}^{[p,q]}\left(\mathbf{D}_{\text{m}}^{[0,p]} = \mathbf{d}_{\text{m}}^{[0,p]}\right)$ and $\text{RFL}\left(\bar{p}_{\text{F}}; t_p, \mathbf{D}_{\text{m}}^{[0,p]} = \mathbf{d}_{\text{m}}^{[0,p]}\right)$ at future times $t_{p,q}$ (with $q = 0, 1, 2, \ldots, \bar{q}$) can then be used as input for risk-based decision making in scheduling and/or updating the maintenance/repair plan for the considered structural system on the basis of a pre-defined maximum acceptable threshold $(\bar{p}_{\text{F}})$ for $P\left[\mathbf{F}_{\text{sys}}^{[p,q]} \middle| \mathbf{D}_{\text{m}}^{[0,p]} = \mathbf{d}_{\text{m}}^{[0,p]}\right]$. A more in-depth discussion on the topic of component and system reliability analysis of monitored structural systems and alternative abstractions of the real structure into a collection of reliability components linked together as a combination of series and parallel subsystems can be found in Gobbato (2011), Gobbato et al. (2011) and Gobbato, Conte, et al. (2012). Based on these considerations, and assuming a number of local and GLSs equal to n_{LLS} and n_{GLS}, respectively, the conditional event of *system failure* at future time $t_{p,q} = t_p + q \cdot \Delta \tau$, with $q \in \{0, 1, 2, ..., \bar{q}\}$, can be defined as

$$\left(F_{\text{sys}}^{[p,q]} \Big| \mathbf{D}_m^{[0,p]} = \mathbf{d}_m^{[0,p]}\right) \triangleq \left[\left(\bigcup_{u=1}^{n_{\text{LLS}}} \left(F_{L,u}^{[p,q]} \Big| \mathbf{D}_m^{[0,p]} = \mathbf{d}_m^{[0,p]}\right)\right) \\ \cup \left(\bigcup_{v=1}^{n_{\text{GLS}}} \left(F_{G,v}^{[p,q]} \Big| \mathbf{D}_m^{[0,p]} = \mathbf{d}_m^{[0,p]}\right)\right)\right] \quad (17.22)$$

where $F_{L,u}^{[p,q]}$ denotes the u-th local failure event and $F_{G,v}^{[p,q]}$ represents the v-th global failure event at time $t_{p,q}$. The probability of the failure event defined in Eqn (17.22) can be evaluated using state-of-the-art simulation-based techniques such as MCMC methods (Gamerman, 1997; Guan et al., 2011, 2012) and transitional MCMC methods (Ching & Chen, 2007) as well as semi-analytical approaches (Gobbato, 2011; Gobbato et al., 2013a, 2013b); both types of methods can in fact be embedded into the proposed SHM-DP framework.

To provide the reader with some practical examples, the simplest (and most logical) local failure event associated with the debonding propagation process evolving at the i-th damage location can be defined as $F_{L,i}^{[p,q]} \triangleq \left\{D_a^{(i,[p,q])} \geq d_{a,\text{crit}}^{(i)}\right\}$, where $d_{a,\text{crit}}^{(i)}$ represent a predefined location-dependent critical debonding length. According to this definition, the corresponding time-varying probability of failure at the local reliability component level can be expressed as

$$P\left[F_{L,i}^{[p,q]} \Big| \mathbf{D}_m^{[0,p]} = \mathbf{d}_m^{[0,p]}\right] = P\left[\left\{D_a^{(i,[p,q])} \geq d_{\text{crit}}^{(i)}\right\} \Big| \mathbf{D}_m^{[0,p]} = \mathbf{d}_m^{[0,p]}\right]$$

$$= \int_{d_{\text{crit}}^{(i)}}^{+\infty} f_{D_a^{(i,[p,q])} | \mathbf{D}_m^{[0,p]}}\left(d_a^{(i,[p,q])} \Big| \mathbf{d}_m^{[0,p]}\right) d\left(d_a^{(i,[p,q])}\right) \quad (17.23)$$

$$= 1 - F_{D_a^{(i,[p,q])} | \mathbf{D}_m^{[0,p]}}\left(d_{a,\text{crit}}^{(i)} \Big| \mathbf{d}_m^{[0,p]}\right)$$

where the term $F_{D_a^{(i,[p,q])} | \mathbf{D}_m^{[0,p]}}\left(d_a^{(i,[p,q])} \Big| \mathbf{d}_m^{[0,p]}\right)$ represents the conditional CDF of the random debonding length $D_a^{(i,[p,q])}$ at time $t_{p,q} = t_p + q \cdot \Delta \tau$. On the other hand, as previously discussed in Sections 17.2 and 17.6, the global performance metric vector can collect a given set of natural frequencies of the structure being monitored, or in some specific aerospace applications the damage extent—dependent flutter velocity of an aircraft wing whose global aeroelastic stability is compromised by the progressive fatigue-driven damage/debonding growth along its critical subcomponents (Gobbato, Conte, et al., 2012). Therefore, according to these two practical examples, the simplest global failure event associated with the r-th global performance metric can be defined as $\mathrm{F}_{\mathrm{G},r}^{[p,q]} \triangleq \left\{ D_g^{(r,[p,q])} \leq d_{g,\mathrm{crit}}^{(i)} \right\}$ where $d_{g,\mathrm{crit}}^{(i)}$ represents a predefined minimum threshold for $D_g^{(r,[p,q])}$. For example, if the random quantity $D_g^{(r,[p,q])}$ represents the flutter velocity of an aircraft wing, $d_{g,\mathrm{crit}}^{(i)}$ defines the critical flutter speed below which the overall aeroelastic stability of the vehicle is considered overly impacted. Consequently, the corresponding time-varying probability of failure, $P\left[\mathrm{F}_{\mathrm{G},r}^{[p,q]} \big| \mathbf{D}_\mathrm{m}^{[0,p]} = \mathbf{d}_\mathrm{m}^{[0,p]} \right]$, can be expressed as (Gobbato et al., 2011; Gobbato, Conte, et al., 2012)

$$\begin{aligned}
P\left[\mathrm{F}_{\mathrm{G},r}^{[p,q]} \big| \mathbf{D}_\mathrm{m}^{[0,p]} = \mathbf{d}_\mathrm{m}^{[0,p]} \right] &= P\left[\left\{ D_g^{(r,[p,q])} \leq d_{g,\mathrm{crit}}^{(i)} \right\} \big| \mathbf{D}_\mathrm{m}^{[0,p]} = \mathbf{d}_\mathrm{m}^{[0,p]} \right] \\
&= \int_0^{d_{g,\mathrm{crit}}^{(i)}} f_{D_g^{(r,[p,q])} | \mathbf{D}_\mathrm{m}^{[0,p]}} \left(d_g^{(r,[p,q])} \big| \mathbf{d}_\mathrm{m}^{[0,p]} \right) d\left(d_g^{(r,[p,q])} \right) \\
&= F_{D_g^{(r,[p,q])} | \mathbf{D}_\mathrm{m}^{[0,p]}} \left(d_{\mathrm{crit}}^{(i)} \big| \mathbf{d}_\mathrm{m}^{[0,p]} \right)
\end{aligned} \qquad (17.24)$$

where the term $F_{D_g^{(r,[p,q])} | \mathbf{D}_\mathrm{m}^{[0,p]}} \left(d_g^{(r,[p,q])} \big| \mathbf{d}_\mathrm{m}^{[0,p]} \right)$ represents the conditional CDF of the global performance metric $D_g^{(r,[p,q])}$ at time $t_{p,q}$.

17.8 Effectiveness of proposed methodology in predicting the remaining time to failure

The overall effectiveness of the SHM-DP framework presented in this chapter depends on the individual performance of each of the five key analysis steps illustrated in Figure 17.2. However, the recursive Bayesian inference and the debonding evolution analysis represent the two most important steps of the overall methodology to provide accurate and reliable RFL predictions. On the one hand, the effectiveness of the recursive Bayesian inference scheme in detecting and quantifying damage is primarily governed by the following factors: detectability and measurement accuracy of the NDE inspection technique used in the monitoring process (see Section 17.3), the number of measurements taken at each inspection, and the inter-inspection time (Gobbato, 2011). On the other hand, the effectiveness, accuracy, and robustness of the predictive debonding propagation analysis step rely on the use of calibrated and validated mechanics-based models capable of capturing the physics of the damage propagation processes and the accurate representation of the statistical correlation between the parameters of such models. These two critical aspects often require a considerable amount of experimental investigations (see Section 17.5). Ultimately, the end user is interested in quantifying the accuracy and reliability of the SHM-DP framework in replicating and predicting the real degradation pattern experienced by the monitored structural or mechanical system. However, in a real-world application, the evolution in time of the true state of damage is unknown and therefore the performance and robustness of any SHM-DP methodology, such as the one proposed in this chapter, must be assessed through component- and system-level validation studies in which the underlying truth is known and used to evaluate the accuracy of the damage prognosis results. Examples of component-level validation studies can be found in Guan et al. (2011, 2012) and Gobbato et al. (2013a, 2013b). Based on the analysis and discussions presented in this chapter, several key performance metrics can be used during the verification and validation phase to ensure that in a real-world application of the proposed methodology, all desired prediction capabilities are achieved. For instance, the end user of the proposed SHM-DP framework is interested in having the posterior conditional mean of the actual debonding length, defined as

$$\mu''\left(D_a^{(i,p)}\Big|\mathbf{D}_m^{[0,p]} = \mathbf{d}_m^{[0,p]}\right) = \int_{D_a^{(i,p)}} d_a^{(i,p)} \cdot f''_{D_a^{(i,p)}\big|\mathbf{D}_m^{[0,p]}}\left(d_a^{(i,p)}\Big|\mathbf{d}_m^{[0,p]}\right) d\left(d_a^{(i,p)}\right)$$

(17.25)

as close as possible to the underlying true debonding length, $d_{\text{true}}^{(i,p)}$, at each detected debonding location and inspection time, i.e., for $i = 1,...,n_{\text{loc}}$ and $p = 0, 1, 2,...$, respectively. Similarly, the end user would like to be assured that the $\bar{p}_F$ percentile of the predicted conditional time at failure, $t_F\left(\bar{p}_F; t_p, \mathbf{d}_m^{[0,p]}\right)$, becomes closer and closer to the true failure time $(t_{F,\text{true}})$ as the time t_p, at which damage prognosis is performed, gets progressively closer to $t_{F,\text{true}}$. In addition, the end user wishes to minimize the odds of not predicting an unexpected failure, i.e., maximize the confidence that the predicted conditional RFL, previously denoted as $\text{RFL}\left(\bar{p}_F; t_p, \mathbf{D}_m^{[0,p]} = \mathbf{d}_m^{[0,p]}\right)$, does not overestimate the true RFL of the monitored system. In other words, he or she wants to minimize the risk that the predicted conditional time at failure, $t_F\left(\bar{p}_F; t_p, \mathbf{d}_m^{[0,p]}\right)$, exceeds the true failure time, i.e., $t_F\left(\bar{p}_F; t_p, \mathbf{d}_m^{[0,p]}\right) \leq t_{F,\text{true}}$. For this purpose, the ideal condition that would maximize the RFL without compromising the minimum safety requirements is met when the function $G_{\text{sys}}\left(t; t_p, \mathbf{d}_m^{[0,p]}\right)$, introduced in Eqn (17.20), is represented by the Heaviside step function $H(t - t_{F,\text{true}})$ defined as

$$H\left(t - t_{F,\text{true}}\right) = \begin{cases} 0 & \text{if } t < t_{F,\text{true}} \\ 1 & \text{if } t \geq t_{F,\text{true}} \end{cases}$$

(17.26)

However, the presence of many sources of uncertainty within the SHM-DP process tends to flatten the function $G_{sys}\left(t; t_p, \mathbf{d}_m^{[0,p]}\right)$ into the characteristic and well-known CDF S-shaped pattern. Furthermore, it can be proven and verified (Gobbato et al., 2013a, 2013b) that higher levels of uncertainty (e.g., measurement uncertainty, load uncertainty) and fewer inspection data lead toward a flatter pattern of the function $G_{sys}\left(t; t_p, \mathbf{d}_m^{[0,p]}\right)$. Figure 17.6 illustrates these concepts by considering two increasing levels of load uncertainty (LU1 and LU2) and three different inspection times (t_0, t_1, and t_p). First, the three plots in each of the two columns of this figure show that the predicted time-varying probability of failure of the monitored system becomes more

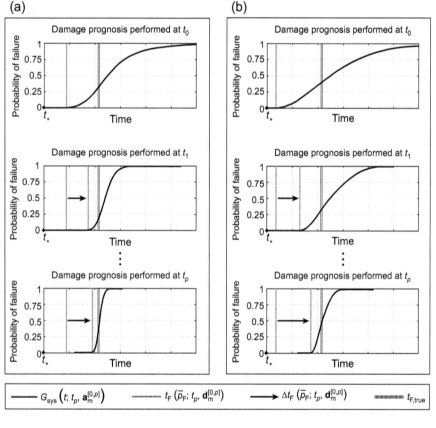

Figure 17.6 Illustrative example of the proposed recursive update of the time-varying probability of system failure and the potential extension of the remaining fatigue life that can be obtained by recursively using the proposed structural health monitoring and damage prognosis framework after each nondestructive evaluation inspection. (a) Load uncertainity LU1; (b) Load uncertainity LU2. Note that $t_0 < t_1 < ... < t_p < t_*$ and the scale of the horizontal axis remains the same for all six subplots.

and more accurate as new inspections are performed and new data are collected and processed. Second, it can be observed that the three functions $G_{\text{sys}}\left(t;t_0,\mathbf{d}_m^{[0,p]}\right)$, $G_{\text{sys}}\left(t;t_1,\mathbf{d}_m^{[0,p]}\right)$, and $G_{\text{sys}}\left(t;t_p,\mathbf{d}_m^{[0,p]}\right)$ shown on the right column of the figure are always less steep than their corresponding counterparts reported on the left column. Finally, the gain in fatigue life at time t_p, defined as $\Delta t_{\text{F}}\left(\overline{p}_{\text{F}};t_p,\mathbf{d}_m^{[0,p]}\right) = t_{\text{F}}\left(\overline{p}_{\text{F}};t_p,\mathbf{d}_m^{[0,p]}\right) - t_{\text{F}}\left(\overline{p}_{\text{F}};t_0,\mathbf{d}_m^{[0,p]}\right)$, is higher for the loading scenario LU2. In other words, the higher the uncertainty in future operational loads, the higher the benefits that can be obtained by deploying a monitoring program and integrating it with the proposed SHM-DP framework presented in this chapter.

To conclude, all quantities mentioned previously are conditional on the particular realization/scenario of NDE inspection results (i.e., $\mathbf{D}_m^{[0,p]} = \mathbf{d}_m^{[0,p]}$). The application of the proposed framework with a different realization (i.e., a different set) of NDE inspection results would lead to different Bayesian updating and damage prognosis results. Therefore, to conduct a more robust statistical performance evaluation of the proposed NDE-based SHM-DP methodology, multiple analyses across an ensemble of NDE inspection results—$\left\{\mathbf{D}_m^{[0,p]} = \mathbf{d}_m^{[0,p]}\right\}_e$ with $e=1,\ldots,n_{\text{ens}}$—should be performed. This approach is referred to as *ensemble scenario analysis* (Gobbato et al., 2013b) and can be used to demonstrate that the developed SHM-DP framework provides unbiased results. For example, it could be shown that the ensemble-average of the quantity $\mu''\left(D_a^{(i,p)}\Big|\mathbf{D}_m^{[0,p]} = \mathbf{d}_m^{[0,p]}\right)$ converges to the true debonding length as n_{ens} increases, i.e., $\lim_{n_{\text{ens}} \to \infty} E_{\text{ens}}\left[\mu''\left(D_a^{(i,p)}\Big|\mathbf{D}_m^{[0,p]} = \mathbf{d}_m^{[0,p]}\right)\right] = d_{\text{true}}^{(i,p)}$ for $i=1,\ldots,n_{\text{loc}}$ and for each $p \geq 0$. Similarly, it could also be shown that as the system gets closer to its true failure time and n_{ens} increases, the ensemble-average of the function $G_{\text{sys}}\left(t;t_p,\mathbf{d}_m^{[0,p]}\right)$ evaluated at $t_{\text{F,true}}$ converges to 0.5 (see Figure 17.6 and Gobbato, 2011), i.e.,

$$\lim_{\substack{n_{\text{ens}} \to \infty \\ t_p \to t_{\text{F,true}}}} E_{\text{ens}}\left[G_{\text{sys}}\left(t_{\text{F,true}};t_p,\mathbf{a}_m^{[0,p]}\right)\right] = 0.5 \qquad (17.27)$$

A practical case study at the local reliability component level, demonstrating the benefits of this type of analysis, is provided in Gobbato et al. (2013b).

17.9 Future trends

There are currently considerable research efforts in the fields of civil, mechanical, naval, and aerospace engineering aimed at improving the integration and field-deployment of SHM and model-based damage prognosis methodologies for RFL predictions and optimal life-cycle cost management (Kim et al., 2013). On the one hand, damage detection and quantification algorithms have reached a good level of maturity and are widely used to assess the current structural integrity and performance level of many monitored systems such as bridges (Wenzel, 2009), tall buildings (Bashor, Bobby, Kijewski-Correa, & Kareem, 2012), composite component and subassemblies in aircraft structures (Lanza di Scalea et al., 2007; Velazquez & Kosmatka, 2010b, 2011), and composite wind turbine and helicopter rotor blades (García Márquez, Tobias, Pinar Pérez, & Papaelias, 2012; Pawar & Ganguli, 2007). On the other hand, being able to efficiently use SHM data to recursively perform model updating, fatigue-driven damage growth predictions, and ultimately compute the time-varying system reliability as damage progresses still requires considerable research efforts. Thus far, verifications and experimental validations of integrated SHM-DP methodologies have been carried out mostly at the local reliability component level with numerically simulated NDE measurement data (Gobbato et al., 2013a, 2013b; Guan et al., 2011, 2012; Rabiei & Modarres, 2013; Vanniamparambil et al., 2012). In these studies, the propagation of a single crack, or debonding front, is considered and the proposed SHM-DP methodology is validated by exploiting experimental datasets collected from coupon-level fatigue tests (e.g., Virkler et al., 1979). To the authors' best knowledge, validation studies at the full system-level—considering multiple damage locations and multiple potential damage mechanisms evolving in time simultaneously—have not yet been performed. These types of validation studies will be part of future research and a concrete example toward this direction is provided by the current efforts of both NASA and the U.S. Air Force Research Laboratory, which are trying to develop a complete computational model of a real aircraft structure (Stargel & Glaessgen, 2012; Tuegel, 2012). Their goal is to be able to update the computational model each time new monitoring data are collected and then use the updated model to forecast future performance levels and maintenance needs of the real structure. Another future research direction, which goes beyond the system-level validation and introduces an additional layer of complexity, will focus on the applicability of integrated SHM-DP methodologies at the global portfolio level, i.e., large civil infrastructure networks, aircraft fleets, wind turbine farms. Finally, being able to significantly reduce the computational time required by these types of SHM-DP algorithms will probably represent one of the most challenging tasks.

17.10 Conclusions, recommendations, and additional sources of information

A reliability-based damage prognosis framework for predicting the time-varying reliability and RFL of adhesively-bonded joints in monitored mechanical/structural systems is presented in this chapter. The methodology relies on periodic NDE inspection results to assess and update the current state of damage of the monitored system and predict the joint probability distribution of the actual debonding extents at future times. This information is then used to estimate the time-varying probability of failure and the corresponding RFL of the monitored system. The proposed framework is formulated to explicitly account for the uncertainties related to: NDE detection capability and measurement accuracy, material model parameters, damage evolution model parameters, and future operational loads (or usage patterns). Furthermore, the SHM-DP framework discussed in this chapter is fairly flexible: it is not restricted to any particular NDE inspection technique, it can assimilate binary (i.e., damage detected or not detected) as well as full-resolution (i.e., damage detected and damage extent measured) inspection results, and, most important, it is not limited to any particular damage mechanism as long as the damage propagation process can be properly modeled and simulated within the predictive module of the framework. All of these desirable attributes allow the user to adopt the framework presented here for a wide variety of practical problems that can be encountered in a real-world scenario.

The success of combined SHM-DP methodologies aimed at providing an optimal structural health management over the entire service life of a given structural, mechanical, aerospace, or automotive system can be quantified through different measures such as total life-cycle cost reduction to the owner, reduction in the number of maintenance-hours per operational hour, reduction in system downtime, extension of system service life, and enhancement of system reliability. With this perspective, the damage prognosis framework proposed in this chapter represents an essential tool to accomplish these objectives through a condition-based and cost-efficient maintenance that uses real-time NDE data collected during the regular operation of the system to prioritize and optimize maintenance resources, i.e., to perform maintenance only upon evidence of need. Further details and additional or complementary sources of information on the topic discussed in this chapter are provided in Inman et al. (2005), Farrar and Lieven (2007), Adams (2007), Boller, Chang, and Fujino (2009), and Farrar and Worden (2010).

References

Adams, D. E. (2007). *Health monitoring of structural materials and components: methods with applications*. Wiley. ISBN-13: 978−0470033135.

Alfano, G., & Crisfield, M. A. (2001). Finite element interface models for the delamination analysis of laminated composites: mechanical and computational issues. *International Journal for Numerical Methods in Engineering, 50*(7), 1701−1736. http://dx.doi.org/10.1002/nme.93.

Bashor, R., Bobby, S., Kijewski-Correa, T., & Kareem, A. (2012). Full-scale performance evaluation of tall buildings under wind. *Journal of Wind Engineering and Industrial Aerodynamics, 104−106*, 88−97. http://dx.doi.org/10.1016/j.jweia.2012.04.007.

Bauchau, O. A., & Loewy, R. G. (1997). *Nonlinear aeroelastic effects in damaged composite aerospace structures*, Technical report. Atlanta, GA, USA: School of Aerospace Engineering, Georgia Institute of Technology.

Berens, A. P. (1989). *NDE reliability analysis. ASM handbook, nondestructive evaluation and quality control* (9th ed.), *17*; (pp. 689−701). Materials Park, OH, USA: ASM International, ISBN 978-0871700230.

Blanco, N., Gamstedt, E. K., Asp, L. E., & Costa, J. (2004). Mixed-mode delamination growth in carbon-fiber composite laminates under cyclic loading. *International Journal of Solids and Structures, 41*(15), 4219−4235. http://dx.doi.org/10.1016/j.ijsolstr.2004.02.040.

Boller, C., Chang, F.-K., & Fujino, Y. (2009). *Encyclopedia of structural health monitoring.* Wiley. http://dx.doi.org/10.1002/9780470061626.

Ching, J., & Chen, Y.-C. (2007). Transitional markov chain Monte Carlo method for Bayesian model updating, model class selection, and model averaging. *Journal of Engineering Mechanics, ASCE, 133*(7), 816−832. http://dx.doi.org/10.1061/(ASCE)0733-9399(2007) 133:7(816).

Degrieck, J., & Paepegem, W. V. (2001). Fatigue damage modeling of fibre-reinforced composite materials: review. *Applied Mechanics Reviews, 54*(4), 279−299. http://dx.doi.org/10.1115/1.1381395.

Deodatis, G., Asada, H., & Ito, S. (1996). Reliability of aircraft structures under non-periodic inspection: a Bayesian approach. *Engineering Fracture Mechanics, 53*(5), 789−805. http://dx.doi.org/10.1016/0013-7944(95)00137-9.

Deodatis, G., Fujimoto, Y., Ito, S., Spencer, J., & Itagaki, H. (1992). Non-periodic inspection by Bayesian method I. *Probabilistic Engineering Mechanics, 7*, 191−204. http://dx.doi.org/ 10.1016/0266-8920(92)90023-B.

Der Kiureghian, A. (2005). Non-ergodicity and PEER's framework formula. *Earthquake Engineering & Structural Dynamics, 34*(13), 1643−1652. http://dx.doi.org/10.1002/eqe.504.

Ditlevsen, O., & Madsen, H. O. (1996). *Structural reliability methods.* West Sussex, England: John Wiley and Sons. ISBN-13: 978−0471960867.

Doebling, S. W., Farrar, C. R., & Prime, M. B. (1998). A summary review of vibration-based damage identification methods. *Shock & Vibration Digest, 30*(2), 91−105.

Farrar, C. R., & Lieven, N. A. J. (2007). Damage prognosis: the future of structural health monitoring. *Philosophical Transactions of the Royal Society A: Mathematical, Physical and Engineering Sciences, 365*(1851), 623−632. http://dx.doi.org/10.1098/ rsta.2006.1927.

Farrar, C. R., & Worden, K. (2010). *Structural health monitoring: a machine learning perspective.* Chichester, UK: John Wiley & Sons, Ltd. ISBN: 978−1119994336.

Farrar, C. R., Worden, K., Lieven, N. A. J., & Park, G. (2010). Nondestructive evaluation of structures. In R. Blockley & W. Shyy (Eds.), *Encyclopedia of aerospace engineering.* John Wiley & Sons, Ltd. ISBN-13: 978−0470754405.

Gamerman, D. (1997). *Markov chain monte carlo: stochastic simulation for bayesian inference.* London, UK: Chapman & Hall. ISBN-13: 978−1584885870.

García Márquez, F. P., Tobias, A. M., Pinar Pérez, J. M., & Papaelias, M. (2012). Condition monitoring of wind turbines: techniques and methods. *Renewable Energy, 46*, 169−178. http://dx.doi.org/10.1016/j.renene.2012.03.003.

Giurgiutiu, V. (2008). *Structural health monitoring with piezoelectric wafer active sensors.* London, UK: Elsevier Inc. ISBN-13: 978−0120887606.

Gobbato, M. (2011). *Reliability-based framework for fatigue damage prognosis of bonded structural elements in aerospace composite structures* (Ph.D. thesis). California, San Diego, USA: Department of structural engineering, University of California.

Gobbato, M., Conte, J. P., & Kosmatka, J. B. (2013a). Remaining fatigue life predictions considering load and model parameters uncertainty. In T. Simmermacher, et al. (Eds.), *Proceeding of the 31st IMAC, a conference on structural dynamics: Vol. 5. Topics in model validation and uncertainty quantification.* http://dx.doi.org/10.1007/978-1-4614-6564-5_2.

Gobbato, M., Conte, J. P., & Kosmatka, J. B. (2013b). Ensemble scenario analysis for the statistical performance evaluation of an NDE-based damage prognosis methodology. In *Proceedings of the 11th international conference on structural safety & reliability*, 16–20 June 2013. New York, NY: Columbia University.

Gobbato, M., Conte, J. P., Kosmatka, J. B., & Farrar, C. R. (2011). *Reliability-based damage prognosis framework for bonded joints in composite unmanned aircrafts*, Structural Systems Research Project, Report No. SSRP-11/03. San Diego, USA: Department of Structural Engineering, University of California.

Gobbato, M., Conte, J. P., Kosmatka, J. B., & Farrar, C. R. (2012). A reliability-based framework for fatigue damage prognosis of composite aircraft structures. *Probabilistic Engineering Mechanics*, 29, 176–188. http://dx.doi.org/10.1016/j.probengmech.2011.11.004.

Gobbato, M., Kosmatka, J. B., & Conte, J. P. (2012). A recursive approach for remaining fatigue life predictions of monitored structural systems. In *Proceedings of the 53rd AIAA/ASME/ASCE/AHS/ASC structures, structural dynamics, and materials conference*, 23–26 April 2012, Honolulu, HI, USA. http://dx.doi.org/10.2514/6.2012-1360.

Guan, X., Jha, R., & Liu, Y. (2011). Model selection, updating, and averaging for probabilistic fatigue damage prognosis. *Structural Safety*, 33(3), 242–249. http://dx.doi.org/10.1016/j.strusafe.2011.03.006.

Guan, X., Jha, R., & Liu, Y. (2012). Probabilistic fatigue damage prognosis using maximum entropy approach. *Journal of Intelligent Manufacturing*, 23(2), 163–171. http://dx.doi.org/10.1007/s10845-009-0341-3.

Heasler, P. G., Taylor, T. T., & Doctor, S. R. (1990). *Statistically based reevaluation of PISC-II round robin test data*, Report NUREG/CR-5410. Washington, DC: U.S. Nuclear Regulatory Commission.

Hoblit, F. M. (1988). *Gust loads on aircraft: concepts and applications*. Washington, DC: AIAA Ed. Series. ISBN-13: 978–0930403454.

Inman, D. J., Farrar, C. R., Lopez, V., Jr., & Steffen, V., Jr. (2005). *Damage prognosis for aerospace, civil and mechanical systems*. West Sussex, England: John Wiley and Sons. ISBN-13: 978–0470869079.

Ito, S., Deodatis, G., Fujimoto, Y., Asada, H., & Shinozuka, M. (1992). Non-periodic inspection by Bayesian method II: structures with elements subjected to different stress levels. *Probabilistic Engineering Mechanics*, 7, 205–215. http://dx.doi.org/10.1016/0266-8920(92)90024-C.

Kim, S., Frangopol, D., & Soliman, M. (2013). Generalized probabilistic framework for optimum inspection and maintenance planning. *Journal of Structural Engineering, ASCE*, 139(3), 435–447. http://dx.doi.org/10.1061/(ASCE)ST.1943-541X.0000676.

Kosmatka, J. B., & Velazquez, E. (2008). Acoustic emission monitoring of interlaminar microcracking and strength. In *Proceedings of the 49th AIAA/ASME/ASCE/AHS/ASC structures, structural dynamics, and materials conference*, 7–10 April 2008, Schaumburg, IL, USA. http://dx.doi.org/10.2514/6.2008-2015.

Kotulski, Z. A. (1998). On efficiency of identification of a stochastic crack propagation model based on Virkler experimental data. *Archives of Mechanics*, 50(5), 829–847.

Kulkarni, S. S., & Achenbach, J. D. (2008). Structural health monitoring and damage prognosis in fatigue. *Structural Health Monitoring*, 7(1), 37–49. http://dx.doi.org/10.1177/1475921707081973.

Lanza di Scalea, F., Matt, H. M., Bartoli, I., Coccia, S., Park, G., & Farrar, C. R. (2007). Health monitoring of UAV skin-to-spar joints using guided waves and macro fiber composite transducers. *Journal of Intelligent Material Systems and Structures, 18*(4), 373−388. http://dx.doi.org/10.1177/1045389X06066528.

Lee, B. H. K., Prince, S. J., & Wong, Y. S. (1999). Nonlinear aeroelastic analysis of airfoils: bifurcation and chaos. *Progress in Aerospace Sciences, 35,* 205−334. http://dx.doi.org/10.1016/S0376-0421(98)00015-3.

Librescu, L., Chiocchia, G., & Marzocca, P. (2003). Implications of cubic physical aerodynamic non-linearities on the character of the flutter instability boundary. *International Journal of Non-linear Mechanics, 38,* 173−199. http://dx.doi.org/10.1016/S0020-7462(01)00054-3.

Lin, K. Y., Du, J., & Rusk, D. (2000). 'Structural design methodology based on concepts of uncertainty', NASA/CR-2000−209847. Hampton, VA, USA: NASA Langley Research Center.

Ling, Y., & Mahadevan, S. (2012). Integration of structural health monitoring and fatigue damage prognosis. *Mechanical Systems and Signal Processing, 28,* 89−104. http://dx.doi.org/10.1016/j.ymssp.2011.10.001.

Nguyen, O., Repetto, E. A., Ortiz, M., & Radovitzky, R. A. (2001). A cohesive model of fatigue crack growth. *International Journal of Fracture, 110*(4), 351−369. http://dx.doi.org/10.1016/j.ijsolstr.2009.01.031.

Ostergaard, D. F., & Hillberry, B. M. (1983). Characterization of the variability in fatigue crack propagation data. In J. M. Bloom, & J. C. Ekvall (Eds.), *Probabilistic fracture mechanics and fatigue methods: applications for structural design and maintenance,* ASTM STP 798 (pp. 97−115). http://dx.doi.org/10.1520/STP33214S.

Paris, P. C., & Erdogan, F. A. (1963). A critical analysis of crack propagation laws. *Journal of Basic Engineering, Transactions of the American Society of Mechanical Engineers, 85*(Series D), 528−534. http://dx.doi.org/10.1115/1.3656900.

Pawar, P., & Ganguli, R. (2007). Fuzzy-logic-based health monitoring and residual-life prediction for composite helicopter rotor. *Journal of Aircraft, 44*(3), 981−995. http://dx.doi.org/10.2514/1.26495.

Quaresimin, M., & Ricotta, M. (2006). Life prediction of bonded joints in composite materials. *International Journal of Fatigue, 28,* 1166−1176. http://dx.doi.org/10.1016/j.ijfatigue.2006.02.005.

Rabiei, M., & Modarres, M. (2013). A recursive Bayesian framework for structural health management using online monitoring and periodic inspections. *Reliability Engineering and System Safety, 112,* 154−164. http://dx.doi.org/10.1016/j.ress.2012.11.020.

Shull, P. (2002). *Nondestructive evaluation: Theory, techniques, and applications.* New York, NY: Marcel Dekker, Inc. ISBN-13: 978−0824788728.

Simola, K., & Pulkkinen, U. (1998). Models for non-destructive inspection data. *Reliability Engineering and System Safety, 60*(1), 1−12. http://dx.doi.org/10.1016/S0951-8320(97)00087-2.

Staat, M. (1993). Sensitivity of and influences on the reliability of an HTR-module primary circuit pressure boundary. In *12th international conference on structural mechanics in reactor technology (SMiRT),* 15−20 August 1993, Amsterdam, The Netherlands.

Stargel, D. S., & Glaessgen, E. H. (2012). Digital twin paradigm for future NASA and U.S. Air force vehicles. In *Proceedings of the 53rd AIAA/ASME/ASCE/AHS/ASC structures, structural dynamics, and materials conference,* 23−26 April 2012, Honolulu, HI, USA.

Styuart, A. V., Demasi, L., Livne, E., & Lin, K. W. (2008). Probabilistic modeling of structural/aeroelastic life cycle for reliability evaluation of damage tolerant composite structures. In *Proceedings of the 59th AIAA/ASME/ASCE/AHS/ASC structures, structural dynamics, and materials conference*, 7−10 April 2008, Schaumburg, IL, USA.

Styuart, A. V., Demasi, L., Livne, E., & Mor, M. (2011). Flutter failure risk assessment for damage-tolerant composite aircraft structures. *AIAA Journal*, *49*, 655−669. http://dx.doi.org/10.2514/1.J050862.

Styuart, A. V., Mor, M., Livne, E., & Lin, K. Y. (2007). Aeroelastic failure risk assessment in damage tolerant composite airframe structures. In *Proceedings of the 48th AIAA/ASME/ASCE/AHS/ASC structures, structural dynamics, and materials conference*, 23−26 April 2007, Honolulu, HI, USA. http://dx.doi.org/10.2514/6.2007-1981.

Tuegel, E. J. (2012). The airframe digital twin: some challenges to realization. In *Proceedings of the 53rd AIAA/ASME/ASCE/AHS/ASC structures, structural dynamics, and materials conference*, 23−26 April 2012, Honolulu, HI, USA. http://dx.doi.org/10.2514/6.2012-1812.

Van Staveren, W.H. (2003). *Analyses of aircraft responses to atmospheric turbulence* (Ph.D. thesis). Delft, The Netherlands: Department of Aerospace Design, Integration & operations, Delft University of Technology.

Vanniamparambil, P. A., Bartoli, I., Hazeli, K., Cuadra, J., Schwartz, E., Saralaya, R., et al. (2012). An integrated structural health monitoring approach for crack growth monitoring. *Journal of Intelligent Material Systems and Structures*, *23*(14), 1563−1573. http://dx.doi.org/10.1177/1045389X12447987.

Velazquez, E., & Kosmatka, J. B. (2010a). Acoustic emission-based structural health monitoring of interlaminar matrix-driven damage in advanced composite structures. In *Proceedings of the 51th AIAA/ASME/ASCE/AHS/ASC structures, structural dynamics, and materials conference*, 12−15 April 2010, Orlando, FL, USA. http://dx.doi.org/10.2514/6.2010-3029.

Velazquez, E., & Kosmatka, J. B. (2010b). Detecting impending bond joint failures in unmanned aircraft composite wing structures. In *Proceedings of the 55th international SAMPE Symposium and Exhibition*, 17−20 May 2010, Seattle, WA, USA.

Velazquez, E., & Kosmatka, J. B. (2011). Detecting shear and tension bond failures in composite aircraft structures. In *Proceedings of the 52th AIAA/ASME/ASCE/AHS/ASC structures, structural dynamics, and materials conference*, 4−7 April 2011, Denver, CO, USA. http://dx.doi.org/10.2514/6.2011-1935.

Velazquez, E., & Kosmatka, J. B. (2012). Acoustic emission structural health monitoring of laminated composite aircraft structures. In *Proceedings of the 53rd AIAA/ASME/ASCE/AHS/ASC structures, structural dynamics, and materials conference*, 23−26 April 2012, Honolulu, HI, USA. http://dx.doi.org/10.2514/6.2012-1788.

Virkler, D. A., Hillberry, B. M., & Goel, P. K. (1979). The statistical nature of fatigue crack propagation. *Transactions of ASME − Journal of Engineering Materials and Technology*, *101*(2), 148−153. http://dx.doi.org/10.1115/1.3443666.

Wang, K., Inman, D. J., & Farrar, C. R. (2005). Crack-induced changes in divergence and flutter of cantilevered composite panels. *Structural Health Monitoring*, *4*(4), 377−392. http://dx.doi.org/10.1177/1475921705057977.

Wen, Y.-K. (1990). *Structural load modeling and combination for performance and safety evaluation*. Elsevier. ISBN-13: 978−0444881489.

Wenzel, H. (2009). *Health monitoring of bridges*. Wiley. ISBN-13: 978−0470031735.

Zhang. Y. (2010). *Fracture and fatigue of adhesively-bonded fiber-reinforced polymer structural joints* (Ph.D. thesis). Lausanne, Switzerland: School of Architecture, Civil and Environmental Engineering, École Polytechnique Fédérale de Lausanne (EPFL).

Zhang, R., & Mahadevan, S. (2001). Fatigue reliability analysis using non-destructive inspection. *Journal of Structural Engineering, ASCE, 127*(8), 957–965. http://dx.doi.org/10.1061/(ASCE)0733-9445(2001)127:8(957).

Zheng, R., & Ellingwood, B. R. (1998). Role of non-destructive evaluation in time-dependent reliability analysis. *Structural Safety, 20*(4), 325–339. http://dx.doi.org/10.1016/S0167-4730(98)00021-6.

Zou, Y., Tong, L., & Steven, G. P. (2000). Vibration-based model-dependent damage (delamination) identification and health monitoring for composite structures — a review. *Journal of Sound and Vibration, 230*(2), 357–378. http://dx.doi.org/10.1006/jsvi.1999.2624.

Index

Note: Page numbers followed by f and t indicate figures and tables respectively.

A
ADCB. *See* Asymmetric double cantilever beam
Adherend
  properties, 50–55, 50f
  shaping, 61–63, 61f
Adherend failure. *See* Substrate failure
Adhesive
  bonding, 94–95
  failure, 301
  joints
    with functionally graded materials, 65–66
    interlaminar failure, 51f
    stick-slip effects on, 126–128
  properties, 48–49
    ductility effect, 49f
    modulus effect, 49f
Adhesively-bonded composite joints, 43, 187
  aluminium-composite joints, 53f
  BL fatigue behavior
    experimental program, 260–269
    experimental results, 269–283
    materials and specimens, 260
  bonded joint applications in F/A-18, 290–291
  closed-form solutions, 44
  design philosophy, 44–47
  durability testing, 22–24
  experimental work, 4–5
  fatigue behavior characterization, 188–190
  fatigue testing, 7–13
    environmental effects determination, 22–24
    modeling behavior, 24–26
    pultruded GFRP profiles, 14–22
    simulation behavior, 24–26
  fleet service, 302–315
  fracture testing
    environmental effects determination, 22–24
    modeling behavior, 24–26
    pultruded GFRP profiles, 14–22
    simulation behavior, 24–26
  hybrid joints, 63–66
  inner-wing full-scale fatigue test, 315–318
  joint designs, 6–7
  joint strength
    factors affecting, 48–59
    methods for increasing, 59–63
  joint types, 6–7
  loading sequence effect, 257–258
  manufacturing and processing, 5–6
  mixed-Mode I/II fracture behavior, 187
  mode partitioning, 188
  objectives and current state of standardization, 3–4
  repair techniques, 66–67, 67f
  software packages, 45t–47t
  stepped-lap joints stress analysis, 291–293
  stress analysis of lap joints, 43–44
  VA fatigue behavior
    experimental program, 260–269
    experimental results, 269–283
    materials and specimens, 260
  WR-SLJ, 293–302
Adhesively-bonded composite structural joints, 226
  CLD, 446–448
  constant amplitude fatigue data, 236t–237t
  constant life diagram formulations, 467–478
  experimental investigation
    experimental setup, 229–230
    material, 227–228
    specimen fabrication, 228–229

Adhesively-bonded composite structural
    joints (Continued)
  specimen geometry, 228–229
  fatigue life, 232–238, 233t–235t
    modeling, 443–446
  fatigue loading, 231–232
  fracture mechanics measurements analysis,
    243–252
  FRP joints behavior, 225
  GFRP, 225–226
  influence of loading parameters, 446
  quasi-static investigation,
    230–231, 230t
  S–N formulations, 448–456
  stiffness degradation, 238–243
American Society for Testing and Materials
    (ASTM), 4
Area method, 405
Asymmetric double cantilever beam
    (ADCB), 11–12, 345
Asymmetric specimens, 188
Asymmetric tapered double cantilever beam
    (ATDCB), 11–12
Asymmetry effect, 345–346
  adherend and adhesive materials,
    348–349
  experimental setup and procedure,
    349–350
  failure modes, 350–351
  fracture data analysis, 351–353
ATDCB. See Asymmetric tapered double
    cantilever beam

B
Basic modes, 5
Bayesian inference analysis, 495–497
Block loading fatigue behavior (BL fatigue
    behavior), 257. See also Variable
    amplitude loading (VA loading)
  BL sequence, 81, 83
  experimental program, 260–269
  failure modes, 269
  load transition effect, 279
  loading sequence effect, 271–279
  materials and specimens, 260
  multi-block, 267t–268t
  two-stage, 263t
Boerstra's constant life diagram (BR CLD),
    471–473, 480

Bolted-bonded joints, 66
Bonded joint applications, 290–291
Bonded stepped-lap joints, 292

C
CA. See Constant amplitude
Carbon fiber-reinforced laminates, 461–463
Carbon-fiber-reinforced polymer (CFRP),
    3–4, 58–59, 93–94, 121–122,
    152–153, 419, 469–470
  adhesive bonding, 94–95
  bonded joints, 96
  fatigue characterization
    by FCG approach, 104–113
    by S–N approach, 102–104, 103f
  mixed-mode characterization, 97–98
  mode I characterization, 96–97
  fatigue behaviour, 95–96
  fracture modes, 113–116
  preparation and testing, 98–102
  mode II characterization, 97
CBT. See Corrected beam theory
C–C fatigue. See Compression–compression
    fatigue
CCfatigue software, 481
CCM. See Compliance calibration method
CCP. See Central cut ply
CDF. See Cumulative distribution function
CDFS. See Compression-dominant fatigue
    spectrum
CDM. See Continuum damage mechanics
Central cut ply (CCP), 97
CFL. See Constant fatigue life
CFRP. See Carbon-fiber-reinforced polymer
Chopped strand mat (CSM), 75–76, 191
Civil engineering, 23–24
CLD. See Constant life diagram
Coefficients of thermal expansion (CTE),
    58–59
Cohesive elements, 355–358
Cohesive failure, 298
Cohesive zone models (CZM), 9, 43–44,
    335, 376–384, 508–509
  cohesive fatigue model, 338–339
  cohesive traction–separation law,
    335–338
  comparison, 391–397
  for cyclic delamination, 421
  Maiti and Guebelle model, 423–424

mixed-mode cohesive area, 430–431
Munoz, Galvanetto, and Robinson
    model, 425
Paris fatigue law, 426–431
Roe and Siegmund model, 422–423
Serebrinsky and Ortiz model, 424–425
Yang, Mall, and Ravi-Chandar model,
    422
fatigue crack growth problems, 373–376
finite element implementation, 381–382
mixed-mode loading, 382–384
strain energy release rate computation,
    380–381
Compliance calibration, 138–139
Compliance calibration method (CCM),
    107–108, 408
Compliance method, 405
Compliance-based beam method, 407–408
Compression-dominant fatigue spectrum
    (CDFS), 296–297
Compression–compression fatigue (C–C
    fatigue), 77, 231–232, 446. *See also*
    Tension–compression fatigue
    (T–C fatigue)
Constant amplitude (CA), 154, 259
Constant fatigue life (CFL), 470
Constant life diagram (CLD), 78–79, 227,
    235, 446–448
    Boerstra's, 471–473
    formulations, 467–478
    Harris, 469–470
    Kassapoglou's, 473
    Kawai's, 470–471
    linear, 468
    piecewise linear, 468–469
    piecewise nonlinear, 474–476
    polynomial, 477–478
Continuum damage mechanics (CDM), 323
Corrected beam theory (CBT), 107–108
Crack growth laws, integration of, 333–334
Crack growth rate calculation, 168
Crack growth rate curves, waviness effect in,
    139–141
Crack length, 194, 414
    measurements
    compliance and, 161–166
    load and, 166–167
    number of cycles *vs.*, 199–201
Crack propagation modelling, 327–328

Creep–fatigue interaction, 79–80
CSM. *See* Chopped strand mat
CTE. *See* Coefficients of thermal expansion
Cumulative distribution function (CDF),
    499
"Cycle mix" effect, 82
Cyclic loading, 419
Cyclic mixed-mode delamination modeling
    cyclic MMB experiments, 434–439
    one element tests, 431
CZM. *See* Cohesive zone models

D

Damage prognosis analysis, 513–516
DBT. *See* Direct Beam Theory
DCB. *See* Double cantilever beam
Debonding evolution analysis, 498
DIC. *See* Digital image correlation
Digital image correlation (DIC), 12
Direct Beam Theory (DBT), 107–108
Direct cyclic analysis, 386–388
Displacement method, 406
DLJ. *See* Double-lap joints
DMA. *See* Dynamic mechanical analysis
Double cantilever beam (DCB), 10, 96–97,
    96f, 124–125, 149–150, 152–153,
    187, 226, 323, 345, 373, 404
    of bonded joint, 125f
    load–displacement curves, 126f
    specimen fabrication, 157
    specimen geometry, 157
    test procedure, 132–135, 133f
Double-lap joints (DLJ), 14–17, 226–227,
    229f. *See also* Single-lap joints (SLJ)
    failure mode, 231f
    geometry, 260f
    two crack gauges, 265f
Ductile adhesive, 48–49
Durability testing, 22–24
Dynamic compliance, 136–138, 138f
Dynamic mechanical analysis (DMA), 22

E

ECM. *See* Experimental compliance method
Edge-crack torsion (ECT), 10
EGM. *See* Extended global method
Elastic foundation method, 406–407
ELS. *See* End-loaded split
ENBC. *See* End-notched cantilever beam

End-notched flexure (ENF), 10, 97, 149–150, 187, 226, 508–509
End-loaded split (ELS), 10, 97, 226
End-notched cantilever beam (ENBC), 97
Energy release rate (ERR), 132
ENF. *See* End-notched flexure
Ensemble scenario analysis, 500, 519
Epoxy (EP), 6
Equivalent crack length approach, 11–12
ERR. *See* Energy release rate
Evolutionary structural optimisation method (EVOLVE), 63
Existing fatigue models comparison, 456. *See also* Adhesively-bonded structural composite joints
  carbon fiber-reinforced laminates, 461–463
  formulations, 478–486
  glass fiber-reinforced laminates, 457–461
  hybrid glass–carbon fiber-reinforced laminates, 463
  pultruded fiber-reinforced adhesively-bonded joints, 464–465
  short fiber-reinforced laminates, 463–464
Experimental characterization, 150–153
Experimental compliance method (ECM), 201, 353
Experimental fatigue procedure, 194–197
Extended finite element method (XFEM), 324, 328
Extended global method (EGM), 190, 201–204
Extended global model, 21

F

F/A-18 wing root stepped-lap joint, 292
  bonded joint applications, 290–291
  fleet service, 302
    environmental effects, 313–315
    fatigue data, 303
    fatigue test matrix, 303
    fatigue test spectrum, 302–303, 304t–305t
    increasing fatigue damage, 309–312
    inspections, 312–313
    residual strength after, 297–298
    inner-wing full-scale fatigue test, 315–318
  stepped-lap joints stress analysis, 291–293
  WR-SLJ, 293–302

Failure modes, 129
  analysis, 231–232
False-call probability (FCP), 501
Fatigue. *See also* Fracture mechanics (FM)
  behavior characterization, 153–154
  characterization
    of bonded joints, 402
    by FCG approach, 104–113
  FCG approach, 402–403
    by S–N approach, 102–104, 103f
    stress-life approach, 402
  control mode, 411–412
  experiments, 229–230
  failure, 402
  interpretation
    compliance, 161–166, 162t–163t
    crack length measurements, 161–167
    crack tip, 164f
    failure mode analysis, 158–161
    load, 166–167
  life, 232–238, 233t–235t
  modeling, 443–446
  loading, 231–232
    fatigue data, 75
    parameters, 76–85
    tensile fatigue *vs.* compressive fatigue, 75–76
Fatigue crack growth technique (FCG technique), 95–96, 150, 152f, 188–189, 369, 403
  curves, 75, 169–173, 179–181
  determination, 409–410
  fatigue characterization by, 104–106
  da/dN estimation, 110–111
  G-parameter estimation in mode I, 107–109
  hypothetic plot, 105f
  testing results, 111–113
  fatigue control mode, 411–412
  integration of, 410–411
Fatigue testing, 131. *See also* Fracture testing
  adhesively-bonded joints, 8–9
  compliance calibration, 138–139
  crack growth, 135f
  dynamic compliance, 136–138, 138f
  G–N crack growth onset curve, 134f
  from pultruded GFRP profiles, 14–17

Index

standards and test protocols for
 environmental effects determination,
 22–24
modeling behavior, 24–26
simulation behavior, 24–26
test procedure, 132–135, 133f
test set-up, 136
FCG technique. *See* Fatigue crack growth
 technique
FCP. *See* False-call probability
FE. *See* Finite element
FEA. *See* Finite element analysis
FEM. *See* Finite element modeling
FH specimen. *See* Filled hole specimen
Fiber bridging effect, 346–347, 360–361
 adherend and adhesive materials, 348–349
 experimental setup and procedure,
 349–350
 failure modes, 350–351
 fracture data analysis, 351–353
Fiber-breaking, 9–10
Fiber-reinforced polymer (FRP), 3, 73, 82,
 225, 258, 443
Fibre-reinforced polymer composite joints,
 121
 bonded joint configuration, 122–124
 bonded repairs application, 121
 design approach, 141–143
 failure modes, 129
 fatigue testing, 124–126, 131–139
 fractography, 129
 fracture surface analysis, 130f
 simulation approach, 141–143
 stick-slip effects, 126–128
 waviness effect, 139–141
Filled hole specimen (FH specimen), 294
Fillets, 59–61
Finite element (FE), 98, 188
 implementation, 381–382
Finite element analysis (FEA), 9, 323,
 401–402, 412
 mode I strain energy release rate, 412–413
 modelling DCB, 412
 numerical integration of crack growth law,
 414
 validation, 415
Finite element modeling (FEM), 43–44,
 204–207, 204t, 354–363, 388–390
 cohesive elements, 355–358

crack propagation modelling, 327–328
DCB joint, 324–325
material models, 325–326
mesh sensitivity, 359–360
VCCT, 358–359
Fleet service
 environmental effects, 313–315
 fatigue data, 303
 fatigue test
  matrix, 303
  spectrum, 302–303, 304t–305t
 increasing fatigue damage, 309–312
 inspections, 312–313
 residual strength after, 297–298
FM. *See* Fracture mechanics
Four point bending end-notched flexure test
 (4-ENF test), 10
Four-point end-notched flexure (4ENF), 97
Fourier transform infrared (FTIR), 22
FPZ. *See* Fracture process zone
Fractography, 129
Fracture behavior characterization, 153–154
Fracture interpretation, 158–167
 compliance, 161–166, 162t–163t
 crack length measurements, 161–167
 crack tip, 164f
 failure modes analysis, 158–161
 load, 166–167
Fracture mechanics (FM), 323, 403
 crack growth laws, 329–330
 data analysis
  crack growth rate calculation, 168
  FCG curves, 169–173
  SERR calculation, 169
 FE meshing, 330–331
 fracture parameters extraction, 331–333
 measurements analysis, 243–252
 modeling
  FCG curves prediction, 179–181
  total life fatigue model, 173–179, 175t,
  180t
 theory, 150
Fracture process zone (FPZ), 109, 142
Fracture testing
 adhesively-bonded joints, 9–13
 from pultruded GFRP profiles, 18–22
 standards and test protocols for
 environmental effects determination,
 22–24

Fracture testing (*Continued*)
   modeling behavior, 24–26
   simulation behavior, 24–26
FRP. *See* Fiber-reinforced polymer
FTIR. *See* Fourier transform infrared

G
G-parameter, 107–109
Glass fiber-reinforced laminates, 457–461
Glass-fiber-reinforced polymer (GFRP), 3–4, 75, 154, 225–226, 445
   adhesively-bonded
      GFRP joint, 76f
      pultruded GFRP joints, 155–158
   load interaction effect, 83–84
   pultruded GFRP profiles
      adhesively-bonded joints fatigue testing, 14–17, 15t–16t
      adhesively-bonded joints fracture testing, 18–22, 19t–20t
Global method, 187–188

H
Harris constant life diagram (HR CLD), 469–470, 481
High loading rate (HLR), 229
High-cycle fatigue (HCF), 443
Hybrid glass–carbon fiber-reinforced laminates, 463
Hybrid joints, 63–66, 64f. *See also* Lap joints
Hybrid materials, 54–55

I
Incremental polynomial method, 280
Inner mold line (IML), 309–310
Inner-wing full-scale fatigue test, 315–318. *See also* Wing root stepped-lap joint (WR-SLJ)
International Organization for Standardization (ISO), 4
Irwin–Kies method, 405, 408

J
J-integral, 332–333
Joint strength, 48
   factors affecting, 48
      adherend properties, 50–55
      adhesive properties, 48–49
      adhesive thickness, 55

   overlap, 56–58
   residual stresses, 58–59
   methods to increasing
      adherend shaping, 61–63
      fillets, 59–61

K
Kassapoglou's constant life diagram (Kassapoglou's CLD), 473
Kawai's constant life diagram (KW CLD), 470–471, 480
Kenane and Benzeggagh theory (KB theory), 384

L
Lap joints, 43–44
   double-lap joints, 62f, 63–65
   single-lap joints, 55, 59f, 63–65
LCF. *See* Low-cycle fatigue
LCO. *See* Limit cycle oscillations
LEFM. *See* Linear elastic fracture mechanics
LEFM-based techniques, 142
Limit cycle oscillations (LCO), 511–512
Linear constant life diagram model (LR CLD model), 468, 480
Linear elastic fracture mechanics (LEFM), 107–108, 126, 136, 405
Linear phase, 104–106
LLR. *See* Low loading rate
LLS. *See* Local limit state
Load hazard analysis, 498
Load method, 405–406
Load ratio, 340
Load severity factor (LSF), 302–303
Load transition effect, 279
Load-displacement responses, 197–198
Loading sequence effect, 80–83
Local limit state (LLS), 514–515
Local method, 187–188
Low loading rate (LLR), 229
Low-cycle fatigue (LCF), 443
LR CLD model. *See* Linear constant life diagram model
LSF. *See* Load severity factor

M
Maiti and Guebelle model, 423–424
Markov Chain Monte Carlo methods (MCMC methods), 499–500

MBT. *See* Modified beam theory
MC simulations. *See* Monte Carlo simulations
MCC. *See* Modified calibration method
MCMC methods. *See* Markov Chain Monte Carlo methods
Mean load effect, 77–79
Mean stress effect. *See* R-ratio — effect
Mesh sensitivity, 359–360
Microscopy, 129
Mixed adhesive joints, 63–65
Mixed CZM and FM approach, 340–342
Mixed-mode bending (MMB), 9–10, 187, 195t, 420
 asymmetric specimens, 188
 fatigue results, 200t
 specimen compliance, 201
Mixed-mode cohesive area, 430–431
Mixed-mode fatigue and fracture behavior, 97–98, 187
 experimental investigation
  experimental fatigue procedure, 194–197
  material description, 191
  specimen description, set-up, and procedure, 192–194
 experimental results
  load and crack length *vs.* number of cycles, 199–201
  load-displacement responses, 197–198
  observed failure modes, 197
 fatigue and fracture data analysis
  ECM, 201
  EGM, 201–204
  FCG curves, 207–208
  FEM, 204–207
 mode partitioning, 187–188
  in asymmetric specimens, 188
  fatigue behavior characterizations, 188–190
  techniques for characterization, 190
 results and discussion, 209–215
Mixed-mode fatigue delamination
 of composites, 419
 diverse approaches, 420–421
 novelties of, 420
Mixed-mode loading, 382–384
Mixed-mode end-loaded split (MMELS), 373
MMB. *See* Mixed-mode bending

MMELS. *See* Mixed-mode end-loaded split
Mode I fatigue behaviour characterization, 96–97
 CFRP joints
  fracture modes, 113–116
  preparation and testing, 98–102
 G-parameter in, 107–109
Mode I fracture of adhesively-bonded joints, 12
Mode I quasi-static fracture, 18
Mode I strain energy release rate calculation, 404–408
Mode II fatigue behaviour, 97
Mode II quasi-static fracture, 18
Modified beam theory (MBT), 126
Modified calibration method (MCC), 111–113
Monte Carlo simulations (MC simulations), 498
Munoz, Galvanetto, and Robinson model, 425

N
N-SIF. *See* Notch-stress intensity factor
NDE techniques. *See* Non-destructive evaluation techniques
NDI. *See* Non-destructive inspection
NH specimen. *See* No hole specimen
NL. *See* Nonlinear
No hole specimen (NH specimen), 294
No-growth design strategy, 141–142
Node release, 327
Non-destructive inspection (NDI), 121, 293
Non-destructive evaluation techniques (NDE techniques), 493–494
Non-linear (NL), 111–113
Notch-stress intensity factor (N-SIF), 104–106
Number of transitions (NT), 271
Numerous methods, 153

O
Observed failure modes, 197
One element tests, 431
One-Side Tacky (OST), 415
Onset curve, 134
Open hole specimen (OH specimen), 294
Outer mold line (OML), 309–310
Overlap, 56–58

P

Paris fatigue law, 426–431
PCs. *See* Personal computers
PDF. *See* Probability distribution function
Piecewise linear model (PWL model), 468–469, 480
Piecewise nonlinear CLD model (PNL model), 474–476, 480
PLANE182 element, 354
PND. *See* Probability of nondetection
PNL model. *See* Piecewise nonlinear CLD model
POD. *See* Probability of detection
Polynomial constant life diagram (POLY CLD), 477–478, 480
Polynomial method, 408–409
Polyurethane (PUR), 6
Probabilistic load hazard analysis, 505–507
Probabilistic mechanics-based debonding evolution analysis, 507–511
Probability distribution function (PDF), 493–494
Probability of detection (POD), 501
Probability of nondetection (PND), 501
Pultruded fiber-reinforced adhesively-bonded joints, 464–465
Pultruded GFRP joints. *See* Pultruded glass fiber-reinforced polymer composite joints
Pultruded glass fiber-reinforced polymer composite joints (Pultruded GFRP joints), 155
  experimental investigation
  experimental set-up, 158
  fiber architecture, 156f, 156t
  loading, 158
  material, 155–156
  specimen fabrication, 157
  specimen geometry, 157
  fatigue interpretation, 158–167
  fracture interpretation, 158–167
  load interaction effect, 83–84
  mean load effects, 78
Pultrusion manufacturing, 6
PUR. *See* Polyurethane
PWL model. *See* Piecewise linear model

Q

Quasi-static investigation, 230–231, 230t

R

R-curve. *See* Resistance curve
R-ratio, 132, 467
  effect, 173–179, 226–227, 232–238
RBIM. *See* Reliability-based inspection and maintenance
Re-meshing, 327
Recursive Bayesian characterization, 500–505
Reliability-based inspection and maintenance (RBIM), 493–494
Remaining fatigue life (RFL), 493–494, 516–519
Repair techniques, 66–67, 67f
Residual stress, 58–59
Resistance curve (R-curve), 346–347
RFL. *See* Remaining fatigue life
Risk-based inspection and maintenance (RBIM). *See* Reliability-based inspection and maintenance (RBIM)
Rivet-bonded joints, 66
RMS. *See* Root mean square
Roe and Siegmund model, 422–423
Root mean square (RMS), 506
Roving bridging, 197

S

SBT. *See* Simple beam theory
Scanning electron microscopy (SEM), 113–115, 129
Secant method, 409–410
SEM. *See* Scanning electron microscopy
Serebrinsky and Ortiz model, 424–425
SERR. *See* Strain energy release rate
SFH. *See* Spectrum fatigue hours
SHM-DP. *See* Structural health monitoring and damage prognosis
Short fiber-reinforced laminates, 463–464
Simple beam theory (SBT), 128
Simulation of fatigue delamination, 369
  CZM, 373–384
  VCCT, 370–373
Single leg bending (SLB), 11–12
Single-lap joints (SLJ), 226, 395–396.
  See also Double-lap joints (DLJ)

Index

SLB. *See* Single leg bending
SLERA. *See* Strength Life Equal Rank Assumption
SLJ. *See* Single-lap joints
S–N formulations, 102–104, 103f, 448–456, 465–466
   disadvantages, 453–456
   hybrid, 451–453
S–N curves, 142, 188–189
Specimen
   asymmetric, 188
   compliance, 201
   fabrication, 98–99, 228–229
   FH, 294
   geometry, 228–229
   NH, 294
   OH, 294
   pristine, 296–297
   testing, 100–102
   for VA fatigue behavior, 260
Spectrum fatigue hours (SFH), 302–303
Stable growth, 104–106
Standard test methods for experimental fatigue
   adhesively-bonded composite joints, 7–13
   pultruded GFRP profiles, 14–22
Stepped-lap joints stress analysis, 291–293. *See also* Wing root stepped-lap joint (WR-SLJ)
Stick-slip effects, 126–128
Stiffness degradation, 238–243
Stiffness-based curves, 241–242
Stiffness-controlled curves, 241–242
Strain energy release rate (SERR), 370
   calculation, 169
   computation, 380–381
   evaluation for VCCT, 370–373, 412–413
Strength Life Equal Rank Assumption (SLERA), 445–446
Stress-life approach, 402
Structural health monitoring and damage prognosis (SHM-DP), 493–494
   damage prognosis analysis, 513–516
   probabilistic load hazard analysis, 505–507
   probabilistic mechanics-based debonding evolution analysis, 507–511
   recursive Bayesian characterization, 500–505
   reliability-based, 495–500
   RFL, 516–519

Substrate failure, 301
Symmetric adhesively-bonded double-lap joints, 260
System performance analysis, 498–499, 511–513

T

Tapered double cantilever beam (TDCB), 8, 96–97, 323
T–C fatigue. *See* Tension–compression fatigue
TDCB. *See* Tapered double cantilever beam
TDFS. *See* Tension-dominant fatigue spectrum
Temperature-dependent fatigue testing, 23
Tension-dominant fatigue spectrum (TDFS), 296–297
Tension–compression fatigue (T–C fatigue), 77, 78f, 231–232, 446
Tension–Tension sector (T–T sector), 467
Test protocols
   environmental effects determination, 22–24
   experimental fatigue, 7–13
   fatigue behavior
      modeling, 24–26
      simulation, 24–26
   fracture behavior
      modeling, 24–26
      simulation, 24–26
   fracture testing, 7–13
   pultruded GFRP profiles, 14–22
Total fatigue life model, 154, 189
Total life fatigue model, 173–179, 175t, 180t
Total probability theorem (TPT), 494
Traction–separation
   cohesive law model, 355–357
   cohesive model, 206t
Transition effect, 258
T–T sector. *See* Tension–Tension sector

U

UAV. *See* Unmanned aerial vehicle
UCL. *See* Ultimate compressive load
UCS. *See* Ultimate compressive stress
UEL. *See* Userdefined element
Ultimate compressive load (UCL), 230–231
Ultimate compressive stress (UCS), 447, 468
Ultimate tensile load (UTL), 230–231

Ultimate tensile stress (UTS), 447, 468
Ultraviolet radiation (UV radiation), 22
Understanding fatigue loading conditions.
   See also Adhesively-bonded
      composite joints
   composite bridge deck, 74f
   fatigue data, 75
   FRP composites, 73
   parameters, 76–85
   Pontresina composite bridge, 73f
   tensile fatigue *vs.* compressive fatigue,
      75–76
   WISPERX spectrum, 85f
Unmanned aerial vehicle (UAV), 506
User-defined field subroutine (USDFLD),
   376
Userdefined element (UEL), 375–376
UTL. *See* Ultimate tensile load
UTS. *See* Ultimate tensile stress
UV radiation. *See* Ultraviolet radiation

V
Variable amplitude loading (VA loading),
   83–85, 258–259, 266
   crack propagation rate calculation, 280
   fatigue behavior
      crack propagation rate, 281, 282f
      experimental program, 260–269
      failure modes, 269
      incremental polynomial method, 280
      materials and specimens, 260
      WISPERX spectrum, 280t, 283f
      two-stage block loading, 113–115
Virtual crack closure technique (VCCT), 18,
   98, 142, 384–388
   comparison with cohesive zone, 391–397
   direct cyclic analysis, 386–388
   FE models, 388–390
   for fracture parameter
      calculation, 207, 358–359
      extraction, 331–332
   mode partition, 188
   SERR evaluation, 370–373, 412–413
Virtual crack extension technique (VCET),
   107–108

W
Wear-out model, 466
Wing root stepped-lap joint (WR-SLJ),
   293
   end-of-life residual strength evaluation
      antibuckling fixtures, 297f
      experimental procedure, 296
      failure modes, 298–302
      after fleet service, 297–298
      instrumentation, 296
      pristine specimens, 296–297
      static test results, 299t–300t
      test matrices, 294–296
      test specimen geometry, 293–294,
         295t
      with strain gauge locations, 293f

X
XFEM. *See* Extended finite element method

Y
Yang, Mall, and Ravi-Chandar model, 422

Lightning Source UK Ltd.
Milton Keynes UK
UKHW021825010421
381394UK00003B/59